Anaerobic Reactors for Sewage Treatment

Design, Construction and Operation

Anaerobic Reactors for Sewage Treatment

Design, Construction and Operation

Edited by
Carlos Augusto de Lemos Chernicharo and
Thiago Bressani-Ribeiro

IWA PUBLISHING

Published by **IWA Publishing**
Alliance House
12 Caxton Street
London SW1H 0QS, UK
Telephone: +44 (0)20 7654 5500
Fax: +44 (0)20 7654 5555
Email: publications@iwap.co.uk
Web: www.iwapublishing.com

First published 2019

British Library Cataloguing in Publication Data
A CIP catalogue record for this book is available from the British Library

ISBN: 9781780409221 (Paperback)
ISBN: 9781780409238 (eBook)
ISBN: 9781780409245 (ePub)

Contents

Chapter 9

Chapter 10

About the Editors

Professor Carlos Chernicharo is a pioneer in the field of anaerobic sewage treatment. In 1990 he obtained a Ph.D. in Environmental Engineering from the University of Newcastle (UK) and in 2008 he held a post-doctoral position at the University of New South Wales (Australia). He is a full professor at the Federal University of Minas Gerais (UFMG - Brazil), researcher of the Brazilian Council for Scientific and Technological Development and IWA Fellow. He is also Coordinator of the National Institute of Science and Technology on Sustainable Sewage Treatment Plants (INCT Sustainable STPs). Since 2013 he has been a guest lecturer at the IHE Delft Institute for Water Education. Overall, he has authored/coauthored 5 books, 15 book chapters, more than 100 papers published in scientific journals and more than 200 conference papers. For more than 30 years he has been working together with consultancy and sanitation companies, gaining extensive experience in the design, construction and operation of full-scale STPs. Such experience has fostered the development of an internationally recognized group on applied research on anaerobic sewage treatment.

Thiago Bressani-Ribeiro is an environmental engineer with a M.Sc. in environment, sanitation and water resources from the Federal University of Minas Gerais (UFMG - Brazil). Since 2010 he has been working in several projects in the sanitation sector, bringing together stakeholders from academy, industry and society, in Brazil and abroad. As a member of the National Institute of Science and Technology on Sustainable Sewage Treatment Plants (INCT Sustainable STPs), he was involved in structuring the strategic pillars on basic and applied research, and in training workshops for plant operators. Overall, he has authored/coauthored 2 book chapters, 17 papers published in national and international scientific journals and more than 25 conference papers. He is currently a joint PhD researcher between the UFMG and Ghent University (Belgium), focusing on post-treatment of anaerobic effluents.

List of Contributors

Paulo G. S. Almeida
Department of Sanitary and Environmental Engineering (DESA), Universidade Federal de Minas Gerais (UFMG), Belo Horizonte, Brazil

Sérgio F. Aquino
Environmental and Chemical Technology Group, Department of Chemistry, Universidade Federal de Ouro Preto (UFOP), Ouro Preto, Brazil

Juliana C. Araújo
Department of Sanitary and Environmental Engineering (DESA), Universidade Federal de Minas Gerais (UFMG), Belo Horizonte, Brazil

Iemke Bisschops
Lettinga Foundation (LeAF) Wageningen, The Netherlands

Emanuel M. F. Brandt
Department of Sanitary and Environmental Engineering, Universidade Federal de Juiz de Fora (UFJF), Juiz de Fora, Brazil

Thiago Bressani-Ribeiro
Department of Sanitary and Environmental Engineering (DESA), Universidade Federal de Minas Gerais (UFMG), Belo Horizonte, Brazil

Carolina B. G. Cabral
Rotária do Brasil, Florianópolis, Brazil

Carlos A. L. Chernicharo
Department of Sanitary and Environmental Engineering (DESA), Universidade Federal de Minas Gerais (UFMG), Belo Horizonte, Brazil

Thomas P. Curtis
School of Civil Engineering and Geosciences, Newcastle University, Newcastle upon Tyne, United Kingdom

Mustafa E. Ersahin
Environmental Engineering Department, Istanbul Technical University, Istanbul, Turkey

Fernando Fdz-Polanco
Chemical Engineering and Environmental Technology Department, University of Valladolid, Valladolid, Spain

Eugênio Foresti
Department of Hydraulic and Sanitation, University of São Paulo (USP), São Carlos, Brazil

Xiangjun Huang
Cranfield Water Science Institute, Cranfield University, Cranfield, United Kingdom

Bruce Jefferson
Cranfield Water Science Institute, Cranfield University, Cranfield, United Kingdom

Darja Kragić Kok
Lettinga Foundation (LeAF) Wageningen, The Netherlands

André Lermontov
Grupo Águas do Brasil, Niterói, Brazil

Jules B. van Lier
Department of Water Management, Delft University of Technology, Delft, The Netherlands

Lívia C. S. Lobato
Department of Sanitary and Environmental Engineering (DESA), Universidade Federal de Minas Gerais (UFMG), Belo Horizonte, Brazil

Leonardo Luis Francisco
Paques Brasil Sistemas para Tratamento de Efluentes Ltda, Piracicaba, Brazil

Ewan J. McAdam
Cranfield Water Science Institute, Cranfield University, Cranfield, United Kingdom

Hélinah C. Moreira
Deutsche Gesellschaft für Internationale Zusammenarbeit (GIZ) GmbH, Brasília, Brazil

César R. Mota
Department of Sanitary and Environmental Engineering (DESA), Universidade Federal de Minas Gerais (UFMG), Belo Horizonte, Brazil

Priscilla P. N. Neves
Department of Sanitary and Environmental Engineering (DESA), Universidade Federal de Minas Gerais (UFMG), Belo Horizonte, Brazil

Adalberto Noyola
Instituto de Ingeniería, Universidad Nacional Autónoma de México, México D.F., México

Fabiana Passos
Department of Sanitary and Environmental Engineering (DESA), Universidade Federal de Minas Gerais (UFMG), Belo Horizonte, Brazil

Rodrigo A. S. Pereira
Grupo Águas do Brasil, Rua Luiz Leopoldo Fernandes Pinheiro, Centro, Niterói, Brazil

Christoph Platzer
Rotária do Brasil, Florianópolis, Brazil

Gustavo R. C. Possetti
Water and Sanitation Company of Paraná State (Sanepar), Curitiba, Brazil

Higher Institute of Administration and Economy of Mercosur (ISAE), Curitiba, Brazil

Julio C. Rietow
Department of Hydraulics and Sanitation, Universidade Federal do Paraná (UFPR), Curitiba, Brazil

Angel Robles
Departament d'Enginyeria Química,
Universitat de València,
València, Spain

Aurora Seco
Departament d'Enginyeria Química,
Universitat de València,
València, Spain

Lucas Seghezzo
Instituto de Investigaciones en Energía No Convencional (INENCO),
Universidad Nacional de Salta (UNSa),
Salta, Argentina

Ana Soares
Cranfield Water Science Institute,
Cranfield University,
Cranfield, United Kingdom

Marcos von Sperling
Department of Sanitary and Environmental Engineering (DESA),
Universidade Federal de Minas Gerais (UFMG),
Belo Horizonte, Brazil

Grietje Zeeman
Lettinga Foundation (LeAF)
Wageningen, The Netherlands

Foreword

For over one hundred years the effective treatment of sewage has been dominated by aerobic fixed film and suspended growth processes. The activated sludge process, the leading technology for larger urban sewage treatment works, has been very successfully adapted over time to meet increasingly stringent effluent standards. Activated sludge plants, however, require substantial energy inputs and hence come at a significant economic and environmental cost.

In the resource constrained world we now live in, a compelling case can be made to fundamentally change how we treat sewage by switching to anaerobic sewage treatment. In warmer tropical and sub-tropical climates, anaerobic sewage treatment is already successfully applied. However, recent developments and the establishment of firm design principles, as described in this book, offer the exciting prospect of not only delivering high effluent treatment standards, but achieving truly energy neutral if not energy positive sewage treatment in both cool and warm climates.

Anaerobic Reactors for Sewage Treatment: Design, Construction and Operation consolidates 20 years of academic and practical industry experience gained by the leading proponents of anaerobic sewage treatment to detail best practice in optimising anaerobic sewage treatment processes and managing liquid, solid and gaseous phases. The introduction covers the fundamentals of anaerobic sewage treatment and how it can deliver more sustainable sanitation. Subsequent chapters discuss the design, construction, operation and full-scale experience of Upflow Anaerobic Sludge Blanket (UASB) reactors. Invaluable information is provided, based on a large survey of academic and industrial experts in Brazil, highlighting potential pitfalls in UASB design and operation and how to avoid them. Later chapters cover the wider treatment system; energy recovery, controlling fugitive greenhouse gas emissions and post-treatment of anaerobic effluents to achieve the desired final effluent quality. The incorporation of ultra-filtration membranes to the UASB flowsheet, as covered in Chapter 11, is particularly relevant for applications in more temperate climates, where the control of solids in the reactor is a key factor. The book is brought to a conclusion by exploring the considerable opportunity that exists with anaerobic flowsheets for a more circular approach to sewage treatment, for example in recovering nutrients.

This book will be a valuable reference for water industry professionals - engineers, designers, operators, and managers, as well as researchers, undergraduates and graduates in the fields of civil engineering, environmental engineering and process engineering.

Peter Vale
Technical Lead Innovation
Severn Trent Water

Are UASB reactors indeed a proven and effective technological option for sewage treatment? What is the role of anaerobic sewage treatment in the circular economy? Should the reported constraints of UASB reactors be faced as inherent limitations or could they be seen as an advantage? Is toxicity an issue in anaerobic sewage treatment? What are the key design parameters to be considered and how to effectively manage them? How are full-scale UASB reactors currently performing? How can critical construction and operational issues be tackled? How can recurring problems with sludge and scum management be addressed? What are the best technological options for biogas recovery? How can odour and methane emissions in UASB reactors be controlled? Which post-treatment technology should be applied? Are membrane-based systems an alternative for temperate climate regions?

This is a glimpse of the issues that *Anaerobic Reactors for Sewage Treatment: Design, Construction and Operation* seeks to address. The use of UASB reactors for sewage treatment is relatively recent (around 30 years), especially when compared to processes such as activated sludge. Additionally, a significant part of the development of anaerobic technology for sewage treatment comes from the work of research groups and professionals associated with institutions in developing countries, notably Brazil, Colombia, Mexico, and India, who have been developing applied research in demo- and full-scale reactors, and working together with water authorities. Therefore, there was a latent demand for the consolidation of experiences in the design, construction, and operation of UASB reactors for sewage treatment. The advantages of the anaerobic process make it a more sustainable option for sewage treatment, especially in warm climate regions, but there are still important constraints related to the design, construction, and operation of anaerobic reactors when applied to sewage treatment. On the other hand, there is enough knowledge, experience and proven technology that can be used to improve all these aspects, which are now consolidated in this book. It is launched at a time of particular interest, in which temperate climate countries (e.g., England) have been carrying out research to better understand and overcome the current limitations associated with the applicability of the anaerobic process under low temperatures.

In *Anaerobic Reactors for Sewage Treatment: Design, Construction and Operation*, a detailed view is provided on the particularities of UASB reactors treating sewage, in which the fundamentals of the anaerobic process are discussed under the typical sewage flows and composition, especially to elucidate occasional limitations. The coordinated set of the 12 chapters, gathering worldwide recognized professionals with different and cross-connected experiences, can effectively contribute to solving almost all problems related to the design, construction, and operation of UASB reactors. Additionally, a simple and ready-to-use spreadsheet for the design of UASB reactors for sewage treatment is available for download through the IWA Publishing website (http://10.2166/9781780409238_0412). Such a design tool is based on the concept of standardized gas-liquid-solid (GLS) separators, which is presented and discussed in Chapter 4.

We sincerely believe this book can significantly contribute to the responsible expansion of the anaerobic technology worldwide, and especially for the universalization of sewage treatment in warm climate regions.

Carlos Augusto de Lemos Chernicharo and Thiago Bressani-Ribeiro

Dedication and Acknowledgments

We dedicate this book to the very special and beloved Adriana, Marina, Júlia, Pedro, Maristela, Jurandir and Gabriel. Without your kind support, this book would not have materialized.

We take this opportunity to thank the many people and institutions that made valuable contributions to this book. Marise Daniele provided crucial support in adapting and preparing many of the figures included in this book. The National Institute of Science and Technology on Sustainable Sewage Treatment Plants (INCT Sustainable STPs) provided financial support for creating the visual identity of the book and for professional grammar/spelling review. The following Brazilian research institutions support the INCT Sustainable STPs: *Conselho Nacional de Desenvolvimento Científico e Tecnológico – CNPq; Coordenação de Aperfeiçoamento de Pessoal de Nível Superior – CAPES; Fundação de Amparo à Pesquisa do Estado de Minas Gerais – FAPEMIG*. The sanitation companies of the state of Minas Gerais (Companhia de Saneamento de Minas Gerais - COPASA) and of the municipality of Itabira (Serviço Autônomo de Água e Esgoto - SAAE Itabira) permitted us to use many pictures across the book, including the one on the book cover. Such a UASB-based STP is one of the largest anaerobic plants in the world.

Carlos Augusto de Lemos Chernicharo and Thiago Bressani-Ribeiro

Chapter 1

Introduction to anaerobic sewage treatment

Carlos A. L. Chernicharo, Thiago Bressani-Ribeiro and Marcos von Sperling

Department of Sanitary and Environmental Engineering (DESA), School of Engineering, Universidade Federal de Minas Gerais (UFMG), Av. Antônio Carlos 6.627, Belo Horizonte/MG, 31270-901, Brazil

Corresponding author: calemos@desa.ufmg.br

1.1 INTRODUCTION

From the experience with anaerobic filters reported in the classic publication of Young and McCarty (1969), a group of researchers from the University of Wageningen in the Netherlands, led by Professor Gatze Lettinga, suggested replacing the support medium by inserting a phase separator (liquid, solid, and gaseous) at the top of the reactor. This concept was considered the initial stage in the development of the UASB (Upflow Anaerobic Sludge Blanket) reactor (Lettinga, 2014). The first investigations with UASB reactors were related to the treatment of high-strength wastewater from potato processing for starch extraction and from beet processing to obtain sugar. In the mid-1970s, the first tests with UASB reactors treating more diluted wastewater, such as sewage, were carried out at laboratory scale (30–120 L) and pilot scale (6 m^3) at the village of Bennekom (the Netherlands). The difficulties associated with the need to heat the sewage to simulate the temperature in tropical countries motivated the search for partnerships in the effort to establish UASB reactors in developing countries. A project between the Dutch and Colombian governments, with the participation of the University of Wageningen and the consulting company Royal Haskoning, enabled the implementation of a demonstration-scale UASB reactor (64 m^3) in the city of Cali in Colombia (Lettinga, 2014). The satisfactory results obtained have led to the definition of the initial basis for the criteria and

Editor(s): Carlos Augusto de Lemos Chernicharo and Thiago Bressani-Ribeiro
doi: 10.2166/9781780409238_0001

parameters for the design, operation, and maintenance of UASB reactors applied to sewage treatment. From this experience, the implementation of full-scale reactors (~1,000 m^3) was carried out in the city of Cali, and research projects were conducted in India, Ghana, the Middle East, and in Latin America in Brazil, Colombia, and Mexico.

In Brazil, specifically, the use of UASB reactors for sewage treatment was introduced in the 1980s, when projects were initiated by several national research groups and engineers working in the sector. The designers of the *Companhia de Saneamento do Paraná* (Sanepar) in the state of Paraná visited the Netherlands in the 1980s, interacting with Professor Gatze Lettinga. This led to the design of a variant of the UASB reactor, named RALF, widely used by Sanepar. During the initial stages, the still incipient design and operation of the UASB reactors led to difficulties in respect of the credibility of the system, culminating in the rejection of anaerobic technology by some sanitation companies and environmental agencies. However, confidence has been re-established in recent decades because of the intensification of studies and research in this field, as well as the experience gained from the operation of full-scale reactors. Undoubtedly, an important contribution to the consolidation and dissemination of anaerobic technology for the treatment of sewage in Brazil was made by the Basic Sanitation Research Program (PROSAB), developed between 1997 and 2007, involving several universities, research institutes and sanitation companies (Campos, 1999; Chernicharo, 2001).

Sewage treatment plants (STPs), comprising UASB reactors followed by aerobic post-treatment, generally enable a reduction of 20–50% in implementation costs (CAPEX) and more than 50% in operational costs (OPEX) compared with conventional activated sludge plants (Alem Sobrinho & Jordão, 2001; Chernicharo, 2006; Silva, 1993; von Sperling & Chernicharo, 2005). This cost reduction is one of the leading factors for the increase in the levels of sewage treatment in Latin America (Chernicharo *et al.*, 2015).

In addition to the operational simplicity of the UASB reactors compared with that of conventional aerobic (e.g. activated sludge) or physicochemical processes, as well as the financial aspects previously mentioned, the concentration of biomass (2–6 g L^{-1}) and the sludge age (>30 days) in the reactor are typically high, which significantly reduces its volume compared with those of the other sewage treatment processes. Other advantages of UASB reactors are the absence of a support medium because of the dispersed biomass growth, the low production of sludge, and the reduced (or zero) energy expenditure, factors that are typical of anaerobic systems. Additionally, the excess sludge that must be withdrawn is thickened and digested already and can therefore be routed directly to the dewatering unit. Finally, an important factor is the generation of biogas that can be used for energy recovery. The results of the Brazil–Germany Project to Promote the Use of Biogas in Brazil (Probiogás) show that the average methane content in biogas from UASB reactors treating sewage varies between 70 and 81% (Cabral *et al.*, 2016). This means that the biogas has high calorific value, which opens possibilities for different uses, as further addressed in Chapters 2, 8 and 12 of this book.

1.2 UASB REACTORS AND POST-TREATMENT SYSTEMS IN BRAZIL

Given the favourable climatic conditions in Brazil, the anaerobic treatment of sewage and post-treatment of anaerobic effluents have been implemented widely. After successful and unsuccessful experiences in the 1980s and early 1990s, a concentrated research effort that started in the late 1990s and continues today has created significant impetus. In this regard, the aforementioned PROSAB should be emphasized. This initiative facilitated the consolidation of knowledge on several aspects of anaerobic treatment and post-treatment, paving the way for the applied research to be converted into full-scale units.

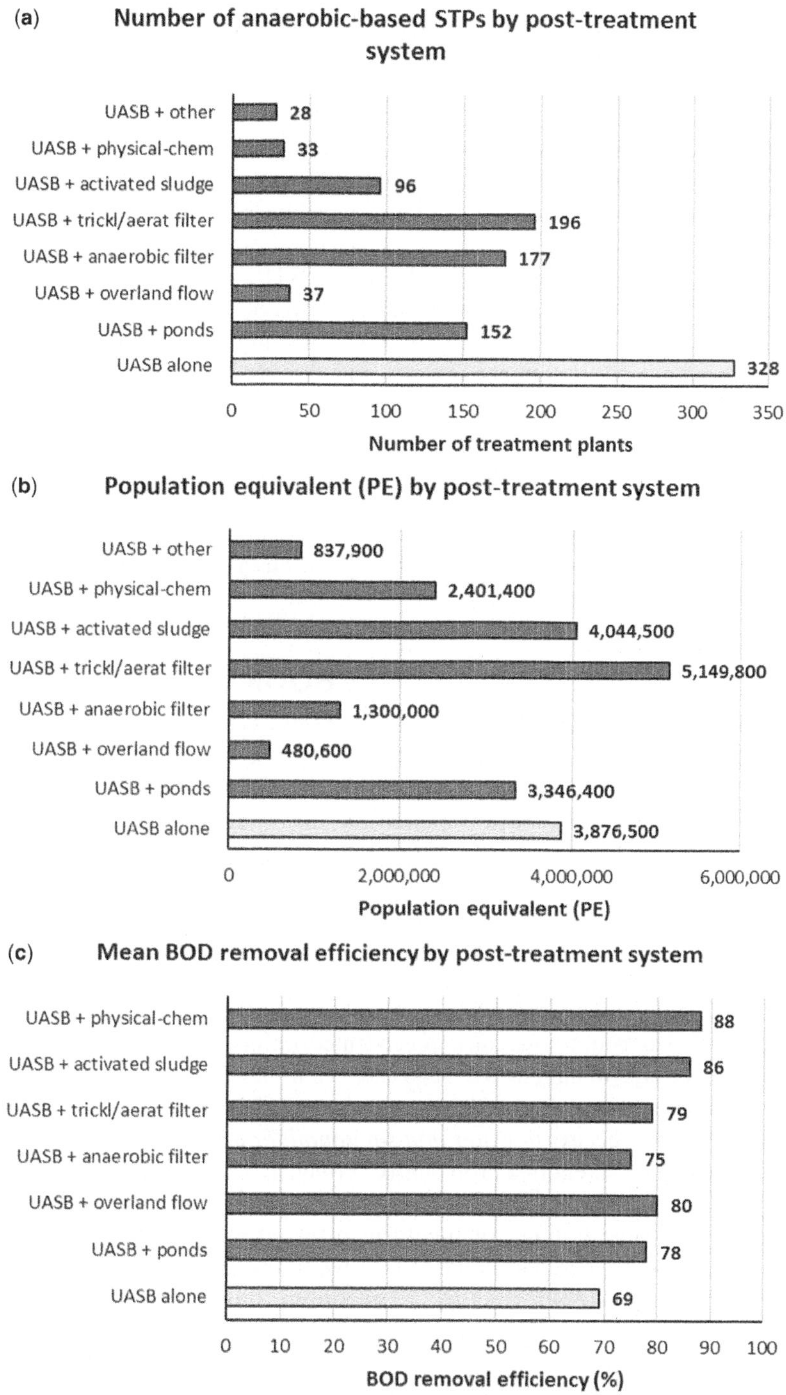

Figure 1.1 (a–c) Data on UASB reactors and post-treatment technologies applied in sewage treatment in Brazil (Source: adapted from ANA, 2017).

A survey undertaken by the National Water Agency (ANA, 2017), listing all the urban STPs in Brazil, reinforces this statement. The survey collected data only from the urban seats of all the 8,570 municipalities in Brazil, excluding treatment plants from small settlements, rural communities, industries, institutions, condominiums, hotels, restaurants, and the like. The number of identified STPs totalled 2,768, treating the sewage produced by approximately 71.7 million inhabitants, using several treatment processes. Altogether, the 1,047 existing anaerobic-based STPs (38% of the total STPs) treat the wastewater from about 21.5 million inhabitants (30% of all wastewater presently being treated in Brazil).

Figure 1.1(a–c) concentrates only on the anaerobic-based STPs and shows the data related to UASB reactors (alone) and UASB reactors followed by different forms of post-treatment technologies. In terms of the number of treatment plants (Figure 1.1a) and the population equivalent (Figure 1.1b), there is a wide distribution of the main post-treatment options, but the prevailing ones involve ponds, trickling filters and activated sludge.

Figure 1.1c shows the mean BOD removal efficiencies obtained from each process variant involving UASB reactor and post-treatment. These data need to be regarded with caution, as they do not reflect all the treatment plants surveyed and, moreover, are based on self-declared information; however, they do present an initial overview of the supposed actual performance under real conditions. In most instances, the mean BOD removal efficiencies are close to 70% without any post-treatment, and in the range of 80 to 90% when post-treatment is applied.

1.3 EXPECTED CONTRIBUTION OF THIS BOOK

Despite the recognised advantages, using UASB reactors for sewage treatment still presents challenges that prevent their maximum operational performance. Notably, the problems associated with these reactors originated from defects in the initial phases of the design. However, inadequate construction processes and operational problems have also been reported in the literature and by sanitation companies (Chernicharo *et al.*, 2018; van Lier *et al.*, 2010; Miki, 2010). In addition to the technical problems of design, construction and operation, operational problems can arise from situations beyond the limits of the STP. These include system management issues that are reflected in administrative and financial failures, which often prevent the proper execution of essential operational procedures. Therefore, neglecting the existence of problems and postponing the solutions can jeopardise the significant advances made to date, thereby detrimentally affecting the future of anaerobic technology for treating sewage. Such negative effects could have an unprecedented influence on the sanitation sector, with the risk of the stagnation of sewage treatment coverage indexes in developing countries, i.e. those countries that lack financial resources and that need progress to reach the United Nations Sustainable Development Goal (SDG) 6, namely "*Ensure access to water and sanitation for all*".

It must be emphasised that most reported problems (Chernicharo *et al.*, 2018) can be corrected in a simple and effective way, without significantly affecting reactor costs, as more attention is being paid to the design and construction phases, and the training of treatment plant operators is being improved. The aim of Chapters 4, 5, and 6 of this book is to contribute to solving almost all the problems related to the design, construction, and operation of UASB reactors. Furthermore, progress is required on managing the sanitation services in developing countries, since management issues are associated with STPs regardless of the treatment technology used.

The main topics related to the need for improvements in the design, construction, and operation of UASB reactors are represented schematically in Figure 1.2, based on a large survey conducted with experts from the academic sector and sanitation companies in Brazil.

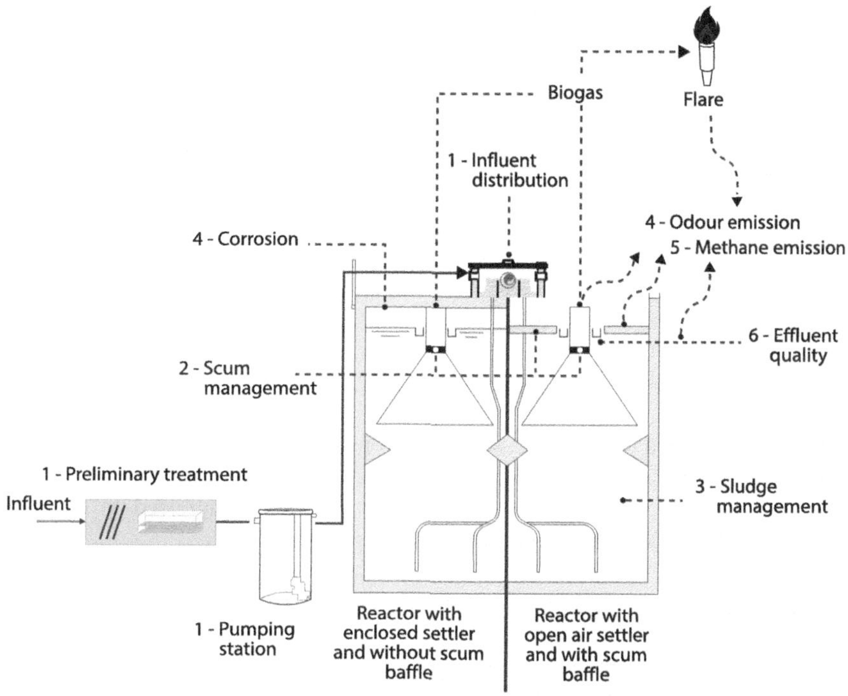

Figure 1.2 Topics of interest for improving design, construction and operation of UASB reactors treating sewage.

The aspects reported in Figure 1.2 are addressed in detail in the following chapters of this book, as set out in Table 1.1.

Throughout this book, STPs are categorised into three different sizes (scales), namely small-, medium- and large-scale STPs. Small-scale STPs are considered to be those with a population equivalent (PE) of over 2,000 but less than 10,000 inhabitants. Medium-scale STPs are those with PE between 10,000 and 100,000 inhabitants, and large-scale STPs are those with PE higher than 100,000 inhabitants. Such size definition based on population scales is used in several chapters of this book, with the purpose being facilitating the grouping of technological solutions or arrangements feasible for a certain STP size (e.g. the use of biogas for cogeneration of heat and electricity, as discussed in Chapters 8 and 12). Obviously, this STP classification is not a narrow concept and a technological solution suitable for a specific STP scale can be applied to the other scales as well. The details are discussed in depth in the chapters of this book, particularly the technological arrangements presented in Chapters 8 and 12.

It must be noted that this book is dedicated particularly to the use of UASB reactors for the treatment of the liquid phase of sewage (mainstream treatment). Occasionally, depending on the sewage treatment flowsheet adopted, a UASB reactor could also receive excess sludge from an aerobic post-treatment stage to perform the anaerobic digestion and thickening steps together with the continuous treatment of the incoming raw sewage. The particularities of this process are discussed in Chapters 6 and 7 and mentioned briefly in Chapter 10. In various more recent proposals for sewage treatment flowsheets, e.g. UASB coupled with high-rate algae ponds, the possibility was considered of sending the biomass (solid phase) for co-digestion with the influent raw sewage. Further details are discussed in Chapter 12.

Table 1.1 Relationship between the main topics of interest for the improvement of UASB reactors and the chapters of this book.

Topic of Interest	Chapters
Preliminary treatment, pumping, and influent flow distribution	Chapters 4 and 6
Management of scum Management of sludge	Chapters 4, 5, 6 and 7
Corrosion and odorous emissions	Chapters 4, 5 and 9
Biogas	Chapters 2, 4, 5, 8 and 12
Effluent quality	Chapters 7, 10 and 11

For the sake of standardisation in this book, sewage is considered as wastewater from domestic activities, which has, on average, a fairly constant and balanced composition in terms of COD and nutrients, as is discussed further in Chapter 3. Additionally in this book, sewage is referred to as domestic sewage; however, it should be understood as a synonym for municipal wastewater.

1.4 FINAL REMARKS

In addition to the recognised advantages of anaerobic processes, particularly lower operating costs per kilogram of COD removed and per cubic metre of treated sewage, it should be emphasised that anaerobic processes, in their different modalities, are possibly the best way for the most deprived populations to benefit from readily available sewage treatment. This implies that employing anaerobic technology enables progress in the democratisation of access to sanitation and in the fulfilment of SDG 6. This is an obvious observation, as many deprived populations live in developing countries, most of which are located in tropical and subtropical regions – exactly those regions where the climatic conditions favour the application of anaerobic technology.

When financial resources are scarce, even for the most important priorities, finding low-cost and more effective sanitation solutions is mandatory (Foresti *et al.*, 2006), while keeping in mind that a more effective solution must, necessarily, also be more sustainable. In this context, anaerobic processes are even more advantageous, as they not only enable the removal of organic matter but also the recovery of energy (methane gas) and compounds in the sewage (e.g. nitrogen, phosphorus, sulphur and water itself). These factors confer high scores in sustainability analyses, taking into account economic, technical, environmental, and social criteria, as discussed in Chapter 2. It is implicit to anaerobic-based STPs that post-treatment systems are important not only to improve the quality of the effluent for health and environmental protection but also to achieve the recovery of resources and to re-establish the link between sanitation and agriculture. The possibilities of closing cycles from the use of anaerobic sewage treatment and from the recovery of water, nutrients, and energy resources are discussed in detail in Chapters 2, 8, and 12 of this book.

REFERENCES

Alem Sobrinho P. and Jordão E. P. (2001). Pós-tratamento de efluentes de reatores anaeróbios – uma análise crítica (Post-treatment of anaerobic reactor effluents – a critical analysis). In: C. A. L. Chernicharo (coordenador). Pós-Tratamento de Efluentes de Reatores Anaeróbios. FINEP/PROSAB, Rio de Janeiro, Brazil, 544pp.

ANA (Agência Nacional de Água) (2017). Atlas Esgotos. Despoluição de Bacias Hidrográficas (Sewage Atlas. Remediation of Watersheds). Agência Nacional de Água, Brasilia, Brazil. Executive summary, 88 pp. Available online at atlasesgotos.ana.gov.br (accessed 6 March 2019).

Cabral C. B. G., Chernicharo C. A. L., Hoffmann H., Neves P. N. P., Platzer C., Bressani-Ribeiro T. and Rosenfeldt S. (2016). Resultados do projeto de medições de biogás em reatores anaeróbios/Probiogás (Results of the biogas measurement project in anaerobic reactors). Ministério das Cidades and Deutsche Gesellschaft für Internationale Zusammenarbeit. Brasília, DF, Brazil.

Campos J. R. (coordenador) (1999). Tratamento de esgotos sanitários por processo anaeróbio e disposição controlada no solo (Sewage treatment through anaerobic process and controlled land application). São Carlos: Reatores anaeróbios. FINEP, Belo Horizonte, Brazil, 464 pp.

Chernicharo C. A. L. (coordenador) (2001). Pós-Tratamento de Efluentes de Reatores Anaeróbios (Post-Treatment of Effluents from Anaerobic Reactors). FINEP/PROSAB - Programa de Pesquisa em Saneamento Básico, Belo Horizonte, 544pp.

Chernicharo C. A. L. (2006). Post-treatment options for the anaerobic treatment of domestic wastewater. *Reviews in Environmental Science and Bio/Technology*, **5**, 73–92.

Chernicharo C. A. L., van Lier J. B., Noyola A. and Bressani-Ribeiro T. (2015). Anaerobic sewage treatment: state of the art, constraints and challenges. *Reviews in Environmental Science and Bio/Technology*, **14**, 649–679.

Chernicharo C. A. L., Bressani-Ribeiro T., Pegorini E., Possetti G. R. C., Miki M. K. and Nonato S. (2018). Contribuição para o aprimoramento de projeto, construção e operação de reatores UASB aplicados ao tratamento de esgoto sanitário – Parte 1: Tópicos de Interesse (Contribution for improving the design, construction and operation of UASB reactors treating sewage – Part 1: Topics of Interest). *Revista DAE* – edição especial.

Foresti E., Zaiat M. and Vallero M. (2006). Anaerobic processes as the core technology for sustainable domestic wastewater treatment: consolidated applications, new trends, perspectives, and challenges. *Reviews in Environmental Science and Bio/Technology*, **5**, 1–2.

Lettinga G. (2014). *My Anaerobic Sustainability Story*. Lettinga Foundation, Wageningen, Netherlands.

Miki M. K. (2010). Dilemas do UASB (Dilemmas of the UASB reactor). *Revista DAE*, **183**(1504), 25–37.

Silva S. M. C. P. (1993). Desenvolvimento de uma nova concepção de tratamento de águas residuárias pela associação de processos anaeróbios e aeróbios (Development of a new concept for wastewater treatment combining anaerobic and aerobic processes). PhD Thesis. Escola Politécnica da Universidade de São Paulo, São Paulo, Brazil.

van Lier J. B., Vashi A., van der Lubbe J. and Heffernan B. (2010). Anaerobic sewage treatment using UASB reactors: engineering and operational aspects. In: Environmental Anaerobic Technology; Applications and New Developments, H. H. P. Fang (ed.), Imperial College Press, London, UK, pp. 59–90.

von Sperling M. and Chernicharo C. A. L. (2005). Biological Wastewater Treatment in Warm Climate Regions (2 vols). London, UK: IWA Publishing. Available at http://www.iwapublishing.com/open-access-ebooks/3567.

Young J. C. and McCarty P. L. (1969). The anaerobic filter for waste treatment. *Journal Water Pollution Control Federation*, **41**, R160–R173.

Chapter 2

Anaerobic treatment as core technology for more sustainable sanitation

Iemke Bisschops[1], Darja Kragić Kok[1], Lucas Seghezzo[2] and Grietje Zeeman[1]

[1]LeAF, PO BOX 500, 6700 AM Wageningen, The Netherlands
[2]Instituto de Investigaciones en Energía No Convencional (INENCO), Universidad Nacional de Salta (UNSa), Avenida Bolivia 5150, A4408FVY Salta, Argentina

2.1 INTRODUCTION

Anaerobic treatment of sewage is preferably applied at ambient temperatures above 20°C mainly due to the biochemistry of anaerobic processes and the low concentration of sewage (see Chapter 3 for an in-depth evaluation of the processes involved). As a result, tropical and sub-tropical countries are most suitable for its application. In the majority of the industrialised world, sewage treatment is already implemented and the systems of choice have been defined. Many industrialised countries have a climate that is too cold to allow for efficient anaerobic sewage treatment without having to spend energy on heating. Some low-temperature anaerobic systems have been researched but are not yet implemented in practice (e.g. Mahmoud *et al.*, 2004; McKeown *et al.*, 2009; Zhang *et al.*, 2018). For successful anaerobic treatment in colder climates, other collection and transport systems are recommended (see Section 2.5). Anaerobic sewage treatment at psychrophilic environments is also addressed in Chapter 11. Developing countries are often still struggling with the provision of sustainable sanitation, including sewage treatment. Sewage treatment projects often failed, as the chosen technologies and their management were not appropriate for the local situation. As many developing countries are in the warmer climate zones, the option of applying anaerobic sewage treatment is available, and evaluations of technological suitability often show that systems including anaerobic reactors score higher than others (Seghezzo, 2004; van Buuren, 2010). The lack of application of anaerobic technologies in developing countries is not only related to technical factors, but could also be explained by the fact that water and sanitation management systems in many of

 Anaerobic Reactors for Sewage Treatment: Design, Construction and Operation
Editor(s): Carlos Augusto de Lemos Chernicharo and Thiago Bressani-Ribeiro
doi: 10.2166/9781780409238_0008

those countries tend to be outdated and governed by short-term political or professional interests (Iribarnegaray *et al.*, 2012; Iribarnegaray & Seghezzo, 2012).

This chapter discusses sustainability in sewage treatment, the role that anaerobic reactors can play in achieving sustainable sanitation and how to evaluate the sustainability of different sewage treatment concepts by applying multi-criteria analysis (MCA).

2.2 THE CONCEPT OF SUSTAINABLE DEVELOPMENT

Over the last half century the world population has more than doubled, the per capita consumption of natural resources in industrialised and some developing countries has rocketed, and the pressure on our planet's resources has never been this large. Experts from different disciplines are working on ways to reduce environmental pollution, demand for resources and greenhouse gas (GHGs) emissions. At the same time, when explaining complex environmental systems, scientific disciplines use different concepts and languages. A common framework to organize knowledge and findings is required to allow knowledge to accumulate (Ostrom, 2009). The concept of sustainable development, despite some criticisms (Seghezzo, 2009), has been generally endorsed as the way forward, and the United Nations (UN) confirmed this by launching seventeen Sustainable Development Goals (SDGs) for the period 2015–2030.

The essence of sustainable development can be found in the 1987 definition by the Brundtland Commission of the United Nations: "*sustainable development is development that meets the needs of the present without compromising the ability of future generations to meet their own needs*" (WCED, 1987). In more than thirty years that have since passed, the advance towards sustainable development has been slow. Although the message is clear, and still very much valid, it has proven difficult to develop the approaches to be followed. On the one hand, establishing seventeen different SDGs clearly shows that sustainable development is a wide-reaching concept, comprising all aspects of our lives. In addition, the individual "smaller" goals seem more reachable. On the other hand, however, there is a risk of fragmented actions, lacking integration with other goals and therefore lowering the chances to achieve their full potential. Still, sustainable development is firmly placed on the agenda.

Since roughly the 1960s the concept of sustainable development has been evolving. In the last decades, the 3P model was widely used: People, Planet, Profit (or Prosperity). The most recent representation includes two additional "Ps": Partnership and Peace (Figure 2.1), explicitly recognising that peace, partnership and sustainable development are reciprocal (UN, 2015).

Figure 2.1 Schematic representation of sustainable development integrating the five dimensions.

Sustainable wastewater treatment is one of sustainable development's vital aspects. Most people would consider wastewater treatment to be something inherently sustainable. However, a local improvement in one place (e.g. a cleaner river) can have less favourable implications elsewhere (e.g. GHG emissions or sludge management problems). The choices made when implementing sanitation concepts and including sewage treatment technologies – in new but also in existing systems – greatly affect the sustainability of the overall system. It seems therefore necessary to take sustainability into account during the decision-making process.

2.3 INTERNATIONAL DEVELOPMENTS IN SANITATION

2.3.1 Sanitation in MDGs and SDGs

International development policies were a trigger for starting international monitoring of access to sanitation. During the last decades, this monitoring was executed by the World Health Organization (WHO), and now jointly by WHO and the United Nations Children's Fund (UNICEF) through the Joint Monitoring Programme (JMP) (Bartram *et al.*, 2014).

In the year 2000 the leaders of 189 countries adopted the Millennium Development Goals (MDGs) (SDGF, 2018). Goal number 7 addressed environmental sustainability. One of its specific targets (7c) was "*to halve, by 2015, the proportion of the population without sustainable access to safe drinking water and basic sanitation*" (UN, 2018). The JMP – for the purpose of monitoring the progress of sanitation – coined the terms "improved" and "unimproved sanitation". Under improved sanitation, the following systems were listed: flush toilets connected to sewers or septic tanks, flush/pour flush toilets connected to pit latrines, pit latrines with a slab, VIP latrines and composting toilets. Unimproved sanitation included: public or shared latrines, pour/flush toilets connected to elsewhere (i.e., other than septic tank, pit or sewer), pit latrines without a slab, bucket latrines, hanging toilets, and no facilities (open defecation).

In response, a massive construction of pit latrines with slabs and VIP latrines took place worldwide. However, counting toilets as the main way to show success may be misleading, as it does not pay sufficient attention to whether the latrines are used or not (WELL, 2004) and consequently, if they are properly maintained and if faecal sludge is managed well. The target for drinking water was reached but the sanitation target was missed by about 700 million people (WHO/UNICEF, 2015).

Based on the lessons learned with the MDGs, the new goals (the SDGs) adopted for the period 2015–2030 include for the first time the word "management" in the sanitation targets in the form of SDG Goal 6: "*Ensure availability and sustainable management of water and sanitation for all*". The associated indicator for target 6.2 – universal access to sanitation by 2030 – refers to people using "safely managed" sanitation services. This was a big step forward, as for the first time sanitation targets include a more systemic approach. However, the issue of shared sanitation facilities remained a challenge. Since shared sanitation is classified as a "limited" sanitation service under JMP definitions (WHO/UNICEF, 2017), some governments are hesitant to invest in it. Concern has been raised (Evans *et al.*, 2017) that leaving out shared sanitation may exclude the poorest citizens of dense slum areas from sanitation service improvements, as decision makers may be directing their efforts to other areas.

2.3.2 Appropriate technologies

In traditional engineering approaches, a specific problem is analysed, technological solutions are designed, and final products are presented to the client. End-users and clients are usually not involved in all steps of the process, which is one of the reasons why traditional engineering approaches often failed (Murphy *et al.*,

2009). Understanding the local context and involving end-users and stakeholders in all steps of the decision-making process is key when working on projects involving appropriate technologies.

The appropriate technology concept emerged in the 1960s after observing many problems with technologies that had been implemented in developing countries. It became obvious that, in retrospect, the chosen technologies too often did not fit the local situation. Through the decades, the term appropriate technology has been reserved for use in a low-income country context (Murphy *et al.*, 2009).

The importance of incorporating local context in decision making was mentioned by the World Bank and published some decades ago. As a result of a research project in the 1970s, the World Bank published a series of reports on Appropriate Technology for Water Supply and Sanitation (appearing in the early 1980s) where next to the technical and economic factors, assessments were made of environmental, public health, institutional, and social constraints (see Figure 2.2 for an overview).

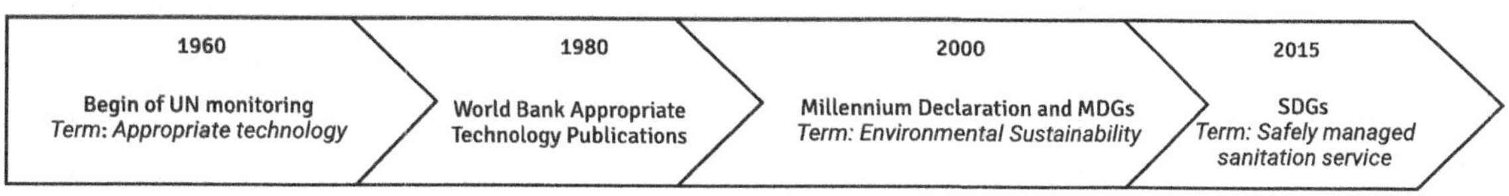

Figure 2.2 International developments and evolution of terminology used in the sanitation field.

Recently, the concept of the "circular economy" has gained attention as an approach towards sustainable development. The research of Schroeder *et al.* (2018) shows that circular economy practices can potentially contribute to achieving several SDG targets and possibly strongly contribute to SDG 6.

2.4 ANAEROBIC TREATMENT: CORE TECHNOLOGY IN THE CIRCULAR ECONOMY?

In many countries, faecal matter and urine used to be collected, transported and composted, or sold for farming purposes without further treatment. In this way, raw materials in faeces and urine, originating from food consumption and production, were recycled for food production. Nowadays, in many developing countries, raw sewage or black (toilet) waste is still used in agriculture. Already in the early 1900s, Franklin Hiram King was severely concerned about the industrialization of agriculture as introduced in the industrialised world. In 1911, King wrote the book *Farmers of Forty Centuries – Permanent Agriculture in China, Korea and Japan* (King, 1911). Any loss was seen as a loss of money. Not only the nutrients were valued in the past for agriculture but also the organics, necessary for soil structure (Zeeman, 2012). Most nutrients in domestic waste (water) originate from faeces and urine. Table 2.1 represents the nutrients produced per person per day (Kujawa-Roeleveld & Zeeman, 2006). The given ranges or concentrations of wastewater components are a compilation of mainly European data (Eriksson *et al.*, 2002; Hellström & Kärrman, 1995; Jönsson *et al.*, 1997; Kujawa-Roeleveld *et al.*, 2003; STOWA, 2001; Vinnerås, 2002).

Nitrogen is abundantly present in the atmosphere and therefore not a finite resource. However, large-scale production of nitrogen fertilizers from the air occurs via the Haber–Bosch process, consuming a considerable amount of energy (37–45 kJ gN^{-1}) (Maurer *et al.*, 2003). About 1% of the world's total annual energy supply is used for artificial nitrogen production (Smith, 2002). Use of nitrogen from sewage therefore stands for reduction of energy consumption. Phosphorus and potassium, however, are finite resources derived from minerals and reserves are very limited (Jönsson, 2016; Zeeman, 2012). Sewage also contains different types of microorganisms (bacteria, viruses, helminths) and organic matter (usually expressed as Volatile Solids, Biochemical Oxygen Demand (BOD) or Chemical Oxygen Demand (COD)).

Table 2.1 Volume and composition of urine and faeces.

Parameter	Unit	Urine	Faeces
Volume	kg or L p^{-1} d^{-1}	1.25–1.5	0.07–0.17
Nitrogen	gN p^{-1} d^{-1}	7–11	1.5–2
Phosphorus	gP p^{-1} d^{-1}	0.6–1.0	0.3–0.7
Potassium	gK p^{-1} d^{-1}	2.2–3.3	0.8–1.0
COD	gO_2 p^{-1} d^{-1}	10–12	45.7–54.5

By applying anaerobic treatment, a large fraction of the COD in sewage can be converted to CH_4 gas, which can be used as an energy source, while nutrients are retained for the largest part in the liquid effluent and partly in the sludge. The UASB reactor, developed in the 1970s by Gatze Lettinga for the treatment of industrial wastewater (Lettinga *et al.*, 1980), is by far the most widely used anaerobic technology for sewage treatment (Chernicharo *et al.*, 2015; Seghezzo *et al.*, 1998). This technology can produce energy and is characterised by a small footprint, low sludge production, low evaporation (compared to open systems such as activated sludge, pond and wetland systems), and low operation and maintenance costs. It is also applied for black water (BW) treatment at several locations in The Netherlands (Hernández Leal *et al.*, 2017). After anaerobic treatment, the nutrients in sewage are retained in the effluent which can, after a proper post-treatment, be used for irrigation/fertilisation in agriculture (also discussed in Chapters 11 and 12). Technologies that are used for converting domestic waste streams into reclaimed water and ecological, hygienic, and reliable fertilisers, can be seen as a figurative bridge between cities and agriculture, as indicated in Figure 2.3 (Zeeman, 2012).

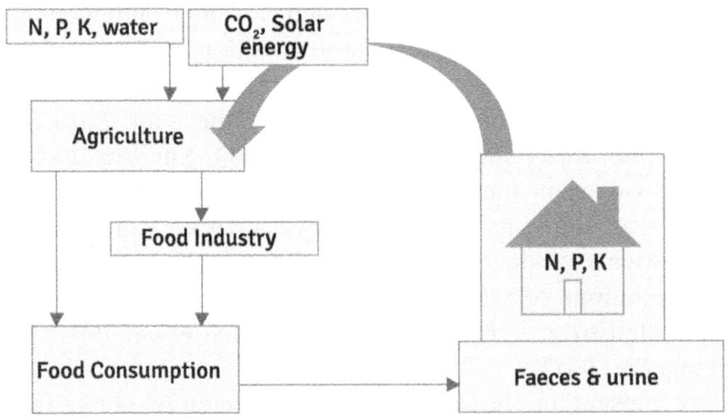

Figure 2.3 Circular approach for the fertilisers that are used in agriculture for food production and are excreted via the urine and faeces after food consumption (Zeeman, 2012).

Nutrient recovery from anaerobically treated sewage is being examined from different perspectives, looking at obtaining both inorganic and organic fertilisers. For example, work is progressing on a UASB reactor treating BW that, next to methanisation of COD, achieves precipitation of calcium phosphate granules in the reactor (Cunha *et al.*, 2018). In addition, within the EU project RUN4LIFE (2017–2021) currently the hyper-thermophilic anaerobic treatment of highly concentrated BW (ultralow flush vacuum toilets) is being researched, with the aim to achieve digestion and disinfection in a one-step reactor system.

2.5 ANAEROBIC TREATMENT IN DIFFERENT SANITATION CONCEPTS

When designing sanitation systems, all components of the chain should be considered: collection (toilet), transport, treatment and recovery, and reuse. Each component influences the design and functioning of the ones that follow, as well as the quantity and quality of the generated products and residues. In this chapter, sanitation chains, including anaerobic treatment followed by recovery and reuse of nutrients for agricultural purposes, will be discussed for different environments. A wrong decision at the beginning of the chain, e.g. in sewage collection, can result in malfunctioning of the whole treatment concept. For example:

(i) In a school in Ghana, new flush toilets were installed but after installation it turns out that the school could not pay for the water bill. Therefore, the toilets are not in use and children go outside for defecation and urination – an unhealthy situation.

(ii) In a housing estate in the Netherlands, anaerobic treatment of source-separated toilet water (conventional flush toilets) was foreseen. The Netherlands have low temperatures in winter, therefore toilets with low flush volumes are needed, otherwise too much energy for heating of the anaerobic reactor is used. The anaerobic reactor was never built. At four other locations in The Netherlands, vacuum toilets were installed, connected to successfully functioning anaerobic treatment (UASB) reactors (Hernández Leal *et al.*, 2017).

Ambient temperature is an important parameter when selecting a sanitation system, including anaerobic treatment followed by irrigation/fertilization or recovery and reuse of nutrients, for a certain area. While anaerobic treatment of conventionally collected sewage is not applied in countries with cold winters, large-scale UASB systems are applied for treatment and sometimes reuse of the treated effluent in agriculture in tropical areas (Chernicharo *et al.*, 2015). The cropping system determines whether reuse of diluted sewage for irrigation and fertilization is possible (addressed further in Chapter 12). In many tropical areas, continuous cropping systems prevail, enabling such reuse and matching water supply and demand. Separation of concentrated and diluted streams, i.e. black (toilet) and grey (bath, shower, kitchen and laundry) water, enable separate fertilization and irrigation, which could also be profitable in some tropical countries. Firmansyah *et al.* (in preparation) describe and compare (using MCA) several sanitation concepts for tropical regions, including anaerobic treatment and the reuse of nutrient-rich effluents.

In low-temperature countries no crops grow in winter. Fertilisation in that period is therefore not environmentally friendly and generally forbidden. Recovery of nutrients from concentrated BW enables storage for use in summer (De Graaff *et al.*, 2010; Vasconcelos Fernandes *et al.*, 2015). Several sanitation concepts in combination with either continuous or discontinuous crop systems are presented below.

2.5.1 Tropical areas

Example a: flush toilets for BW collection followed by house-on-site anaerobic treatment in either a conventional Septic Tank (ST) or a UASB-ST (Lettinga *et al.*, 1993), connected via a small-bore sewer system (Mara, 2006) with a community-on-site post-treatment, such as a Trickling Filter (TF), for fertilisation in agriculture. Conventional STs do not allow for the collection and use of the produced biogas, but UASB-STs do. Biogas could be used for lighting (Buysman, 2009). In Brazil for example, for each household (5 residents) 3 m^3 ST systems with an HRT of 20 days are applied in rural areas and liquid effluents are reused in local agriculture (Lopez da Silva, 2014).

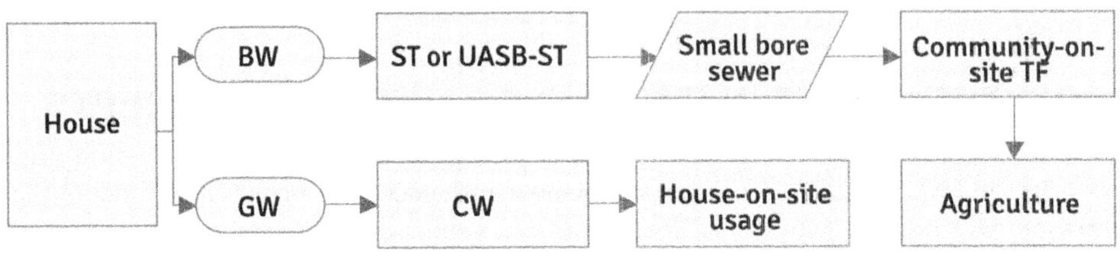

Example b: flush toilets connected to a conventional sewer to transport BW to a community-on-site UASB or UASB-ST followed by, e.g., a TF for post-treatment for fertilisation in agriculture. Biogas could be used for lighting (Buysman, 2009).

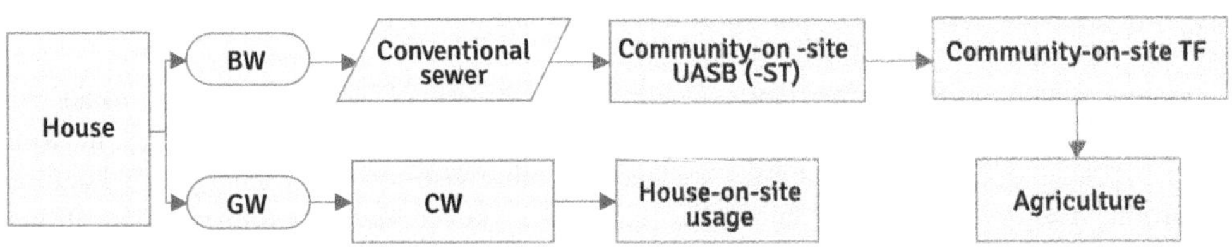

In both Examples a and b, Grey Water (GW) is used house-on-site, after treatment in, for example, a constructed wetland (CW).

Example c: BW of flush toilets and GW are collected together and transported via a conventional sewer to a community-on-site or centralised UASB followed by, for example, a TF (Chernicharo & Almeida, 2011) for irrigation/fertilisation in agriculture. Biogas could be used for lighting (Buysman, 2009) or electricity production.

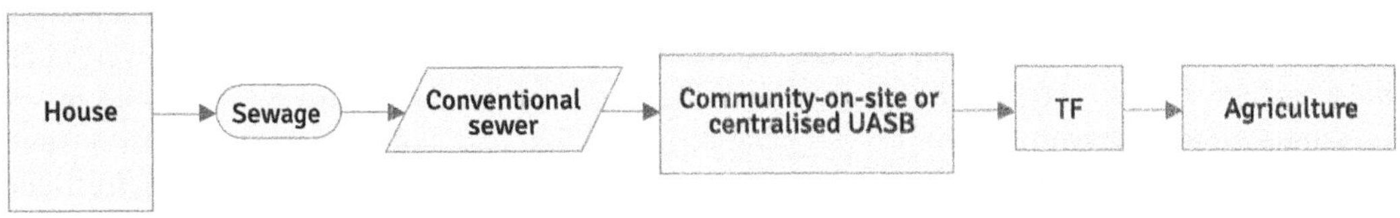

2.5.2 Low/moderate temperature areas

Example d: vacuum toilet (≤1 L per flush) and sewer for the collection and transport of concentrated BW and kitchen waste (KW) connected with a heated (25–35°C) community-on-site UASB reactor (De Graaff *et al.*, 2010; Hernández Leal *et al.*, 2017) followed by a struvite precipitation reactor for P-recovery and an OLAND reactor for nitrogen removal. UASB-sludge can be co-composted with garden waste for reuse in agriculture. Struvite can be used in agriculture as a P-fertiliser. GW is separately collected, transported and treated on-site in an activated sludge plant preceded (or not) by a bioflocculation unit for anaerobic treatment of the produced non-stabilised sludge in the BW-UASB (Hernández Leal *et al.*, 2010).

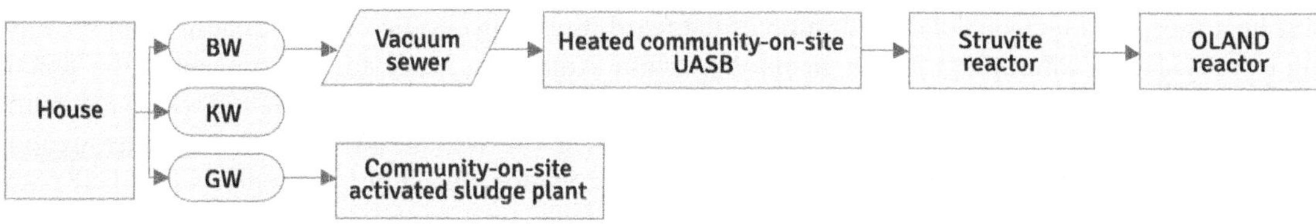

This system of source separation of BW and GW and community-on-site treatment/recovery is applied in a 250 house estate in Sneek, and in 3 office buildings in the Netherlands (Hernández Leal *et al.*, 2010). Several new projects are under development, including one for 550 houses in Amsterdam, currently under construction.

2.6 CHOOSING APPROPRIATE TECHNOLOGIES AND SUSTAINABILITY ANALYSIS

As stated before, the sustainability analysis of sewage treatment options is increasingly gaining importance, and rightly so. Furthermore, the internet has made it possible to publically share information on the topic. For example, the well known water and sanitation platform Akvopedia has a sub-section on sustainability (Akvo, 2018).

In the literature, there are several examples of studies that compared different sewage treatment options in order to choose the most sustainable option. High-rate anaerobic sewage treatment is not always included as a pre-selected option in these analyses. However, in the cases where it was part of the evaluation, its sustainability scores are high. An important point in MCA is that none of the technologies alone receives a sustainability score, but always as part of a technology chain. The choice of pre-treatment, post-treatment and by-product management options strongly influences the outcomes.

To be able to fairly judge the sustainability of different sewage treatment options, it is very important to analyse the whole system and not just the main treatment technology. Similarly to Life Cycle Analysis, the choice of system boundaries greatly influences the outcome. Limits should be chosen carefully, to encompass the widest possible boundaries without making the analysis unfeasible with regard to the practical constraints set by the available time and budget.

A key aspect in choosing sustainable sewage treatment options is a thorough understanding of the concepts under study. It is not sufficient to include only (a part of) the treatment chain in the analysis, as this can lead to erroneous conclusions on the impact of the chosen system as a whole. The entire concept should be evaluated, from sewage collection, via transport and treatment to the final destination of products, emissions and residues. In cases where a sewage collection system is already in place, the evaluation may be limited to treatment and reuse options, unless the existing sewerage infrastructure needs to be updated or replaced.

A thorough review of MCAs including UASBs is outside the scope of this publication, but the following examples are illustrative of the points presented in this chapter.

Research by van Buuren (2010) offers a screening aid for community-scale sewage treatment in warm climates and lists a number of criteria for the comparison of sewage treatment technologies. This study proposes a method for involving stakeholders in sanitation planning and decision making using an elaborate list of criteria, including 2 site-specific criteria to aid the screening of options, and 23 technology-specific criteria. Analysis of the technology chains showed that the ones that include UASB reactors followed by different post-treatment options score the best for tropical countries.

In his PhD research, Seghezzo (2004) demonstrated the necessity of a thorough sustainability analysis. Despite a successful technical demonstration of the feasibility of anaerobic sewage treatment for the city of Salta, located in a subtropical region in north-western Argentina, other factors prevented the practical implementation of this technology. An assessment was performed of three different sewage treatment concepts that were selected based on their availability in the region: (a) aerobic high-rate treatment system (primary and secondary settling, trickling filters, sludge digestion, and chlorination); (b) waste stabilisation ponds (WSP); and (c) UASB reactors with post-treatment in polishing ponds. The comparison was done using four basic criteria, subdivided into 20 operational indicators that were weighted for their importance in the local context and given a performance score. For all the indicators, the importance and performance were multiplied and the results were then added up to calculate a so-called sustainability index (SI). The best score was obtained for concept (c). Based on the SI results, the author listed strengths and weaknesses for each sewage treatment concept considered; Figure 2.4 focuses on concept (c).

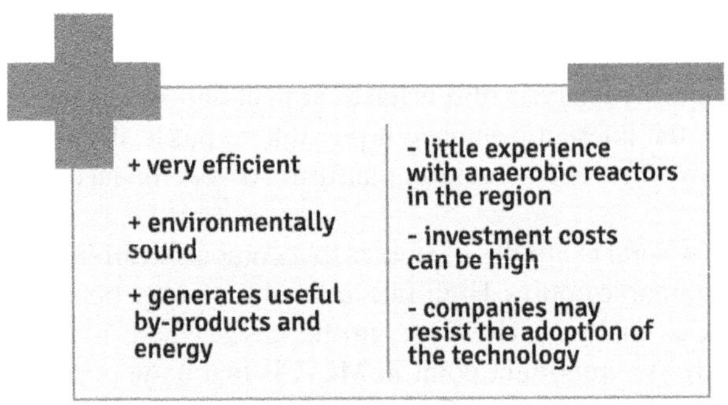

Figure 2.4 Identified strengths and weaknesses of the sewage treatment concept UASB + polishing ponds for a case study in Argentina (Adapted from Seghezzo (2004)).

2.6.1 Evaluating the sustainability of sewage treatment options

When discussing and comparing sewage treatment technologies, the most obvious criterion for many is usually investment costs. However, many other criteria and parameters are also relevant in the selection of appropriate technologies. Examples include operation and maintenance (costs and complexity), footprint, possibility for up-scaling, availability of spare parts, emissions, resource recovery, institutional aspects, and by-product management, among others.

Sustainability is often equated to the maximum reduction of energy use or emissions, but those are only two aspects out of many. The difficulty in selecting sustainable technologies is that a broad range of factors should be taken into account, and that their nature and importance can vary widely depending on local conditions. It is generally recognised that adequate management of excreta is a key factor in improving and maintaining human health. Proper excreta management can take place in many different ways, ranging from individual house-on-site solutions to large-scale centralised systems. Ideally, the best solution for each case should be chosen taking into account all relevant variables within the local context. In practice, however, technology selection is often based on less holistic approaches. Examples are the thoughtless expansion of existing systems to include new neighbourhoods, positive or negative biases towards certain systems that influence the decision-making process, and rigid legislative guidelines that prescribe specific systems for certain cases. The purpose of protecting people from exposure to their excreta is usually met, at least initially. However, for the sustainable, long-term protection of public health and the environment, it is vital that technology selection is done with a more comprehensive approach, taking into account local circumstances and consulting local actors. Examples of important factors in sewage treatment sustainability are:

- Acceptance of the system by the society it is operating in; this includes not only the technologies used, but also their products, emissions and by-products
- Affordable running costs and a business plan that makes sure they can be paid, including in the long term
- System management should be possible within local management, governing, operating and maintenance capacities
- Availability of spare parts, chemicals, etc.
- Reuse and/or discharge requirements should be met
- Resilience to future changes
- Acceptable emission levels (odour, GHG)
- Small footprint
- Enabling legislation.

2.6.2 Sustainability index method

The sustainability index (SI) method first proposed by (Seghezzo, 2004) and later modified and applied by Iribarnegaray *et al.* (2012, 2015) and LeAF (2005–current) to different case studies, is a systematic approach to compare sewage treatment technologies or water and sanitation management systems using a weighted-scale checklist. The method is simple and transparent and it can be used as a framework that can be adapted to different situations. Prior to the start of evaluation of the technologies and concepts, the system boundaries need to be set. The potential viable technological options can be pre-screened by using developed screening aid (van Buuren, 2010). The original SI method encompasses four criteria – environmental, technical, social, and economic – which are divided into operational indicators, which are all assigned weights and performance in relation to technology. Subsequent versions of the method included other criteria and indicators for the assessment of specific technologies and management systems.

The procedure can be divided into five steps:

- Step 1. Deciding on the criteria and indicators to be used
- Step 2. Assigning importance to criteria
- Step 3. Assigning importance to indicators under each criterion
- Step 4. Assigning performance to each indicator under each concept
- Step 5. Calculating the sustainability index of different concepts

An objectively performed MCA is a useful tool that allows decision makers to perform a proper system selection in a structured way. Different stakeholders should be involved to be able to take into account all relevant views, and to be able to select and adjust the indicators and their weights to the local circumstances. The investigated system may include only sewage treatment technologies, or can include different wastewater conveyance systems and resource recovery options. Selecting indicators to be assigned under each of the criteria is the first step in the process. Decision on inclusion of different indicators should always be a two-way process with decision makers (Valentin & Spangenberg, 2000).

When the framework is ready, the next steps of evaluation can take place. In the second step the criteria are weighted according to their relative contribution to the whole environmental and social system in the local context. In the third step, the assignment of a (subjective) importance to each indicator is done by experts, independently of the technologies compared. In step four, the performance of each indicator is assigned, in relation to the each evaluated technological concept. In the last step, the SI is calculated as the average of the score obtained by each indicator, calculated as the importance of the indicator multiplied by the performance given to the technology for this particular indicator and divided by the maximum possible performance (e.g. 100, assuming that a centesimal scale allows a more intuitive and fast interpretation of the results). The SI obtained for different technological options can be divided into ranges or categories such as: very low sustainability if $SI < 25$; low sustainability if $25 < SI < 50$; medium sustainability if $50 < SI < 75$; and high sustainability if $SI > 75$. The SI is a relative but also "relatively absolute" account of sustainability. These categories allow for a quick comparison of alternatives, but can also be used to discard technologies that score below a certain mutually agreed threshold or "band of equilibrium" (e.g. 50, or 75) (Bell & Morse, 2008).

2.6.3 Experiences with the SI method

Over recent years the SI method for evaluating the sustainability of technologies was applied in several contexts, ranging from an urban area in Argentina (scale: city), to rural areas in Jordan (scale: village) and Kenya (scale: school). Table 2.2 shows results of the importance of criteria for evaluated technologies.

Table 2.2 Different results of importance of criteria given by local stakeholders at different case studies.

	Importance of Criteria from Highest (left) to Lowest (right)			
Kenya – rural	Economic	Technical	Environmental	Social
Jordan – rural	Technical	Economic	Environmental	Social > Institutional
Argentina – urban	Environmental	Economic	Technical	Social

The indicators that were rated as the least important within social/institutional criteria per case study were: social acceptability (Kenya), scientific support (Jordan) and regulatory framework (Argentina). Remarkably, the social criterion scored the lowest on importance in each case study, while past experience shows that it plays a key role in sustainable applications of technologies.

The SI method can help identify the strengths and weaknesses of the investigated systems and technologies in any given context (see, e.g., Figure 2.4). The method also offers a communication platform for decision makers which can contribute to developments in local policy making. Achievement of the new SDG set out by the United Nations will require a combination of accurate monitoring methods and adequate governance processes that need to include more social participation. Assessment methodologies such as the SI can only have impact on governance processes if there is adequate communication between researchers and policy makers (Iribarnegaray *et al.*, 2015). A way to improve this could be to include social participation as a more fundamental component in the assessment methods and to ensure that all actors have equal opportunities to take part in the decision-making process, including the implementation phase (OECD, 2015). Water institutions also need to play a more active role in organizing interaction processes and facilitating the formulation of more sustainable water and sanitation policies.

2.6.4 Case study: decentralized collection, wastewater treatment and reuse in rural area in Jordan

Bringing new technologies to a region where they have not been applied earlier calls for inclusion of international experts who have experience with the technology in question in the decision making. This situation arose when developing sanitation options for two villages in rural Jordan. An initial feasibility study identified the following four potentially interesting technological concepts for sewage collection and treatment:

(1) Conventional and simplified sewer in combination with waste stabilization ponds;
(2) Conventional and simplified sewer in combination with French-type constructed wetlands;
(3) Conventional sewer in combination with anaerobic main treatment and post-treatment in constructed wetlands; and
(4) Solids-free sewer in combination with constructed wetlands.

Two anaerobic reactor types were considered in option 3: the Upflow Anaerobic Sludge Blanket (UASB) reactor and the Anaerobic Baffled Reactor (ABR). Both systems are proven technology for the treatment of domestic wastewater, although the experience in Jordan is limited.

The four concepts were compared using the SI method. Experts from different Jordanian institutions and organisations (including the governmental, academic, water and development sectors) adapted the list of criteria and indicators from Seghezzo (2004) to the local situation, and assigned weights to them. The motivation to include only national experts in this step was that they are well aware of the situation in the rural areas of the country and could thus significantly contribute to the evaluation. Both national and international technical experts subsequently participated in assigning performance values to each indicator for the four different sanitation concepts. The motivation to include international experts in the performance evaluation was the fact that some of the considered technologies were not yet implemented in Jordan. After the importance was assigned to each criterion and indicator, the SI of the four concepts was calculated.

Concept 3, with anaerobic treatment as a core technology, obtained not only the overall highest SI score, but also the highest score for the technical, environmental and social criteria (three out of five). Concept 4 included solids-free sewers, an alternative sewer system not yet applied in Jordan. It scored highest for economic aspects and lowest with respect to institutional aspects, since its application would demand an adaptation of the national standards.

Despite the higher costs for a UASB as compared to an ABR, the UASB was selected as the main treatment step in concept 3, as it allows for easier collection of biogas and GHG emissions were considered an important factor.

2.7 FINAL REMARKS

Over the last half of a century many international developments took place aiming at reducing the pressure on our planet's resources. Environmental sustainability as a goal was endorsed and included in the UN MDGs. International organizations such as WHO and UNICEF joined forces to monitor progress towards MDG targets. Improving sanitation worldwide was demonstrated to be a challenge, and as an era of MDGs was coming to an end, it was clear that the sanitation target will not be met. In addition, a need for an integrated approach in sanitation was identified, rather than assuring only access to toilets (Tilley *et al.*, 2014). SDG goal number 6 took a more holistic view for tackling sustainable sanitation challenges and the circular economy approach is acknowledged to potentially play an important role in achieving the sanitation targets for 2030.

Anaerobic reactors provide a robust technology that can be employed in different configurations of sanitation chains, and obtain high scores in sustainability analyses. Anaerobic treatment is a valuable tool in reaching the SDGs on sanitation in (sub)tropical countries. It allows for efficient removal of organic matter while facilitating resource recovery, allowing the link between sanitation and agriculture to be restored, thus creating circularity. Certainly, also in developing countries where the resources water, nutrients and energy are in short supply, it is vital that these are recovered and cycles are closed.

The local situation does not always provide sufficient agricultural land to allow reuse of (all) effluent, which usually means that the water needs further treatment, to be discharged to the surface water. In many countries, legislation effectively prevents treated sewage to be used in agriculture, by prohibiting the reuse of all treated sewage or employing very strict limits on the concentration of nitrogen and phosphorous compounds, pathogens and/or other components. The use of sewage sludge is also very restricted in several countries. There are different developments in this field, investigating the recovery of inorganic fertilisers from anaerobically treated sewage and the production of hygienically safe sludge for agricultural application. See also Chapters 11 and 12. Moreover, adjustment of irrigation practices are being applied to avoid pathogen transmission to humans through crops (Halalsheh *et al.*, 2008). Social, cultural, and political aspects can dramatically affect the acceptability and effective use of certain technologies and need to be taken into account by both researchers and policy makers interested in the sustainability of water and sanitation management systems. For truly sustainable sanitation it is vital that the worlds of sanitation and agriculture are once again connected and working together in the circular economy; that is the future.

REFERENCES

Akvo (2018). Akvopedia – Sustainability Portal. See https://akvopedia.org/wiki/SustainabilityPortal (accessed 6 March 2019).

Bartram J., Brocklehurst C., Fisher M. B., Luyendijk R., Hossain R., Wardlaw T. and Gordon B. (2014). Global monitoring of water supply and sanitation: history, methods and future challenges. *International Journal of Environmental Research and Public Health*, **11**, 8137–8165.

Bell B. and Morse S. (2008). Sustainability Indicators: Measuring the Immeasurable? Routledge, London, UK.

Buysman E. (2009). Anaerobic digestion for developing countries with cold climates. MSc thesis, Wageningen University, Netherlands.

Chernicharo C. A. L. and Almeida P. G. S. (2011). Feasibility of UASB/trickling filter systems without final clarifiers for the treatment of domestic wastewater in small communities in Brazil. *Water Science and Technology*, **64**, 1347–1354.

Chernicharo C. A. L., van Lier J. B., Noyola A. and Bressani-Ribeiro T. (2015). Anaerobic sewage treatment: state of the art, constraints and challenges. *Reviews in Environmental Science and Bio/Technology*, **14**(4), 649–679.

Cunha J. R., Schott C., van der Weijden R. D., Hernández Leal L., Zeeman G. and Buisman C. (2018). Calcium addition to increase the production of phosphate granules in anaerobic treatment of black water. *Water Research*, **130**, 333–342.

De Graaff M. S., Temmink H., Zeeman G. and Buisman C. J. N. (2010). Anaerobic treatment of concentrated black water in a UASB reactor at a short HRT. *Water*, **2**, 101–119.

Eriksson E., Auffarth K., Henze M. and Ledin A. (2002). Characteristics of grey wastewater. *Urban Water*, **4**(1), 85–104.

Evans B., Hueso A., Johnston R., Norman G., Pérez E., Slaymaker T. and Trémolet S. (2017). Editorial: Limited services? The role of shared sanitation in the 2030 Agenda for Sustainable Development. *Journal of Water Sanitation and Hygiene for Development*, **7**, 349–351.

Firmansyah I., Spiller M., Carsjens G. J., de Ruijter F. J. and Zeeman G. (in preparation). An approach for a holistic evaluation of the performance of sanitation concepts.

Halalsheh M., Ghunmi L. A., Al-Alami N. and Fayyad M. (2008). Fate of Pathogens In Tomato Plants and Soil Irrigated With Secondary Treated Wastewater. In: Efficient Management of Wastewater, I.A. Baz, R. Otterpohl and C. Wendland (eds), Springer, Berlin, Heidelberg.

Hellström Kärrman D. E. (1995). Nitrogen and phosphorus in fresh and stored urine. *Proceedings of the 2nd International Conference on Ecological Engineering for Wastewater Treatment. Environmental Research Forum, Waedenswil, Switzerland.* Zurich, Switzerland: Transtec.

Hernández Leal L., Temmink H., Zeeman G. and Buisman C. J. N. (2010). Bioflocculation of grey water for improved energy recovery within decentralized sanitation concepts. *Bioresource Technology*, **101**, 9065–9070.

Hernández Leal L., Tervahaut T. and Zeeman G. (2017). Resource recovery from source separated domestic wastewater; energy, water, nutrients and organics. In: Innovative Wastewater Treatment & Resource Recovery Technologies: Impacts on Energy, Economy and Environment, J. M. Lema and S. Suarez Martinez (eds), IWA Publishing, London, UK.

Iribarnegaray A. M. and Seghezzo L. (2012). Governance, sustainability and decision making in water and sanitation management systems. *Sustainability*, **4**(11), 2922–2945.

Iribarnegaray M. A., Copa F. R., Gatto D., Andrea M. L., Arredondo M. F., Cabral J. D., Correa J. J., Liberal V. I. and Seghezzo L. (2012). A comprehensive index to assess the sustainability of water and sanitation management systems. *Journal of Water Sanitation and Hygiene for Development*, **2**(3), 205–222.

Iribarnegaray A. M., D'Andrea L. M., Rodriguez-Alvarez S. M., Hernández E. M., Brannstrom C. and Seghezzo L. (2015). From indicators to policies: open sustainability assessment in the water and sanitation sector. *Sustainability*, **7**(11), 14537–14557.

Jönsson H. (2016). Recovery and reuse of resources from source separated domestic wastewater. Presentation at Farewell Symposium for Grietje Zeeman, Wageningen University, Netherlands.

Jönsson H., Stenström T.-A., Svensson J. and Sundin A. (1997). Source separated urine-nutrient and heavy metal content, water saving and faecal contamination. *Water Science and Technology*, **35**(9), 145–152.

King F. H. (1911). Farmers of Forty Centuries, Or Permanent Agriculture in China, Korea and Japan, Organic Gardening Press. Emmaus, Pennsylvania, USA.

Kujawa-Roeleveld K., Elmitwalli T., van Leeuwen M., Tawfik A., de Mes T. and Zeeman G. (2003). Anaerobic digestion and physiological waste and kitchen refuse towards resource management in the DESAR concept. 2nd IWA/GTZ International Symposium on Ecological Sanitation, Lübeck, Germany.

Kujawa-Roeleveld K. and Zeeman G. (2006). Anaerobic treatment in decentralised and source-separation-based sanitation concepts. *Reviews in Environmental Science and Bio/Technology*, **5**, 115–139.

LeAF (2005–current). Non-public reports involving MCA of wastewater management concepts.

Lettinga G., de Man A., van der Last A. R. M., Wiegant W., van Knippenberg K., Frijns J. and van Buuren J. C. L. (1993). Anaerobic treatment of domestic sewage and wastewater. *Water Science and Technology*, **27**(9), 67–73.

Lettinga G., van Velsen A. F. M., Hobma S. W., de Zeeuw W. and Klapwijk A. (1980). Use of the upflow sludge blanket (USB) reactor concept for biological wastewater treatment, especially for anaerobic treatment. *Biotechnology and Bioengineering*, **22**, 699–734.

Lopez da Silva W. T. (2014). Saneamento básico rural. Brasília, DF, Empresa Brasileira de Pesquisa Agropecuária (Rural basic sanitation, Brasília, DF: Brazilian Agricultural Research Company). Embrapa Instrumentação Ministério da Agricultura, Pecuária e Abastecimento.

Mahmoud N., Zeeman G., Gijzen H. and Lettinga G. (2004). Anaerobic sewage treatment in a one-stage UASB reactor and a combined UASB-digester system. *Water Research*, **38**, 2348–2358.

Mara D. (2006). Water Sector in Small Urban Centres. Water Supply and Sanitation Options for Small Urban Centres in Developing Countries, Geneva, Switzerland: UN-Habitat.

Maurer M., Schwegler P. and Larsen T. A. (2003). Nutrients in urine: energetic aspects of removal and recovery. *Water Science and Technology*, **48**(1), 37.

McKeown R. M., Scully C., Enright A.-M., Chinalia F. A., Lee C., Mahony T., Collins G. and O'Flaherty V. (2009). Psychrophilic methanogenic community development during long-term cultivation of anaerobic granular biofilms. *ISME Journal*, **3**, 1231.

Murphy H. M., McBean E. A. and Farahbakhsh K. (2009). Appropriate technology – A comprehensive approach for water and sanitation in the developing world. *Technology in Society*, **31**(2), 158–167.

OECD (2015). *Stakeholder Engagement for Inclusive Water Governance*, OECD, Paris, France.

Ostrom E. (2009). A general framework for analyzing sustainability of social-ecological systems. *Science*, **325**(5939), 419.

RUN4LIFE. (2017–2021). EU H2020 project Recovery and Utilization of Nutrients 4 Low Impact Fertilizer. See http://run4life-project.eu/

Schroeder P., Anggraeni K. and Weber U. (2018). The relevance of circular economy practices to the Sustainable Development Goals. *Journal of Industrial Ecology*, **23**, 77–95.

SDGF (2018). From MDGs to SDGs. See http://www.sdgfund.org/mdgs-sdgs (accessed April 2018).

Seghezzo L. (2004). Anaerobic treatment of domestic wastewater in subtropical regions. PhD, Wageningen University.

Seghezzo L. (2009). The five dimensions of sustainability. *Environmental Politics*, **18**(4), 539–556.

Seghezzo L., Zeeman G., van Lier J. B., Hamelers H. V. M. and Lettinga G. (1998). A review: The anaerobic treatment of sewage in UASB and EGSB reactors. *Bioresource Technology*, **65**(3), 175–190.

Smith B. E. (2002). Nitrogenase reveals its inner secrets. *Science*, **297**(5587), 1654–1655.

STOWA (2001). Separate Urine Collection and Treatment. Options for Sustainable Wastewater Systems and Mineral Recovery, STOWA, Utrecht, Netherlands.

Tilley E., Strande L., Lüthi C., Mosler H.-J., Udert K. M., Gebauer H. and Hering J. G. (2014). Looking beyond technology: an integrated approach to water, sanitation and hygiene in low income countries. *Environmental Science & Technology*, **48**(17), 9965–9970.

UN (2015). Transforming our World: The 2030 Agenda for Sustainable Development, United Nations. See https://sustainabledevelopment.un.org/post2015/transformingourworld/publication

UN (2018). *Goal 7: Ensure environmental sustainability*. See http://www.un.org/millenniumgoals/environ.shtml (accessed April 2018).

Valentin A. and Spangenberg J. H. (2000). A guide to community sustainability indicators. *Environmental Impact Assessment Review*, **20**(3), 381–392.

van Buuren J. C. L. (2010). Sanitation Choice Involving Stakeholders – A Participatory Multi-Criteria Method for Drainage and Sanitation System Selection in Developing Cities Applied in Ho Chi Minh City, Vietnam. PhD, Wageningen University.

Vasconcelos Fernandes T., Shrestha R., Sui Y., Papini G., Zeeman G., Vet L. E. M., Wijffels R. H. and Lamers P. (2015). Closing domestic nutrient cycles using microalgae. *Environmental Science & Technology*, **49**(20), 12450–12456.

Vinnerås B. (2002). Possibilities for sustainable nutrient recycling by faecal separation combined with urine diversion, Department of Agricultureal Engineering. Swedish University of Agricultural Sciences.

WCED (1987). Our Common Future, UN World Commission on Environment and Development Oxford University Press, Oxford.

WELL (2004). Water Supply and Sanitation Sector Monitoring A new decade to measure. Leicestershire: WEDC.

WHO/UNICEF (2015). 2015 ANNUAL REPORT of the WHO/UNICEF Joint Monitoring Programme for Water Supply and Sanitation, WHO/UNICEF. See https://washdata.org/sites/default/files/documents/reports/2017-07/JMP-2015-Annual-Report.pdf

WHO/UNICEF (2017). Progress on Drinking Water, Sanitation and Hygiene: 2017 Update and SDG Baselines. World Health Organization (WHO) and the United Nations Children's Fund (UNICEF), Geneva.

Zeeman G. (2012). New Sanitation – Inaugural Speech, Wageningen University, Netherlands. Available from: https://www.wur.nl/en/Publication-details.htm?publicationId=publication-way-343335303333 (accessed 14 March 2019).

Zhang L., De Vrieze J., Hendrickx T. L. G., Wei W., Temmink H., Rijnaarts H. and Zeeman G. (2018). Anaerobic treatment of raw domestic wastewater in a UASB-digester at 10°C and microbial community dynamics. *Chemical Engineering Journal*, **334**, 2088–2097.

Chapter 3

Fundamentals of anaerobic sewage treatment

Sérgio F. Aquino[1], Juliana C. Araújo[2], Fabiana Passos[2], Thomas P. Curtis[3] and Eugênio Foresti[4]

[1]Environmental and Chemical Technology Group, Department of Chemistry, Universidade Federal de Ouro Preto (UFOP), Ouro Preto, Minas Gerais, Brazil
[2]Department of Sanitary and Environmental Engineering (DESA), Universidade Federal de Minas Gerais (UFMG), Av. Antonio Carlos 6627, Belo Horizonte, Brazil
[3]School of Civil Engineering and Geosciences, Newcastle University, Newcastle upon Tyne, United Kingdom
[4]Department of Hydraulic and Sanitation, School of Engineering, University of São Paulo (USP), São Carlos, SP, Brazil

3.1 INTRODUCTION: SEWAGE CHARACTERIZATION

Domestic sewage is a diluted wastewater ($\leq$1,000 mgCOD L^{-1}), with complex characteristics, and a relatively high suspended solids content, i.e., a low $COD_{soluble}/COD_{total}$ ratio (Van Lier *et al.*, 2010). The composition of the sewage, with regard to soluble and particulate organic matter and biodegradable, toxic, and/or recalcitrant compounds, influences the reactor performance, process kinetics, and by-products outcome. Accordingly, detailed knowledge of the characteristics of such effluent facilitates understanding of the microbial metabolism and the reactor performance. This section contains information on the influent, effluent and biomass characteristics to enhance the reader's knowledge and understanding of anaerobic sewage treatment.

The purpose of most studies on the characterization of domestic wastewater is the modelling and controlling of the biological treatment of sewage and, in particular, the activated sludge model (ASM). In such studies, domestic sewage is defined as a complex mixture of molecules and heterobiopolymers represented by lipids, proteins, sugars, nucleic acids and humic substances. The main parameters related

Editor(s): Carlos Augusto de Lemos Chernicharo and Thiago Bressani-Ribeiro
doi: 10.2166/9781780409238_0023

to sewage characteristics are the organic matter content, particle size and biodegradability. Organic matter may be described in terms of biochemical oxygen demand (BOD), total organic carbon (TOC) and chemical oxygen demand (COD). However, COD is the most established parameter, as employing it enables more-appropriate correlation of the influent substrate, biomass and dissolved oxygen in terms of electron equivalence (Henze *et al.*, 2002; Orhon & Çokgör, 1997). Accordingly, sewage is characterized generally as COD and subsequently divided further into various categories, namely, particulate and soluble fractions (Figure 3.1), and biodegradable, inert or recalcitrant fractions (Figure 3.2). COD characterization also enables performing mass balances.

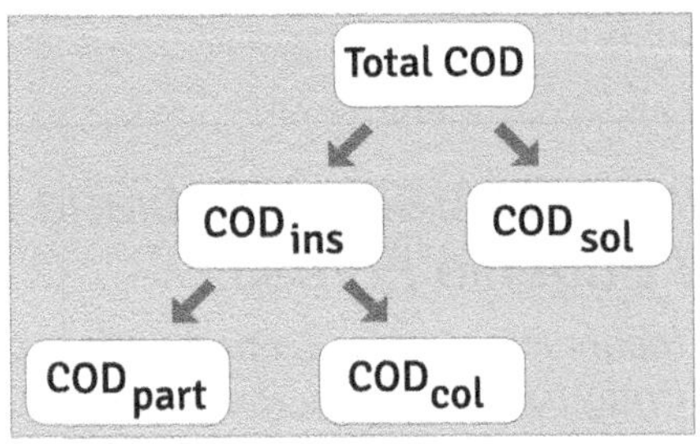

Figure 3.1 Classification of sewage COD according to its particle size (Chernicharo, 2007). COD_{ins}: COD of material retained by 1.2 μm pore-size membranes; COD_{sol}: COD of material passing through 1.2 μm pore-size membranes; COD_{part}: COD related to particulate matter; COD related to colloidal matter.

The biodegradable organic matter present in raw sewage is estimated by the amount of substrate which is degraded biologically in anaerobic conditions, fraction of the sewage contents available for microbial growth. Experimental analyses have indicated that the biodegradable COD is approximately 80–90% of the total COD present in the sewage. However, in practice, only approximately 70% of the COD is removed in an anaerobic reactor (Chernicharo, 2007). This difference is related to the amount of organic matter not removed during the process because of either the operational/environmental conditions of the reactor (e.g., short circuiting, which reduces the contact time of the substrate and biomass, or a drop in temperature, which lowers the rates of kinetic biodegradation), or the biodegradable material produced during anaerobic metabolism (such as volatile fatty acids [VFA], extracellular polymeric substances [EPS], and other microbial products).

Biodegradable COD can be divided further into readily biodegradable, rapidly hydrolysable and slowly hydrolysable, as shown in Figure 3.2. The readily biodegradable COD comprises the soluble compounds present in the sewage as simple carbohydrates, alcohols, amino acids and VFA (produced along the sewerage system), and other compounds that can be used directly for microbial synthesis (Orhon & Çokgör, 1997). However, most biodegradable COD needs to be hydrolyzed in the reactor, including rapidly and slowly hydrolysable COD. The rapidly hydrolysable COD is expected to be degraded in the reactor and represents larger compounds that are converted easily into readily soluble biodegraded COD. However, the slowly hydrolysable COD may be or not degraded in the reactor depending on various factors, such as hydraulic retention time (HRT), sludge retention time (SRT), temperature and organic loading rate (OLR). This slowly hydrolysable fraction consists of particulate COD that cannot pass through the microbial cell wall and needs to undergo extracellular hydrolysis prior to being taken up. Therefore, it is the most difficult fraction to determine; however, it is regarded as the most important in terms of the reactor performance, as it describes the microbial growth and the rate-limiting step of the anaerobic process (Henze *et al.*, 2002).

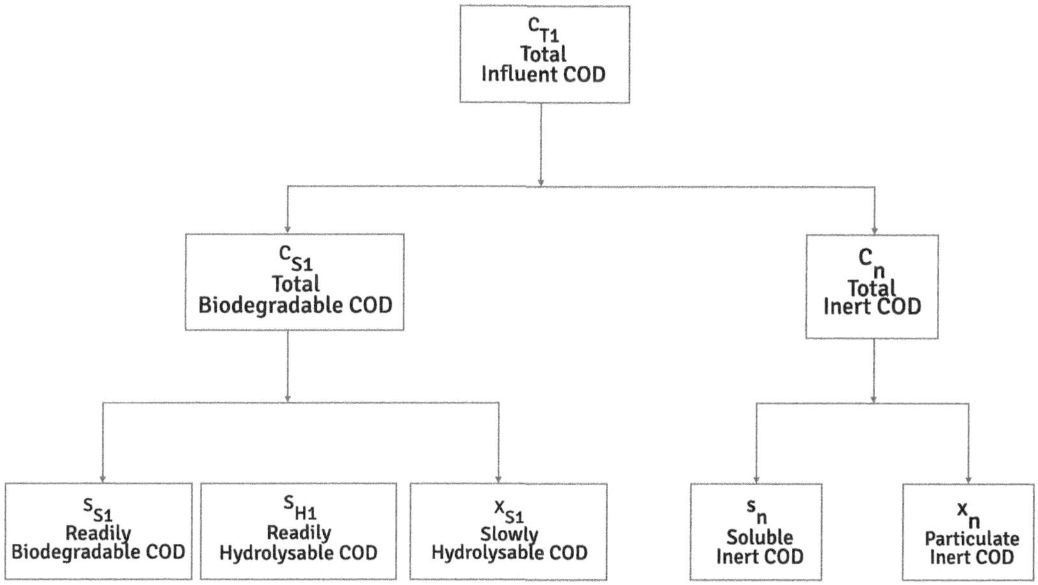

Figure 3.2 Classification of biodegradable and inert COD (Orhon & Çokgör, 1997).

A possible approach to determining biodegradable organic matter is to divide COD in terms of particle size, as shown in Figure 3.1. Previous studies have shown that domestic sewage contains approximately 27–52% particulate COD (settleable solids, >1 μm), 18–31% colloidal COD (0.08–1 μm) and 24–43% soluble COD (<0.08 μm) (Guellil *et al.*, 2001; Mahmoud *et al.*, 2003; Munch *et al.*, 1980). Suspended solids could compromise sewage anaerobic treatment by reducing the biomass activity, limiting substrate transfer, inhibiting granulation and increasing the accumulation of inert material in the reactor. Therefore, pre-treatment methods can be applied to increase particle surface contact through physical cleavage, thereby increasing the biodegradability of the sewage (Van Lier *et al.*, 2001).

In this regard, a study was conducted using forced sieving as a pre-treatment step prior to an upflow anaerobic sludge blanket (UASB) reactor. The results indicated that although the hydrolysis rate was not affected by the smaller particle size, the biomass grown in the reactor showed higher methanogenic activity. Compared with the non-pre-treated control UASB reactor, the specific methanogenic activity (SMA) was 72% higher (0.31 $gCOD \cdot gVS^{-1} \cdot d^{-1}$), COD removal 20% higher (65% COD removal), and the effluent produced had lower particulate organic matter content (48 *versus* 74 $mg\ COD \cdot L^{-1}$) (Teixeira, 2007).

The inert or recalcitrant organic COD fraction cannot be degraded in the reactor, as it is biologically resistant and, therefore, unaffected during anaerobic treatment (Chernicharo, 2007). It must be noted that inert COD can be present in the sewage or could be generated during the biological treatment as a residual soluble fraction of microbial products (SMP), as is discussed further in Section 3.7. SMPs are not quite biodegradable and, therefore, are commonly present in the effluent (Orhon & Çokgör, 1997).

In addition to COD biodegradability, nutrient concentration of the sewage could also affect the anaerobic digestion performance. The nutrient concentration required for methanogenic metabolism is shown in Table 3.1. Nonetheless, all appropriate macro- and micronutrients are generally present at suitable concentrations in domestic sewage. This provides the ideal environment for microbial conversion, with no limitations to the treatment process (Chernicharo, 2007).

Table 3.1 Nutrients required for methanogenic activity (Chernicharo, 2007; Lettinga *et al.*, 1996).

Nutrient	Concentration ($g \cdot kgTSS^{-1}$)
Nitrogen	65
Sulphur	10
Phosphorus	15
Iron	1.8
Cobalt	0.075
Nickel	0.1
Molybdenum	0.06
Selenium	Traces
Riboflavin	Traces
Vitamin B12	Traces

In addition, as pH has a strong controlling effect on the growth rate of anaerobic microorganisms, the alkalinity concentration would have an effect on the microbial selection. However, in general, domestic sewage has relatively high alkalinity and, when combined with a high organic loading rate and upflow velocity, this is sufficient to buffer the pH at values above 7. Therefore, generally, no additional chemicals are required for the process.

Although domestic sewage has, on average, a constant and balanced composition in terms of COD and nutrients, variations in the influent characteristics could occur because of contamination by other waste streams, such as industrial effluents, or from dilution by rain water. This will affect the COD concentration and biodegradability, as well as the nutrient content in the influent. For instance, some industrial streams contain recalcitrant and toxic materials, which could inhibit the metabolism and equilibrium of anaerobic microorganisms. This topic is discussed in detail in Section 3.6. For combined sewer system, rainwater runoff will be collected with domestic sewage, thus increasing influent flow and sewage dilution (reducing the influent organic matter concentration). This will contribute to reduce biogas production rates in anaerobic reactors.

In general, biogas production and composition do not vary in an anaerobic reactor operating in steady state. The biogas composition depends on the environmental conditions and could be affected particularly in the initial start-up phase and when methanogenesis is inhibited. For domestic sewage, the CH_4 and CO_2 concentrations are, approximately, 70–80% and 5–15%, respectively (biogas composition is further addressed in Chapter 8). For biogas production, except for the losses caused by microbial inefficiency, almost all the COD removed is recovered as CH_4 (Chernicharo, 2007). Detailed information on biogas management is presented in Chapter 8.

3.2 MICROBIOLOGY OF ANAEROBIC TREATMENT

In this section, information is presented on the microbiology of anaerobic conversions that occur generally in anaerobic environments (including anaerobic reactors). In addition, specific information on the microbiology of anaerobic sewage treatment (as available in the literature) is provided and discussed.

The anaerobic degradation of organic matter to end-products involves the formation of various metabolites and the interactions of different groups of microorganisms. Under favourable conditions for the interaction of the microbial community, the primary end-products are methane and carbon dioxide.

The anaerobic degradation of organic matter involves four major microbial processes, namely, hydrolysis, acidogenesis, acetogenesis, and methanogenesis (Figure 3.3). These steps are carried out by several types of microorganisms that act in a sequential manner and according to their syntrophic interactions. Malfunctioning in a single metabolic group can, therefore, affect the entire microbial community, resulting in process imbalance or failure (Gujer & Zehnder, 1983). In anaerobic environments and reactors, sulphate-reducing bacteria are always present even when the sulphate concentration is low (in sewage, the sulphate concentration could vary between 30 and 250 mg L^{-1} [WEF, 1995]). Therefore, the scheme of anaerobic degradation of organic matter should include the sulphate-reduction step (Fig. 3.3). Accordingly, the complete mineralization of organic matter under anaerobic conditions will produce methane (CH_4), carbon dioxide (CO_2), ammonium (NH_3), hydrogen sulphide (H_2S) and water (H_2O). The overall reaction is represented in equation 3.1 (Polprasert, 1989).

$$\text{Organic matter} \rightarrow CH_4 + CO_2 + H_2 + NH_3 + H_2S \tag{3.1}$$

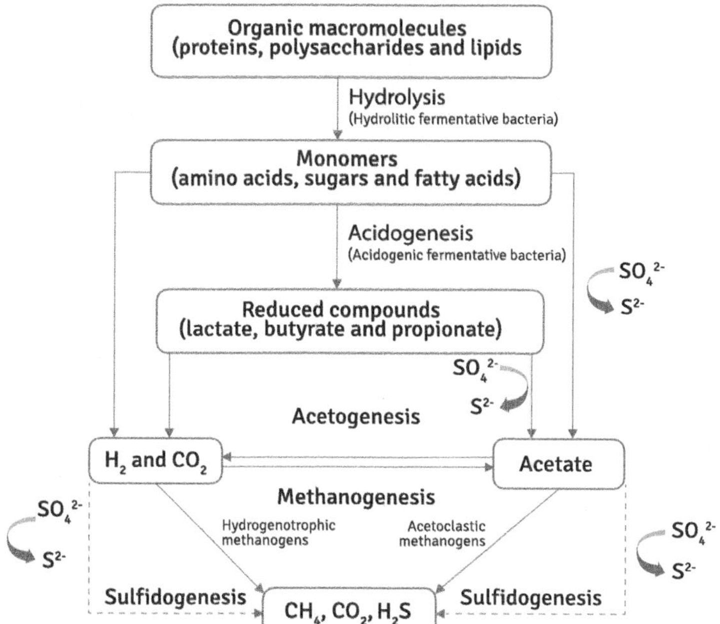

Figure 3.3 Scheme of the anaerobic degradation of organic matter usually present in sewage, with the addition of sulphate reduction pathways (dashed gray lines). Adapted from Muyzer and Stams (2008).

In the first step in anaerobic degradation, macromolecules, such as proteins, polysaccharides and lipids are hydrolyzed by hydrolytic fermentative bacteria to monomers, such as amino acids, sugars and fatty acids. In the second step, the monomers are fermented by acidogenic fermentative bacteria into different fermentation products, such as acetate, propionate, butyrate, lactate and hydrogen. In the third step, hydrogen-producing acetogenic bacteria convert the higher volatile fatty acids, e.g., propionic and butyric acids, to H_2, CO_2, and acetic acid. In the presence of sulphate, even at low concentrations as for domestic sewage (50–200 mg L^{-1}; Barbosa & Sant'Anna, 1989), sulphate-reducing bacteria can consume the fermentation products (as electron donors for sulphate reduction). In this way, hydrogen sulphide is formed and will be present in the biogas and also dissolved in the liquid (5–22 mg L^{-1} H_2S; Azevedo *et al.*, 2018; Garcia *et al.*, 2015; Souza *et al.*, 2012). These sulphide concentrations found in methanogenic reactors treating sewage are lower than are the values reported in the literature for methanogen inhibition (50–250 mgH_2S L^{-1} at pH 7 to 8; O'Flaherty *et al.*, 1998). The fourth and last step is the formation of methane by the methanogens from the consumption of acetate, hydrogen, and carbon dioxide.

3.2.1 Hydrolysis

Polymeric materials such as lipids, proteins, and carbohydrates are hydrolyzed first by the extracellular enzymes (also called "exoenzymes") excreted by fermentative microbes. Hydrolytic enzymes (lipases, proteases, cellulases, amylases and the like) break down their respective substrates (biopolymers) into smaller molecules, primarily monomeric units that are consumed subsequently by microbes. Depending on the type of polymer present in the wastewater (or in the waste to be digested), the hydrolysis can be a slow process and could be a rate-limiting step in the production of simpler bacterial substrates to be used in the subsequent degradation steps.

Lipases convert lipids to long-chain fatty acids, which are degraded further by β-oxidation reactions to produce acetyl CoA. Polysaccharides such as cellulose, starch, and pectin are hydrolyzed by cellulases, amylases, and pectinases, respectively. Proteins are hydrolyzed generally to amino acids by proteases. The amino acids produced are degraded subsequently into fatty acids, such as acetate, propionate, and butyrate, and to ammonia.

A wide variety of bacterial genera within the phyla Firmicutes and Synergistetes, order Synergistales (Vos *et al.*, 2009), Bacteroidetes (Delbès *et al.*, 2000) and Chloroflexi (Ariesyady *et al.*, 2007) are thought to be hydrolytic and fermentative bacteria. They have been detected in anaerobic reactors treating sewage (Li & Lu, 2017) and in full-scale anaerobic digesters (Sundberg *et al.*, 2013). However, it is difficult to separate the fermentative bacteria into hydrolytic and acidogenic groups. For instance, Bacteroidales (Bacteroidetes phylum), affiliated with the class Bacteroidia, has been identified as hydrolyzers and fermenters involved in the degradation of cellulose and polysaccharide into VFAs (Goux *et al.*, 2015). Clostridiales, affiliated with the class Clostridia, has diverse metabolism that can ferment polysaccharide to produce VFAs for methanogenesis (Sundberg *et al.*, 2013). Members of Anaerolineaceae (Chloroflexi) are involved in saccharide degradation (Yamada *et al.*, 2006). Within Firmicutes (class Clostridia), which are fermenting bacteria, the Syntrophomonadaceae is a group of acetogenic bacteria that can oxidize butyrate to acetate (Stams, 1994). The development of metagenome assembly could facilitate the determining of the functions of the different types of bacteria.

3.2.2 Fermentation (acidogenesis)

In this step, the sugars, amino acids, and fatty acids (monomers) produced by the microbial degradation of biopolymers are metabolized successively, by separate groups of acidogenic fermenting bacteria, to acetate, propionate, butyrate, lactate, ethanol, carbon dioxide, and hydrogen. The biological reactions in the acidogenic step are thermodynamically favourable (Table 3.2), and acidogenic bacteria have the highest growth rates (Thauer *et al.*, 1977). Fermenting bacteria affiliated with Firmicutes, Proteobacteria, Bacteroidetes and Chloroflexi correspond to approximately 90% of the total bacterial phyla detected in anaerobic reactors treating sewage (Li & Lu, 2017).

Table 3.2 Acidogenic, acetogenic, sulphate-reducing and methanogenic reactions that usually occur in anaerobic reactors treating domestic sewage.

Reactions	ΔG°′ (kJ/Reaction)*	Microbial Groups
Acidogenic		
$\underset{\text{(Glucose)}}{C_6H_{12}O_6} + 2H_2O \rightarrow \underset{\text{(Acetate)}}{2C_2H_3O_2^-} + 2CO_2 + 2H^+ + 4H_2$	−206	Members of Firmicutes (Bacilli, Clostridia), Bacteroidetes (Bacteroidales), Chloroflexi (Anaerolineae), Synergistetes (*Aminomonas, Aminobacterium*).
$\underset{\text{(Glucose)}}{C_6H_{12}O_6} + 2H_2 \rightarrow \underset{\text{(Propionate)}}{2C_3H_5O_2^-} + 2H_2O + 2H^+$	−358	
$\underset{\text{(Glucose)}}{C_6H_{12}O_6} \rightarrow \underset{\text{(Butyrate)}}{C_4H_7O_2^-} + 2CO_2 + H^+ + 2H_2$	−255	
Acetogenic reactions		
$\underset{\text{(Propionate)}}{C_3H_5O_2^-} + 3H_2O \rightarrow \underset{\text{(Acetate)}}{C_2H_3O_2^-} + \underset{\text{(Bicarbonate)}}{HCO_3^-} + H^+ + 3H_2$	+76.1	*Desulfovibrio, Desulfomicrobium, Smithella,* and species of *Syntrophobacter*
$\underset{\text{(Butyrate)}}{C_4H_7O_2^-} + 2H_2O \rightarrow \underset{\text{(Acetate)}}{2C_2H_3O_2^-} + H^+ + 2H_2$	+48.3	*Syntrophomonadaceae (Syntrophomonas wolfei)*
$\underset{\text{(Lactate)}}{C_3H_5O_3^-} + 2H_2O \rightarrow \underset{\text{(Acetate)}}{C_2H_3O_2^-} + HCO_3^- + H^+ + 2H_2$	−4.2	*Desulfovibrio, Desulfomicrobium*
Homoacetogenic		
$4H_2 + \underset{\text{(Bicarbonate)}}{2HCO_3^-} + H^+ \rightarrow \underset{\text{(Acetate)}}{C_2H_3O_2^-} + 4H_2O$	−104.6	Sulphate reducers
$\underset{\text{(Lactate)}}{C_3H_5O_3^-} \rightarrow \underset{\text{(Acetate)}}{1.5C_2H_3O_2^-} + 0.5H^+$	−56.5	Sulphate reducers

(*Continued*)

Table 3.2 Acidogenic, acetogenic, sulphate-reducing and methanogenic reactions that usually occur in anaerobic reactors treating domestic sewage (*Continued*).

Reactions	$\Delta G^{o\prime}$ (kJ/Reaction)*	Microbial Groups
Sulphate reducing		
$4H_2 + SO_4^{2-} + H^+ \rightarrow HS^- + 4H_2O$ (Sulphate) (Hydrogen sulphide ion)	−151.9	Sulphate reducers
$C_2H_3O_2^- + SO_4^{2-} \rightarrow 2HCO_3^- + HS^-$ (Acetate) (Bicarbonate)	−47.6	Sulphate reducers
Methanogenic		
$4H_2 + HCO_3^- + H^+ \rightarrow CH_4 + 3H_2O]] >$ (Bicarbonate) (Methane)	−135.6	Hydrogenotrophic methanogens Members of *Methanomicrobiales*, *Methanococccales*, and *Methanobacteriales*
$C_2H_3O_2^- + H_2O \rightarrow CH_4 + HCO_3^-]] >$ (Acetate) (Methane)	−31.0	Acetoclastic methanogens Members of *Methanosarcinales*

*Data from Thauer *et al.* (1977).

Cardinali-Rezende *et al.* (2013) have reported that Proteobacteria, Bacteroidetes and Firmicutes (Clostridia class) phyla predominated in 16SrRNA gene libraries of sludge samples from UASB reactor treating sewage at mesophilic temperature. Sinischalchi *et al.* (2015) also reported that most of the identified sequences retrieved from an anaerobic sludge of a UASB reactor treating domestic sewage are related to putative fermentative bacteria belonging to the orders Synergistales (48%) and Clostridiales (9%). Therefore, most of the sequences reported in anaerobic reactors are thought to be fermentative bacteria. Fermentation is not regarded as a limiting step in the anaerobic process (Gujer & Zehnder, 1983). The acidogenesis step will limit the process only if the substrate to be degraded were not hydrolyzed readily (Moraes *et al.*, 2015).

3.2.3 Acetogenesis

In this step, the acidogenesis products are oxidized further to acetate, hydrogen and carbon dioxide, in a process called acetogenesis (Stams, 1994), which is mediated by obligate hydrogen-producing acetogens. The hydrogen-producing acetogenic bacteria can degrade higher fatty acids (e.g., propionic acid and butyric acid) and alcohols (e.g., ethanol) into acetate, hydrogen and carbon dioxide (see equations (3.2) to (3.4)). However, the oxidation of butyric and propionic acids is thermodynamically problematic (close to equilibrium) under standard conditions (1 mol L^{-1} reactants and products, 25°C, and 100 kPa). Even so, such reactions are possible in anaerobic environments because of syntrophic interactions (Stams *et al.*, 1998) between the hydrogen-producing acetogenic bacteria and the hydrogen-consuming microorganisms. The latter can be a hydrogen-consuming methanogen, hydrogen-consuming sulphate-reducing bacteria (SRB), or a homoacetogenic microorganism (Table 3.2).

The conversion of ethanol, propionic acid, and butyric acid into acetic acid is represented in reactions (3.2) to (3.4):

$$\text{(ethanol) } CH_3CH_2OH + H_2O \rightarrow CH_3COOH \text{ (acetic acid)} + 2H_2 \tag{3.2}$$

$$\text{(propionic acid) } CH_3CH_2COOH + 2H_2O \rightarrow CH_3COOH \text{ (acetic acid)} + CO_2 + 3H_2 \tag{3.3}$$

$$\text{(butyric acid) } CH_3CH_2CH_2COOH + 2H_2O \rightarrow 2CH_3COOH \text{ (acetic acid)} + 2H_2 \tag{3.4}$$

Typically, the conversion of fatty acids to acetate is carried out by syntrophic communities of acetogenic bacteria and methanogenic archaea (Muyzer & Stams, 2008). Some examples of acetogenic bacteria able to degrade propionate are sulphate reducers, such as *Desulfovibrio* and *Desulfomicrobium,* and species of *Syntrophobacter* (Muyzer & Stams, 2008). *Syntrophomonas wolfei* is an example of butyrate-degrading bacteria. Species of *Desulfovibrio* and *Desulfomicrobium* grow by fermenting pyruvate to form acetate, carbon dioxide and hydrogen as products. They can oxidize lactate and ethanol to acetate, but only when hydrogen is removed efficiently by hydrogen-consuming methanogens (Muyzer & Stams, 2008).

Syntrophobacter wolinii is a sulphate reducer that, in the presence of a hydrogen-utilizing sulphate reducer, suppresses sulphate reduction and grows as an acetogenic microorganism (Harmsen *et al.*, 1993). It has been reported that the propionate-degrading SRB, *Desulfobulbus propionicus,* could oxidize propionate to acetate more efficiently compared with a syntrophic association of proton-reducing acetogens with hydrogen-utilizing methanogens (Widdel & Pfennig, 1982). Therefore, in anaerobic reactors treating sewage, the oxidation of various organic compounds can occur together with sulphate utilization and, therefore, the presence of sulphate-reducing bacteria in the sludge appears particularly important (Agrawal *et al.*, 1997).

In an anaerobic filter treating real domestic sewage, the genus *Smithella* (which can degrade propionate to acetate) was dominant during winter (3.5%), likely being responsible for the accumulation of the acetate observed in the system (Li & Lu, 2017). In addition, acetate accumulation could be linked to the growth rate of the acetogenic bacteria being much faster than that of the methanogens. The former group has a specific maximum growth rate (μ_{max}) of ~1 h^{-1}, whereas the μ_{max} of the latter group is ~0.04 h^{-1} (Hammer, 1986).

The oxidation of acetate to hydrogen and carbon dioxide can be performed by *Clostridium* sp. in syntrophic association with hydrogenotrophic methanogens (Schnurer *et al.*, 1996). Syntrophic acetate-oxidizing bacteria can be found in both mesophilic (*Clostridium ultunense* and *Syntrophaceticus schinkii*) and thermophilic (*Thermacetogenium phaeum* and *Thermotoga lettingae*) conditions (Hattori *et al.*, 2000; Schink, 1997; Schnurer *et al.*, 1996; Shigematsu *et al.*, 2003).

3.2.4 Methanogenesis

The last step in anaerobic treatment is the formation of methane by the methanogenic archaea that use a limited number of substrates for growth. Hydrogen, carbon dioxide and acetate (produced directly by fermentation or indirectly by acetogenesis) are quantitatively the most important and best-known substrates for the methanogenic archaea. In some instances, methanogenesis may be the controlling step of the process because of the low growth rates of the methanogens (compared with the acidogenic and acetogenic bacteria), and the sensitivity of the methanogenic consortium to various environmental factors, such as pH and ammonia. The optimal pH range for the methanogens is between 6.8 and 7.5 (Gujer & Zehnder, 1983). In relation to ammonia levels, various studies have reported that methanogens could tolerate ammonia levels up to 7 g L^{-1} of total ammonia nitrogen in anaerobic digesters (Schnurer & Nordberg, 2008). Other studies have observed that mesophilic hydrogenotrophic methanogens were more sensitive to ammonia toxicity compared with thermophilic methanogens (Wang *et al.*, 2015). Therefore, an abrupt change in pH (e.g., due to fatty acids accumulation), increase in the salt concentration, change in the organic loading rate, absence of nutrients, or the introduction of toxic compounds often cause system imbalance or failure (Ma *et al.*, 2009). The likelihood of these events occurring in an anaerobic reactor treating domestic sewage is explored further in Section 3.6.

The methanogens belong to domain Archaea and most of the methanogen sequences detected in anaerobic reactors are affiliated with the phylum Euryarchaeota. Compared with the bacterial communities that normally show high diversity (several studies have identified 33 taxonomic categories at phyla level), archaeal communities show low diversity and they have a specialized metabolism. Currently, seven orders of methanogens are known, with at least four having been reported in anaerobic reactors (and digesters), namely: Methanobacteriales, Methanomicrobiales, Methanosarcinales and Methanococcales (Demirel & Scherer, 2008). Li and Lu (2017) have reported that the archaeal community of an anaerobic filter treating real domestic sewage has lower diversity, and most of the sequences are related to Methanosarcinales, Methanobacteriales, and Methanomicrobiales. Methanopyrales, the fifth order, includes only one hyperthermophilic species, which is unlikely to be found in anaerobic reactors (Bapteste *et al.*, 2005). The sixth order, Methanocellales, contains only one genus, *Methanocella*, which is a hydrogenotrophic methanogen that was first isolated from a propionate-degrading culture obtained from rice paddy soils (Sakai *et al.*, 2008). The seventh order, Methanoplasmatales, was proposed by Paul *et al.* (2012) and includes members previously believed to be distantly related to a different lineage in the phylum *Euryarchaeota* (Paul *et al.*, 2012).

The methanogens can be divided into two main groups according to the substrate they utilize (Table 3.2), namely, acetoclastic methanogens (also called aceticlastic methanogens) and hydrogenotrophic methanogens (i.e., hydrogen-utilizing chemolithotrophs). The acetate-utilizing methanogens convert acetate into methane and CO_2 (Table 3.2). This group comprises two main genera, *Methanosarcina* and *Methanosaeta* within the order Methanosarcinales. In this order, Methanosaetaceae is the only family that includes strictly acetoclastic anaerobes (Smith & Ingram-Smith, 2007), whereas the other members (Methanosarcina) are more diverse in terms of metabolism and can metabolize acetate, hydrogen, formate, ethanol, isopropanol and methylated compounds (Kendall & Boone, 2006).

In various anaerobic reactors with low acetate concentrations, the *Methanosaeta* are the dominant acetoclastic methanogens (Zheng & Raskin, 2000). *The Methanosaeta* species are specialists in utilizing acetate, which they can use at concentrations as low as 5–20 μM (or ~0.3 to ~1.2 mg L^{-1}) (Liu & Whitman, 2008). Additionally, the presence of *Methanosaeta* species usually leads to an improved granulation process (presumably because of their filament morphology) and results in bioreactor performance that is more stable (Barbosa & Sant'Anna, 1989). *Methanosarcina*, an acetoclastic methanogen that can also use other substrates, has a higher maximum growth rate but lower affinity for acetate compared with *Methanosaeta* (see Table 3.3).

Table 3.3 Kinetic parameters of microbial growth for acetoclastic methanogens (data from Gujer & Zehnder, 1983; Zinder *et al.*, 1984).

Acetoclastic Methanogens	K_s for Acetate (mg L^{-1})	μ_{max} (day^{-1})
Methanosaeta	30	0.11
Methanosarcina	200	0.3

Methanosarcina sp. have high growth rates (i.e. doubling times in the order of 1.0–1.2 day), are tolerant of sudden changes in pH of approximately 0.8–1.0 units, and are tolerant of high levels of ammonium (up to 7 g L^{-1} total ammonia nitrogen; Schnurer & Nordberg, 2008) compared with the other methanogens (e.g., *Methanosaeta*). Whereas *Methanosaeta sp.* have doubling times of four to six days and tend to be affected by a pH shock of 0.5 units (Shin *et al.*, 2011).

The concentrations of VFAs in domestic sewage are low (according to Pontes *et al.* (2014) the median concentrations of acetic, butyric, and propionic acids in raw sewage are 50.6 mg L^{-1}, 4.0 mg L^{-1}, and 0.54 mg L^{-1}, respectively), the pH is approximately neutral (6.8 to 7.2), and the ammonium concentration varies from 35 to 60 mg L^{-1} (Mota & von Sperling, 2009). Therefore, *Methanosaeta* prevail in anaerobic reactors treating sewage (at mesophilic temperatures), as they can reach the maximum growth velocity with extremely low acetic acid concentration in comparison with *Methanosarcina*.

The hydrogenotrophic methanogens are members of Methanomicrobiales, Methanococcales and Methanobacteriales (such as *Methanobrevibacter, Methanococcus* and *Methanobacterium*, among other genera) that utilize carbon dioxide as electron acceptor and hydrogen as the common electron donor (Table 3.4). However, some species also use formate and/or alcohols, such as ethanol or isopropanol, and members of Methanobacteriales (with the exception of *Methanosphaera*) cannot use acetate or one-single carbon compounds (Bonin & Boone, 2006). In an anaerobic reactor treating real domestic sewage, the hydrogenotrophic genera *Methanobacterium* and *Methanolinea* were the dominant detected species, but other less-abundant hydrogenotrophic genera were also observed, including *Methanospirllum*, *Methanosphaerula* and *Methanosphaera* (Li & Lu, 2017).

Table 3.4 Typical substrates used by methanogenic archaea as electron donors for growth (Adapted from Zehnder, 1988).

Substrate	Microorganism
H_2 (hydrogen)	*Methanobacterium bryantii, M. formicicum, M. thermoautotrophicum, Methanobrevibacter arboriphilus, M. smithii, M. ruminantum, Methanococcus vannielli, M. voltae, Methanomicrobium mobile, Methanogenium cariaci, M. marisnigri, Methanospirillum hungatei, Methanosarcina barkeri, Methanosarcina mazei.*
HCOOH (formic acid)	*Methanobacterium formicicum, Methanobrevibacter ruminantum, M. smithii, Methanomicrobium mobile, Methanogenium cariaci, Methanococcus vannielli, Methanospirillum hungatei, Methanogenium cariaci, M. marisnigri*
CO (carbon monoxide)	*Methanosarcina barkeri*
CH_3OH (methanol)	*Methanosarcina mazei, Methanosarcina barkeri*
CH_3NH_2 (methylamine), $(CH_3)_2NH$ (dimethylamine), $(CH_3)_3N$ (trimetylamine)	*Methanosarcina mazei, Methanosarcina barkeri*
$CH_3CH_2N(CH_3)_2$ (ethylamine)	*Methanosarcina barkeri*
CH_3COOH (acetic acid)	*Methanosarcina mazei, Methanosarcina barkeri, Methanosaeta soehngenii, Methanosaeta concilii*

3.3 MICROBIAL COMMUNITIES IN ANAEROBIC REACTORS TREATING SEWAGE

Several studies have investigated the microbial community of anaerobic reactors treating various types of wastewater or residues. These include synthetic wastewater (Connaughton *et al.*, 2006; Kundu *et al.*, 2012; McHugh *et al.*, 2004; Qiaoying *et al.*, 2013); brewery wastewater (Diaz *et al.*, 2006; Liu *et al.*, 2002; Xing *et al.*, 2009); and papermill (Roest *et al.*, 2005), pharmaceutical (LaPara *et al.*, 2000), and other industrial wastewater (McKeown *et al.*, 2009); dairy-processing (Lee *et al.*, 2008), dairy (Bialeck *et al.*, 2012), and food-processing wastewater (Narihiro *et al.*, 2009; Tang *et al.*, 2007); swine wastewater (Kim *et al.*, 2010; Song *et al.*, 2010); sewage sludge (Hao *et al.*, 2016); food-waste (Guo *et al.*, 2014; Martins *et al.*, 2017); and other residues (De Vrieze *et al.*, 2015; Regueiro *et al.*, 2012).

However, only a few studies have investigated the microbial diversity in anaerobic reactors treating domestic sewage at tropical temperatures and/or at low temperatures, as well as under different operational conditions. These studies include the following, with detail information summarized in Table 3.5: Barbosa and Sant'Anna (1989); Agrawal *et al.* (1997); Zhang *et al.* (2005); Gomec *et al.* (2008); Silva *et al.* (2008); Gomec *et al.* (2009); Bandara *et al.* (2012); Cardinali-Rezende *et al.* (2013); Smith *et al.* (2015); Saia *et al.* (2016); Petropoulos *et al.* (2017) and Li and Lu (2017).

Typically, in the anaerobic treatment of domestic sewage, 70% of the total methane derives from the acetate-degradation pathway; therefore, the acetoclastic methanogens have been considered a methanogenic species more important (Lettinga, 1995) than the hydrogenotrophic methanogens. However, this can vary depending on the temperature, substrate availability, pH, HRT, SRT, among others. This aspect is discussed further based on various studies (Cardinali-Rezende *et al.*, 2003; Li & Lu, 2017; Petropoulos *et al.*, 2017).

Table 3.5 Studies investigating the composition and structure of microbial communities in anaerobic reactors treating sewage.

Reactor Type	Wastewater	Microbial Community	Reference
Pilot-scale UASB reactor	**Synthetic** Low-strength wastewater (containing 15 to 20 mg L^{-1} of sulphate)	Changes in the activity and bacterial populations of various trophic groups and the morphology of the sludge over time were investigated. The viable cell numbers of trophic groups increased by 2 to 4 orders of magnitude for methanogens, 3 orders for acetogens, and 2 to 3 orders for sulphate-reducing bacteria compared with the numbers of viable cells in the seed sludge.	Agrawal *et al.* (1997)
Laboratory-scale UASB reactor	**Synthetic** and real municipal wastewater	DGGE fingerprints showed that the bacterial community was subject to major changes. The sludges for treating raw and pretreated wastewater had quite similar bacterial and archaeal communities (82% and 96% similarity), but differed from that treating the synthetic sewage. Despite a similar percentage of particulate COD in the synthetic and real wastewaters, the two wastewaters were selected for different microbial communities.	Zhang, Y. *et al.* (2005)
Bench-scale anaerobic membrane bioreactor (AnMBR) at psychrophilic temperatures of 15, 12, 9, 6, and 3°C.	**Synthetic** domestic wastewater	Chemical oxygen demand (COD) removal of >95% was achieved at temperatures as low as 6°C, but this declined to 86% at 3°C. As temperature decreased, soluble COD in the bioreactor increased, suggesting a reduction in suspended biomass activity. Sequencing of 16S rRNA suggested that the metabolic diversity of the biofilm increased as temperature decreased in response to a greater flux of complex organics into the biofilm that is attributable to temperature-based suspended biomass inhibition. Hydrogenotrophic methanogenesis was the preferred pathway in the biofilm compared with acetoclastic methanogenesis, but not in the suspended biomass.	Smith *et al.* (2015)
Pilot-scale UASB	**Real** sewage	This study describes granule formation in a pilot-scale UASB reactor treating sewage and investigates the morphology of the microorganisms present in the sludge. After nine months of reactor operation, granular particles up to 6–8 mm in diameter were evident. The granules showed a heavy structure, with an appreciable quantity of inert material and various morphotypes	Barbosa and Sant'Anna (1989)

(Continued)

Table 3.5 Studies investigating the composition and structure of microbial communities in anaerobic reactors treating sewage (*Continued*).

Reactor Type	Wastewater	Microbial Community	Reference
		(coccus, coccus chains, *sarcinas*, and multicellular rod-shaped filaments). These agglomerations of filaments (*Methanosaeta*-like) appear to have a significant role in the structural formation of the granules.	
Two pilot-scale UASB *reactors* at ambient temperature (25–30°C)	**Real** *sewage*	Two UASB *reactors were operated during five* different HRT and upflow velocities. Microbial diversity was investigated by PCR-DGGE (polymerase chain reaction-denaturing gradient gel electrophoresis), revealing that bacterial and archaeal diversities were not altered by the imposition of different HRT and upflow velocities. Microbial diversity did not change with the increase in the upflow velocity, despite the decrease in COD removal when the highest upflow velocity was imposed. Stable microbial diversity remained in the reactor after environmental perturbations caused by changing HRT, upflow velocities, and the size of solids in the influent.	Silva *et al.* (2008)
Laboratory-scale UASB at 13°C	**Synthetic** domestic wastewater	According to FISH results, archaeal cells representing methanogens were dominant in the bottom sampling port of the UASB reactor, and acetoclastic *Methanosaeta* was the abundant methanogen. Other methanogens such as *Methanosarcina*- and *Methanobacterium*-like species were also observed. The abundance of originally mesophilic *Methanosaeta*-related Archaea under low temperature on all sampling days revealed the microbial adaptation to psychrophilic conditions. Members of the order Methanobacteriales constituted the major hydrogenotrophic methanogens present in the psychrophilic UASB reactor, whereas the other hydrogenotrophic methanogens (Methanococcales and Methanogenium) were absent.	Gomec *et al.* (2008)

(*Continued*)

Table 3.5 Studies investigating the composition and structure of microbial communities in anaerobic reactors treating sewage (*Continued*).

Reactor Type	Wastewater	Microbial Community	Reference
Laboratory-scale UASB reactor at 6–31°C	**Real** wastewater	The COD removal efficiency was more than 50% in summer and below 40% in winter, with no evolution of biogas. Analysis of the archaeal community structures of the granular sludge using 16S rRNA gene cloning indicated that after microorganisms had adapted to the low temperatures, the archaeal community had a lower diversity and the relative abundance of acetoclastic methanogens decreased together with an increase in the hydrogenotrophic methanogens.	Bandara *et al.* (2012)
Demonstration-scale UASB reactor at ambient temperature (25–30°C)	**Real** domestic wastewater	Fluctuations in the OLR and F/M (food/microorganisms) ratio affected the abundance and composition of the UASB prokaryotic community, mainly at the species level, as well as the performance of the UASB reactor. This study revealed higher levels of bacterial diversity and metabolic specialization of the microbes present in the UASB reactor during the rainy season (at lower OLR).	Cardinali-Rezende *et al.* (2013)
Pilot-scale upflow reactors with immobilized biomass	**Real** sewage	The microbial community was investigated of two reactors treating domestic sewage to promote anaerobic digestion and denitrification using endogenous electron donors. Pyrosequencing results revealed that in the anaerobic chamber, the microbial community comprised fermentative, syntrophic, and sulphate-reducing bacteria. Most sequences were related to *Aminobacterium* and *Syntrophorhabdus*. In the anoxic chamber, most sequences were related to the mixotrophic and strictly autotrophic denitrifiers *Arcobacter* and *Sulfuricurvum*, respectively, both involved in sulphur-driven denitrification.	Saia *et al.* (2016)

(*Continued*)

Table 3.5 Studies investigating the composition and structure of microbial communities in anaerobic reactors treating sewage (*Continued*).

Reactor Type	Wastewater	Microbial Community	Reference
Pilot-scale three-stage anaerobic wool-felt filter reactor (AWFR)	**Real** domestic wastewater	High-throughput MiSeq sequencing results indicated that seasonal temperature variations of 8–35°C had a significant effect on the microbial community composition. The genera *Methanosaeta, Methanobacterium*, and *Methanolinea* were the predominant methanogens, whereas *Bacillus* was constantly the most-abundant genus, which contributed to the fermentation processes throughout the operation period (summer and winter).	Li and Lu (2017)
Eight laboratory-scale anaerobic batch reactors incubated for 400 days at 4, 8, and 15°C	**Real** domestic wastewater (UV sterilized)	The microbial diversity at 4 and 8°C was distinct from that seen at 15°C and the formation of this distinct community could be a prerequisite for successful treatment at lower temperatures. DGGE results showed that the archaeal taxa were detected in all digesters regardless of the operational temperature and belonged to the Methanomicrobiales, Methanosarcinaceae, and Methanosaetaceae groups. After 102 days, Methanomicrobiales were marginally more abundant than Methanosaetaceae in the 4 and 8°C reactors. In the reactors maintained at 15°C, the abundance of these groups was statistically indistinguishable. The Methanosarcinaceae group was detected in low abundances at all temperatures.	Petropoulos *et al.* (2017)

Generally, these studies (Table 3.5) have observed that changes in bacterial and methanogenic communities occur because of modifications in temperature (and/or seasonal temperature changes), organic loading rate, and pH, among others. In addition, the studies have suggested that at mesophilic temperatures the acetoclastic methanogenesis appears to be the preferred pathway for methane formation, whereas the hydrogenotrophic methanogenesis seems to play a significant role in the low-temperature anaerobic treatment of sewage.

In a mesophilic UASB reactor treating domestic sewage during two different periods of organic loading rate (OLR) and food-to-microorganism (F/M) ratio (Cardinali-Rezende *et al.*, 2013), the archaeal 16S rRNA gene libraries showed a predominance of sequences affiliated with the Methanosarcinales order in rainy and dry seasons. That is, corresponding to low and high OLR, respectively. Members of the Methanosaetaceae family were represented by all the sequences retrieved from the dry-season sample. The emergence of sequences related to the Methanosarcinaceae family was observed in the rainy season (i.e., low OLR) that predominated in relation to the Methanosaetaceae sequences. OTU (operational taxonomic units) affiliated with Methanosaetaceae presented similarity (97%) to *Methanosaeta* sp. and (99%) to *Methanosaeta concilli*. Among the Methanosarcinaceae members, representatives of *Methanomethylovorans* sp. predominated and only one OTU showed 99% similarity to *Methanosarcina horonobensis*. The hydrogenotrophic methanogens were represented by the Methanomicrobiales order, but they were detected only at low OLR, with one OTU with 95% similarity to *Methanolinea* sp.

Several studies have shown an apparent reduction of *Methanosaeta* and an enrichment of hydrogen-utilizing methanogens in psychrophilic anaerobic reactors (Gomec *et al.*, 2008; Li & Lu, 2017; Petropoulos *et al.*, 2017; Syutsubo *et al.*, 2008; Zhang *et al.*, 2012), suggesting that the hydrogenotrophic pathway could have a significant role in low-temperature anaerobic treatment. Under low-temperature conditions, the improved thermodynamics of methane formation from H_2/CO_2, coupled with the enhanced solubility and accessibility of H_2/CO_2 in the bulk liquid, is thought to account for this phenomenon (Lettinga *et al.*, 2001). However, other studies have shown that under low-temperature conditions, acetate represents the main precursor of methanogenesis in both natural (Glissmann *et al.*, 2004) and engineered systems (McHugh *et al.*, 2003). Such a shift is likely the result of enhanced homoacetogenic activity at low temperatures (Schulz & Conrad, 1996), which could contribute to increased acetate accumulation and could decrease the conversion rate of H_2/CO_2 to methane (as observed by Li & Lu, 2017). These authors have observed that the genus *Smithella* (which can degrade propionate to acetate) is dominant in winter.

In anaerobic reactors treating sewage, microbes are present in the scum layer in addition to the microbial community present in the sludge bed (comprising hydrolytic bacteria, fermenting bacteria, acetogens and methanogens). The scum that occurs on the surface of the settler compartment of UASB reactors comprises mainly organic material of low biodegradability (such as hair strands), as well as several bacterial morphotypes (such as filaments and rods) (Souza *et al.*, 2006).

A previous study has shown that the higher upflow velocities imposed on the reactor resulted in a higher production rate of the scum material and in the reduced biodegradability of the scum (Souza *et al.*, 2006; see also Chapter 6). Other studies (Garcia *et al.*, 2012, 2013) have investigated the bacterial diversity present in the scum layer and found sequences related to sulphur-oxidizing bacteria (such as *Thiorhodovibrio* sp, *Thiocapsa* sp., *Thiorhodococcus drewsii*, and *Chlorobium ferrooxidans*) and methane- oxidizing bacteria (such as *Methylacidiphilum fumariolicum*). These results showed that the scum layer could be involved in the biological oxidation of reduced sulphur compounds, as well as the oxidation of the methane produced in the UASB reactor, and remained dissolved in the effluent.

3.4 BIOMASS ORGANIZATION AND STRUCTURE

The efficiency of anaerobic sewage treatment in the instance of UASB reactors depends essentially on the spontaneous formation of a microbial biomass associated with a biological activity and sedimentation velocity. This biomass or sludge can be organized and structured inside the reactor in various ways depending on the reactor hydrodynamics and the substrate characteristics.

The sludge can be aggregated into granules or flocs. Granules are formed by the natural self-immobilisation of bacteria, without the employment of any solid supporting material, and are structured in a dense agglomeration presenting high settling velocities (~60 m h^{-1}) (Pereira, 2012; Schmidt & Ahring, 1995). Because of their immobilisation in the reactor, sludge granulation is associated with several advantages, such as a high rate of COD removal attributable to the higher upflow velocities, improvement of the contact between substrate and biomass and mass transfer, volume decrease (reactors with higher height and lower area), and resistance to eventual hydraulic and organic loading shocks (Hulshoff Pol *et al.*, 2004). Even so, sludge granulation is more likely to occur in reactors fed with concentrated effluents or with high organic loads that facilitate conditions for a more-selective biomass (Hulshoff Pol, 1989).

Generally, in UASB reactors treating sewage, the sludge is developed with a dispersed structure, forming flocs, with a looser arrangement and poor sedimentation, presenting a homogeneous macroscopic layer after settling (Schmidt & Ahring, 1995). Regardless, sludge granulation in reactors treating domestic sewage is probably not a great gain in terms of the system hydrodynamics, as flow velocities cannot be so high because of the presence of particulate COD. As discussed previously, domestic sewage has high content of solids and organic matter in particulate form that could be lost to the effluent, without treatment at high hydraulic loads and upflow velocities. Although sludge granulation in UASB reactors treating sewage has been reported (Pereira, 2012), the topic is controversial (Aiyuk *et al.*, 2006), with the main issues being the low concentrations of rapidly acidified COD and the low rate of development of the methanogens *Methanosaeta*.

Granular cell yields are shown to vary because of the involvement of complex microbiological interactions that involve physical, microbiological, and thermodynamic parameters. For instance, granulation can occur through physical aggregation with solids. This phenomenon could take place if, on one hand, high amounts of particulate matter were present in sewage; and, on the other hand, there is insufficient hydraulic mixture to promote cell contact. Other mechanisms could include chemical coagulation and charge neutralization with bivalent cations (Ca^{2+}, Mg^{2+}, Fe^{2+}, Fe^{3+}) (Liu *et al.*, 2003). However, domestic sewage normally has low concentrations of these cations, possibly hindering the granulation process (Henze *et al.*, 2002).

Granulation can take place in laboratory-scale and pilot-scale reactors treating sewage, as reported by Barbosa and Sant'Anna (1989), as it has plug-flow hydraulics instead of the mixed flow that full-scale systems have. This eases the development of acidogenic bacteria, which promotes the role of EPS production in sludge granulation (Batstone *et al.*, 2005). However, when the reactor is scaled up, the hydrodynamic conditions are no longer favourable. In such instance, low upflow velocities (0.5–0.7 m h^{-1}) and reactor geometry hinder specific methanogenic activity and promote flocculent sludge.

The low content of rapidly hydrolyzed COD in domestic sewage (approximately 10% of the total influent COD) limits the growth of the acidogenic bacteria that are responsible for the production of EPS. The EPS matrix plays a role in the structure and function of the sludge. In actual fact, bacteria in granules are surrounded by EPS, and their composition affects its surface and physical properties.

In terms of sewage characteristics, a higher C/N ratio stimulates the production of polymers, resulting in an improved bacterial attachment to solid surfaces. These polymers are composed of organic debris, lysed cells, and other organic material excreted by the microbial cells, containing polymers of saccharides, proteins, lipids, phenols, and nucleic acids. Experimental studies have demonstrated that after acidifying sewage with 300 mg L^{-1} of sucrose, sludge granulation could be observed, even though the granules had a structure of only 2 mm, were fragile and had low density (Ligero & Soto, 2002).

In general, although specific microbial community is crucial in the formation of the sludge matrix, operational conditions and reactor configuration have also been shown to be important for defining the biomass structure.

3.5 IMPACT OF ENVIRONMENTAL AND OPERATIONAL CONDITIONS ON PROCESS KINETICS AND MICROBIAL COMMUNITIES

Process kinetics is defined by the influencing factor with respect to the rate of biochemical reactions in the reactor. As anaerobic digestion comprises a sequence of reactions, the overall rate is determined by the slowest reaction, i.e., the rate-limiting step. In such instance, substrate degradation can be modelled using first-order kinetics (Vavilin *et al.*, 2008), which means that the reaction rate is correlated positively to the substrate concentration in terms of COD.

The high content of suspended solids in sewage is mainly composed of not readily biodegraded compounds, which lead to a low hydrolysis rate. The hydrolysis step usually occurs at a slow rate in anaerobic conditions and depends on several environmental and operational factors, such as temperature, pH, HRT, substrate composition, particle size, concentration of toxic compounds and concentration of byproducts (e.g., VFAs) (Chernicharo, 2007). This is relevant to particulate organic matter, such as sewage sludge or lignocellulosic biomass, and can occur also in highly concentrated sewage (~2500 mg COD L^{-1}) that has a significant fraction of particulate matter.

For diluted sewage (<1000 mg COD L^{-1}) in tropical conditions, methanogenesis can be the step limiting the overall process; however, hydrolysis has been reported as the rate-limiting step at low temperatures (Petropoulos *et al.*, 2017). In such instance, SRT should always be more than three times the doubling time of methanogens ($T_d = 10$ days; SRT = 30 days) (Van Lier *et al.*, 2010). Therefore, kinetically, the anaerobic treatment of sewage comprises two relatively slow steps. The first step is the decomposition of particulate matter into soluble compounds and the second is the conversion of dissolved organic matter to biogas (Vavilin *et al.*, 2008).

As seen before, fermentative bacteria degrade complex compounds into soluble ones and, subsequently, into short chain VFAs. Afterward, fermented compounds are converted into methane by methanogenic microorganisms (see Figure 3.3). The degradation of propionate and butyrate is not thermodynamically favourable at standard conditions, i.e., $\Delta G_o > 0$ (see Table 3.2), but the bioconversion can occur with low hydrogen partial pressure ($<10^{-4}$ atm; $\sim 10^{-6}$ atm), a common condition in anaerobic reactors in stable equilibrium (Schink, 1997).

Anaerobic sewage treatment occurs at ambient temperatures and, therefore, process kinetics depends on environmental conditions. This could appear a drawback in terms of the kinetics for attaining the best methanogenic growing conditions compared with the anaerobic digestion of sludge that takes place in heated reactors (at mesophilic temperature of 35–38°C). However, the large flowrates of diluted wastewater produced makes it unfeasible to treat it in conditions other than the environment one. For lower temperatures, the biological conversion capacity appears to determine the overall COD removal rather than the prevailing hydrodynamic conditions.

This means that the efficiency and stability of the anaerobic treatment process are influenced by various environmental parameters, including temperature, pH, OLR, hydraulic retention time (HRT), and nutrient availability (Gao *et al.*, 2011; Lew *et al.*, 2003). These affect the kinetics of microbial growth and substrate utilization and cause imbalance in the microbial communities. As regards sewage, nutrients are available, and neutral and stable pH is maintained by the alkalinity present (ranging from 120 to 388 mg L^{-1}; Seghezzo *et al.*, 1998).

3.5.1 Temperature effect

Temperature is a crucial factor because of its significant effects on microbial growth and activity, and substrate conversion (Jung *et al.*, 2011). Most reactions in the biodegradation of organic matter require more energy to proceed at low temperatures than at mesophilic temperature (37°C), because the activity of enzymes are lower at low temperatures compared to high temperatures. Generally, the anaerobic reactors treating sewage are operated at a mesophilic range of 30°–37°C, or at ambient temperatures of 23–34°C in tropical countries (Barbosa & Sant'Anna, 1989; Foresti, 2002; Pereira, 2012; Silva *et al.*, 2008). Currently, full-scale anaerobic treatment of sewage has been limited to tropical areas (Chernicharo *et al.*, 2015), where sewage temperature (~20–30°C) allows the hydrolysis of complex organics and suspended solids at a fast enough rate. In temperate climates, where sewage temperature can vary from 8 to 20°C (Seghezzo *et al.*, 1998), efficient anaerobic sewage treatment needs to be demonstrated before implementation can be considered (see also Chapter 12).

Few studies have investigated the effects of operational parameters, such as OLR, upflow velocities, HRT and temperature variation in anaerobic reactors treating real domestic sewage (Bandara *et al.*, 2012; Cardinali-Rezende *et al.*, 2013; Lew *et al.*, 2003; Li & Lu, 2017; Pereira, 2012; Petropoulos *et al.*, 2017; Silva *et al.*, 2008). As regards the effect of temperature, Li and Lu (2017) have observed that temperature is the key factor in the structures of archaeal and bacterial communities in an anaerobic reactor treating real domestic sewage. Shifts in the methanogenic community were observed during seasonal temperature variations of 8–35°C. High-throughput MiSeq sequencing analyses showed that acetoclastic methanogens predominate in summer, whereas hydrogenotrophic genera (*Methanobacterium* and *Methanolinea*) comprise a substantial portion (>63.7%) of the archaeal community in winter. In relation to the bacteria, four phyla affiliated with Firmicutes, Proteobacteria, Bacteroidetes and Chloroflexi were predominant (89.3%) during the entire operating period. With the reduction in temperature (from summer to winter), the relative abundance of Firmicutes decreased (from 64.9 to 29.4%), whereas the abundances of Proteobacteria and Bacteroidetes increased from 15 to 31.5% and 10.0 to 20.6%, respectively.

In relation to the reactor performance, it has been reported that COD removal efficiency decreased with temperature reduction from 82 to 44% when T changed from 28 to 10°C (Lew *et al.*, 2003), and from 76 to 43% when T changed from 35 to 8°C (Li & Lu, 2017). In addition, higher COD efficiency and biogas production are observed in summer (Bandara *et al.*, 2012; Li & Lu, 2017). Furthermore, the methane formation pathway is affected by temperature variation. At mesophilic temperatures, acetoclastic methanogenesis appears to be the main pathway for methane production (as *Methanosaeta* is the most abundant genus detected in many studies), with some contribution from the hydrogenotrophic methanogens (Cardinali-Rezende *et al.*, 2013; Li & Lu, 2017). At low temperatures (15°C or lower) hydrogenotrophic methanogenesis appears to be the main pathway for methane production and, therefore, Methanomicrobiales and Methanobacteriales are generally more abundant and dominate the community compared with Methanosarcinaceae and Methanosaetaceae (Li & Lu, 2017; Petropoulos *et al.*, 2017; Smith *et al.*, 2015).

At extremely low temperatures (4°C) methanogenic activity is extremely low (methane yield at 4°C is less than 35% of the theoretical potential) and, in a study using adapted mesophilic seed at this temperature, methanogenesis was the rate-limiting step (Bowen *et al.*, 2014). On the other hand, in another study that used arctic seed and evaluated the specific rates of methanogenesis and hydrolysis at 4, 8, and 15°C

(Petropoulos *et al.*, 2017), the hydrolysis step was twice as sensitive to temperature as the methanogenesis, and the lipids (but not the proteins and carbohydrates) failed to degrade. A recent study showed this is probably because of the reduced lipolysis activity at lower temperatures rather than absence of lipases (Petropoulos *et al.*, 2018). However, at a low temperature (4°C), the specific methanogenic activity was lower than was that observed at 15°C (6.3 and 10.3 fmol CH_4 $cell^{-1}$ day^{-1}, respectively).

To overcome the challenge of slow hydrolysis and the slow growth of methanogens at low-temperature anaerobic sewage treatment, Zhang *et al.* (2013) investigated a pilot-scale UASB digester system treating real sewage, with glucose as co-substrate. Glucose was added in the sludge digester to produce additional methanogenic biomass, which was recycled continuously to inoculate the UASB reactor. The results indicated that COD removal efficiency increased from 6 to 23%, specific methanogenic activity at 15°C increased three times, and methane production in the UASB reactor increased by more than 90%. Therefore, the authors concluded that co-digestion is a suitable approach to support a UASB digester for pre-treating low-temperature municipal sewage. The challenge of low-temperature anaerobic sewage treatment is further addressed in Chapter 12.

3.5.2 Organic loading and dilution rate effects

In relation to the impact of different operational conditions on the performance of a UASB reactor treating domestic sewage, Cardinali-Rezende *et al.* (2013) have reported that fluctuations in the OLR and F/M ratio affected the abundance and composition of the UASB prokaryotic community, mainly at the species level, as well as the performance of the reactor. There was a decrease in the bacterial cell numbers during the rainy season (at a lower OLR and F/M ratio). However, the diversity of the bacterial and archaeal communities was higher during this time, suggesting that the community degraded more diversified substrates. In another study (Silva *et al.*, 2008), the authors have reported that microbial diversity (investigated by DGGE analyses) did not change with the increase in the upflow velocity, despite the decrease in COD removal when the highest upflow velocity was imposed. However, this study investigated bacterial and archaeal diversity by PCR-DGGE, which detects the dominant members of a community. Therefore, more studies using high-throughput amplicon sequencing, which provide more-detailed information about microbial communities, should be conducted to confirm previous results.

Regarding the effect of the dilution rate (volumetric hydraulic load) on the methanogenic community, several studies have reported that the microbial community was affected by changes in product distribution (VFA concentration) that occur at various dilution rates (Kim *et al.*, 2010). *Methanosaeta* dominated at a low dilution rate (0.025 d^{-1}), whereas *Methanosarcina* dominated at a high dilution rate (0.6 d^{-1}) in mesophilic acetate-fed reactors (Shigematsu *et al.*, 2003). The dominance of *Methanosaeta* at the low dilution rate could be explained by the lower Ks value for acetate compared with that of *Methanosarcina*. In addition, at a low dilution rate, hydrogenotrophic methanogens (*Methanoculleus*) were detected in significant numbers, suggesting that syntrophic acetate oxidation by hydrogenotrophic methanogens and acetate-oxidizing bacteria could occur. In UASB reactors treating sewage, the dilution rate (volumetric hydraulic loading) is typically between 2 and 4 d^{-1}, considering the recommended design HRT (see Chapter 4).

Based on the data from the literature, it can be concluded that microbial communities in anaerobic reactors treating sewage are affected by operational conditions, such as temperature, OLR and HRT, among others. At mesophilic temperatures, methane formation derives from the acetoclastic methanogenesis pathway, whereas at low temperatures the hydrogenotrophic pathway appears to be more important. Ample work has been done on anaerobic digesters relating to the response of the microbial communities to the operational parameters (De Vrieze *et al.*, 2012; Martins *et al.*, 2017; Regueiro *et al.*, 2012). However, there is a lack of knowledge on anaerobic reactors treating real domestic wastewater. From the data presented in Table 3.5, it is clear that only two studies have conducted in-depth investigation (using high-throughput sequencing analyses) of the microbial community structure in anaerobic reactors treating sewage. Therefore, future studies should focus on the characterization of microbial communities, including the genomes of these organisms, to determine which bacteria adopt which role in the degradation of polymers. This, in turn, could lead to comprehension of metabolic networks (and how these networks are affected by operational parameters) and a rational basis for the improvement of anaerobic sewage treatment efficiency. Augmented knowledge on the influence of environmental variables (temperature, OLR, HRT, acetate concentration and the like) on the microbial community structure could inform community-based strategies to increase the stability of anaerobic reactors.

3.6 IS TOXICITY AN ISSUE IN ANAEROBIC SEWAGE TREATMENT?

Microorganisms of the anaerobic consortia, especially those with low diversity and low growth rates, such as the methanogenic and acetogenic microorganisms, can be inhibited by distinct classes of compounds at different concentrations. This factor is of great concern when anaerobic biotechnology is applied to industrial wastewater treatment.

In developing countries, sewage and non-domestic wastewater are rarely collected separately. Therefore, commonly, landfill leachates, as well as raw or partially treated industrial wastewater, such as laundry, dairy, silk screen and dyeing effluents are discharged into sewers. The question is whether the presence of inhibiting chemicals and/or recalcitrant or slow biodegradable compounds in these effluents could influence the anaerobic sewage treatment? In this section, we discuss the probability of the various contaminants usually present in non-domestic contributions causing adverse effects in anaerobic reactors during sewage treatment.

3.6.1 Heavy metals and complex organics

Silk screen and textile effluents contain pigments and dyes, formulated with metals, such as Cr, Cu, Ni, and Co, as well as complex organics that have aromatic rings, azo bounds, condensed rings, and N ring compounds, which might exhibit some toxicity toward anaerobic microbial communities. Landfill leachates also contain inhibiting agents, such as ammonia and metals, as well as countless organic compounds of different classes. In this regard, Maria (2010) characterized four sanitary landfill leachates, detecting Cr (0.39–0.61 mg L^{-1}), Ni (0.53–0.65 mg L^{-1}), Zn (0.07–1.38 mg L^{-1}) and up to 169 compounds of different classes (e.g., acid, alcohol, amine, ketone, ether, aldehyde, and amide), some of which (e.g., chlorinated, phenolic, benzene) are well known toxicants to anaerobic microorganisms.

As regards heavy metals, the likelihood of their affecting anaerobic processes is low, even for low-dilution factors, because of the side effects of sulphide precipitation and the low solubility of most metal sulphide precipitates, which renders an extremely low metal concentration in the reactor. For instance, the sulphide concentration in effluents from anaerobic reactors treating domestic sewage (5–22 mgS L^{-1}; see section 3.2), would allow a metal concentration as low as $\sim 10^{-6}$ ng L^{-1}, considering the most-soluble sulphide (K_{sp} Mn $= 7 \times 10^{-16}$).

As regards industrial organic contaminants, they are normally found in industrial wastewater and leachate in the μg L^{-1} to ng L^{-1} range and, although some of them exhibit toxicity toward the methanogens in this range (Blum & Speece, 1991), the dilution factor is normally high enough to ensure the concentration entering the anaerobic reactor is below the threshold levels.

3.6.2 Sulphate and sulphide

Sulphate is an important contaminant present in several industrial effluents because of the use of sulphuric acid and other sulphur compounds in various manufacturing processes. Sulphate *per se* is not considered toxic to anaerobic microorganisms at the concentrations usually found in sewage (Speece, 1996) but, in anaerobic environments sulphate is reduced to sulphide, which is toxic at much lower levels depending on the pH (Paula Jr. & Foresti, 2009). Sulphide concentrations from 100 to 150 mg L^{-1} have been considered severely inhibitory to methanogens at pH approximately neutral (Speece, 1983).

In addition to the non-competitive sulphide inhibition on methanogens, sulphide production from sulphate decreases the amount of electron donors (carbon equivalents) available to methanogens in a competitive inhibition process. The competition between acetoclastic methanogenic archaea (AMA) and sulphate-reducing bacteria (SRB) has been found to be dependent on the carbon/sulphate ratio. This means that at a given carbon/sulphate ratio, SRB can outcompete AMA, thereby decreasing the net amount of biogas and lowering the methane concentration in anaerobic reactors. This could be a problem particularly in coastal or arid areas because of the higher presence of sulphate in brackish water entering the sewerage system.

The sulphate concentration in Brazilian raw sewage varies from 40 to 70 mg L^{-1} (13.3 to 23.3 mgS L^{-1}), whereas dissolved sulphide ranges between 5 and 22 mgS L^{-1}, as aforementioned. This implies that the expected sulphide concentration in anaerobic reactors is well below the range considered toxic to the methanogens. Therefore, sulphide inhibition will not likely occur in anaerobic reactors treating domestic sewage unless other sulphur-enriched effluents are being discharged into the sewers.

In this regard, it is important to point out that landfill leachates have been increasingly discharged into sewers in some Brazilian cities, which can contribute to sulphur increasing in the sewage, depending on the dilution factor. In a leachate characterization study, Maria (2010) found that landfills of different ages had average sulphate concentrations varying from 67.1 to 95.2 mg L^{-1}. Considering only a tenfold dilution factor, the leachate discharge would cause an average increase in sewage sulphur concentration lower than 15%.

Industrial wastewater could also contribute to sulphur input, as sulphuric acid, which is the main industrial sulphur compound, is used widely by various large water consumers such as the electroplating industry. However, such chemical industries are usually under surveillance by environmental agencies, even in developing countries, which reduces the chances of high sulphate concentrations reaching the sewers. In addition to sulphuric acid, sulphur can be found in several chemical products, such as dyes, pigments, and detergents. Detergents are formulated mainly with alkyl benzene sulfonates that release sulphate upon biological degradation. However, the sulphur concentration from these chemicals is expected to be negligible because of the low mass proportion of sulphur in these compounds.

3.6.3 Surfactants

Surfactants, used in various household cleaning products, are significant contaminants naturally present in sewage in concentrations varying from 1 to 20 mg L^{-1} (Clara *et al.*, 2007; Holt *et al.*, 1995; Jepsen *et al.*, 2007; Painter & Zabel, 1989). Furthermore, the discharge of laundry effluents in the sewers could contribute to an increase in such compounds in the sewage.

Anionic surfactants, mainly linear alkylbenzene sulfonate (LAS), are the most important class of detergents in domestic sewage. LAS is a mixture of isomers and homologues formed by a linear alkyl chain, with 10–13 carbons bound to a sulfonated aromatic ring. Although LAS has been introduced in the market as a 'readily biodegradable' compound to replace the poorly biodegradable branched alkylbenzene sulfonate (ABS), several studies have shown that LAS is fairly recalcitrant, mainly in anaerobic environments. The anaerobic recalcitrance of LAS has been reported by various research groups (Carosia *et al.*, 2014; Garcia *et al.*, 2005; Souza *et al.*, 2016), particularly in relation to the inhibition of the methanogenic process (Garcia *et al.*, 2006; Garcia-Morales *et al.*, 2001).

Wagener and Schink (1987) have suggested that the action of the surfactant in cells is related to fluidization of the membrane and solubilization of structural components. These authors found that methane production was inhibited by 18 and 84% at LAS concentrations of 10 and 100 mg L^{-1}, respectively. Souza *et al.* (2016) investigated the effect of LAS concentration on the specific methanogenic activity (SMA) and total methane production (TMP). The exposure of anaerobic sludge to LAS at 10 mg L^{-1} led to 30% reduction in SMA and TMP compared with the control (without LAS). Further exposure to concentrations up to 100 mg L^{-1} led to SMA reduction from 44 to 97%, whereas TMP decreased from 30 to 90%. On the other hand, Garcia *et al.* (2006) concluded that LAS concentrations of less than 10 mg L^{-1} did not inhibit methane production. This concentration (10 mg L^{-1}) was the highest found by Tavares (2008) in raw sewage of Belo Horizonte city, Minas Gerais state, Brazil. Therefore, it appears that the LAS concentrations typically found in raw sewage could have only a slight effect on methane production and methanogenic activity.

Although not causing inhibition in anaerobic reactors, the presence of LAS in anaerobic effluents causes problems during their discharge in water bodies. Problems, such as foam formation, have been reported in anaerobic sewage treatment plants, even with aerobic post-treatment units, particularly, when there is a significant difference in height between the treatment plant and the point of discharge.

3.6.4 Ammonia

Ammonia is another pollutant that can be detrimental to anaerobic growth. Ammonia is the main product of anaerobic degradation of nitrogenous matter, of which proteins are the main representatives in sewage. As ammonia is a weak base, it reacts with water to form ammonium ion, thereby releasing hydroxide ions that increases pH (Equation (3.5)). This is a reversible reaction and the relative concentrations of the ionic (NH_4^+) and molecular (NH_3) forms are pH dependent (Equation (3.6)). The acidity constant (K_a) of NH_4^+ is 5.6×10^{-10} ($pK_a = 9.25$), which implies that at pH approximately neutral, nearly all the ammonia is in ionic form (Figure 3.4). The molecular form, also known as free ammonia (FAN), prevails only at higher pH values. At pH $= pK_a$, equal amounts of NH_3 and NH_4^+ will be present, at pH 10, ~80% of NH_3 will be present and, at pH 11.8, nearly all the ammonia will be present as FAN (Figure 3.4).

$$NH_3 + H_2O \Leftrightarrow NH_4^+ + OH^- \quad (3.5)$$

$$NH_4^+ + H_2O \Leftrightarrow NH_3 + H^+ \quad (3.6)$$

Both ionic and molecular forms can directly and indirectly cause inhibition in an anaerobic digestion system (Yenigün & Demirel, 2013). It has been acknowledged in the literature that FAN is a powerful inhibitor in an anaerobic digester at a concentration exceeding toxic threshold levels. The classical study by McCarty and McKinney (1961) showed that 150 mg L^{-1} FAN completely inhibited anaerobic digestion. Shanmugam and Horan (2009) found that 500 mg L^{-1} FAN inhibited methane production during the mesophilic anaerobic digestion of meat and bone meal. El Hadj *et al.* (2009), studying the anaerobic digestion of the organic fraction of municipal solid waste, found inhibitory FAN concentrations in this range, as, at 215 mg L^{-1} and 468 mg L^{-1}, 50% inhibition of methane production occurred during mesophilic and thermophilic treatment, respectively.

Assuming a digester operated at pH 7, the ammonium ion concentration would have to be ~25,860 mg L^{-1} to attain the FAN concentration (150 mg L^{-1}) deemed to cause methanogens toxicity. This is much higher than is the expected ammonium concentration in anaerobic reactors treating domestic sewage, which contains from 35 to 60 mg L^{-1} total nitrogen (Mota & von Sperling, 2009).

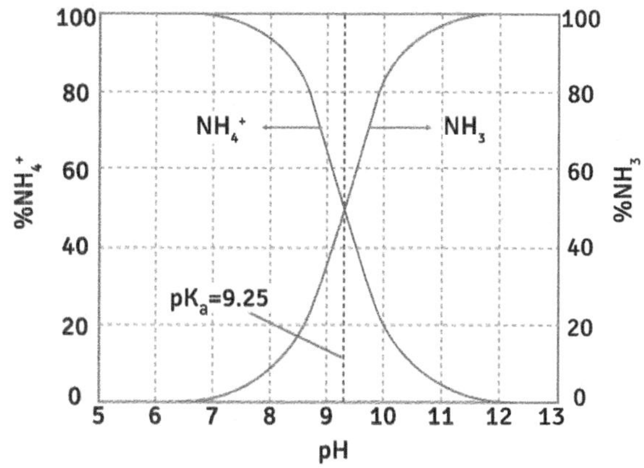

Figure 3.4 Distribution of ammonium and free ammonia forms as a function of pH.

As stated before, small sewers can receive non-domestic contributions that could contain high ammonia levels, such as leachates. The sanitary landfill leachates characterized by Maria (2010) showed average total nitrogen concentrations that varied from 1,331 to 1,803 mg L^{-1}, which is ~15 times lower than is the concentration needed to produce toxic levels of FAN at neutral pH.

Obviously, as pH increases, the less would be the ionic ammonia needed to yield a given FAN concentration (Figure 3.4). Therefore, an increase of one pH unit would reduce the threshold ionic ammonia concentration by 10 times.

In this way, considering the given example, in a digester operated at pH 9, only 258.6 mg L^{-1} ammonia would be required to yield a FAN concentration of 150 mg L^{-1}. Although ammonium ions can also exhibit some toxicity, it is clear from the literature (Yenigün & Demirel, 2013) that FAN is the most toxic form. Accordingly, pH control in anaerobic reactors is crucial to preventing ammonia toxicity and process failure.

3.7 ANAEROBIC EFFLUENT CHARACTERIZATION

The effluents from anaerobic reactors fed with raw sewage contain soluble and insoluble matter, with the latter accounting for nearly half of the organic content leaving the reactor (i.e., 57–63% total COD and 59–67% total BOD; Aquino *et al.*, 2009). The effluent insoluble matter is mainly organic, with a volatile suspended solids to total solids ratio (VSS/TSS) averaging 65% (Bressani-Ribeiro, 2015), and including anaerobic microorganisms (active and dead cells or cell fragments), as well as influent particulate matter (e.g., cellulose fibres, hair, fats and proteins). Most of these insoluble compounds have low density and a slow hydrolysis rate, which, for some reason, cannot be entrapped in the reactor.

The so-called insoluble COD is obtained by the difference between the total COD (sample neither filtered nor centrifuged) and the COD given by the sample after filtration through 1.2 μm nominal pore-size membranes or centrifuged at 5000 rpm for 5 min (Chernicharo, 2007). As not all colloidal material, which causes turbidity, is retained by 1.2 μm filtration membranes or can be centrifuged at 5000 rpm, the insoluble COD includes colloidal (e.g., proteinaceous material and extracellular polymers) and particulate (insoluble *stricto sensu*) matter, as shown in Figure 3.1.

In turn, the dissolved fraction of anaerobic effluents contains countless organic compounds, some present at higher amounts (mg L^{-1}) that contribute to soluble or dissolved BOD and COD. Others, are present at low concentration (ng L^{-1} to μg L^{-1}), i.e., the micropollutants, such as pharmaceutical and hygiene/personal care products, could have a disproportionate ecological impact relative to their mass. The former includes non-biodegradable compounds of raw sewage, transformation products from the incomplete (partial) degradation of influent substrates, such as the volatile fatty acids (VFA), as well as other compounds produced by microorganisms during their metabolism or decay. Notably, biological effluents are characterized poorly from a chemical point of view, which means that the identity of the compounds and their relative contribution to the residual COD or BOD are barely known.

It is known that volatile fatty acids (VFA), from C1 (formic acid) to C5 ((iso)valeric acid), are the main intermediate products of anaerobic digestion and that they normally do not accumulate in reactors treating low-strength wastewater, such as sewage, unless a stressful condition is present (Aquino & Chernicharo, 2005). Therefore, in well operated bioreactors treating domestic sewage, the main constituents of anaerobic effluent are not VFA, but non- or slowly biodegradable compounds (inert from an operational point of view). These could originate from raw sewage or be produced during treatment, as has been demonstrated by various research groups (Aquino & Stuckey, 2008; Laspidou & Rittmann, 2002; Petropolous *et al.*, 2018) using simpler substrates or synthetic wastewaters. The latter are referred to as soluble microbial products (SMPs).

Figure 3.5 shows a COD balance throughout the anaerobic digestion process that encompasses all the possibilities for electron channelling. As there is a lack of published information on the characterization of biological effluents from reactors treating real sewage, it is difficult to pinpoint the contribution of each of the aforementioned classes to the residual COD.

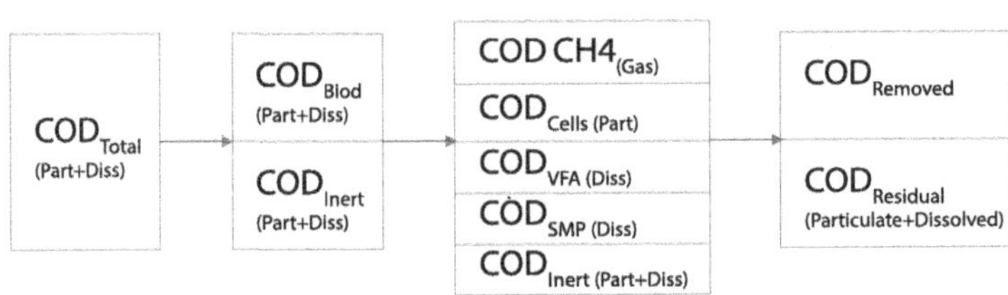

Figure 3.5 Diagram of the COD balance in a typical anaerobic reactor.

One of the few studies on effluent characterization from anaerobic reactors treating sewage was carried out by Aquino *et al.* (2009). The authors studied effluents from demonstration-scale UASB reactors treating real domestic sewage and estimated that 9.2 ± 1.3% of the influent soluble COD could be considered inert to anaerobic treatment. In addition, it was estimated that the amount of soluble COD produced by biomass varied from 30 to 70 mg L^{-1}, accounting for nearly half of the soluble COD that left the UASB reactor. Owing to the analytical difficulties of chemically identifying the compounds that cause the residual COD or BOD in biological effluents, most studies present effluent characterization in terms of global (e.g., protein/carbohydrate/lipid content, molecular weight distribution) or functional (e.g., biodegradability, toxicity) parameters.

Analyses of the main chemical groups, such as protein, carbohydrate, and lipids of raw and treated sewage have been conducted by a few authors (Pontes & Chernicharo, 2009; Pontes *et al.*, 2014). Pontes *et al.* (2014) studied the influence of returning excess aerobic sludge from a trickling filter on the degradation kinetics of organic matter in UASB reactors. These authors found that anaerobic effluents contain 1.4–3.9 mg L^{-1} lipids; 5.5–13.1 mg L^{-1} carbohydrates, and 31.7–57.7 mg L^{-1} proteins, which implies that this class of compound constitutes the majority of organics in the dissolved fraction.

In their study (Pontes & Chernicharo, 2009), considering that the COD effluent of the UASB reactors varies from 120 to 140 mg L^{-1} and that the total suspended solids (TSS) varies from 25 to 30 mg L^{-1}, the authors estimated that the soluble COD contributed approximately 70% of the residual COD. Therefore, the amount of carbohydrate, lipid, and mainly protein, would represent 85–95% of the soluble COD that leaves the UASB reactor. Notably, VFAs are found in the anaerobic effluent in concentrations lower than the limit of detection of the methodology employed, estimated at 4 mg L^{-1} for acetic acid and 1 mg L^{-1} for propionic and butyric acids (Pontes & Chernicharo, 2009).

The protein-like material found in anaerobic effluents could derive from raw sewage, as well as be formed during the treatment, as the microbial metabolism and death-release proteins might not be biodegraded promptly by the biomass from which they originate (Aquino & Stuckey, 2008; Zhang *et al.*, 2016). One example of such material is extracellular polymeric substances (EPS), also known as extracellular polymers (ECP). Various research groups have found that EPS or ECP, which are important components of biofilms and microbial arrangements, can be sheared and can contribute to the effluent colloidal COD, or can even be hydrolyzed to integrate the pool of soluble COD of the effluent. Unfortunately, not enough published data are available to facilitate determining the main source of protein in anaerobic effluents from reactors treating real sewage.

Extremely little is known about the exact composition of the residual COD from biological reactors, especially from those treating real sewage, and only a few researchers have focused on the chemical characterization of anaerobic effluents by using sophisticated instruments. Such instruments and methods include gas chromatography and mass spectrometry (GC-MS) (Aquino, 2004; Zhou *et al.*, 2009), high-resolution liquid chromatography ion-trap time-of-flight mass spectrometry (LC-IT-TOF-MS) (Mesquita *et al.*, 2010), and matrix-assisted laser desorption ionization time-of-flight mass spectrometry (MALDI-TOF-MS) (Aquino, 2004; Mesquita *et al.*, 2010). Zhou *et al.* (2009) have investigated the residual COD in the effluent of a full-scale UASB reactor (3100 m^3) treating real sewage using GC-MS. These authors reported that 27 compounds could be identified comprising long-chain carbohydrates and esters, such as N-hexadecanoic acid (10.5%), heneicosane (8.6%), and hentriacontane (5.1%). According to the authors, the identified compounds derive mostly from the metabolism of the anaerobic bacteria, are not related directly to the raw pollutants, and account for 55–65% of the organic matter leaving the anaerobic reactor.

The origin of the residual COD from biological reactors, if produced or originally present in the raw sewage, is a matter of debate in the literature, and more studies are needed on this topic. However, the characterization of such COD is most important from a practical operational point of view, as it would ease designing post-treatment units to polish anaerobic effluents. In addition, published information (Aquino & Stuckey, 2007; Barber & Stuckey, 1999; Mesquita *et al.*, 2010) has shown that the COD produced during treatment can be minimized by changing operational (e.g., SRT, HRT) and environmental (e.g., temperature, toxic compounds, nutrients) conditions of both aerobic and anaerobic reactors. Therefore, knowing the origin of the residual COD could help engineers to improve bioreactors and enhance the efficiency of sewage treatment.

REFERENCES

Agrawal L., Harada H. and Okui H. (1997). Treatment of dilute wastewater in a UASB reactor at a moderate temperature: performance aspects. *Journal of Fermentation and Bioengineering*, **83**, 179–184.

Aiyuk S., Forrez I., Lieven D., van Haandel A. and Verstraete W. (2006). Anaerobic and complementary treatment of domestic sewage in regions with hot climates – a review. *Bioresource Technology*, **97**, 2225–2241.

APHA–AWWA–WEF (1995). Standard methods for the examination of water and wastewater, 19th edn. APHA-AWWA-WEF, Washington, DC.

Aquino S. F. (2004). Formation of soluble microbial products (SMP) in anaerobic reactors during stress conditions. PhD thesis, Imperial College, London, UK.

Aquino S. F. and Chernicharo C. A. L. (2005). Build up of volatile fatty acids (VFA) in anaerobic reactors under stress conditions: causes and control strategies. *Engenharia Sanitária e Ambiental*, **10**, 152–161.

Aquino S. F. and Stuckey D. C. (2007). Bioavailability and toxicity of metal nutrients during anaerobic digestion. *Journal of Environmental Engineering (New York)*, **133**, 28–35.

Aquino S. F. and Stuckey D. C. (2008). Integrated model of the production of soluble microbial products (SMP) and extracellular polymeric substances (EPS) in anaerobic chemostats during transient conditions. *Biochemical Engineering Journal*, **38**, 138–146.

Aquino S. F., Glória R. M., Silva S. Q. and Chernicharo C. L. (2009). Quantification of the inert COD of raw sewage and evaluation of SMP production in demo-scale UASB reactors under different operational conditions. *Water Environmental Research*, **81**, 608–616.

Ariesyady H. D., Ito T. and Okabe S. (2007). Functional bacterial and archaeal community structures of major trophic groups in a full-scale anaerobic sludge digester. *Water Research*, **41**, 1554–1568.

Azevedo L. S., Castro I. M. P, Leal C. D., Araujo J. C. and Chernicharo C. A. L. (2018). Performance and bacterial diversity of bioreactors used for simultaneous removal of sulphide, solids and organic matter from UASB reactors effluents. *Water Science and Technology*, **78**, 1312–1323.

Bandara W. M., Kindaichi T., Satoh H., Sasakawa M., Nakahara Y., Takahashi M. and Okabe S. (2012). Anaerobic treatment of municipal wastewater at ambient temperature: analysis of archaeal community structure and recovery of dissolved methane. *Water Research*, **46**, 5756–5764.

Bapteste É., Brochier C. and Boucher Y. (2005). Higher-level classification of the archaea: evolution of methanogenesis and methanogens. *Archaea*, **1**, 353–363.

Barbosa A. and Sant'Anna G. L., Jr. (1989). Treatment of raw domestic sewage in an UASB reactor. *Water Research*, **23**, 1483–1490.

Barker D. J. and Stuckey D. C. 1999. A review of soluble microbial products (SMP) in wastewater treatment systems. *Water Research*, **33**, 3063–3082.

Batstone D., Hernandez J. and Schmidt J. (2005). Hydraulics of laboratory and full-scale upflow anaerobic sludge blanket (UASB) reactors. *Biotechnology and Bioengineering*, **91**, 387–391.

Bialeck K., Kumar A., Mahony T., Lens P. and O'Flaherty V. (2012). Microbial community structure and dynamics in anaerobic fluidized-bed and granular sludge-bed reactors: influence of operational temperature and reactor configuration. *Microbial Biotechnology*, **5**(6), 738–752.

Blum D. J. W. and Speece R. E. (1991). A database of chemical toxicity to environmental bacteria and its use in interspecies comparisons and correlations. *Research Journal WPCF*, **63**(3), 198–207.

Bonin A. S. and Boone D. R. (2006). The order Methanobacteriales. In: The Prokaryotes, M. Dworkin, S. Falkow, E. Rosenberg, K.-H. Schleifer and E. Stackebrandt (eds), Springer, New York, USA, pp. 231–243.

Bowen E., Dolfing J., Davenport R. J., Read F. and Curtis E. P. (2014). Low-temperature limitation of bioreactor sludge in anaerobic treatment of domestic wastewater. *Water Science and Technology*, **69**, 1004–1013.

Bressani-Ribeiro T. (2015). Sistema UASB-FBP submetido a hidrograma típico de vazão: avaliação do uso de meio suporte baseado em espuma de poliuretano e operação sem decantadores secundário (UASB/TF system submitted to a typical hydrograph flow: assessment of sponge-based media and operation without secondary systems). MSc thesis, Federal University of Minas Gerais, Belo Horizonte, Brazil.

Cardinali-Rezende J., Araujo J. C., Almeida P. G. S., Chernicharo C. A. L., Sanz J. L., Chartone-Souza E. and Nascimento A. M. A. (2013). Organic loading rate and food to microorganism ratio shape prokaryotic diversity in a demo-scale up-flow anaerobic sludge blanket reactor treating domestic wastewater. *Antonie van Leeuwenhoek*, **104**(6), 993–1003.

Carosia M. F., Okada D. Y., Sakamoto I. K., Silva E. L. and Varesche M. B. A. (2014). Microbial characterization and degradation of linear alkylbenzene sulfonate in an anaerobic reactor treating wastewater containing soap powder. *Bioresource Technology*, **167**, 316–323.

Chernicharo C. A. L. (2007). Anaerobic Reactors. (Biological Wastewater Treatment Series, Volume 4). IWA Publishing, London, UK. ISBN electronic: 9781780402116

Chernicharo C. A. L., van Lier J. B., Noyola A. and Bressani-Ribeiro T. (2015). Anaerobic sewage treatment: state of the art, constraints and challenges. *Reviews in Environmental Science and Bio/Technology*, **14**, 649–679.

Clara M., Scharf S. and Gans O. (2007). Occurrence of surfactants in untreated and treated. *Water Research*, **41**, 4339–4348.

Connaughton S., Collins G. and O'Flaherty V. (2006). Development of microbial community structure and activity in a high-rate anaerobic bioreactor at 18°C. *Water Research*, **40**, 1009–1017.

Delbès C, Moletta R. and Godon J. J. (2000). Monitoring of activity dynamics of an anaerobic digester bacterial community using 16S rRNA PCR-single-strand conformation polymorphism analysis (SSCP). *Environmental Microbiology*, **2**, 506–15.

Demirel B. and Scherer P. (2008). The roles of acetotrophic and hydrogenotrophic methanogens during anaerobic conversion of biomass to methane: a review. *Reviews in Environmental Science and Biotechnology*, **7**(2), 173–190.

Diaz E. E., Stams A. J. A., Amils R. and Sanz J. L. (2006). Phenotypic properties and microbial diversity of methanogenic granules from a full-scale upflow anaerobic sludge bed reactor treating brewery wastewater. *Applied and Environmental Microbiology*, **72**, 4942–4949.

De Vrieze J., Hennebel T., Boon N. and Verstraete W. (2012). *Methanosarcina*: the rediscovered methanogen for heavy duty biomethanation. *Bioresource Technology*, **112**, 1–9.

De Vrieze J., Saunders A. M., He Y., Fang J., Nielsen P. H., Verstraete W. and Boon N. (2015). Ammonia and temperature determine potential clustering in the anaerobic digestion microbiome. *Water Research*, **75**, 312–23.

El Hadj T. B., Astals S., Gali A, Mace S. and Mata-Alvarez J. (2009). Ammonia influence in anaerobic digestion of OFMSW. *Water Science and Technology*, **59**, 1153–1158.

Foresti E. (2002). Anaerobic treatment of domestic sewage: established technologies and perspectives. *Water Science and Technology*, **45**(10), 181–186.

Gao W., Leung K., Quin W. and Liao B. (2011). Effects of temperature and temperature shock on the performance and microbial community structure of a submerged anaerobic membrane bioreactor. *Bioresource Technology*, **102**, 8733–8740.

Garcia G. P. P., Souza C. L., Gloria R. M., Silva S. Q. and Chernicharo C. A. L. (2012). Biological oxidation of sulphides by microorganisms present in the scum layer of UASB reactors treating domestic wastewater. *Water Science and Technology*, **66**, 1871–1878.

Garcia G. P. P., Araujo J. C., Franco V. A., Oliveira R. C., Souza C. L. and Chernicharo C. A. L. (2013). *Book of abstracts of the 13th World Congress on Anaerobic Digestion AD13*, Santiago de Compostela, 25–28.

Garcia G. P. P., Oliveira R. C., Bicalho S. K., Franco V., Gontijo E. M., Toscano R. A., Canhestro K. O., Santos M. R., Carmo A. L. D., Lobato L. C., Brandt E. M., Chernicharo C. A. L. and Araujo J. C. (2015). Biological sulphide removal from anaerobically treated domestic sewage: reactor performance and microbial community dynamics. *Environmental Technology*, **36**, 1–26.

Garcia M. T., Campos E., Ribosa I., Latorre A. and Sanchez-Leal J. (2005). Anaerobic digestion of linear alkyl benzene sulfonates: biodegradation kinetics and metabolite analysis. *Chemosphere*, **60**, 1636–1643.

Garcia M. T., Campos E., Sanchez-Leal J. and Ribosa I. (2006). Effect of linear alkylbenzenesulphonates (LAS) on the anaerobic digestion of sewage sludge. *Water Research*, **40**, 2958–2964.

Garcia-Morales J. L., Nebot E., Romero L. I. and Sales D. (2001). Comparison between acidogenic and methanogenic inhibition caused by linear alkylbenzenesulfonate (LAS). *Chemical and Biochemical Engineering Quartely*, **15**, 13–19.

Glissmann K., Chin K. J., Casper P. and Conrad R. (2004). Methanogenic pathway and archaeal community structure in the sediment of eutrophic Lake Dagow: effect of temperature. *Microbial Ecology*, **48**(3), 389–399.

Gomec C. Y., Letsiou I., Ozturk I., Eroglu V. and Wilderer P. A. (2008). Identification of archaeal population in the granular sludge of an UASB reactor treating sewage at low temperatures. *Journal of Environmental Science and Health A*, **43**, 1504–1510.

Gomec C. Y., Eroglu V. and Wilderer P. A. (2009). Identifying aceticlastic and hydrogenotrophic methanogens in psychrophilic and mesophilic granular sludges treating synthetic sewage by means of FISH and CSLM. *Journal of Rapid Methods & Automation Microbiology*, **17**(2), 135–153.

Goux X., Calusinska M., Lemaigre S., Marynowska M., Klocke M., Udelhoven T., Benizri E. and Delfosse P. (2015). Microbial community dynamics in replicate anaerobic digesters exposed sequentially to increasing organic loading rate, acidosis, and process recovery. *Biotechnology Biofuels*, **8**, 122, 18pp.

Guellil A., Thomas F., Block J.-C., Bersillon J. L. and Ginestet P. (2001). Transfer of organic matter between wastewater and active sludge flocs. *Water Research*, **35**, 143–150.

Gujer W. and Zehnder A. (1983). Conversion processes in anaerobic-digestion. *Water Science Technology*, **15**(8–9), 127–167.

Guo X., Wang C., Sun F., Zhu W. and Wu W. (2014). A comparison of microbial characteristics between the thermophilic and mesophilic anaerobic digesters exposed to elevated food waste loadings. *Bioresource Technology*, **152**, 420–428.

Hammer M. J. (1986). Water and wastewater technology, 2nd edn. United States.

Hao L., Bize A., Conteau D., Chapleur O., Courtois S., Kroff P., Queménér E. D., Bouchez T. and Mazéas L. (2016). New insights into the key microbial phylotypes of anaerobic sludge digesters under different operational conditions. *Water Research*, **102**, 158–169.

Harmsen H., Wullings B., Akkermans A. D. L., Ludwig W. and Stams A. J. M. (1993). Phylogenetic analysis of *Syntrophobacter wolinii* reveals a relationship with sulphate-reducing bacteria. *Archives of Microbiology*, **160**(3), 238–240.

Hattori S., Kamagata Y., Hanada S. and Shoun H. (2000). *Thermacetogenium phaeumgen. Nov.*, sp nov., a strictly anaerobic, thermophilic, syntrophic acetate oxidizing bacterium. *International Journal of Systematic and Evolutionary Microbiology*, **50**, 1601–1609.

Henze M., Gujer W., Mino T. and van Loosdrecht M. (2002). Activated Sludge Models ASM1, ASM2, ASM2d and ASM3. IWA Publishing, London, UK.

Henze M., Harremoës P., la Cour Jansen J. and Arvin E. (2002). Wastewater Treatment: Biological and Chemical Processes, 3rd edn. Springer-Verlag, Berlin.

Holt M. S., Waters M. H., Comber R., Armitage G., Morris G. and Newbery C. (1995). AIS/CESIO environmental surfactant monitoring programme. SDIA sewage treatment pilot study on linear alkylbenzene sulphonates (LAS). *Water Research*, **29**, 2063–2070.

Hulshoff Pol L. (1989). The phenomenon of granulation of anaerobic sludge. PhD thesis, Wageningen Agricultural University, The Netherlands.

Hulshoff Pol L. W., De Castro Lopes S. I., Lettinga G. and Lens P. N. L. (2004). Anaerobic sludge granulation. *Water Research*, **38**, 1376–1389.

Jepsen J., Smith S. R., Krogh P. H. and Versteeg A. T. (2007). European risk assessment of LAS in agricultural soil revisited: species sensitivity distribution and risk estimates. *Chemosphere*, **69**, 880–892.

Jung K. W., Kim D. H., Kim S. H. and Shin H. S. (2011). Bioreactor design for continuous dark fermentative hydrogen production. *Bioresource Technology*, **102**, 8612–8620.

Kendall M. and Boone D. (2006). The order Methanosarcinales. In: The Prokaryotes, M. Dworkin, S. Falkow, E. Rosenberg, K.-H. Schleifer and E. Stackebrandt (eds), Springer, NewYork, USA, pp. 244–256.

Kim W., Lee S., Shin S. G., Lee C., Hwang K. and Hwang S. (2010). Methanogenic community shift in anaerobic batch digesters treating swine wastewater. *Water Research*, **44**, 4900–4907.

Kundu K., Sharma S. and Sreekrishnan T. (2012). Effect of operating temperatures on the microbial community profiles in a high cell density hybrid anaerobic bioreactor. *Bioresource Technology*, **118**, 502–511.

LaPara T. M., Nakatsu C. H., Pantea L. and Alleman J. E. (2000). Phylogenetic analysis of bacterial communities in mesophilic and thermophilic bioreactors treating pharmaceutical wastewater. *Applied Environmental Microbiology*, **66**, 3951–3959.

Laspidou C. S. and Rittmann B. E. (2002). A unified theory for extracellular polymeric substances, soluble microbial products, and active and inert biomass. *Water Research*, **36**, 2711–2720.

Lee C., Kim J., Shim S. and Hwang S. (2008). Monitoring bacterial and archaeal community shifts in a mesophilic anaerobic batch reactor treating a high-strength organic wastewater. *FEMS Microbiology Ecology*, **65**, 544–554.

Lettinga G. (1995). Anaerobic digestion and wastewater treatment systems. *Antonie van Leeuwenhoek*, **67**(1), 3–28.

Lettinga G., Hulshoff Pol L. W. and Zeeman G. (1996). Biological Wastewater Treatment. Part I: Anaerobic Wastewater Treatment. Lecture Notes. Wageningen Agricultural University, Netherlands.

Lettinga G., Rebac S. and Zeeman G. (2001). Challenge of psychrophilic anaerobic wastewater treatment. *Trends Biotechnology*, **19**(9), 363–370.

Lew B., Belavski M, Admon S., Tarre S. and Green M. (2003). Temperature effect on UASB reactor operation for domestic wastewater treatment in temperate climate regions. *Water Science and Technology*, **48**(3), 25–30.

Ligero P. and Soto M. (2002). Sludge granulation during anaerobic treatment of pre-hydrolysed domestic wastewater. *Water SA*, **28**, 307–311.

Liu J. and Lu X. (2017). Effect of Seasonal Temperature on the Performance and on the Microbial Community of a Novel AWFR for Decentralized Domestic Wastewater Pretreatment. *Applied Science*, **7**(6), 1–18.

Liu Y. and Whitman W. B. (2008). Metabolic, phylogenetic, and ecological diversity of the methanogenic archaea. *Annals of the New York Academy of Sciences*, **1125**, 171–189.

Liu W. T., Chan O. C. and Fang H. H. P. (2002). Characterization of microbial community in granular sludge treating brewery wastewater. *Water Research*, **36**, 1767–1775.

Liu Y., Xu H.-L., Yang S.-F. and Tay J.-H. (2003). Mechanisms and models for anaerobic granulation in upflow anaerobic sludge blanket reactor. *Water Research*, **37**, 66–673.

Ma J. X., Mungoni L. J., Verstraete W. and Carballa M. (2009). Maximum removal rate of propionic acid as a sole carbon source in UASB reactors and the importance of the macro- and micro-nutrients stimulation. *Bioresource Technology*, **100**(14), 3477–3482.

Mahmoud N., Amarneh M. N., Al-Sa'Ed R., Zeeman G., Gijzen H. and Lettinga G. (2003). Sewage characterization as a tool for the application of anaerobic treatment in Palestine. *Environmental Pollution*, **126**, 115–122.

Maria M. A. (2010). Caracterização de lixiviados de aterros Sanitário e industrial da Região Metropolitana de Belo Horizonte (Characterization of leachate from sanitary and industrial landfills of the Metropolitan Region of Belo Horizonte). MSc thesis, Federal University of Minas Gerais, Belo Horizonte, Brazil.

Martins A. S., Ferreira B. O., Ribeiro N. C., Martins R., Rabelo-Leite L., Oliveira G., Colturato L. F., Chernicharo C. A. L. and Araújo J. C. (2017). Metagenomic analysis and performance of a mesophilic anaerobic reactor treating food waste at various load rates. *Environmental Technology*, **38**, 2153–2163.

McCarty P. L. and McKinney R. E. (1961). Salt toxicity in anaerobic digestion. *Journal Water Pollution Control Federation*, **33**, 399–415.

McHugh S., Carton M., Mahony T. and Oâ Flaherty V. (2003). Methanogenic population structure in a variety of anaerobic bioreactors. *FEMS Microbiology Letter*, **219**(2), 297–304.

McHugh S., Carton M., Collins G. and O'Flaherty V. (2004). Reactor performance and microbial community dynamics during anaerobic biological treatment of simple and complex wastewaters at 16–37°C. *FEMS Microbial Ecology*, **48**, 369–378.

McKeown R. M., Scully C., Enright A., Chinalia F. A., Lee C., Mahony T., Collins G. and O'Flaherty V. (2009). Psychrophilic methanogenic community development during long-term cultivation of anaerobic granular biofilms. *ISME Journal*, **3**, 1231–1242.

Mesquita P. L., Aquino S. F., Xavier A. L. P., da Silva J. C. C., Afonso R. C. F. and Silva S. Q. (2010). Soluble Microbial Product (SMP) characterization in bench-scale aerobic and anaerobic CSTR under different operational conditions. *Brazilian Journal of Chemical Engineering*, **27**(1), 101–111.

Moraes B. S., Triolo J. M., Lecona V. P., Zaiat M. and Sommer, S. G. (2015). Biogas production within the bioethanol production chain: Use of co-substrates for anaerobic digestion of sugar beet vinasse. *Bioresource Technology*, **190**, 227–234.

Mota S. B. and von Sperling M. (2009). Nutrientes de esgoto sanitário: utilização e remoção (Nutrients of municipal sewage: utilization and removal). ABES, Rio de Janeiro, Brazil.

Munch R., Hwang C. P. and Lackie T. H. (1980). Wastewater fractions add to total treatment picture. *Water and Sewage Works*, **127**, 49–54.

Muyzer G. and Stams A. (2008). The ecology and biotechnology of sulphate-reducing bacteria. *Nature Reviews Microbiology*, **6**, 441–454.

Narihiro T., Terada T., Kikuchi K., Iguchi A., Ikeda M., Yamauchi T., Shiraishi K., Kamagata Y., Nakamura K. and Sekiguchi Y. (2009). Comparative analysis of bacterial and archaeal communities in methanogenic sludge granules from upflow anaerobic sludge blanket reactors treating various food-processing, high-strength organic wastewaters. *Microbes Environments*, **24**(2), 88–96.

O'Flaherty V., Lens P., Leahy B. and Colleran E. (1998). Long-term competition between sulphate-reducing and methane-producing bacteria during full-scale anaerobic treatment of citric acid production wastewater. *Water Research*, **32**, 815–825.

Orhon D. and Çokgör E. U. (1997). COD fractionation in wastewater characterization – The state of the art. *Journal of Chemical Technology & Biotechnology*, **68**(3), 283–293.

Paul K., Nonoh J. O., Mikulski L. and Brune A. (2012). "Methanoplasmatales," thermoplasmatales-related Archaea intermite guts and other environments, are the seventh order of methanogens. *Applied Environmental Microbiology*, **78**, 8245–8253.

Paula D. R., Jr. and Foresti E. (2009). Sulfide toxicity kinetics of a UASB reactor. *Brazilian Journal of Chemical Engineering*, **26**(4), 669–675.

Painter H. A. and Zabel T. (1989). The behaviour of LAS in sewage treatment. *Tensid Surfactant Detergent*, **26**(2), 108–115.

Pereira J. (2012). Desenvolvimento de reator UASB com duplo estagio de coleta de biogás (DECB) aplicado ao tratamento de esgoto sanitário objetivando o controle da escuma e a granulação do lodo. (Development of a double-stage biogas collection UASB reactor (DEBC) applied to the treatment of domestic wastewater aiming at scum control and sludge granulation). PhD thesis, Federal University of Minas Gerais, Belo Horizonte, Brazil.

Petropoulos E., Dolfing J., Davenport R. J., Bowen E. J. and Curtis T. P. (2017). Developing cold-adapted biomass for the anaerobic treatment of domestic wastewater at low temperatures (4, 8 and 15°C) with inocula from cold environments. *Water Research*, **112**, 100–109.

Petropoulos E., Dolfing J., Yu Y., Wade M. J., Bowen E. J., Davenport R. J. and Curtis T. P. (2018). Lipolysis of domestic wastewater in anaerobic reactors operating at low temperatures. *Environmental Science: Water Research and Technology*, **4**, 1002–1013.

Polprasert C. 1989. Organic Wastes Recycling. John Wiley & Sons, Chichester, UK.

Pontes P. P. and Chernicharo C. A. L. (2009). Effect of aerobic sludge return on the characteristics of anaerobic biomass present in UASB reactors treating domestic sewage. *Engenharia Sanitaria e Ambiental*, **14**(2), 223–234.

Pontes P. P., Chernicharo C. A. L. and Von Sperling M. (2014). Kinetics of concentration decay of specific organic matter in UASB reactors operating with and without return of aerobic sludge. *Environmental Technology*, **35**(16), 2046–2054.

Qiaoying B., Li J. M., Zhang L., Kumar Jha A., Zhang Y. and Ai B. (2013). Syntrophic Propionate Degradation Response to Temperature Decrease and Microbial Community Shift in an UASB Reactor. *Journal Microbiology Biotechnology*, **23**(3), 382–389.

Regueiro L., Veiga P., Figueroa M., Alonso-Gutierrez J., Stams A. J. M., Lama J. and Carballa M. (2012). Relationship between microbial activity and microbial community structure in six full-scale anaerobic digesters. *Microbiology Resource*, **167**, 581–589.

Roest K., Heilig H. G. H. J., Smidt H., de Vos W. M, Stams A. J. M. and Akkermans A. D. L. (2005). Community analysis of a full-scale anaerobic bioreactor treating paper mill wastewater. *Systematic Applied Microbiology*, **28**(2), 175–185.

Saia F. T., Souza T. O., Duarte R. T. D., Gianotti E. P., Fonseca D. and Foresti E. (2016). Microbial community in a pilot-scale bioreactor promoting anaerobic digestion and sulphur-driven denitrification for domestic sewage treatment. *Bioprocess and Biosystems Engineering*, **39**(2), 341–352.

Sakai S., Imachi H., Hanada S., Ohashi A., Harada H. and Kamagata Y. (2008). *Methanocellapaludicola* gen. nov., sp.nov., a methane-producing archaeon, the first isolate of the lineage "Rice Cluster I", and proposal of the new archaeal order Methanocellales ord. nov. *International Journal Systematic Evolutionary Microbiology*, **58**, 929–936.

Schmidt J. E. and Ahring B. K. (1995). Granulation in thermophilic upflow anaerobic sludge blanket (UASB) reactors. *Antonie van Leeuwenhoek*, **68**, 339–344.

Schink B. (1997). Energetics of syntrophic cooperation in methanogenic degradation. *Microbiology and Molecular Biology Reviews*, **61**(2), 262–280.

Schnurer A. and Nordberg A. (2008). Ammonia, a selective agent for methane production by syntrophic acetate oxidation at mesophilic temperature. *Water Science and Technology*, **57**, 735–740.

Schnurer A., Schink B. and Svensson B. H. (1996). *Clostridium ultunense* sp. nov., a mesophilic bacterium oxidizing acetate in syntrophic association with a hydrogenotrophic methanogenic bacterium. *International Journal of Systematic Bacteriology*, **46** (4), 1145–1152.

Schulz S. and Conrad R. (1996). Influence of temperature on pathways to methane production in the permanently cold profundal sediment of Lake Constance. *FEMS Microbiology Ecology*, **20**(1), 1–14.

Seghezzo L., Zeeman G., Van Lier J. B., Hamelers H. V. M. and Lettinga G. (1998). The anaerobic treatment of sewage in UASB and EGSB reactors Reviews in. *Bioresource Technology*, **65**(3), 175–190.

Shanmugam P. and Horan N. J. (2009). Optimising the biogas production from leatherfleshing waste by co-digestion with MSW. *Bioresource Technology*, **100**(18), 4117–4120.

Shigematsu T., Tang Y. Q., Kawaguchi H., Ninomiya K., Kijima J., Kobayashi T., Morimura S. and Kida K. (2003). Effect of dilution rate on structure of a mesophilic acetate-degrading methanogenic community during continuous cultivation. *Journal of Bioscience and Bioengineering*, **96**(6), 547–558.

Shin S. G., Zhou B. W., Lee S., Kim W. and Hwang S. (2011). Variations in methanogenic population structure under overloading of pre-acidified high-strength organic wastewaters. *Process Biochemistry*, **46**(4), 1035–1038.

Silva S. Q., Abreu E. F., Zerbini A. M. and Chernicharo C. A. L. (2008). Stability of microbial diversity in UASB bioreactors under different operational conditions. Proc. IX Latin American Workshop and Symposium on Anaerobic Digestion, Easter Island, Chile.

Sinischalchi L. A. B., vale I. C., Dell' Isola J., Chernicharo C. A. and Araujo J. C. (2015). Enrichment and activity of methanotrophic microorganisms from municipal wastewater sludge. *Environmental Technology*, **36**(12), 1563–1575.

Smith K. S. and Ingram-Smith C. (2007). *Methanosaeta*, the forgotten methanogen? *Trends Microbiology*, **15**, 150–155.

Smith A. L., Skerlos S. J. and Raskin L. (2015). Anaerobic membrane bioreactor treatment of domestic wastewater at psychrophilic temperatures ranging from 15°C to 3°C. *Environmental Science Water Resource Technology*, **1**, 56–64.

Song M., Shin A. G. and Hwang S. (2010). Methanogenic population dynamics assessed by real-time quantitative PCR in sludge granule in upflow anaerobic sludge blanket treating swine wastewater. *Bioresource Technology*, **101**, S23–S28.

Souza C. L., Silva S. Q., Aquino S. F. and Chernicharo C. A. L. (2006). Production and characterization of scum and its role in odour control in UASB reactors treating domestic wastewater. *Water Science and Technology*, **54**(9), 201–208.

Souza C. L., Chernicharo C. A. L. and Melo G. C. B. (2012). Methane and hydrogen sulphide emissions in UASB reactors treating domestic wastewater. *Water Science and Technology*, **65**, 1229–1237.

Souza L. F. C., Florencio L., Kato M. T. and Gavazza S. (2016). Methanogenic activity inhibition by increasing the linear alkylbenzene sulfonate (LAS) concentration. *Journal of Environmental Science and Health. Part A, Toxic Hazardous Substances and Environmental Engineering*, **51**(8), 1–5.

Speece R. E. (1983). Anaerobic biotechnology for industrial wastewater treatment. *Environmental Science Technology*, **17**(9), 416–427.

Speece R. E. (1996). Anaerobic biotechnology for industrial wastewaters. Archae Press, Nashville.

Stams A. J. M. (1994). Metabolic interactions between anaerobic bacteria in methanogenic environments. *International Journal General Molecular Microbiology*, **66**(1–3), 271–294.

Stams A. J. M., Dijkema C., Plugge C. M. and Lens P. (1998). Contribution of 13C-NMR spectroscopy to the elucidation of pathways of propionate formation and degradation in methanogenic environments. *Biodegradation*, **9**(6), 463–473.

Sundberg C., Al-Soud W. A., Larsson M., Alm E., Yekta S. S., Svensson B. H., Sørensen S. J. and Karlsson A. (2013). 454 pyrosequencing analyses of bacterial and archaeal richness in 21 full-scale biogas digesters. *FEMS Microbiology Ecology*, **85** (3), 612–626.

Syutsubo K., Yoochatchaval W., Yoshida H., Nishiyama K., Okawara M., Sumino H., Araki N., Harada H. and Ohashi A.. (2008). Changes of microbial characteristics of retained sludge during lowtemperature operation of an EGSB reactor for low-strength wastewater treatment. *Water Science and Technology*, **57**, 277–281.

Tang Y-Q., Fujimura Y., Shigematsu T., Morimura S. and Kidal K. (2007). Anaerobic treatment performance and microbial population of thermophilic upflow anaerobic filter reactor treating Awamori Distillery wastewater. *Journal of Bioscience and Bioengineering*, **104**, 281–287.

Tavares F. V. F. (2008). Remoção de surfactantes aniônicos em filtros biológicos percoladores com diferentes meios suporte aplicados ao pós-tratamento de efluentes de reatores UASB (Removal of anionic surfactants in trickling filters with different media applied to the post-treatment of UASB reactor effluents). MSc thesis, Federal University of Minas Gerais, Belo Horizonte, Brazil.

Teixeira A. R. (2007). Influência da alteração da distribuição do tamanho de partículas no desempenho de reatores UASB tratando esgotos doméssticos (Influence of changing particle size distribution on the performance of UASB reactors treating domestic wastewater). PhD thesis, Federal University of Minas Gerais, Belo Horizonte, Brazil.

Thauer R. K., Jungermann K. and Decker K. (1977). Energy conservation in chemotrophic anaerobic bacteria. *Bacteriology Reviews*, **41**, 100–180.

van Lier J. B., Tilche A., Ahring B. K., Macarie H., Moletta R., Dohanyos M., Hulshoff Pol L. W., Lens P. and Verstraete W. (2001). New perspectives in anaerobic digestion. *Water Science and Technology*, **43**(1), 1–18.

van Lier J. B., Vashi A., van der Lubbe J. and Heffernan B. (2010). Anaerobic sewage treatment using UASB reactors: engineering and operational aspects. In: Environmental Anaerobic Technology; Applications and New Developments, H. H. P. Fang (ed.), Imperial College Press, London, UK, pp. 59–90.

Vavilin V. A., Fernandez B., Palatsi J. and Flotats X. (2008). Hydrolysis kinetics in anaerobic degradation of particulate organic material: an overview. *Waste Management*, **28**, 939–51.

Vos P. D., Garrity G. M., Jones D., Krieg N. R., Ludwig W., Rainey F. A., Schleifer K. H. and Whitman W. B. (2009). Bergey's Manual of Systematic Bacteriology, Vol. Three, The Firmicutes, 2nd edn. Springer, New York, USA, pp. 1422.

Wagener S. and Schink B. (1987). Anaerobic degradation of nonionic and anionic surfactants in enrichment cultures and fixed-bed reactors. *Water Resource*, **21**, 615–622.

Wang H., Fotidis I. A. and Angelidaki I. (2015). Ammonia effect on hydrogenotrophic methanogens and syntrophic acetate-oxidizing bacteria. *FEMS Microbiology Ecology*, **91**(11), 1–8.

Widdel F. and Pfennig N. (1982). Studies on dissimilatory sulphate reducing bacteria that decompose fatty acids. II. Incomplete oxidation of propionate by *Desulfobulbus propionicus* gen. nov., sp. nov. *Archives of Microbiology*, **131**, 360–365.

Xing W., Zuo J.-e., Dai N., Cheng J. and Li J. (2009). Reactor performance and microbial community of an EGSB reactor operated at 20 and 15 C. *Journal of Applied Microbioogy*, **107**, 848–857.

Yamada T., Sekiguchi Y., Hanada S., Imachi H., Ohashi A., Harada H. and Kamagata Y. (2006). *Anaerolinea thermolimosa* sp. nov., *Levilinea saccharolytica* gen. nov., sp. nov. and *Leptolinea tardivitalis* gen. nov., sp. nov., novel filamentous anaerobes, and description of the new classes *Anaerolineae* classis nov. and *Caldilineae* classis nov. in the bacterial phylum *Chloroflexi*. *International Journal of Systematic and Evolutionary Microbiology*, **56**, 1331–1340.

YenigÃ¼n O. and Demirel B. (2013). Ammonia inhibition in anaerobic digestion: A review. *Process Biochemistry*, **48**, 901–911.

Zehnder J. B. A. (ed.) (1988). Biology of Anaerobic Microorganisms. Wiley, New York, 872 p.

Zhang D., Trzcinski A. P., Kunacheva C., Stuckey D. C., Liu Y., Tan S. K. and Ng W. J. (2016). Characterization of soluble microbial products (SMPs) in a membrane bioreactor (MBR) treating synthetic wastewater containing pharmaceutical compounds. *Water Research*, **102**, 594–606.

Zhang D. D., Zhu W. B., Tang C., Suo Y. L., Gao L. J., Yuan X. F., Wang X. F. and Cui Z. J. (2012). Bioreactor performance and methanogenic population dynamics in a low-temperature (5–18°C) anaerobic fixed-bed reactor. *Bioresource Technology*, **104**, 136–143.

Zhang L., Hendrickx T. L. G., Kampman C., Temmink H. and Zeeman G. (2013). Co-digestion to support low temperature anaerobic pretreatment of municipal sewage in a UASB-digester. *Bioresource Technology*, **148**, 560–566.

Zhang Y., Aiyuk S., Xu H., Chen G. and Verstraete W. (2005). Study of microbial community structures in UASB sludge treating municipal wastewaters by denaturing gradient gel electrophoresis of 16S rRNA. *Science China Life Science*, **48**, 128–135.

Zheng D. and Raskin L. (2000). Quantification of *Methanosaeta* species in anaerobic bioreactors using genus- and species-specific hybridization probes. *Microbial Ecology*, **39**, 246–262.

Zhou W. L., Wu B. T., She Q. H., Chi L. and Zhang Z. J. (2009). Investigation of soluble microbial products in a full-scale UASB reactor running at low organic loading rate. *Bioresource Technology*, **100**, 3471–3476.

Zinder S. H., Cardwell S. C., Anguish T., Yee M. and Koch M. (1984). Methanogenesis in a thermophilic (58°C) anaerobic digestor: *methanothrix* sp. as an important aceticlastic methanogen. *Applied and Environmental Microbiology*, **47**, 796–807.

Chapter 4

Design of UASB reactors for sewage treatment

Thiago Bressani-Ribeiro, Carlos A. L. Chernicharo, Lívia C. S. Lobato and Priscilla N. P. Neves

Department of Sanitary and Environmental Engineering (DESA), Universidade Federal de Minas Gerais (UFMG), Av. Antonio Carlos, 6.627, Campus Pampulha, Belo Horizonte 31270-901, Brazil

**Corresponding author: calemos@desa.ufmg.br*

4.1 INTRODUCTION

As discussed in Chapter 1, UASB reactors were originally designed to treat industrial wastewater. Nevertheless, countries like Brazil, Colombia, Mexico, and India have been systematically implementing and adapting UASB reactors for sewage treatment. Therefore, the anaerobic process can be considered a consolidated technology for such purpose. Mainstream anaerobic sewage treatment at low temperatures has not yet been fully demonstrated, but looks promising, as addressed in Chapters 3 and 11. This is important because apart from the majority of developed countries (in temperate zones) that have already defined their sewage treatment systems (see Chapter 2) as aerobic-based (i.e. activated sludge), retrofitting to less energy-intensive, carbon-neutral processes is a current objective.

The design of UASB reactors for sewage treatment can be simple and does not require sophisticated equipment or packing media for biomass retention and growth (although innovative configurations have been using packing media, as discussed in Chapter 7). However, inadequate design assumptions that do not consider the sewage treatment specificities could hamper anaerobic reactor performance, thereby jeopardizing the acceptance of the technology. The main design criteria are well explained in consolidated chapters and textbooks, such as those by van Haandel and Lettinga (1994), Chernicharo (2007), van Lier *et al.* (2010), and Chernicharo *et al.* (2015). Nonetheless, the improved design of the integrated management of the liquid, solid, and gaseous phases remains of great importance.

Therefore, this Chapter describes critical aspects for the comprehensive design of UASB reactors treating sewage based on the last 20 years of experience in design, construction, and operation of such units.

Editor(s): Carlos Augusto de Lemos Chernicharo and Thiago Bressani-Ribeiro
doi: 10.2166/9781780409238_0056

Significant attention has been called to the design of the preliminary treatment (PT) step. As presented in Chapter 6, the satisfactory operation of UASB reactors depends on the accurate operation of the PT units, which in turn is associated with proper design. This Chapter also presents an innovative design of UASB reactors for sewage treatment based on the concept of standard gas–liquid–solid (GLS) separators.

4.2 PRELIMINARY TREATMENT: A CRUCIAL STEP BEFORE UASB REACTORS

The main role of the PT units (screens, sieves, and grit chambers) is the removal of solids that could compromise the proper functioning of UASB reactors, mainly through pipe obstructions, electromechanical equipment failures, grit accumulation inside the reactors, and scum formation. These problems usually also extend to equipment and by-product processing units, such as for sludge and scum, thereby requiring frequent shutdowns, which can range from simple cleaning and clearing to the stoppage of equipment for time-consuming maintenance in the context of developing countries (e.g. centrifuges).

Several characteristics of the sewage treatment plant (STP) contribution basin influence the quantity and quality of the material removed in the PT stage, such as the type of occupation, size of the area, and habits and sanitary education of the population, among others. The alternatives associated with the management of the material retained in the screens and grit chambers are discussed in Chapter 12. Especially for UASB-based STPs, the receipt of sludge from cesspool cleaning trucks, non-domestic effluents (NDEs), waste from other STPs (such as scum and sludge), and sludge from water treatment plants (WTPs) leads to an increase in sludge production and the formation and accumulation of scum in UASB reactors, as discussed in Chapters 6 and 7.

4.2.1 Main configurations of the preliminary sewage treatment before UASB reactors

The units that comprise the PT stage can be implanted, buried (see Figure 4.1(a)), on the ground, or elevated (Figure 4.1(b)) in relation to the terrain level. Buried solutions may facilitate the PT enclosure, as shown in Figure 4.1(a) for the grit chamber and Parshall flume; however, they may imply greater operational difficulty associated with the cleaning routine of the units, and in most cases require a pumping station (PS) within the STP. In this configuration, the PT units upstream of the PS may play an important role in protecting the pumping equipment, but there is no way to ensure that this arrangement compensates for the greater difficulty associated with the operation of the buried units. Aspects associated with the containment, exhaustion, and treatment of odours in the PT step are discussed in Chapter 9.

In turn, the elevated (above-ground) solution may incur a higher deployment cost compared with the solution on the ground; however, it has the advantage of allowing discharge (by gravity) from the grit chambers, which may simplify or eliminate the need for electromechanical equipment (e.g. screw classifier). Additionally, the bucket for collecting the screened material can be strategically placed near the screens at a level below the PT units. This contributes to the improvement in the ergonomic conditions associated with the manual cleaning of the bar racks. Eventually, the screening step may be associated with the raw sewage PS considering the need to adopt a structure to control the incoming flows (overflow) to prevent flooding of the screens. In this case, considering that the screens are components of the PS, the PT units would be reduced to the implantation of the grit chambers, and if necessary, fine screens and sieves. In many situations, the elevation of the PT units eliminates the need for a new PS within the STP because the level increase can come directly from the pressurised mains, thereby making the hydraulic profile in favour of gravity.

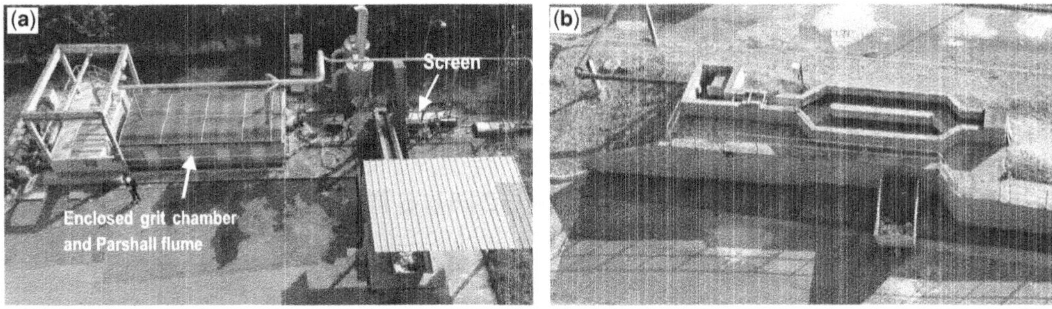

Figure 4.1 Preliminary treatment units of UASB-based sewage treatment plants: (a) buried and enclosed solution and (b) elevated (above the ground) solution.

4.2.2 Main design criteria for preliminary sewage treatment before UASB reactors

The main design logic of the PT units is to ensure proper flow velocities for the removal of coarse solids; to avoid obstruction, wear and tear, and locking problems in the electromechanical equipment; and to avoid problems with the sensors and submerged speed meters. However, it must be ensured that such velocities do not cause the sedimentation of particulate organic matter, which would lead to the generation of odours. The main parameters and sizing criteria of the PT components are reported in Table 4.1. The different arrangements recommended for the PT units of UASB-based STPs are presented in Section 4.2.3 depending on the population equivalent (or influent flow) of the STP. Such arrangements were obtained from operational experience in STPs in Brazil, and have been extensively discussed with specialists in the country's sanitation sector.

Table 4.1 Main parameters and sizing criteria of the preliminary treatment components.

Parameter	Value
Screen (bar racks)	
Maximum velocity through the bars for the final (end of plan) flow rate[a]	1.20 m s^{-1}
Inclination of the bars in relation to the horizontal	45° to 60° – manual cleaning 60° to 90° – mechanised cleaning
Minimum head loss to be considered (upstream flow)	0.15 m – manual cleaning 0.10 m – mechanised cleaning
Grit Chamber[b]	
Flow velocity	0.25–0.40 m s^{-1}
Minimum depth for accumulation of settled material	0.20 m
Hydraulic loading rate	600–1,000 m^{3} · m^{-2} · d^{-1}

Note: The flow rate used for sizing the preliminary treatment components must be the maximum influent flow rate.
[a]The upstream and downstream channels of coarse solids removal devices shall ensure, at least once per day from the start of the operation, a velocity equal to or greater than 0.40 m s^{-1}. The minimum flow velocity in these channels should ideally be greater than 0.30 m s^{-1} in order to avoid grit deposition in the channels.
[b]Criteria valid for grit chambers of the horizontal flow and rectangular section.

The main aspects to be considered in the design of the components that comprise the PT unit in UASB-based STPs are summarised in Table 4.2. We define the degree of importance associated with the observance of each aspect according to the following criteria:

- Medium importance – aspects that, if not observed, can compromise the proper operation and maintenance of the system.
- High importance – aspects whose observation is crucial for adequate operation and maintenance of the system.

It is noted that most of the aspects listed in Table 4.2 were categorised as of high significance. This fact reflects the importance of PT for adequate conditioning of the sewage before entering the UASB reactors. In summary, an adequate PT unit will positively influence: (i) the maintenance of the flow distribution structures and effluent collecting channels, thereby avoiding obstructions; (ii) the reduction in deposition/accumulation of grit in the bottom of UASB reactors; (iii) the management of the sludge because it will be possible to obtain a solid with lower contents of debris and grit and better aesthetic quality; (iv) the management of scum mainly owing to the fact that when not properly removed in the PT unit, some materials (e.g. plastics and light debris in general) tend to float inside the reactor and accumulate in the upper part of the GLS separator; and (v) the significant reduction in common problems associated with the operation of mechanised sludge and scum processing units, such as stoppages caused by pipe obstructions, breakage, and loss of efficiency due to equipment wear and tear.

4.2.3 Recommended technological arrangements for preliminary treatment before UASB reactors

For small-scale STPs (population equivalent (PE) < 10,000), operational experience has pointed to the sequential use of two fine screens with manual cleaning and 20 and 10 mm spaces between the bars, respectively (Figure 4.2). It should be noted that the option to use manual or mechanised cleaning screens for STPs of this size is the responsibility of the service provider given the volume of material to be retained or operational difficulties. In the case of plants with a maximum influent flow rate of 100 L s^{-1} or greater, the screens should be mechanically cleaned.

It is essential to use a coarse solids removal device with openings of 12 mm or smaller for STPs with a maximum flow rate of up to 100 L s^{-1}. Operational experience using static sieving units (3–5 mm mesh opening) in small-scale STPs, namely post screen units, has shown their importance in avoiding obstructions in the UASB reactor feeding structure (e.g. channels, weirs and pipes), as well as in the treated effluent collection channels. For plants whose maximum flow rate is greater than 100 L s^{-1}, it is essential to use a sieve with openings equal to or smaller than 6 mm. In this case, the operational experience has indicated the preferential use of sieves with a mesh opening of 3 mm or 4 mm.

The quality of the sludge and scum removed from UASB reactors in terms of the presence of coarse solids is also reflected in the reliability of operation of the mechanised dewatering systems, thereby reducing the need for maintenance due to the obstruction of pump and centrifuge feeding pipes, locking of stators and rotors of pumps and mixers, rupture of mechanical seals, accumulation of residues in equalisation tanks, etc. In addition, a more efficient removal of coarse solids leads to more aesthetic sludge, which may favour agricultural use, as well as reduced formation of scum inside the anaerobic reactors.

In relation to the grit chambers, for small-scale STPs (PE < 10,000), horizontal and rectangular section units (channel type; Figure 4.1(b)) are usually adopted, with a constant flow velocity guaranteed by means of a control section implanted downstream (usually a Parshall flume) and non-mechanised grit removal. A widely used option for small and medium-scale STPs (PE < 50,000)

Table 4.2 Main aspects to be observed in the design of the components that comprise the preliminary treatment (PT) unit.

Aspect	Degree of Importance	
General Aspects	**High**	**Medium**
Size the units using the maximum influent flow rate and ensure minimum velocities to prevent unwanted deposition of organic solids.	X	
Check the hydraulic profile of the PT components. The head losses must be considered for all components (channels, locks, and equipment) from the downstream flow control point (e.g. Parshall flume) to the upstream point (e.g. screens) for the definition of the water depths (upstream and downstream of each unity).	X	
Provide an overflow structure for flow rates higher than the design capacity of the STP. Note that in STPs in which the PS is located downstream of the PT units and where there is a risk of interruption of electric power supply, two overflow devices should be deployed, namely one before the PT units and the other associated with the suction well of the PS. This measure aims to prevent the drowning (or flooding) of units installed upstream of PSs.	X	
In the STP overflow channel, implement flow meters to ensure the recording of the frequency and magnitude of the inflow directed to the water body.		X
Provide a flap valve when there is risk of drowning of the overflow exit; this must be installed closer to the PT unit to prioritise access for maintenance of the valve.	X	
The by-pass upstream of the PT units shall be provided with a grit removal device.	X	
For medium and large-scale STPs (flow rate (Q) $\geq 20\ L\ s^{-1}$ or population equivalent (PE) $\geq$ 10,000), adopt redundancy (duplication) of the PT units (screens, grit chambers, and sieves). Evaluate modulation alternatives to reduce the initial investment, but ensure the maintenance of minimum velocities in order to avoid unwanted solids deposition.	X	
For small-scale STPs ($Q < 20\ L\ s^{-1}$ or PE $<$ 10,000), adopt a parallel (by-pass) channel equipped with a screen of 20 mm bar aperture.	X	
Verify the technical–economic feasibility of using a mechanised screen and/or static screen for small and medium-scale STPs ($Q < 100\ L\ s^{-1}$ or PE $<$ 50,000).		X
Evaluate the possibility of implementing sequential course, medium, and fine screens if it is not appropriate to use mechanised screens.		X
Provide a collection, storage, transport, and final disposal system for the solid waste retained in the PT units considering the operational routine (removal frequency and destination) and disturbance to the neighbourhood (odours).	X	
	X	

(Continued)

Table 4.2 Main aspects to be observed in the design of the components that comprise the preliminary treatment (PT) unit (*Continued*).

Aspect	Degree of Importance	
Provide gates upstream and downstream of the units to allow the admission, isolation, cleaning, and by-pass of the incoming sewage.		
In the specification of gates and equipment, consider the use of materials resistant to corrosion.	X	
For PSs, observe the depth of the suction well, which is preferably limited to 8 m. In the case of wells over 8 m, access conditions for maintenance and cleaning (intermediate platforms, ramp, ladder, electric hoist, etc.) and guardrails in the surroundings should be provided. However, depending on the type of screen cleaning, even at depths of less than 8 m, adequate conditions of access and cleaning must be guaranteed.	X	
Provide smaller buckets in small STPs in order to avoid accumulating waste for long periods of time, which would generate odours.		X
Evaluate the need to provide coverage devices for the solids buckets as a function of local rainfall conditions and the need for odour control.	X	
Ensure that access is compatible with the vehicles used by the operation for the removal of buckets/containers, including proper inclination of access ramps, a sufficient manoeuvring area suitable for truck traffic, and a base, sub-base, and pavement compatible with vehicular traffic.	X	
Provide adequate access for operators to perform routine activities and for electromechanical maintenance of the units (e.g. lubrication, removal of motors, and threads).	X	
Consider collecting and directing the leachate from the buckets to the headwork of the STP.	X	
In STPs that receive effluents from cesspool cleaning trucks, and scum and sludge from other STPs and PSs, the installation of a specific unit to receive and control these (e.g. balance and homogenisation tank) should be considered and take into account the impact on the production of sludge in the UASB reactor (see Chapters 6 and 7).	X	
Screens	**High**	**Medium**
For the estimation of retained coarse solids and grit, use operational data related to the local reality (check the characteristics of the contribution basin, such as occupation and economic activities).	X	
In the case of mechanised units, provide a parallel channel with manual cleaning screen with the same apertures between bars.	X	
The screens should be manufactured with rectangular section bars. Do not use circular section bars or horizontal bars to aid the fixing of the main bars because they make free passage of the rake difficult.	X	
For PT units implanted downstream of the PS, consider the use of a coarse screen upstream of the PS (note that the spacing between bars must be smaller than the diameter of the pump rotor to avoid passage	X	

(*Continued*)

Table 4.2 Main aspects to be observed in the design of the components that comprise the preliminary treatment (PT) unit (*Continued*).

Aspect	**Degree of Importance**	
of large coarse solids). In addition, the possibility of implantation of medium and fine screens upstream of the PS should be evaluated.		
For STPs with mechanised sludge and/or scum dewatering, provisions should be made for a sieving step with a maximum aperture of 6 mm (preferably 3 mm or 4 mm) in order to ensure the proper operation and minimisation of stoppages of the dewatering components (e.g. mixers and feeding pumps).	X	
In large-scale STPs with mechanised screens, evaluate the possibility of using compacting devices for the removed residues to reduce the volume and the release of leachate in the buckets and containers, reduce odours, and improve the characteristics of the material for final disposal.		X
Ensure compatibility between the equipment considered in the design and that actually acquired to avoid adjustments in the civil and electromechanical structures of the PT units during the construction and operation phases.	X	
Grit Chambers	**High**	**Medium**
When the width of the grit chamber is less than 30 cm, evaluate and propose alternatives, such as a vortex-flow-type grit chamber, customised cleaning equipment compatible with the smaller channel dimensions, or the adoption of prefabricated solutions manufactured with alternative materials (e.g. glass fibre and polypropylene).		X
For manual cleaning of grit chambers, check that there is adequate depth for grit accumulation according to the cleaning frequency adopted.	X	
Where the topography allows, adopt grit removal through the bottom in order to reduce the expenses of cesspool cleaning trucks.		X
For mechanised square section grit chambers, the moulding of the scraper circumference must be conducted within the square in which it is inscribed (contemplating chamfers at the vertices) using the scraper arm of the equipment itself.	X	
For square-section grit chambers, ensure uniform distribution of the flow along the width of the unit. In addition to having a suitable hydraulic sizing in the access channel, it should be possible to regulate the flow at each entry point.	X	
For square-section grit chambers, ensure the possibility of access for the cleaning of the influent channel because grit can be deposited at the ends of this channel owing to the occurrence of velocities below the minimum predicted in the design.	X	
For square-section grit chambers, evaluate the possibility of having a bottom discharge to drain the unit in order to allow cleaning and maintenance of the scraper arm and screw classifier.	X	

is the vortex flow grit chamber (cyclone type) with removal of grit by means of air lift. For larger STPs (PE > 50,000), horizontal flow units and square sections are commonly adopted. In this case, the removal of the retained grit is conducted by means of scraper arms and a screw classifier. Larger units require special attention to ensure uniform distribution of the influent flow across the entire section of the channel as well as attentive operation to identify the occurrence of preferential flows and/or solids deposition at specific points of the channel, notably upstream of the screens, especially during the first years of operation of the STPs when the flows can be lower than those considered in the design.

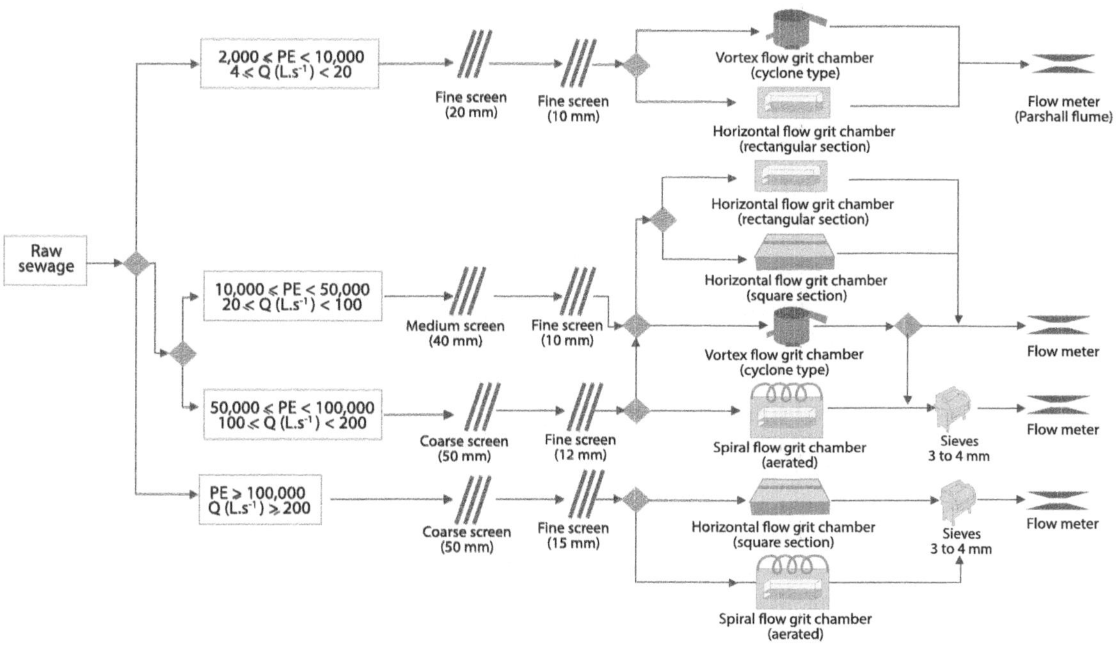

Figure 4.2 Recommended configurations of preliminary treatment (PT) units for UASB-based STPs of different sizes. Note: The equivalences between Q (flow rate) and PE (population equivalent) were determined from $Q_{average}$. The sizing of the PT units must be performed using Q_{max}.

In relation to the aerated grit chambers (spiral flow), although they are recognised as efficient equipment for grit removal, the Brazilian experiences reported in the literature indicate low efficiency of this equipment for the removal of oils and greases. Among the possible reasons given, the difference in viscosity of the liquid as a function of temperature can be an important factor, as such equipment is mainly designed for temperate-climate conditions. The possibility of removal of oils and greases in the PT stage may be interesting for the reduction in scum formation in UASB reactors, which, especially in larger STPs, represents one of the main operational problems, as discussed in Chapter 6.

In addition to the configurations recommended in Figure 4.2, the modulation and redundancy possibilities of the PT units, specifically for medium and large-scale STPs, must be verified. For small-scale STPs (PE < 10,000), it is recommended to implement a parallel by-pass channel with a fine screen (bar spacing equal to 20 mm).

For small-scale STPs the dimensions of the PT unit, especially the influent and effluent channels to the screens and grit chambers, become minute to meet the ideal flow conditions. Therefore, the use of reinforced concrete structures may not be an ideal solution given the difficulties associated with construction. In this case, precast solutions in alternative materials can be used instead, such as glass fibre and polypropylene.

4.3 TYPICAL CONFIGURATIONS/GEOMETRY AND MODULATION OF UASB REACTORS

Through the adaptation process of UASB reactors to treat low-concentration wastewater (such as sewage), an important outcome has been identified, which is the control of excessive upflow velocities in the digestion and settler compartments, especially during peak flows. Consequently, the height of the reactor is reduced (compared to reactors treating industrial wastewater) and the cross section is increased, as the design is mainly ruled by the hydraulic loading criteria, as discussed in Section 4.4.

Additionally, in an effort to prevent higher upflow velocities in the settler compartment and to enable sludge to return to the digestion compartment, larger cross sections close to the settler compartment have been tested. In this case, the reactor adopts a variable section that is smaller close to the digestion compartment and larger close to the settler compartment. A UASB variation called RALF (mentioned in Chapter 1) was developed in Brazil following this assumption. Nevertheless, this significantly enhances the construction complexity (e.g. inclined walls), and drawbacks associated with sludge withdrawal and biogas collection have been widely reported by sanitation companies. An alternative has been recently developed in Brazil and is further addressed in Section 4.7.

Mounting of the GLS separator and reduced upflow velocities prevent the need for enlarged surfaces at the top (van Lier *et al.*, 2010). Section 4.4.5 addresses the possibility of using standardised GLS separators for UASB reactors. The design example provided in Section 4.6 follows this approach. Additionally, Section 4.7 presents different configurations of UASB reactors that have been applied for sewage treatment, but without detailing design aspects.

Full-scale UASB reactors treating sewage are typically circular (Figure 4.3(a–c)) or rectangular (Figure 4.4(a–c)). Circular reactors are more economical from a structural standpoint (as they occupy a smaller perimeter than rectangular reactors with the same surface area), and they are used more often for small-scale STPs that generally have a single UASB reactor. Rectangular units are more suitable for large-scale STPs, especially when modulation is necessary, where a wall can serve two contiguous modules. Modulation is a crucial aspect for UASB reactors treating sewage, as the influent to the STP undergoes no equalisation, thereby exposing the UASB reactor to flow variations that may be extremely high. For sewage collected in combined sewers (a typical condition in developed countries), large flow fluctuations (e.g. a 5–6 fold increase in flow during rain events) can be verified. Even when using sewerage systems exclusively for sewage collection, excessive rainwater can enter the system through both infiltration and the irregular cross-connections of rainwater (a typical condition in developing countries), as discussed in Chapters 6 and 8. Therefore, modulation aims to couple with these markedly average flow variations, in which one or more UASB reactors would only be used during peak flows to assure the recommended hydraulic retention times (HRTs) (see also Section 4.4). From an economic standpoint, implementing larger reactors may decrease costs compared to implementing several smaller units keeping the same total volume. This is because shared walls can be removed for larger reactors. However, from an operational standpoint, modulation is also advisable as it allows a reactor to be taken out of operation without losing treatment capacity.

Figure 4.3 Cylindrical UASB reactors using different materials: (a) glass fibre, (b) polypropylene and (c) ferrocement.

Figure 4.4 Rectangular UASB reactors: (a) large-scale STP (PE $>$ 1,000,000); (b) medium-scale STP (PE$\approx$80,000); and (c) small-scale STP (PE$\approx$20,000).

For small-scale STPs, alternative materials can be used instead of reinforced concrete, such as glass fibre (Figure 4.3(a)), polypropylene (Figure 4.3(b)) and ferrocement (Figure 4.3(c)). As discussed in Chapter 5, the proper selection of construction materials is crucial to ensure reliable operation.

Open-air UASB reactors (without a top slab covering the settler compartment) can be used if strict odour control is not required, as discussed in Chapter 9. In this case, a baffle for scum retention should be provided in the settler surface. Applied studies have indicated the beneficial effects provided by sulphur-oxidizing microorganisms that grow in scum (Souza *et al.*, 2006; Garcia *et al.*, 2012).

4.4 CURRENT DESIGN CRITERIA OF UASB REACTORS FOR SEWAGE TREATMENT

A UASB-based STP typically comprises a preliminary treatment (PT) step (see Section 4.2) followed by the UASB reactor itself, a sludge dewatering system (see Section 4.4.6), and a biogas management system (see Section 4.4.9 and Chapter 8). Even when following strict design, construction, and operation guidelines, the typical organic matter removal efficiencies of UASB reactors (see Section 4.5) usually do not comply with strict discharge standards (e.g. COD removal efficiencies over 80%). Important experiences have been gained in the last 20 years regarding post-treatment options for anaerobically treated sewage, as discussed in Chapter 10 and further addressed for membrane technologies in Chapter 11.

The process principles of UASB reactors treating sewage are consolidated (the basics can be found in Chernicharo (2007)). A UASB reactor comprises four functional units in a single tank (van Lier *et al.*, 2010), namely (i) a primary settler for removing suspended solids from the influent, (ii) a biological reactor (secondary treatment) for carbon removal through its conversion into methane, (iii) a secondary settler for clarifying the effluent in the settling zone at the top of the reactor, and (iv) a sludge digester for thickening and digestion (or stabilisation) of the retained organic solids and the sludge contributions from aerobic post-treatment systems. The main components of a UASB reactor for sewage treatment are shown in Figure 4.5.

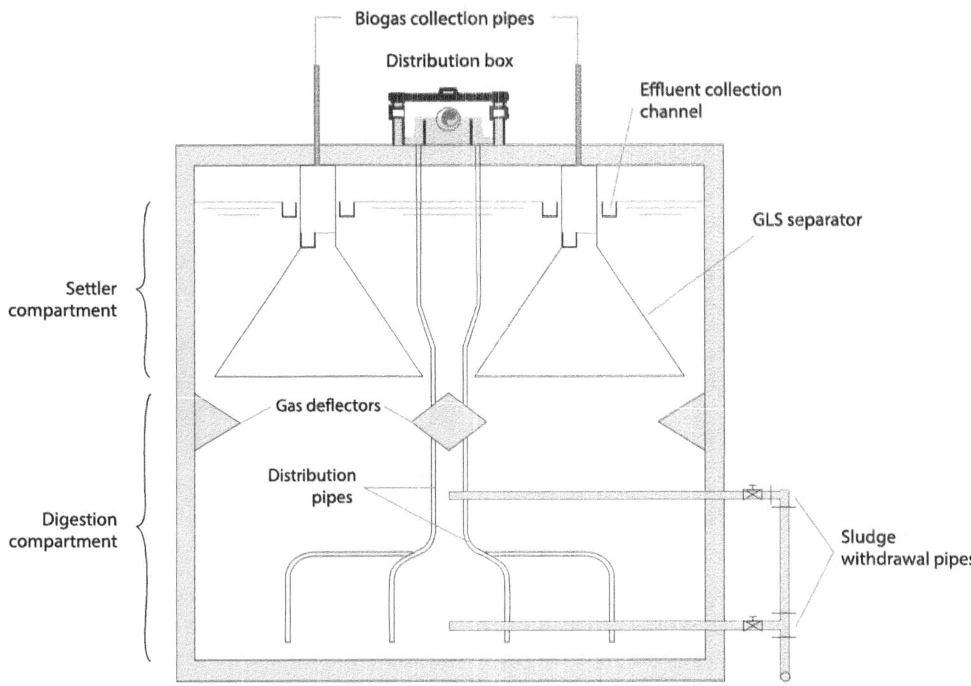

Figure 4.5 Schematic representation of the main components of a UASB reactor treating sewage.

An apparent concern of design engineers is associated with the volume occupied by the internal structures of the UASB reactor (e.g. gas deflectors, pillars and beams, feeding and sludge pipes, and effluent channels). In a survey conducted on several designs of small, medium and large-scale reactors treating sewage, it was noticed that such structures occupy a relative volume usually less than 5% of the useful volume of the reactors (Figure 4.6(a)), mainly ascribed to gas deflectors and pillars and beams. In practice, there is no need to increase the useful volume of UASB reactors because of the volume occupied by internal structures.

The start-up of the system can be slow (3 months to 5 months; $T_{average} \approx 20°C$), but only in situations in which seed sludge is not used. Nevertheless, full-scale experiences in the past few years have shown that the use of well based start-up methodologies and the establishment of appropriate operational routines significantly decrease the start-up period of the systems and minimise operational problems in this phase. As demonstrated by Chernicharo *et al.* (2009), the application of high sludge (biological) loading rates (e.g. 0.50–0.75 kgCOD $kgVS^{-1}$ d^{-1}) can significantly reduce the amount of seed sludge necessary for the start-up of full-scale UASB reactors. In this case, the loss in reactor efficiency could be compensated by the post-treatment unit. In this assessment, remarkable differences in reactor performances during start-up were only observed for the first 40 days of operation. After 80 days all reactors presented similar behaviour. Hence, the start-up of UASB reactors with reduced amounts of seed sludge (or even none) is an appealing alternative owing to the difficulties and costs involved in the inoculation of large full-scale UASB reactors. The lower efficiencies observed during the first months can be compensated by the post-treatment unit if it is implemented in the flowsheet treatment. Nevertheless, the quality of the biomass to be developed (e.g. settleability and specific methanogenic activity) in the system will depend on the appropriate operational routine, as further addressed in Chapter 6. It should be emphasised that the maximum biological loading rate depends on the methanogenic activity of the sludge. Therefore, for reliable operation, the limit for the biological load should not exceed 0.3–0.4 kgCOD $kgVS^{-1}$ d^{-1}, which seems to correspond to the maximum methanogenic activity that can be achieved in UASB reactors treating sewage. This minimum sludge mass (amount of biomass sufficient to digest the influent organic load) is further discussed in Section 6.3 of Chapter 6.

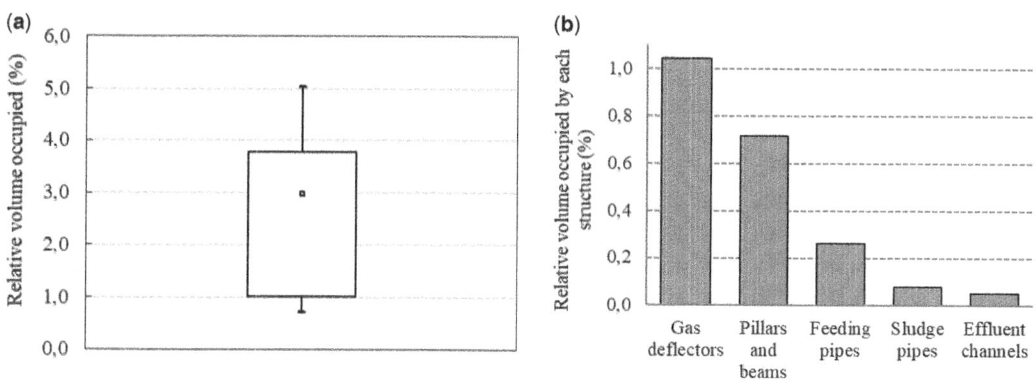

Figure 4.6 Relative volume occupied by different structures inside the UASB reactor: (a) total relative volume occupied, and (b) relative volume occupied by each structure.

Strategies for effluent recirculation during the start-up phase to enhance mixing (better contact between the substrate and biomass) are no longer recommended, as they increase operational complexity. Additionally, operational experience has shown that the development of sludge granules in full-scale UASB reactors treating sewage should not be expected, even if anaerobic granular sludge is utilised as inoculum. This is probably because of the low concentration and proportion of soluble COD in the influent, and hydraulic conditions (as addressed in Chapter 3). However, a proper operational routine is sufficient for selecting high-quality anaerobic biomass.

4.4.1 Input data: flows, concentrations, and loads

The design of UASB reactors treating domestic sewage should consider the average influent sewage flow rate; however, maximum flow checks should be performed in relation to the HRT, upflow velocity in the digestion compartment, and surface loading rate in the settler compartment. The recommended values for these design parameters are described in the following section (4.4.2).

In the case of using pumping stations (PSs) to feed UASB reactors, the maximum pumping flow rate must not exceed more than 25% of the maximum influent sewage flow rate. The use of pumps with rotation speed control or a minimum of three pumps, with one as spare and for switch, is recommended.

In the definition of the average design flow, apart from the per capita water consumption, rainfall infiltration coefficients that reflect local realities should be considered in the sewerage network. As previously discussed, even when using sewerage systems exclusively for sewage collection, excessive rainwater can enter the system through both infiltration and the irregular cross-connections of rainwater. The design examples presented in this Chapter (Section 4.6) consider an incremental per capita water consumption between the design PE of 20,000 (QPC = 150 L PE^{-1} d^{-1}; example 4.5) and 200,000 inhabitants (QPC = 165 L PE^{-1} d^{-1}; example 4.6), as well as an infiltration coefficient of 0.1 L s^{-1} km^{-1} of sewerage (sewerage density = 2.0 m PE^{-1} or 1.5 m PE^{-1} for small- and large-scale STPs, respectively).

The maximum influent flow rates are typically obtained using coefficients in relation to the average flow rate (e.g. hourly and daily variation coefficients; K_1 or K_2), as discussed in von Sperling and Chernicharo (2005). Another approach, which is used for regions where the sewage collection system is integrated into the drainage system, refers to dry weather flow (DWF) conditions. In these cases, the maximum flow can be considered up to six times higher than the average flow in the dry period. Therefore, it may be important to design the UASB reactors to operate in a modular way, as discussed in Section 4.3 of this Chapter. Alternatively, the reactors can be designed to operate under influent flow conditions of up to 2 DWF, with higher flows being directed to an equalisation tank and then routed to the UASB reactors.

The design examples presented in this Chapter (Section 4.6) consider the calculation of the influent COD concentration from the definition of a per capita load that represents the average contribution of each individual. In this case, a per capita load of 100 gCOD PE^{-1} d^{-1} was used. The same logic has been adopted for the other chapters of this book that present design examples (e.g. Chapters 6, 8, 10 and 12). This approach is useful even for the determination of incoming COD concentrations under limited water supply conditions, as discussed in Section 4.4.2.

4.4.2 Sizing parameters

As previously presented in this Chapter and reported by van Haandel and Lettinga (1994), Chernicharo (2007), and van Lier *et al.* (2010), the design of UASB reactors treating low-strength wastewater such as sewage (COD < 1000 mg L^{-1}) under tropical temperatures ($T > 20$°C) is controlled by the hydraulic loading criteria and not by the organic loading criteria. Therefore, the upflow velocity in the digestion and settler compartments is a crucial parameter; excessive velocities result in solids washout with the effluent, while low velocities can jeopardise the system performance owing to the worse contact between the influent substrate and microorganisms.

A number of parameters need to be evaluated for proper UASB reactor design. To cope with this, the Brazilian Association of Technical Standards has published a guide that comprises hydraulic and process engineering design criteria for UASB reactors treating sewage: NBR 12209 (ABNT, 2011). The main design criteria are summarised in Table 4.3. These criteria are consonant with those that have been typically used for UASB reactors in India (van Lier *et al.*, 2010). The following sections of this Chapter further address the design particularities. Specifically, Section 4.6 presents two design examples of UASB reactors treating sewage of 20,000 PE and 200,000 PE considering a design approach based on modular GLS separators (standardised dimensions; see Section 4.4.5). This is of utmost importance, as the GLS separator is a crucial aspect for proper UASB reactor operation.

Table 4.3 Main design criteria recommended for UASB reactors based on Brazilian experiences.

Parameter	Unit	Value		Comment
		T (°C)	HRT	
Hydraulic retention time (HRT)	h	>25	>6	The maximum peak flows should not last more than 2–4 h to not harm the buffer capacity of the settler compartment for solids retention.
		22–25	7–9	
		18–21	8–10	
		15–17	10–14	Adopting values close to the upper limits of each range ensures a larger capacity for sludge storage in the digestion compartment. This can also provide a better operational flexibility for sludge withdrawal (see Chapter 6).
Upflow velocity at average flow	$m\ h^{-1}$	0.5–0.7		Less than 1.1 $m\ h^{-1}$ for the maximum peak flow.
Useful depth	m	4–6		The minimum useful depths of digestion and settler compartments are 2.5 m and 1.5 m, respectively. For the recommended upflow velocities and HRT, the UASB reactor depths should be between 4 m and 6 m.
Feed inlet density	m^2 per feed point	2.0–3.0		The minimum inlet pipe internal diameter should be 75 mm, as discussed in Section 4.4.3.
Angle of gas collector	degrees	≥50		See Section 4.4.5 for the standard GLS separator.
Depth of the settler compartment	m	≥1.5		
Velocity in the apertures to the settler compartment	$m\ h^{-1}$	<2.0–2.3		Less than 4.0–4.2 $m\ h^{-1}$ for the maximum peak flow.
Surface loading rate in the settler compartment	$m\ h^{-1}$	0.6–0.8		Less than 1.2 $m\ h^{-1}$ for the maximum peak flow. This parameter is easier to determine than the HRT in the settler compartment, which should be between 1.5 and 2.0 m for the average flow.
Overlap of the gas deflectors over the apertures for the settler compartment	m	0.15–0.20		

4.4.2.1 Design based on the solids retention time (SRT)

The prime design criterion for UASB reactors, even with dilute sewage, is apparently the reactor solids retention time (SRT). As discussed in Chapter 3, for typical diluted sewage in tropical conditions (COD $< 1{,}000$ mg L^{-1} and $T > 20$°C), methanogenesis can be the step limiting the overall process; however, hydrolysis has been reported as the rate-limiting step at low temperatures (Petropoulos *et al.*, 2017). In such instances, the SRT should always be more than three times the doubling time of methanogens ($T_d = 10$ d; SRT $= 30$ d) (van Lier *et al.*, 2010). With dilute sewage under tropical conditions, the minimum required SRT will always be met. An exception is made for arid climate countries with limited water supply where the sewage concentration can reach up to 2,500 mg COD L^{-1}, whereas winter temperatures may drop to 15°C (Chernicharo *et al.*, 2015). Therefore, for such exceptional conditions (especially for sewage temperatures below 20°C), the SRT should be the main design criterion for determining the reactor volume.

In this case, as briefly presented in Section 4.7 of this Chapter and further addressed in Chapter 11, for low-temperature operation, different approaches can be implemented, such as applying primary settling prior to the anaerobic reactor, integrating the UASB reactor with a sludge digester (the UASB–digester concept (Mahmoud *et al.*, 2004)), or adopting anaerobic membrane bioreactors (AnMBR).

The prevailing SRT depends on various sewage characteristics, as discussed in van Lier *et al.* (2010). The estimation of the volume of a UASB reactor based on the SRT concept is concisely shown in Example 4.1.

EXAMPLE 4.1

Design of a UASB reactor for 20,000 population equivalent (PE) based on the solids retention time (SRT) criteria.

Input data

- Population equivalent: PE = 20,000 inhabitants
- Average influent sewage flow rate: $Q_{s-avg} = 38.7$ L s^{-1} (139.4 m^3 h^{-1})
- Temperature of the liquid: 15°C
- Minimum adopted SRT: 40 d
- For practical purposes, the same excess sludge withdrawal used in Example 4.3 was adopted: $Mass_{sludge\text{-}total} = 400$ kg TS d^{-1}
- Assumed sludge mass concentration in the digestion compartment of the UASB reactor: $C_{sludgeâ\ avg} = 4\%$ (40 kg TS m^{-3})
- Assumed TS concentration in the effluent (TS_{effl}): 100 mg L^{-1} (0.1 kg m^{-3})

Solution

(1) Estimate the solids load in the effluent (L_{TS})

$$L_{TS} = Q_{s\text{-}avg} \times TS_{effl} = 139.4\ m^3\ h^{-1} \times 24\ (h/d) \times 0.1\ kg\ m^{-3} = 334.6\ kg\ TS\ d^{-1}$$

(2) Determine the reactor volume ($V_{reactor}$)

$$V_{reactor} = \frac{SRT \times (L_{TS} + Mass_{sludge-total})}{C_{sludge-avg}} = \frac{40\ d \times (334.6\ kgTS\ d^{-1} + 400\ kg\ TS\ d^{-1})}{40\ kg\ TS\ m^{-3}} = 735\ m^3$$

4.4.3 Influent distribution system

As stated by Chernicharo (2007) and van Lier *et al.* (2010), an even distribution of the influent is a crucial aspect in UASB reactors to ensure a better mixing regime (better contact between the incoming substrate and the biomass) and reduced occurrence of dead zones in the sludge bed. Nonetheless, no direct correlation can be made between the feed inlet density and the overall treatment performance of UASB reactors because their overall performance is affected by multiple parameters, as discussed through this book. Conversely, computational fluid dynamics (CFD) simulations have recently been applied to UASB reactors treating sewage (Das *et al.*, 2018) to identify possible improvements in the mixing regime and to enhance the reactor performance for organic matter removal.

The contact time of the incoming substrate and the biomass inside the reactor is of utmost importance, and short circuiting should be prevented, as also addressed in Chapter 3. As presented in Table 4.3, the ideal influence area of each distributor should be between 2.0 and 3.0 m^2 (Chernicharo, 2007). The number of distribution pipes is determined according to the influence area adopted for each distributor, as addressed in Section 4.6. In the particular case of trunk-conical reactors, the influence area of the distribution pipes is not uniform over the height of the digestion compartment because the cross section of the reactor increases with height (see Section 4.7). In these cases, the calculations should consider the cross section close to the deepest part of the reactor (where the sludge bed, which is more concentrated, is located), i.e. close to the first metre of depth of the reactor, in order to ensure an influence area suitable for the flow distributors (Chernicharo, 2007).

The equal division of the influent flow to the several distributing pipes should be conducted using small compartments (boxes) fed by channels. Each box feeds a single distribution pipe extending to the bottom of the reactor. These compartments, which are installed in the upper part of the reactor, ensure the uniform distribution of sewage throughout the bottom of the UASB reactor in addition to enabling the visualisation of occasional increments in the head loss in each distributor. Once an increased head loss is detected in a distributor, the tube can be easily unblocked using appropriate rods or pressurized devices (see Chapter 6). Examples of influent distribution structures in UASB reactors are presented in Figure 4.7(a–h).

Specifically, regarding these influent distribution boxes it is common to notice structures with a flat bottom and/or with a low slope, which favours the accumulation of grit and/or debris, especially when there is a failure in the PT step. Therefore, it is recommended to adopt boxes with a sloping bottom (Figure 4.7(d)). In addition, structures made of materials that allow better finishing (e.g. glass fibre) as well as the correct design and verification of the water levels in the V-notch for the maximum pumping rates are important aspects to guarantee uniform flow distribution into the compartments feeding the distribution pipes. In turn, these pipes connected to the bottom of the distribution boxes must have an internal diameter of 75 mm. While this ensures an adequate downward velocity, such a diameter is also large enough to prevent frequent obstructions (considering a satisfactory operation of the PT units). It is important to ensure that the pipes are correctly fixed to the distribution structures, as they can be released under the slab without the operator noticing that the pipe is disconnected (see Chapter 5). In this case, the influent sewage would enter the settler compartment of the UASB reactor rather than being distributed to the bottom of the unit.

In STPs that employ aerobic post-treatment units and where there is a designed sludge return for thickening and digestion in UASB reactors, it is important to provide reception/distribution boxes for the excess aerobic sludge flow (Figure 4.8). This strategy allows flexibility of manoeuvres during the feeding of the aerobic sludge between the modules of the UASB reactors.

For the influent distribution system, between the distribution boxes and the bottom of the reactors, polyvinyl chloride (PVC) and high-density polyethylene (HDPE) pipes are typically employed. For the pipes that feed the distribution boxes, cast-iron pipes (especially for diameters greater than 200 mm) are typically employed. Table 4.4 summarises the main design criteria regarding the influent distribution system.

Figure 4.7 Examples of influent distribution structures: (a) concrete distribution boxes on the top of a UASB reactor; (b) distribution box with several compartments linked to the pipes extending to the bottom of the reactor; (c) distribution channels on the top of a UASB reactor; (d) distribution box with a sloping bottom; (e) glass fibre distribution boxes on the top of a UASB reactor (circular section); (f) detail of one distribution box shown in (e); (g) glass fibre distribution boxes on the top of a UASB reactor (rectangular section); (h) concrete side distribution box on the top of a UASB reactor; and (i–l) views of different configurations of the exit of the distribution pipes.

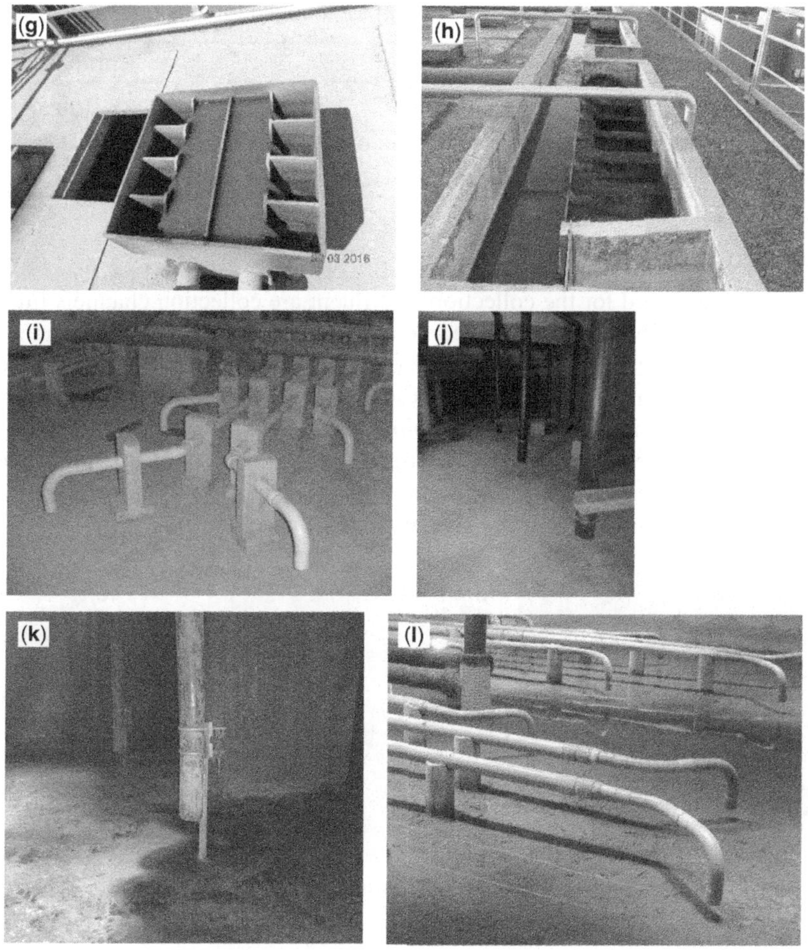

Figure 4.7 (***Continued***).

Figure 4.8 Reception/distribution boxes for the excess aerobic sludge.

Table 4.4 Summary of the main design criteria for the influent distribution system.

Criteria	Range in Values
Diameter of the influent feeding pipe (mm)	75–100
Distance between the pipe exit and the bottom of the reactor (m)	0.10–0.20
Influence area of each distribution pipe (m^2)	2.0–3.0

4.4.4 Effluent collection

The devices that are usually used for the collection of effluent are collection channels fitted with V-notch weirs (Figure 4.9(a–b)) and submerged perforated pipes (Figure 4.9(c)) (with a minimum slope of 1% so they can be self-cleaned). Special care should be taken with levelling of effluent collection channels, as addressed in Chapter 5, as small misalignments can lead to significant variation in the flow collected at different points (van Lier *et al.,* 2010). The submerged collection of the effluent can lead to the formation of scum in the settler compartment, the removal of which should be predicted in the design stage if this alternative of effluent collection is implemented. In the case of the adoption of V-notches, it is recommended to adopt one or two units per square metre of reactor.

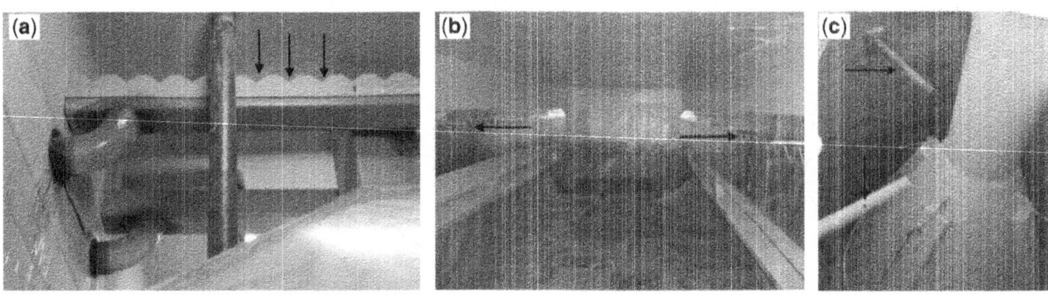

Figure 4.9 (a, b) Effluent collection channel fitted with V-notch weirs, and (c) perforated pipes for submerged effluent collection.

The collection and transport of effluents from UASB reactors should prevent turbulence points to minimise the release of dissolved gases. In cases where the removal of dissolved gases in the effluent and waste gases from the settler compartment is desired, a device must be provided to desorb and collect these gases for further treatment. The anaerobic reactor effluent transport system must be corrosion resistant. In addition, the areas on the settler compartment can be covered or uncovered. In cases where they are covered, the whole structure must be protected against corrosion above the water level. A more in-depth discussion is presented in Chapter 9.

4.4.5 Gas–liquid–solid (GLS) separator

In the improvement evolution of UASB reactors for sewage treatment, the most recent designs have incorporated collection channels inside the GLS separator (Figure 4.10(a, b)) to allow the hydrostatic removal of scum. Some designs also incorporate sprinklers with pressurised water jets in order to assist the transport of the scum into the collection channels. Such sprinklers may be dispensed with if scum is removed more frequently (two to three times per week), since it is a fluid material, as also mentioned in Chapter 6.

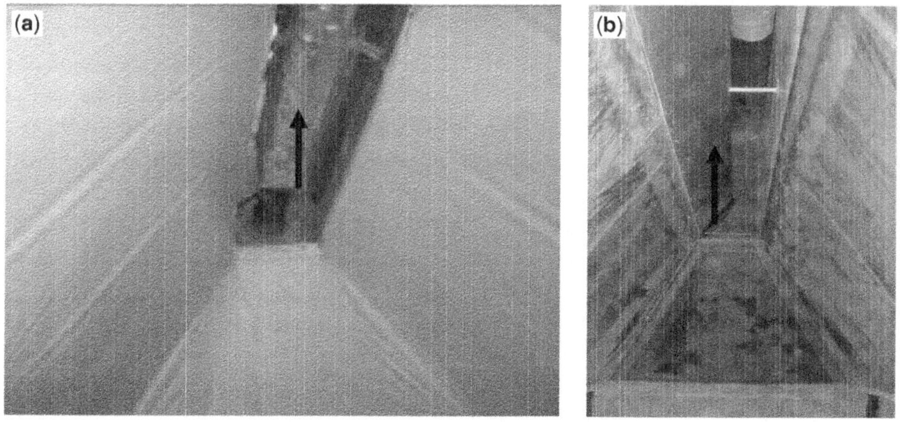

Figure 4.10 (a, b) Gas–liquid–solid separators equipped with internal scum removal channels.

An alternative to the use of contiguous GLS separators (rectangular trunk-pyramidal; see Figure 4.9 and Figure 4.10) equipped with long collection channels is the use of smaller, quadrangular trunk-pyramidal GLS separators equipped with bulge-shaped collectors on the inside, as shown in Figure 4.11.

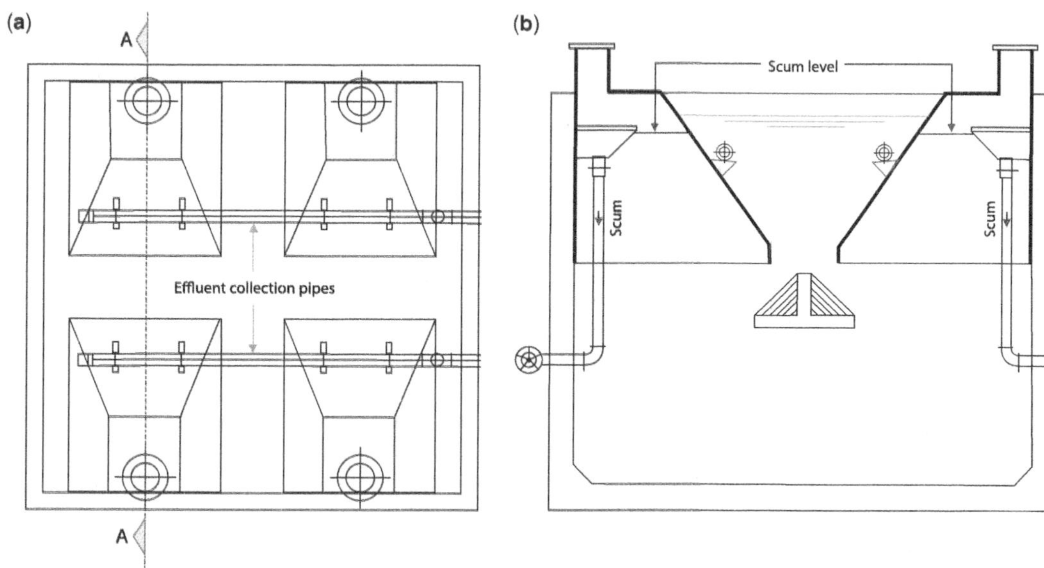

Figure 4.11 A quadrangular trunk-pyramidal gas–liquid–solid separator equipped with an internal scum collector: (a) plan view, (b) cross-section.

The hydrostatic removal procedure is based on the change in the water level inside the GLS separator in order to allow the scum to flow to an internal channel and then to a discharge pipe located outside the reactor. Figure 4.12 shows the schematic representation of the hydrostatic scum removal device and its working principle. The operation procedures related to this device are described in Chapter 6.

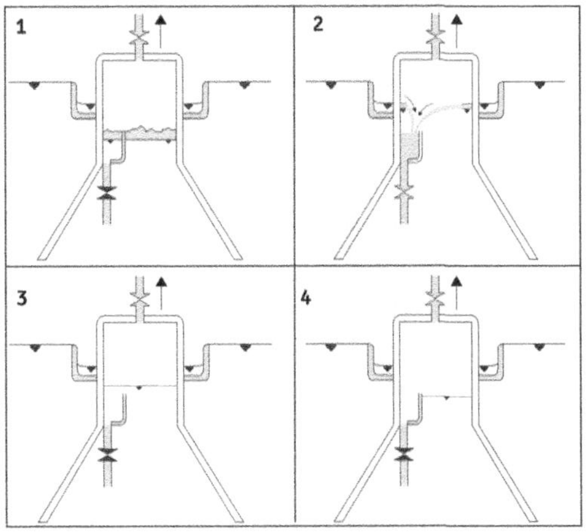

Sequence of procedures:

(1) Level of scum during normal operation of the reactor (pressurised gas line to maintain the scum level 2 cm to 3 cm below the upper edge of the inner channel).
(2) Elevation of the level of scum achieved with the relief of the pressure of the gas line in order to allow scum to enter the internal channel.
(3) Closure of the drain valve of the inner channel.
(4) Augmentation of the gas line pressure to return to the operating condition shown in 1.

Figure 4.12 Schematic representation of the system for hydrostatic removal of scum. (Source: Chernicharo, 2007)

This device for hydrostatic removal was tested with good results in UASB reactors at the demonstration scale and full scale with removal efficiencies between 75% and 90%, as mentioned in Chapter 6. Table 4.5 shows the main design guidelines for the adequate functioning of a GLS separator equipped with hydrostatic scum removal devices. Note that responsibilities are often shared between design, construction, and operation, as presented in Section 6.4.3 of Chapter 6.

Table 4.5 Main guidelines for the design of hydrostatic scum removal devices.

- The GLS separator must be integrated with a water seal to allow the control of internal pressures and the adjustment of water levels in relation to the scum collecting channel (see Chapter 6).
- The scum collection channels should preferably be positioned adjacent to one of the walls of the GLS separator and not at the centre. The positioning of the channel next to one of the sides allows a higher water level and superior scum removal efficiency. A scum collection channel in the triangular section is shown in Figure 4.10b and Section 6.4.3 of Chapter 6.
- Inspection points and valves should be provided on the scum outlet pipes of each GLS separator aiming to identify clogging points and avoid uneven hydraulic head loss (see Chapter 6).
- Extraction nozzles (drain points) should be provided along the length of the scum collection channel to allow more adequate hydraulic flow conditions (see Figure 4.13b).
- Gas compartment tightness should be ensured to allow the establishment of small pressures in the biogas line (5–10 cm.w.c).
- Whenever possible, motorised valves should be provided to facilitate scum removal.

4.4.5.1 Modular gas–liquid–solid separator: Étsus 1000

The Étsus 1000 modular GLS separator (Figure 4.13(a, b)) is in the final stage of development, and is the result of a partnership between the Universidade Federal de Minas Gerais (UFMG, Brazil) through the National Institute of Science and Technology on Sustainable Sewage Treatment Plants (INCT Sustainable

STPs) and Brazilian sanitation companies, and glass fibre manufacturers. This product, which is made of glass fibre, contributes to the solution of several operational problems of UASB reactors, namely (i) adequate management of the formation and removal of scum, (ii) efficient collection of biogas without losses by leakages, and (iii) adequate collection of the treated effluent while avoiding the release of waste gases and the occurrence of hydraulic short circuits. The main advantages associated with the product are as follows:

- A modular GLS separation structure incorporating lightness, strength, watertightness, and ease of transport and installation.
- Incorporation of a scum removal system that directly contributes to the improvement in operational aspects of UASB reactors and reduction/elimination of odorant emissions and biogas losses.
- Incorporation of an effluent collection system coupled to the GLS separator body with a level checking device.

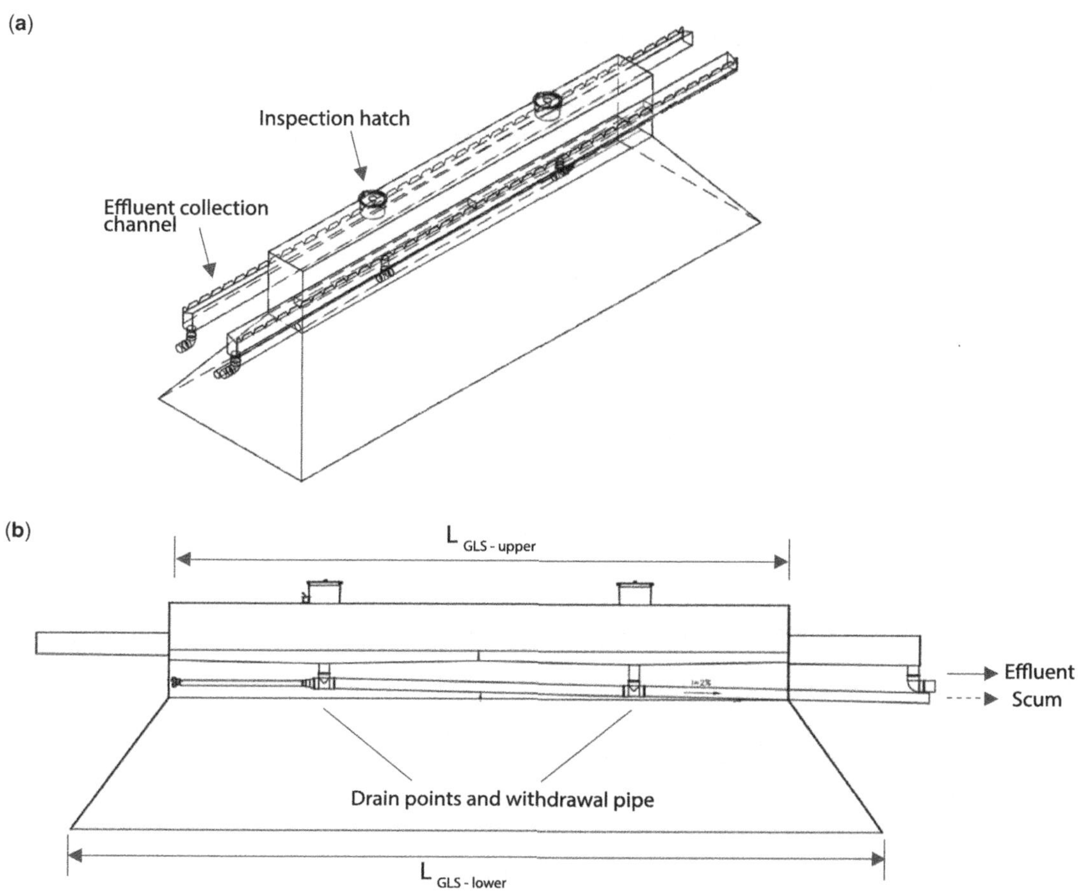

Figure 4.13 Schematic view of the Étsus 1000 modular gas–liquid–solid separator: (a) perspective view, and (b) longitudinal profile.

Therefore, as stated in Section 4.4.2, the design of the UASB reactor can be based on the standardised dimensions of the GLS separator. This allows optimisation of the design of such reactors considering the crucial point to guarantee adequate operation. It is important to mention that such standardised GLS separators typically fulfil the requirement of the minimum biogas release rate ($1.0\ m^3{}_{biogas}\ m^{-2}\ h^{-1}$; Souza *et al.*, 2006) to overcome the scum layer without dragging sludge (verification is shown in the design examples in Section 4.6). Example 4.2 demonstrates the determination of the standardised dimensions of a GLS separator based on the development of the Étsus 1000. A schematic cross-section of such a GLS separator is presented in Figure 4.14.

EXAMPLE 4.2

Determination of a standard gas–liquid–solid (GLS) separator module (based on the Étsus 1000).

(1) Adopted depth for the inclined GLS separator wall (GLS_H)

$GLS_H = 1.3$ m

(2) Adopted depth for the vertical GLS separator wall (GLS_V)

$GLS_V = 0.4$ m

(3) Adopted slope for the inclined GLS separator wall (GLS_S)

$GLS_S = 53.8°$

(4) Width of the inclined GLS separator wall (GLS_W)

$GLS_W = GLS_H/\tan(GLS_S) = 1.3\ m/\tan(53.8°) = 0.95$ m

(5) Adopted width of the scum removal channel inside the GLS separator ($W_{channel}$)

$W_{channel} = 0.15$ m

(6) Adopted free space between the scum removal channel and the GLS separator wall (W_{free})

$W_{free} = 0.25$ m

(7) Width of the GLS separator gas hood ($W_{gashood}$)

$W_{gashood} = W_{channel} + W_{free} = 0.15\ m + 0.25\ m = 0.40$ m

(8) Adopted width for the aperture between two contiguous GLS separators ($GLS_{aperture}$) (see Figure 4.21 – Section 4.6)

$GLS_{aperture} = 0.70$ m

Therefore, the standardised centre-to-centre distance between two modular GLS separators would be 3.0 m.

From the definition of the standardised dimensions of the GLS separator, different GLS models could be proposed as a function of their length ($L_{GLS-upper}$ or $L_{GLS-lower}$ identified in Figure 4.13(b)). Therefore, the UASB reactor would be designed considering available GLS models, as further exemplified in Section 4.6.

Table 4.6 presents the modulation proposal elaborated for the Étsus 1000 GLS separator considering the manufacturing of three different models. In practice, these proposed models have the capacity to be applied to UASB reactors designed for any population equivalents (small, medium, and large-scale STPs).

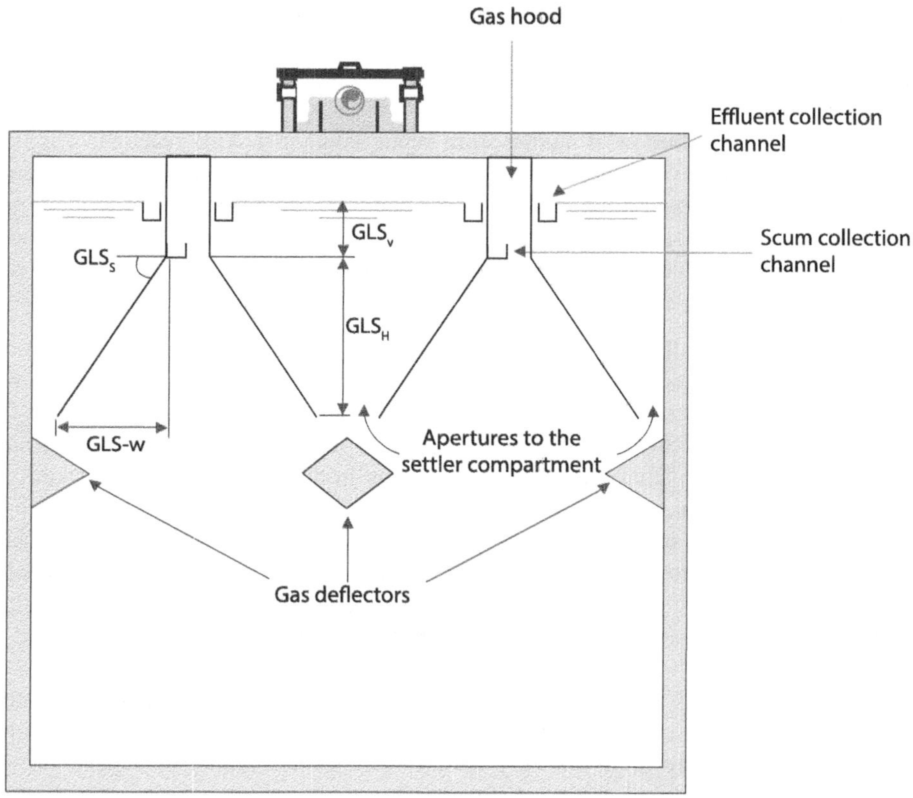

Figure 4.14 Schematic cross-section of a UASB reactor equipped with the modular gas–liquid–solid separator (Étsus 1000).

Table 4.6 Proposed gas–liquid–solid (GLS) separator (Étsus 1000) modulation.

GLS Separator Model	Number of Scum Withdrawal Points[a] ($N_{scum\text{-}points}$)	Length of the GLS Separator (upper part)[b] $L_{GLS\text{-}upper}$ (m)	Length of the GLS Separator (lower part)[a] $L_{GLS\text{-}lower}$ (m)
1	1	3	4.9
2	2	3–6	4.9–7.9
3	3	6–9	7.9–10.9

[a]The length of the scum withdrawal channel should be the same as $L_{GLS\text{-}upper}$ but with convergent divisions to the $N_{scum\text{-}points}$ (see Figure 4.13(b)).
[b]The values between the range limits can be incremented every 0.5 m up to the upper limit of the range.

4.4.6 Sludge production, withdrawal, and processing

Usually, estimates of sludge production are made from the range in sludge yields (Y_{sludge}) reported in the literature, which is from 0.10 to 0.20 kgTS $kgCOD_{app}^{-1}$. However, this range refers only to the production of sludge in UASB reactors exclusively applied to sewage treatment. In cases where the reactor receives other types of contributions (e.g. aerobic sludge produced in the post-treatment unit, sludge from septic tanks, landfill leachate, and NDEs), the value of Y_{sludge} will be higher. Therefore, it is imperative that the design defines the value of Y_{sludge} with the maximum possible safety because it will directly affect (i) the production of sludge from the reactor, (ii) the sizing of the dewatering unit, and (iii) the establishment of sludge withdrawal routines.

When considering the contribution of sludge from septic tanks in UASB reactors, it is necessary to rigorously evaluate the expected volumes and the possibility of implementing a specific unit to receive this type of contribution, thereby allowing the separation and sending of the supernatant to the UASB reactor and the thickened sludge to a complementary digester or, if well stabilised, directly to the dewatering unit.

Regarding the reception of landfill leachate and non-domestic effluents (NDEs) in UASB reactors, the following main issues should be addressed:

- Leachates from young landfills (5 years to 10 years) usually have very high concentrations of BOD, COD, sulphides, and ammonia, so any impacts on stability and reactor performance will depend intrinsically on the degree of dilution of the leachate in relation to the sewage influent flow rate to the STP. Higher concentrations of BOD and COD in the mixture of raw sewage and leachate contribute directly to the increase in sludge production in the reactor, but also make it possible to increase biogas production, which is of interest in cases where the energy recovery is practiced. In turn, high concentrations of sulphides and ammonia in the mixture of raw sewage and leachate could eventually cause toxicity to anaerobic microorganisms, notably methanogenic archaea. However, since the dilution factor is usually very high, the concentrations of sulphides and ammonia in the mixture result in values well below the levels that are considered toxic (see Chapter 3).
- Leachates from old landfills (>10 years) tend to have high COD/BOD ratios, so they contribute little to increase the production of sludge and biogas in the reactor. The concentrations of sulphides and ammonia are still typically high, and any toxicity problems continue to depend on the dilution factors of these constituents, as discussed above.
- Regarding NDEs, the ability of the anaerobic reactor to treat them together with sewage depends intrinsically on the type of NDE. NDEs of essentially organic nature and with low concentrations of suspended solids (e.g. brewery effluents and soft-drink factories) can usually be received without major problems, and the designer should pay particular attention to aspects of increased sludge production (but also biogas) and risk of acidification of the reactor if the alkalinity of the sewage is not high enough to neutralise the organic acids formed from these NDEs (which acidify very rapidly). For other types of NDE of organic nature, an even more careful analysis should be conducted with special attention to the presence of high concentrations of suspended solids and potentially toxic elements (e.g. sulphides, ammonia, and heavy metals). NDEs of inorganic nature (e.g. from metal finishing industries) should not be referred for treatment in STPs that employ biological processes. In addition, it would be desirable for the STPs to have minimally available sensors for online pH measurement upstream of the arrival of raw sewage in order to allow the identification of sudden changes in the values of this parameter. Although remediation actions are difficult when detecting pH values outside the usual sewage range (6.5 to 7.5), recording measurements could help in identifying the source of the problem and in defining the measures to be taken.

Regarding the chemical sludge from water treatment plants (WTPs), it should not be sent to UASB reactors, since its characteristics may impair the functioning and efficiency of the reactor. Sludge from cleaning flocculators and settlers usually has a high content of inorganic solids of 0.5–1.0% total solids (TS) (notably silt and clay) and, if launched into the sewage collection and transport network, will most likely accumulate at the bottom of the reactors, thereby reducing their useful volume. In addition, these solids are difficult to remove with the sludge withdrawal system used in UASB reactors. If the possibility of routing chemical sludge from WTPs to UASB-based STPs is considered, then a technical study should be conducted to verify that the elements contained in the chemical sludge will not cause the formation of sediment banks at the bottom of the reactor. Consideration should also be given to possible risks of damage to pumps and other equipment (e.g. centrifuges). Receiving WTP chemical sludge in STPs can also impact on microbial activity, as observed by Scalize (2003). This study assessed the effects of receiving WTP sludge (with the use of aluminium sulphate as a coagulant) in the primary settlers of a STP. In this case, the methanogenic activity tests showed higher methane production in the control flask (not containing WTP sludge) and decreasing methane production as the fraction of WTP sludge in the flasks increased.

4.4.6.1 Prediction of an adequate sludge sampling system along the height of the reactor

In order to establish withdrawal strategies (quantity and frequency), it is necessary to continuously monitor the sludge present in the reactor from the solids profile and solids mass determination, as discussed in Chapter 6. Therefore, it is essential that the UASB reactor has a sampling system that allows the proper identification of the height and concentration of the sludge inside the reactor digestion compartment. The main guidelines for the design of the sludge sampling system are described in Table 4.7 and exemplified in Figure 4.15(a–d).

Table 4.7 Main guidelines for the design of the sludge sampling system.

- Vertical sampling points of sludge should be provided along the entire height of the reactor digestion compartment, not more than 0.50 m apart. The first point should be at the same height as the bottom withdrawal pipe (0.20–0.30 m) and the last should be at the interface of the digestion and settler compartments. The other sampling points should be evenly distributed between the first and last sampling point, and there should ideally be a minimum of four sampling points (Figure 4.15(a)).
- There should be provision of a manoeuvring box for the sludge sampling valves that allows the proper work of the operator, taking into account the need to handle the collection flasks and cleaning of the installations (Figure 4.15(b)).
- The valves should be ball type, and the use of metal valves followed by PVC valves in all sludge sampling points is recommended (Figure 4.15(c) and 4.15(d)).
- Pipes and sampling valves must have a minimum internal diameter of 50 mm.

4.4.6.2 Prediction of an adequate sludge withdrawal system

In order to provide greater operational flexibility, sets of sludge withdrawal pipes at two different depths inside the reactor digestion compartment should be considered: a lower set should be located between 0.20 and 0.30 m above the reactor bottom and an upper set should be located between 1.00 and 1.50 m above the bottom of the reactor. The lower set of pipes allows the withdrawal of the more concentrated

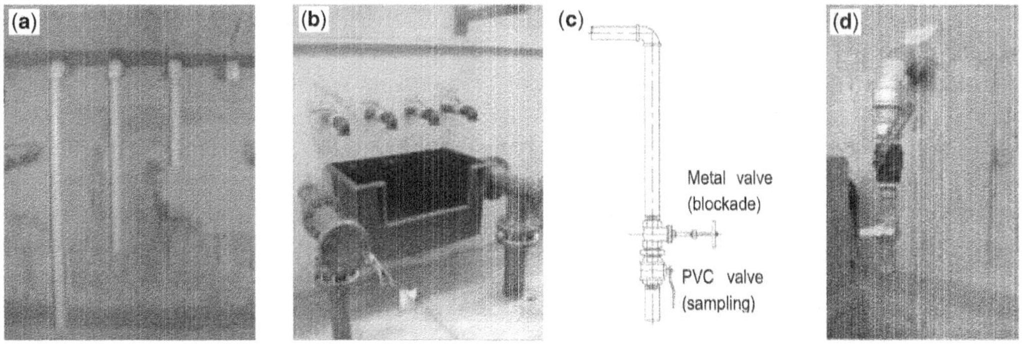

Figure 4.15 Sludge sampling system of a UASB reactor: (a) internal view of vertical sampling points within the digestion compartment spaced 50 cm apart, (b) external view of the sampling points and the manoeuvring box, and (c) and (d) details of the metal valve followed by a polyvinyl chloride (PVC) valve, respectively.

sludge, which presents better settling characteristics, while the upper set allows the removal of the less concentrated sludge, which has worse settling characteristics. This upper and lower withdrawal system configuration makes it possible to withdraw the worst quality (upper, less concentrated) sludge and preserve the best quality (lower, more concentrated) sludge in order to achieve the following: (i) a greater sludge age and the withdrawal of a more stabilised sludge; (ii) the maintenance of a dense sludge bed near the bottom of the reactor, which can act as a "filter" to assist in the retention of suspended solids long enough for them to be hydrolysed; and (iii) lower solids loss to the settler compartment, thereby directly reflecting the improvement in reactor effluent quality. In addition, the amount of sludge discarded from the bottom allows the partial removal of inert solids (grit) that have entered the reactor. The sludge withdrawal system shall present the main characteristics described in Table 4.8 and is exemplified in Figure 4.16(a, b).

Examples 4.3 and 4.4 illustrate the calculation of the required capacities of natural and mechanised dewatering systems for processing the withdrawn sludge as a function of sludge production in the

Table 4.8 Main design guidelines for the sludge withdrawal system.

- Sludge withdrawal pipes should be provided at two different heights in the reactor digestion compartment: lower pipes located between 0.20 and 0.30 m from the bottom, and upper pipes located between 1.00 and 1.50 m from the bottom. Ensure that the upper sludge withdrawal pipes are positioned at least 0.50 m below the gas deflector in order to effectively remove the less concentrated sludge. According to the Brazilian standards (NBR 12209), the upper sludge withdrawal pipes should be positioned at least 1.00 m below the lower edge of the GLS separator (ABNT, 2011) (Figure 4.16(a)).
- At least one bottom sludge withdrawal pipe should be provided for every 20–30 m^2 of reactor area. For the withdrawal of the higher, less concentrated sludge, larger areas of influence (30–50 m^2) can be used, or the possibility of using a single large-diameter pipe connected to collector branches of 100 mm or 150 mm can be evaluated. The large-diameter piping must be designed in order to achieve near zero head loss, considering the total flow resulted from the sum of the collecting branches connected to it.
- Pipes and valves for withdrawal of the lower sludge must have a minimum internal diameter of 150 mm. Pipes and valves with an internal diameter of 100 mm can be adopted for withdrawal of the upper sludge. The manoeuvring valves should preferably be knife type.

anaerobic reactor. The sequence of calculations that should be conducted by the operator of the plant to estimate the mass and volume of sludge to be removed by considering the strategy of weekly removals and the cyclic use of each drying bed for three weeks are addressed in Section 6.5 of Chapter 6.

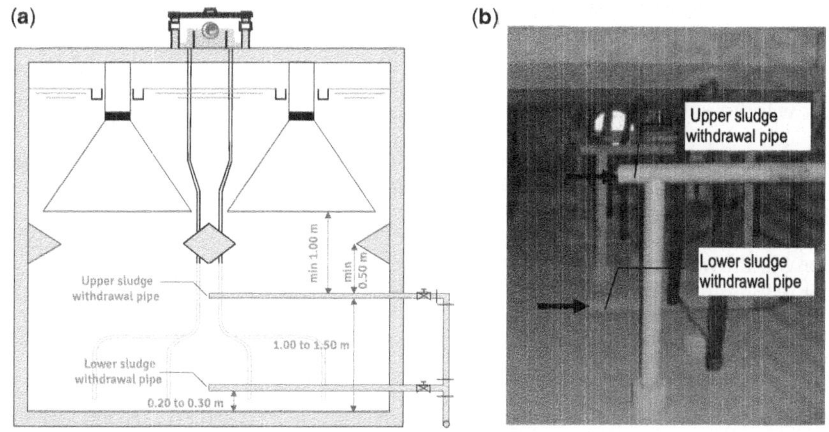

Figure 4.16 Sludge withdrawal system of UASB reactors: (a) schematic representation of the positioning of the lower and upper sludge withdrawal pipes and (b) internal view of the lower and upper sludge withdrawal pipes.

EXAMPLE 4.3

Design of a natural dewatering system to process the sludge produced in a UASB reactor considering the following design data:

- Population equivalent: PE = 20,000 inhabitants
- Per capita COD contribution: $QPC_{COD} = 0.10$ kgCOD PE^{-1} d^{-1}
- Applied organic load: $OL_{app} = 2{,}000$ $kgCOD_{app}$ d^{-1}
- Sludge yield (usually 0.10–0.20 kgTS $kgCOD_{app}^{-1}$): $Y_{sludge} = 0.20$ kgTS $kgCOD_{app}^{-1}$
- Sludge specific mass (usually 1,020–1,040 kgTS m^{-3}): $\gamma = 1{,}020$ kgTS m^{-3}

Important note: This example only considers the sludge production coefficient of the UASB reactor. If the STP receives other types of contributions (e.g. sludge from septic tanks, landfill leachate, and NDEs) and/or secondary (aerobic) sludge from the post-treatment unit (e.g. trickling filters and activated sludge) for thickening and digestion in the UASB reactors, then the increase in the value of the sludge production coefficient (Y_{sludge}) from these contributions should be included in the calculation of the dewatering system.

Solution

(1) Estimate the daily sludge production (P_{sludge})

- $P_{sludge} = Y_{sludge} \times OL_{app} = 0.20$ kgTS $kgCOD_{app}^{-1} \times 2{,}000$ $kgCOD_{app}$ $d^{-1} = 400$ kgTS d^{-1}

(2) Define the cycle time (C_t) of the drying beds

- $C_t = 21$ d (the cycle time of the drying beds should be defined for each individual STP based on local climatic conditions and operational aspects)

(3) Sludge concentrations at different reactor heights (adopted based on usual concentrations verified in full-scale UASB reactors)

- Upper withdrawal pipe (see Figure 4.16): $C_{sludge-upper} = 2.0\%$ (20,000 mgTS L^{-1})
- Lower withdrawal pipe (see Figure 4.16): $C_{sludge-lower} = 5.0\%$ (50,000 mgTS L^{-1})

(4) Percentages of sludge mass to be removed from the different reactor heights (adopted considering the recommendations to preserve the lower (more concentrated) sludge; see Chapter 6)

- Upper withdrawal pipe 60%
- Lower withdrawal pipe 40%

(5) Mass of sludge to be removed from the different reactor heights (upper and lower pipes; see Figure 4.16)

- $Mass_{sludge-upper} = P_{sludge} \times C_t \times 60\% = 400\ kgTS\ d^{-1} \times 21\ d \times 0.60 = 5{,}040$ kgTS every 21 d
- $Mass_{sludge-lower} = P_{sludge} \times C_t \times 40\% = 400\ kgTS\ d^{-1} \times 21\ d \times 0.40 = 3{,}360$ kgTS every 21 d

(6) Total mass of sludge to be removed from the reactors ($Mass_{sludge-total}$)

- $Mass_{sludge-total} = Mass_{sludge-upper} + Mass_{sludge-lower} = 5{,}040\ kgTS + 3{,}360\ kgTS =$ 8,400 kgTS every 21 d

(7) Volume of sludge to be removed from the different reactor heights (upper and lower pipes; see Figure 4.16)

- $V_{sludge-upper} = Mass_{sludge-upper}/(\gamma \times C_{sludge-upper}) = 5{,}040\ kgTS/(1{,}020\ kgTS\ m^{-3} \times 0.02) =$ 247.1 m^3 every 21 d
- $V_{sludge-lower} = Mass_{sludge-lower}/(\gamma \times C_{sludge-lower}) = 3{,}360\ kgTS/(1{,}020\ kgTS\ m^{-3} \times 0.05) =$ 65.9 m^3 every 21 d

(8) Total volume of sludge to be removed from the reactors ($V_{sludge-total}$)

- $V_{sludge-total} = V_{sludge-upper} + V_{sludge-lower} = 247.1\ m^3 + 65.9\ m^3 = 313.0\ m^3$ every 21 d

(9) Calculate the area of drying beds necessary to process the mass and volume of sludge withdrawal from the reactors ($A_{drying-bed}$)

- Sludge height on the drying beds: $H_{drying-bed} = 0.30$ m (should not exceed 0.35 m)
- $A_{drying-bed} = V_{sludge-total}/H_{drying-bed} = 313\ m^3/0.30\ m = 1{,}043.3\ m^2$

(10) Geometry of the sludge drying beds

- Adopt five drying beds of 21 m × 10 m ($A_{drying-bed-unit} = 210\ m^2$ and $A_{drying-bed-total} = 1{,}050\ m^2$)

(11) Verify the solids load on the drying beds (SL_{app})

- $SL_{app} = Mass_{sludge-total}/A_{drying-bed-total} = 8{,}400\ kgTS/1{,}050\ m^2 = 8.0\ kgTS\ m^{-2}$ (expected to be lower than 15 kgTS m^{-2} in each operation cycle)

EXAMPLE 4.4

Design of a mechanised dewatering system to process the sludge produced in a UASB reactor considering the following design data:

- Population equivalent: PE = 200,000 inhabitants
- Per capita COD contribution: $QPC_{COD} = 0.10$ kgCOD PE^{-1} d^{-1}
- Applied organic load: $OL_{app} = 20{,}000$ kgCOD$_{app}$ d^{-1}
- Sludge yield (usually 0.10–0.20 kgTS kgCOD$_{app}^{-1}$): $Y_{sludge} = 0.20$ kgTS kgCOD$_{app}^{-1}$
- Sludge specific mass (usually 1,020–1,040 kgTS m^{-3}): $\gamma = 1{,}020$ kgTS m^{-3}

Important note: This example only considers the sludge production coefficient of the UASB reactor. If the STP receives other types of contributions (e.g. sludge from septic tanks, landfill leachate, and NDEs) and/or secondary (aerobic) sludge from the post-treatment unit (e.g. trickling filters and activated sludge) for thickening and digestion in the UASB reactors, then the increase in the value of the sludge production coefficient (Y_{sludge}) from these contributions should be included in the calculation of the dewatering system.

(1) Estimate the daily sludge production (P_{sludge})

- $P_{sludge} = Y_{sludge} \times OL_{app} = 0.20$ kgTS kgCOD$_{app}^{-1}$ $\times$ 20,000 kgCOD$_{app}$ d^{-1} = 4,000 kgTS d^{-1}

(2) Define the sludge frequency of withdrawal (F_w)

- Frequency of withdrawals: $F_w = 5$ d wk^{-1} (days per week)

(3) Sludge concentrations at the different reactor heights (adopted based on usual concentrations verified in full-scale UASB reactors)

- Upper withdrawal pipe (see Figure 4.16): $C_{sludge\text{-}upper} = 2.0\%$ (20,000 mgTS L^{-1})
- Lower withdrawal pipe (see Figure 4.16): $C_{sludge\text{-}lower} = 5.0\%$ (50,000 mgTS L^{-1})

(4) Percentage of sludge mass to be removed from the different reactor heights (adopted considering the recommendations to preserve the lower (more concentrated) sludge; see Chapter 6)

- Upper withdrawal pipe: 60%
- Lower withdrawal pipe: 40%

(5) Mass of sludge to be removed from the different reactor heights (upper and lower pipes; see Figure 4.16)

- $Mass_{sludge\text{-}upper} = [(P_{sludge} \times 7$ d wk$^{-1})/F_w] \times 60\%$ = [(4,000 kgTS d^{-1} × 7 d wk^{-1})/5 d wk^{-1}] × 0.60 = 3,360 kgTS d^{-1} (5 d wk^{-1} – Monday to Friday)
- $Mass_{sludge\text{-}lower} = [(P_{sludge} \times 7$ d wk$^{-1})/F_w] \times 40\%$ = [(4,000 kgTS d^{-1} × 7 d wk^{-1})/5 d wk^{-1}] × 0.40 = 2,240 kgTS d^{-1} (5 d wk^{-1} – Monday to Friday)

(6) Total mass of sludge to be removed from the reactors ($Mass_{sludge\text{-}total}$)

- $Mass_{sludge\text{-}total} = Mass_{sludge\text{-}upper} + Mass_{sludge\text{-}lower}$ = 3,360 kgTS d^{-1} + 2,240 kgTS d^{-1} = 5,600 kgTS d^{-1} (5 d wk^{-1} – Monday to Friday)

(7) Volume of sludge to be removed from the different reactor heights (upper and lower pipes; see Figure 4.16)

- $V_{sludge\text{-}upper} = Mass_{sludge\text{-}upper}/(\gamma \times C_{sludge\text{-}upper}) = 3{,}360$ kgTS $d^{-1}/(1{,}020$ kgST $m^{-3} \times 0.02) = 164.7$ m^3 d^{-1} (5 d wk^{-1} – Monday to Friday)
- $V_{sludge\text{-}lower} = Mass_{sludge\text{-}lower}/(\gamma \times C_{sludge\text{-}lower}) = 2{,}240$ kgTS $d^{-1}/(1{,}020$ kgST $m^{-3} \times 0.05) = 43.9$ m^3 d^{-1} (5 d wk^{-1} – Monday to Friday)

(8) Total volume of sludge to be removed from the reactors ($V_{sludge\text{-}total}$)

- $V_{sludge\text{-}total} = V_{sludge\text{-}upper} + V_{sludge\text{-}lower} = 164.7$ m^3 $d^{-1} + 43.9$ m^3 $d^{-1} = 208.6$ m^3 d^{-1} (5 d wk^{-1} – Monday to Friday)

(9) Average concentration of the sludge pumped to the centrifuge ($C_{sludge\text{-}avg}$)

- $C_{sludge\text{-}avg} = [(C_{sludge\text{-}upper} \times V_{sludge\text{-}upper}) + (C_{sludge\text{-}lower} \times V_{sludge\text{-}lower})]/V_{sludge\text{-}total}$
- $C_{sludge\text{-}avg} = [(0.02 \times 164.7\ m^3\ d^{-1}) + (0.05 \times 43.9\ m^3\ d^{-1})]/208.6\ m^3\ d^{-1} = 0.026$ (2.6%)

(10) Sludge load processed by the dewatering unit ($Solids_{load\text{-}dew\text{-}unit}$)

- $Solids_{load\text{-}dew\text{-}unit} = V_{sludge\text{-}total} \times \gamma \times C_{sludge\text{-}avg} = 208.6$ m^3 $d^{-1} \times 1{,}020$ kgTS $m^{-3} \times 0.026 = 5{,}532.1$ kgTS d^{-1} (5 d wk^{-1} – Monday to Friday)

(11) Number of dewatering units ($N_{dew\text{-}unit}$)

- $N_{dew\text{-}unit} = 1 + 1$ (reserve equipment). The number of dewatering units in operation and in reserve is a function of the size of the plant as well as the maintenance structure to be implemented and the availability of alternative routes for sludge disposal (e.g. spare drying beds).

(12) Operational routine

- Number of hours of operation of the dewatering unit ($N\text{-}hours_{design}$) = 12 h d^{-1}
- Number of days per week of operation of the dewatering unit ($N\text{-}days_{design}$) = 5 d wk^{-1}

(13) Volumetric and massic capacities of the dewatering unit

- $V_{cap\text{-}dew\text{-}unit} = V_{sludge\text{-}total} = 208.6$ m^3 d^{-1} (5 d wk^{-1} – Monday to Friday)
- $M_{cap\text{-}dew\text{-}unt} = Solids_{load\text{-}dew\text{-}unit} = 5{,}532.1$ kgTS d^{-1} (5 d wk^{-1} – Monday to Friday)
- Considering 12 h of daily operation, the centrifuge must have the capacity to process 17.4 m^3 h^{-1} or 461 kgTS h^{-1}.
- It is important to note that the total volume (and mass) of sludge to be processed daily must be divided by the number of reactors that comprise the STP in order to keep the sludge mass in each reactor above the recommended minimum (see Chapter 6). In addition, the volume to be withdrawn from each reactor must be divided by the number of upper and lower withdrawal pipes/valves by first discarding the upper sludge line and then the lower sludge line.

4.4.6.3 Dewatering systems

For the design of the sludge dewatering system, whether natural or mechanised, in addition to estimating the sludge production in the system considering all possible contributions that may impact the sludge production coefficient (Y_{sludge}), it is essential to account for the specificities of each location (e.g. climatic conditions, operational conditions, and logistical and administrative issues). It is also crucial to consider the sludge withdrawal at two different depths inside the reactor digestion compartment, as discussed previously in this Chapter and in Chapter 6, which results in a larger volume of withdrawn sludge. This apparently disadvantage can be minimised with the design of a balance and homogenisation tank.

4.4.6.3.1 Natural dewatering

In STPs with drying beds, sludge withdrawal must be conducted in batches. The frequency should be compatible with the sludge storage capacity of the UASB reactors, the average drying time (according to local climatic conditions), and the contract with the service provider regarding the removal and transportation of the dewatered sludge.

Regarding the design of the drying beds, attention needs to be paid to the modulation (the quantity and dimensions of the beds). This should be strictly associated to the planning (frequency and volume) of sludge withdrawal and cleaning of the drying beds in order to avoid overlapped sludge discharges (i.e. overloaded sludge), which can accelerate the clogging of drying beds.

4.4.6.3.2 Mechanised dewatering

In STPs where mechanised dewatering processes are used, sludge withdrawal should be performed daily and the magnitude of the discharges should correspond to the daily production of sludge considering all the implemented UASB reactors. However, this operational routine should be adopted only when the reactors are in a normal situation in relation to the sludge level, that is, when the biomass in the reactors is below the maximum value adopted or the TS concentration at the upper sampling point is less than 0.5% (see Chapter 6). In situations with excess sludge stored in the UASB reactors, the withdrawal operation and functioning of the mechanised dewatering units must be intensified.

The definition of the operational sludge withdrawal routine should be compatible with the following main aspects: (i) the operators' working hours; (ii) contracts of the service providers regarding the transport of the dewatered sewage sludge and the preventive and corrective equipment maintenance; (iii) the possibility of shutdowns; and (iv) the availability of chemicals and spare parts. Additionally, the possibility of implementing spare critical equipment and/or the provision of drying beds (in specific cases, even spare ones) should be evaluated as an alternative in the case of shutdowns and/or maintenance of the equipment of the dewatering system. This would prevent excess sludge accumulation within the UASB reactors, which can lead to severe loss of solids with the effluent.

Table 4.9 summarises the main design criteria regarding the sludge handling system.

Table 4.9 Summary of the main design criteria regarding the sludge handling system.

Criteria	Range in Values	Comments
Solids production yield (kgTS $kgCOD_{app}^{-1}$)	0.10–0.20	UASB reactors exclusively applied to sewage treatment
Expected solids concentration in the excess sludge (%)	2–5	Dependent on the sludge withdrawal height
Sludge density (kg m^{-3})	1,020–1,040	
Diameter of the sludge discharge pipes (mm)	100–150	Bottom sludge withdrawal: minimum internal diameter of 150 mm
Diameter of the sludge sampling pipes (mm)	50	Minimum internal diameter

4.4.7 Scum production, withdrawal, and processing

A major operational limitation reported in most full-scale UASB reactors treating sewage is the removal of the scum that accumulates within the GLS separator, as addressed further in Chapters 6 and 7. The scum formation, accumulation, and type of scum that occur inside UASB reactors depend essentially on the composition of the raw sewage, particularly in relation to the presence of debris. Different scum accumulation coefficients have been reported for the inner part of the GLS separator and for the surface of the settler compartment, as summarised in Table 4.10. As can be noticed, values may differ substantially for various studies, possibly due to the absence of a consolidated methodology for evaluating scum accumulation and also because of inherent differences in each reactor (e.g. hydraulic short circuits).

Table 4.10 Scum accumulation coefficients for the inner GLS separator and settler compartment.

Type of scum	Accumulation Coefficient		Reference
	Value	Unit	
Settler compartment	1.0–2.4	$gTS\ kgCOD_{app}^{-1}$	Chernicharo (2007)[a]
Inner GLS separator	0.11–1.26		Souza *et al.* (2006)
	0.15–1.32		Pereira (2012)
	5.0–12.3		Flórez (2016)
Settler compartment plus inner GLS separator	4.82–105.3		Ross (2015)

[a]UASB reactors operating with upflow velocities between 0.5 and 0.7 m h^{-1}.

Advances in scum removal from UASB reactors have been made through the design of hydrostatic removal devices, as discussed previously in this Chapter. Indeed, the GLS separator is a crucial aspect for proper scum handling in UASB-based STPs. Construction and operation issues are addressed further in Chapters 5 and 6.

Specifically, in relation to the scum formation in the settler compartment of UASB reactors, it is typically very concentrated, but the accumulation coefficients are markedly lower. Therefore, it is possible to design UASB reactors without scum retention devices in the settler compartment, allowing the supernatant material to be conveyed with the effluent. As discussed in Chapter 6, the impact of adopting such a strategy on the deterioration of the effluent quality is negligible (approximately 1% in terms of effluent COD), as exemplified in Chernicharo (2007). However, attention should be paid to reactors operating at higher average upflow velocities (around 1.1 m h^{-1} or above), as this can lead to higher scum production coefficients. Chapter 6 also discusses the main aspects of the control over the formation of scum on the surface of the settler compartment.

4.4.7.1 Scum management

After the withdrawal of scum from UASB reactors, proper management and the final destination of the scum are of utmost importance. In Brazil, some anaerobic-based STPs have considered drying beds as a crucial management step. However, it is necessary to include a screening step (static sieve) in order to separate the high amount of coarse material present in the scum, as schematically represented in Figure 4.17 (see also the full-scale example in Chapter 7). Because of the high volumes of liquid discharged along with the scum, the sieved fraction can be directed to a thickening tank. The scum is separated in three phases: (i) supernatant fraction (usually about 5% of the sieved scum), (ii) clarified (liquid) fraction (approximately 90% of the sieved scum), and (iii) sedimented fraction (about 5% of the sieved scum). The clarified fraction typically presents average concentrations of TS, COD, BOD, and oils and greases within the usual ranges for the raw sewage (Flórez, 2016). As such, this liquid fraction can be returned to the UASB reactor itself, without damaging its operation and efficiency. Therefore, only the supernatant and sedimented fractions, together or separately, are sent for dewatering in drying beds, contributing to a substantial reduction of the required area of these units. These supernatants and sedimented fractions typically have a remarkably good drainage rate in the drying bed, resulting in a virtually complete dewatering in less than three days without any apparent detrimental effect of the oils and greases in terms of clogging the drying beds.

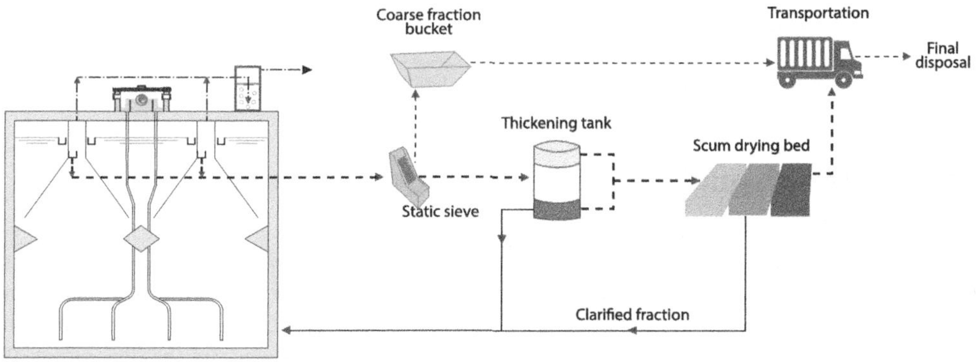

Figure 4.17 Schematic flowsheet of the scum removal system.

Therefore, it is recommended that UASB reactor designs include a screening unit (static sieve) for the scum to be withdrawn from the inner GLS separators and, subsequently, a tank for thickening, as follows:

(i) A screening unit (# 1.0 mm) to allow separation of the coarse material (debris) contained in the scum, which is routed to a final disposal site (e.g. landfill).

(ii) A thickening unit located after the sieving unit, to allow the separation of the clarified fraction from the other fractions (supernatant and sedimented fractions), with return of the clarified fraction to the preliminary treatment unit, or to the UASB reactor itself, and the other fractions to the drying beds. It has been observed that one hour of thickening time is sufficient to promote good separation of the clarified fractions relative to the supernatant and sedimented material.

(iii) The experience with drying beds has shown good results for a scum height around 5–10 cm.

The scum management could be integrated to the sludge management, as schematically represented in Figure 4.18. Once the ideal scum removal protocol is established (see Chapter 6), the routine operation of the hydrostatic removal device consists simply of opening the external withdrawal valve, according to the time and degree of valve opening defined during the start-up period, and allowing the *fluid scum* to flow to the drying bed or to a sludge balance and homogenisation tank for processing in the mechanised dewatering unit. In addition, as addressed in Chapters 8 and 12, the biogas produced can be utilised for sanitisation and/or thermal drying of the sludge and scum.

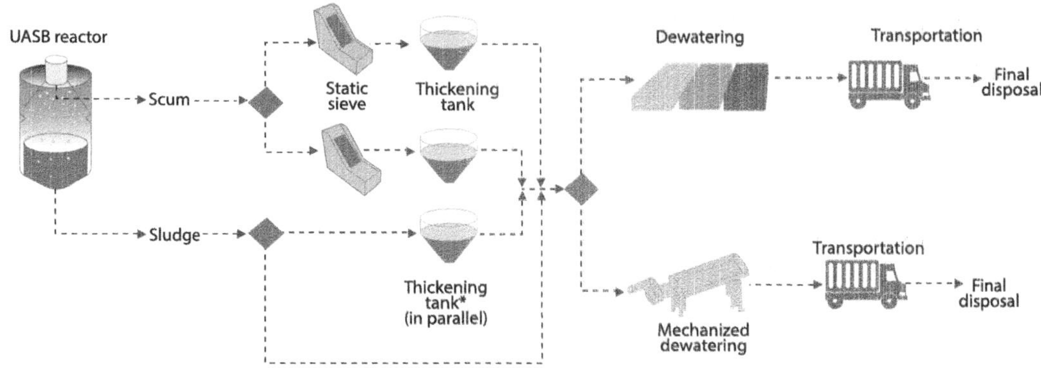

* The clarified fraction from the thickening tanks, static sieves, and dewatering units should be returned to the UASB reactor

Figure 4.18 Schematic representation of the alternatives for integrated management of scum and sludge.

4.4.8 Proper design guidelines for the biogas collection and transport system

The design of a biogas collection system in UASB reactors treating sewage requires specificities due to its differentiated characteristics of low flow rate and low pressure, presence of moisture and particulate matter, and potential for corrosion. This makes the design of the biogas management system different from that of usual gas applications. Special attention should be directed to the design of the GLS separators, as addressed in Section 4.4.5 and also in Chapter 8. In particular, gas chambers must be impermeable to gas and protected from and resistant to corrosion. The biogas pipeline must be designed with a maximum velocity of 5 $m.s^{-1}$ for the average gas flow and a minimum diameter of 50 mm. The biogas transport pipes and their special parts should preferably be aerial to maintain linearity and runoff of the condensate inside the pipes. The biogas collection and transport system should have safety features, including at least, condensate and sediment withdrawal in the lowest piping points, pressure and vacuum relief valves, and flame arresters. The implementation of individualised biogas metering is recommended for each UASB reactor.

To minimise corrosion problems in the biogas collection lines, the pipes and fittings must be made of stainless steel or a suitable plastic. Because of its higher resistance to corrosion, AISI 316Ti steel (EN 1.4571) is recommended. Pipes and fittings made from high-density polyethylene (HDPE) are considered suitable for conducting biogas, unlike pipes and connections made from polyvinyl chloride (PVC). Regarding the use of HDPE, conduction pipes with electrical resistance lower than 109 Ω and coefficient of thermal expansion of 0.20 mm m^{-1} K^{-1} must be used to minimise the risk of fire. Chapter 5 presents some examples of biogas pipelines.

For large-scale STPs (PE > 100,000) in which no biogas energy utilisation is considered, provision should be made for at least two biogas flares, one of which is a reserve unit. The flares should be fitted with a flame arrestor and an automatic ignition system. Large-scale STPs must also have an automatic flame control panel. In cases in which the biogas is used for energy recovery, in addition to the recovery units themselves, at least one flare must be provided as a safety unit. In cases in which the collected biogas is not used for energy recovery, it should be burned, preferably with enclosed flares (see Chapter 8).

In situations in which it is not possible to guarantee a minimum continuous biogas flow rate, an automatic or pilot ignition system, powered by, e.g., liquefied petroleum gas (LPG), must be provided. As described in

Section 4.4.5 and Section 8.3.2 of Chapter 8, a minimum pressure of 1,500 Pa (0.15 mca (metres of water column)) within the reactor gas chambers should be ensured by the use of a water seal or throttle valve pressure.

For estimation of the biogas production in UASB reactors treating sewage, the unit relations reported in Chapter 8 should be considered. Table 4.11 summarises the main design criteria of the biogas collection system.

Table 4.11 Summary of the main design criteria for the biogas collection system.

Criterion	Values/ Range of Values	Comments
Width of the GLS gas hood (m)	0.25–0.50	See Section 4.4.5 for the standard GLS separator.
Minimum biogas release rate ($m^3\ m^{-2}\ h^{-1}$)	1.0	A proper scum withdrawal protocol should be established to prevent mechanical stresses in the GLS separator and biogas release through the settler compartment.
Maximum biogas release rate ($m^3\ m^{-2}\ h^{-1}$)	3.0–5.0	
Methane concentration in the biogas (%)	70–80	See also Chapter 8.

4.5 PERFORMANCE OF UASB REACTORS TREATING SEWAGE

As mentioned in Chapter 3, the BOD and COD removal efficiencies of UASB reactors treating sewage in warm climate countries are in the range of 60–75% and 55–70%, respectively without a post-treatment step. A review of efficiencies of full-scale UASB reactors for carbon and TSS removal (Chernicharo *et al.* 2015) was further complemented, as shown in Table 4.12. Some of the implemented UASB-based STPs were observed to be underperforming. Their deficient performance could be ascribed to engineering flaws incurred during the construction stage (see Chapter 5) as well as operational shortcomings, as elaborated in the Chapter 6. Full-scale experiences are also covered in Chapter 7.

Mathematical modelling based upon the Anaerobic Digestion Model No 1 (ADM1) has been widely applied and validated for simulating the anaerobic digestion of several organic wastes, such as sewage sludge. The ADM1 was published in 2002 (Batstone *et al.*, 2002) and it stands as the most comprehensive model of anaerobic digestion process in the present time. However, predicting the efficiency of full-scale UASB reactors for treating sewage is not an undemanding process. This is so since actual complex flow behaviour is significantly different from that observed in a constant-volume, completely mixed system as assumed by ADM1. In addition, construction aspects and operational protocols (mainly in terms of sludge and scum withdrawal) are of utmost importance to ensure proper effluent quality, as discussed further in Chapters 5 and 6.

Therefore, the carbon removal efficiency of UASB reactors treating sewage is estimated mainly by means of empirical relations which are obtained from experimental results of STPs in operation. Chernicharo (2007) proposed two equations to estimate the COD and BOD removal efficiencies of UASB reactors treating sewage under tropical conditions (sewage temperature within 20–27°C), as a function of the HRT.

Table 4.12 Performance of full-scale UASB-based STPs.

Location	Effluent Concentration			Removal Efficiency			Population Equivalent (inhabitants)	Reference
	COD ($mg\ L^{-1}$)	BOD ($mg\ L^{-1}$)	TSS ($mg\ L^{-1}$)	COD (%)	BOD (%)	TSS (%)		
India	309–605	175–251	229–454	51	56	54	–	Gaur *et al.* (2017)
India	145	63	72	61	66	73	–	Gani *et al.* (2016)
India	177	56	53	56	61	74	280,000	Okubo *et al.* (2015)
India	202	60	150	63	67	70	93,500	Pandey and Dubey (2014)
India	139–567	57–159	72–452	29–75	45–79	40–70	–	Khan *et al.* (2014)
Brazil	283	–	132	58	–	49	3,047	Silva *et al.* (2013)
Brazil	114	38	132	79	84	59	70,000	Rosa *et al.* (2012)
Brazil	251	98	85	65	74	71	24,000	Oliveira and von Sperling (2011)
India	515	115	113	41	50	47	–	Mungray and Patel (2011)
India	405	153	167	44	40	36	–	Mungray and Patel (2011)
India	145–250	55–75	160–240	45	60	34	–	Walia *et al.* (2011)
Colombia	–	60	–	–	77	–	320,000	WERF (2010)
Brazil	170	66	75	58	68	56	544,000	Franco (2010)
Brazil	247	97	112	62	67	54	–	Van Lier *et al.* (2010)
India	285	121	357	46	41	49	–	Van Lier *et al.* (2010)
Brazil	190	70	60	60	65	61	1,000,000	Chernicharo *et al.* (2009)
Colombia	144	–	81	58	–	65	–	Peña *et al.* (2006)
Brazil	181	75	127	64	74	51	24,719	Baréa and Alem Sobrinho (2006)
Brazil	106	69	–	72	72	–	150,000	Carraro (2006)
Brazil	161	66	–	77	78	–	–	Tachini *et al.* (2006)
India	403	130	380	47	50	7	55,000–570,000	Sato *et al.* (2006)
Middle East	221	83	63	71	70	85	–	Nada *et al.* (2006)
India	–	–	–	61	61	66	–	Khalil *et al.* (2006)
Jordan	632	–	180	58	–	62	–	Halalsheh *et al.* (2005)
Brazil	237	64	127	60	69	52	3,808	Busato (2004)
Brazil	202	–	80	67	–	61	18,000	Florencio *et al.* (2001)
Colombia	177	69	72	66	78	69	9,000	Peña *et al.* (2000)
Mexico	–	–	–	70–80	–	–	–	Monroy *et al.* (2000)
Median values	*221*	*73*	*127*	*60*	*67*	*58*	3,808–1,000,000	

Source: Adapted from Chernicharo *et al.* (2015).

The original database included 16 full-scale UASB reactors with influent COD ranging from 300 mg L^{-1} to 1,400 mg L^{-1} and influent BOD varying between 150 mg L^{-1} and 850 mg L^{-1}. This database was expanded after consideration of the other nine full-scale reactors (Figure 4.19(a–b)) operating within the same influent concentration ranges and temperature to obtain a reviewed equation for estimating the efficiency of UASB reactors treating sewage. Efficiencies estimated using the proposed equations seemed to represent typical trends which could be observed in well operated full-scale systems, especially in the operational HRT range of 8–12 h.

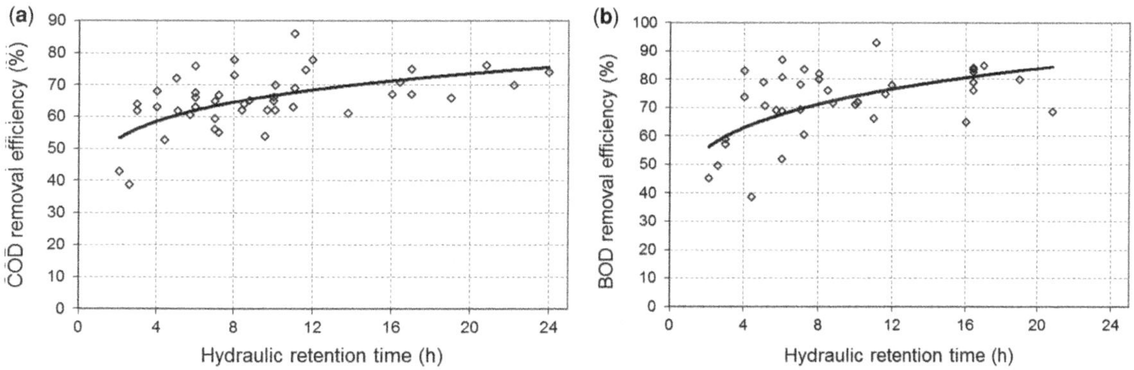

Figure 4.19 Proposed efficiency curves for estimating COD (a) and BOD (b) removal efficiencies in UASB reactors treating sewage.

Therefore, the following efficiency equations are recommended for estimating COD and BOD removal efficiencies of UASB reactors treating sewage:

$$E_{COD} = 48 \times (HRT^{0.14}) \tag{4.1}$$

where:
E_{COD} = UASB reactor efficiency for COD removal (%)
HRT = hydraulic retention time (h)
48 = empirical constant
0.14 = empirical constant.

$$E_{BOD} = 49.2 \times (HRT^{0.18}) \tag{4.2}$$

where:
E_{BOD} = UASB reactor efficiency for BOD removal (%)
HRT = hydraulic retention time (h)
49.2 = empirical constant
0.18 = empirical constant.

The TSS effluent concentration of UASB reactors treating sewage depends on a series of factors including the sludge and scum withdrawal routines, proper GLS separator and adequate upflow velocities in the digestion and settler compartments, as discussed throughout this Chapter and Chapter 6. Due to inaccurate and unavailable data, only 13 out of the 25 aforementioned STPs were considered (Figure 4.20) for deriving a reviewed equation to estimate the TSS concentration in the effluent of UASB reactors treating sewage. The equation was obtained on the basis of the operational HRT.

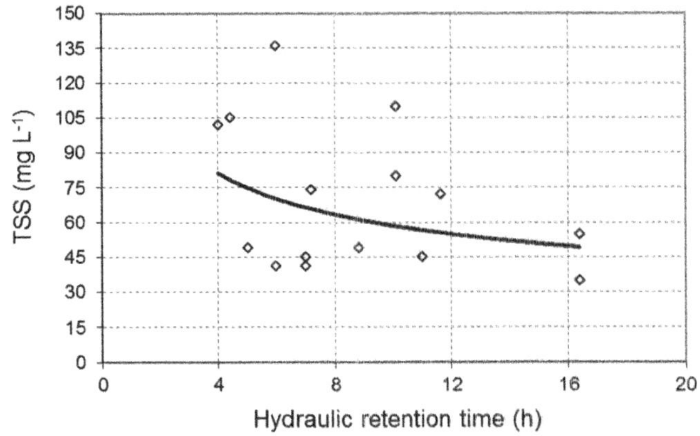

Figure 4.20 TSS concentrations in the effluent of UASB reactors treating sewage.

Therefore, the following equation is recommended for estimating the TSS concentration in the effluent of UASB reactors treating sewage. Such equation should be applied with particular care, because even within a proper HRT condition, the absence of an adequate sludge withdrawal routine can jeopardize the effluent quality, as previously addressed in this Chapter.

$$TSS = 144.8 \times (HRT^{-0.41}) \tag{4.3}$$

where:

TSS = TSS concentration in the effluent of UASB reactors (mg L^{-1})
HRT = hydraulic retention time (h)
144.8 = empirical constant
0.41 = empirical constant.

4.6 DESIGN EXAMPLES

EXAMPLE 4.5

Design of a UASB reactor for 20,000 PE, based on the concept of modular GLS separators (see topic 4.4.5).

Input data

- Population equivalent: PE = 20,000 inhabitants
- Average influent sewage flowrate: $Q_{s\text{-}avg} = 38.7\ L\ s^{-1}$ ($139.4\ m^3\ h^{-1}$)
- Maximum hourly sewage influent flowrate: $Q_{s\text{-}max} = 66.5\ L\ s^{-1}$ ($239.4\ m^3\ h^{-1}$)
- *Per capita* COD contribution: $QPCP_{COD} = 0.10$ kg $inhab^{-1}\ d^{-1}$
- Average influent COD concentration: $C_{COD} = 598$ mg L^{-1} (0.598 kg m^{-3})
- Average influent BOD concentration: $C_{BOD} = 333$ mg L^{-1} (0.333 kg m^{-3})
- Temperature of the liquid: 20°C

Solution

(1) Adopt a hydraulic retention time ($HRT_{adopted}$)

- $HRT_{adopted} = 9.0$ h (see Table 4.3)

(2) Determine the total reactor volume (V_t)

- $V_t = Q_{s\text{-}avg} \times HRT = 139.4\ m^3\ h^{-1} \times 9.0\ h = 1{,}255\ m^3$

(3) Adopt a number of UASB reactors to be implemented (N_r)

- $N_r = 2$ reactors

(4) Determine the volume of each UASB reactor (V_r)

- $V_r = V_t/N_r = 1{,}255\ m^3/2$ reactors $= 627\ m^3$

(5) Adopt a reactor useful height (H_u)

- $H_u = 4.6$ m
  (N.B. adopting lower values leads to larger areas and, consequently, lower upflow velocities)

(6) Determine the area of each UASB reactor (A_r)

- $A_r = V_r/H_u = 627\ m^3/4.6\ m \approx 137\ m^2$

(7) Determine the dimensions of each UASB reactor based on the available GLS separator dimensions (see topic 4.4.5).

(7.1) Adopt a UASB reactor length (L_{UASB}) (preferably a multiple of 3, because of the GLS separator standard dimension)

- $L_{UASB} = 15.0$ m

(7.2) Determine the number of GLS separators to be implemented (N_{GLS})

- $N_{GLS} = L_{UASB}/W_{GLS}$ (for W_{GLS} standard dimension see Section 4.4.5) $= 15.0\ m/3.0\ m = 5$

(7.3) Determine the required UASB reactor width ($W_{UASB\text{-}req}$)

- $W_{UASB\text{-}req} = A_r/L_{UASB} = 137\ m^2/15.0\ m = 9.1$ m

(7.4) Select a GLS separator length based on Table 4.6

- Selected GLS separator: Model 3→ Length of the bottom GLS separator part (considering the inclined walls) ($L_{GLS\text{-}lower}$) = 8.4 m

(7.5) Adopt a width for the aperture between the GLS separator and the UASB reactor wall ($W_{aperture\text{-}wall}$) (see Figure 4.14 – Section 4.4.5 and Figure 4.21)

- $W_{aperture\text{-}wall} = 0.35$ m

(7.6) Determine the resulting UASB reactor width (W_{UASB})

- $W_{UASB} = L_{GLS\text{-}lower} + 2W_{aperture} = 8.4\ m + 2 \times 0.35\ m = 9.1$ m

(7.7) Verify the area and volume of each UASB reactor

- $A_r = W_{UASB} \times L_{UASB} = 9.1\ m \times 15.0\ m = 137\ m^2$
- $V_r = A_r \times H_u = 137\ m^2 \times 4.6\ m = 630\ m^3$

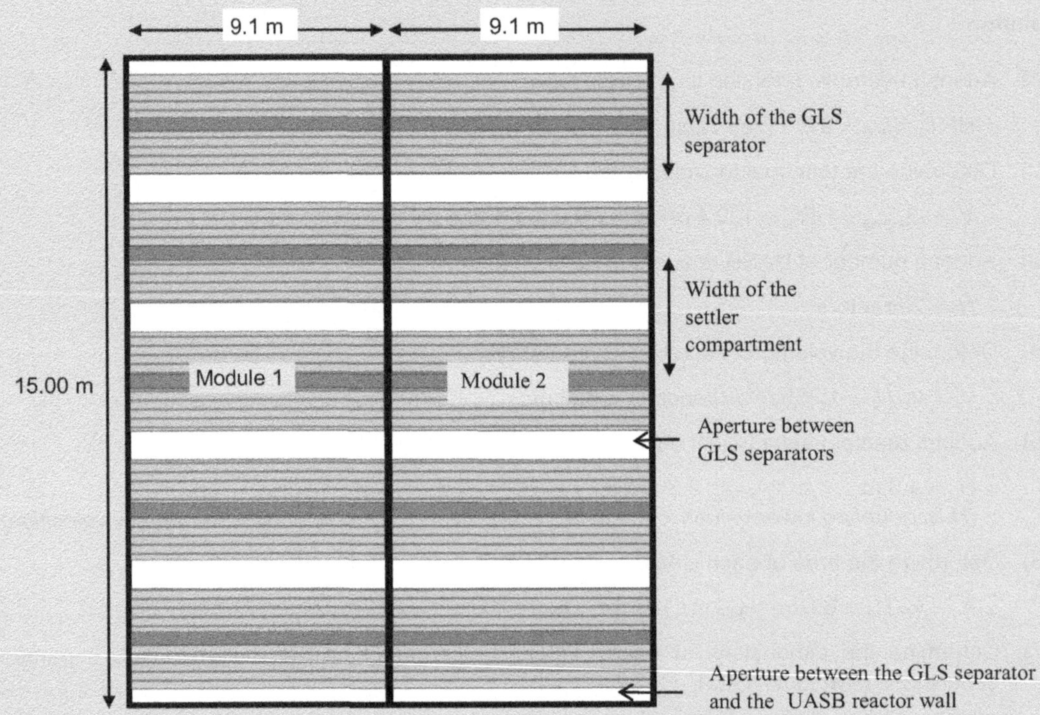

Figure 4.21 Schematic representation of the UASB reactors dimensions (plant view) and the GLS separators (on the top of the reactors).

(7.8) Verify the hydraulic retention time (HRT)

- $HRT = V_r \times N_r / Q_{s\text{-}avg} = 630\ m^3 \times 2/139.4\ m^3\ h^{-1} \approx 9.1\ h$ (The resulting HRT is practically the same of the previously adopted HRT)

Note: the variables H_u, L_{UASB} *(as long as it is a multiple of 3.0 m)*, $W_{aperture\text{-}wall}$ (usually between 0.30 and 0.40 m) and the design HRT can be managed to obtain different UASB reactor configurations.
A schematic representation of the UASB reactors dimensions (plant view) and the GLS separators configuration is shown in Figure 4.21.

(7.9) Verify the upflow velocities

- For $Q_{s\text{-}avg}$: $v = Q_{s\text{-}avg}/(A_r \times N_r) = 139.4\ m^3\ h^{-1}/(137\ m^2 \times 2) = 0.5\ m\ h^{-1}$
- For $Q_{s\text{-}max}$: $v = Q_{s\text{-}max}/(A_r \times N_r) = 239.4\ m^3\ h^{-1}/(137\ m^2 \times 2) = 0.9\ m\ h^{-1}$

(7.10) Verify the velocities through the apertures to the settler compartment

- Number of GLS separators (N_{GLS}): 10 (5 in each module)
- Number of apertures between GLS separators ($N_{apertureGLS\text{-}GLS}$): 8 (4 in each module)
- Number of apertures between the GLS separator and the reactor wall ($N_{apertureGLS\text{-}wall}$): 4 (2 in each module)
- Adopted width for the aperture between two contiguous GLS separators ($GLS_{aperture}$): 0.70 m (see example 4.2)

- Adopted width for the aperture between the GLS separator and the UASB reactor wall ($W_{aperture\text{-}wall}$): 0.35 m
- Length of each aperture: $L_{aperture} = 9.1$ m (along the width of the reactor)
- Total area of the apertures: $A_{apertures} = L_{aperture} \times (N_{apertureGLS\text{-}GLS} \times GLS_{aperture} + N_{apertureGLS\text{-}wall} \times W_{aperture\text{-}wall}) = 9.1\ m \times (8 \times 0.70\ m + 4 \times 0.35\ m) = 63.7\ m^2$
  - o Verification of the velocities through the apertures:
    - For $Q_{s\text{-}avg}$: $v = Q_{s\text{-}avg}/A_{apertures} = 139.4\ m^3\ h^{-1}/63.7\ m^2 = 2.2\ m \cdot h^{-1}$
    - For $Q_{s\text{-}max}$: $v = Q_{s\text{-}max}/A_{apertures} = 239.4\ m^3\ h^{-1}/63.7\ m^2 = 3.8\ m \cdot h^{-1}$

(7.11) Verify the surface loading rates on the settler compartment

- Number of settler compartments ($N_{settler}$): 10 (5 in each reactor)
- Length of each settler (the same of $L_{GLS\text{-}lower}$) = 8.4 m
- Total length of the settlers ($L_{tsettlers}$) = 10 × 8.4 m = 84 m
- Width of each GLS separator gas hood ($W_{gashood}$): 0.40 m (see example 4.2). Note: the thickness (e) of the gas hood wall can be disregarded, due to the use of glass fibre GLS separator ($e = 0.005$ m)
- Useful width of each settler compartment: $W_{settler} = [L_{UASB} - (N_{GLS} \times W_{gashood})]/N_{settler} = [15.0\ m - (5 \times 0.40\ m)]/5 = 2.6\ m$
- Total area of the settlers: $A_{settler} = L_{tsettlers} \times W_{settler} = 84\ m \times 2.6\ m = 218.4\ m^2$
  - o Surface loading rates on the settler compartment
    - For $Q_{s\text{-}avg}$: $v = Q_{s\text{-}avg}/A_{settler} = 139.4\ m^3\ h^{-1}/218.4\ m^2 = 0.6\ m\ h^{-1}$
    - For $Q_{s\text{-}max}$: $v = Q_{s\text{-}max}/A_{settler} = 239.4\ m^3\ h^{-1}/218.4\ m^2 = 1.1\ m \cdot h^{-1}$

(7.12) Verify the HRT on the settler compartment ($HRT_{settler}$)

- Adopted depth for the inclined GLS separator wall (GLS_H): 1.3 m (see example 4.2)
- Adopted depth for the vertical GLS separator wall (GLS_V): 0.40 m (see example 4.2)
- Adopted slope for the inclined GLS separator wall (GLS_S): 53.8° (see example 4.2)
- Width of the inclined GLS separator wall (GLS_W): 0.95 m (see example 4.2)
- Adopted width for the aperture between two contiguous GLS separators ($GLS_{aperture}$) 0.70 m (see example 4.2)
  - o Determine the volume of the settler compartment ($V_{settler}$)
    - Triangular area; between the inclined walls of the GLS settler (see Figure 4.14): $A_{settler1} = (GLS_W \times GLS_H)/2 = (0.95\ m \times 1.3\ m)/2 = 0.62\ m^2$
    - Rectangular area; between the triangular areas (see Figure 4.14): $A_{settler2} = GLS_H \times GLS_{aperture} = (1.3\ m \times 0.70\ m) = 0.91\ m^2$
    - Rectangular area; between the vertical walls of the GLS settler (see Figure 4.14): $A_{settler3} = GLS_V \times W_{settler} = 0.40\ m \times 2.6\ m = 1.04\ m^2$
    - Total area along the settler compartment depth ($A_{settler}$): $A_{settler} = (A_{settler1} \times 2) + A_{settler2} + A_{settler3} = (0.62\ m^2 \times 2) + 0.91\ m^2 + 1.04\ m^2 = 3.19\ m^2$
    - Total volume of the settler compartment ($V_{settler}$): $V_{settler} = N_{settler} \times (L_{GLS\text{-}lower} + GLS_{aperture}) \times A_{settler} = 10 \times (8.4\ m + 0.70\ m) \times 3.19\ m^2 = 290.3\ m^3$
  - o Determine the HRT of the settler compartment ($HRT_{settler}$):
    - For $Q_{s\text{-}avg}$: $HRT_{settler} = V_{settler}/Q_{s\text{-}avg} = 290.3\ m^3/139.4\ m^3\ h^{-1} = 2.1\ h$
    - For $Q_{s\text{-}max}$: $HRT_{settler} = V_{settler}/Q_{s\text{-}max} = 290.3\ m^3/239.4\ m^3\ h^{-1} = 1.2\ h$

(7.13) Determine the influent distribution system

- Adopting an influence area (A_i) of 2.25 m^2 per distribution pipe (according to Table 4.4), the number of distribution pipes is:
- $N_d = A_r/A_i = 137\ m^2/2.25\ m^2 = 61$ distributors. Due to symmetry, adopt 60 distributors, as follows:
    - o Along the length of each module (15.0 m): 10 pipes
    - o Along the width of each module (9.1 m): 6 pipes
- Therefore, each module will have 50 (10 × 5) distribution pipes, each one with an influence area equivalent to: $A_i = 137\ m^2/60 = 2.28\ m^2$.

(8) Estimate the COD removal efficiency (E_{COD})

o According to Section 4.5: $E_{COD} = 48 \times (HRT^{0.14})$
o $E_{COD} = 48 \times (9.1^{0.14}) = 65\%$
Note: a COD removal efficiency of 80% (in terms of filtered COD) will be adopted for the methane production estimation (see example 1 in Chapter 8)

(9) Estimate the BOD removal efficiency (E_{BOD})

o According to Section 4.5: $E_{BOD} = 49.2 \times (HRT^{0.18})$
o $E_{BOD} = 49.2 \times (9.1^{0.18}) = 73\%$

(10) Estimate the effluent COD and BOD concentrations (C_{effl})

o $C_{efflCOD} = C_{COD} - (E_{COD} \times C_{COD})/100 = 598\ mgCOD\ L^{-1} - (65\% \times 598\ mgCOD\ L^{-1})/100 = 209\ mgCOD\ L^{-1}$
o $C_{efflBOD} = C_{BOD} - (E_{BOD} \times C_{BOD})/100 = 333\ mgBOD\ L^{-1} - (73\% \times 333\ mgBOD\ L^{-1})/100 = 90\ mgBOD\ L^{-1}$

(11) Estimate the biogas flow rate (Q_g)

o $Q_{CH4} = 208\ m^3CH_4\ d^{-1}$ (see example 2 in Chapter 8)
o Adopting a methane content of 75% (see Chapter 8): $Q_g = Q_{CH4}/0.75 = 277.3\ m^3biogas \cdot d^{-1}$

(12) Verify the biogas release rate (K_g) in the GLS gas hood

o Total length of the gas collector ($L_{gascollector}$) $= N_{GLS} \times L_{GLS\text{-}upper} \times N_r = 5 \times 6.5\ m \times 2 = 65\ m$
o Adopted free space between the scum removal channel and the GLS separator wall (W_{free}): 0.25 m (see example 4.2)
o Total area of the gas collectors ($A_{gascollector}$) $= L_{gascollector} \times W_{free} = 65\ m \times 0.25\ m = 16.3\ m^2$
o Biogas release rate in the gas collectors (K_g) $= Q_g/A_{gascollector} = (277.3\ m^3biogas\ d^{-1}/24\ h\ d^{-1})/16.3\ m^2 = 0.71\ m^3m^{-2}h^{-1}$
Note: For the biogas production estimation, the methane yield considers the losses of methane, such as leaks and dissolution in the liquid effluent (see Chapter 8). Therefore, it is difficult to meet the minimum biogas release rate (see Table 4.11).

(13) Determine the biogas piping diametre (D_{gas})

o Adopted velocity (v_{gas}): 2.0 m s^{-1}
$D_{gas} = ((4 \times Q_g/86{,}400)/v_{gas} \times Ï)^{0.5} \times 1{,}000)/N_r = ((4 \times 277.3\ m^3biogas\ d^{-1}/86{,}400)/2\ m\ s^{-1} \times Ï)^{0.5} \times 1{,}000)/2 = 23\ mm$
$D_{gas} = 23$ mm. Adopt a minimum diameter of 50 mm.

The sludge management system was addressed in example 4.3.

Note: As mentioned in the book foreword, a design spreadsheet can be found on the IWA Publishing website. Additionally, the conventional design method for UASB reactors treating sewage can be found in Chernicharo (2007).

EXAMPLE 4.6

Design of a UASB reactor for 200,000 PE, based on the concept of modular GLS separators (see topic 4.4.5).

Input data

- Population equivalent: PE = 200,000 inhabitants
- Average influent sewage flowrate: $Q_{s\text{-}avg} = 411.9\ L\ s^{-1}$ ($1{,}483\ m^3\ h^{-1}$)
- Maximum hourly sewage influent flowrate: $Q_{s\text{-}max} = 717.5\ L\ s^{-1}$ ($2{,}583\ m^3\ h^{-1}$)
- *Per capita* COD contribution: $QPCP_{COD} = 0.10$ kg inhab^{-1} d^{-1}
- Average influent COD concentration: $C_{COD} = 562\ mg\ L^{-1}$ ($0.562\ kg\ m^{-3}$)
- Average influent BOD concentration: $C_{BOD} = 309\ mg\ L^{-1}$ ($0.309\ kg\ m^{-3}$)
- Temperature of the liquid: 20°C

Solution

(1) Adopt a hydraulic retention time ($HRT_{adopted}$)

- $HRT_{adopted} = 9.0$ h (see Table 4.3)

(2) Determine the total reactor volume (V_t)

- $V_t = Q_{s\text{-}avg} \times HRT = 1{,}483\ m^3\ h^{-1} \times 9.0\ h = 13{,}347\ m^3$

(3) Adopt a number of UASB reactors to be implemented (N_r)

- $N_r = 8$ reactors

(4) Determine the volume of each UASB reactor (V_r)

- $V_r = V_t/N_r = 13{,}347\ m^3/8$ reactors $= 1{,}668\ m^3$

(5) Adopt a reactor useful height (H_u)

- $H_u = 4.7$ m (adpting lower values lead to larger areas and, consequently, lower upflow velocities)

(6) Determine the area of each UASB reactor (A_r)

- $A_r = V_r/H_u = 1{,}668\ m^3/4.7\ m = 355\ m^2$

(7) Determine the dimensions of each UASB reactor based on the available GLS separator dimensions (see topic 4.4.5)

(7.1) Adopt a UASB reactor length (L_{UASB}) (preferably a multiple of 3, because of the GLS separator standard dimension)

- $L_{UASB} = 30.0$ m

(7.2) Determine the number of GLS separators to be implemented (N_{GLS})

- $N_{GLS} = L_{UASB}/W_{GLS}$ (for W_{GLS} standard dimension see Section 4.4.5) = 30.0 m/3.0 m = 10

(7.3) Determine the required UASB reactor width ($W_{UASB\text{-}req}$)

- $W_{UASB\text{-}req} = A_r/L_{UASB} = 355\ m^2/30.0\ m = 11.8\ m$

(7.4) Select a GLS separator length based on Table 4.6

- Selected GLS separator: Model 3→ Length of the bottom GLS separator part (considering the inclined walls) ($L_{GLS\text{-}lower}$) = 10.9 m

(7.5) Adopt a width for the aperture between the GLS separator and the UASB reactor wall ($W_{aperture\text{-}wall}$) (see Figure 4.14 – Section 4.4.5 and Figure 4.21 – Example 4.5)

- $W_{aperture\text{-}wall} = 0.35\ m$

(7.6) Determine the resulting UASB reactor width (W_{UASB})

- $W_{UASB} = L_{GLS\text{-}lower} + 2 \times W_{aperture} = 10.9\ m + 2 \times 0.35\ m = 11.6\ m$

(7.7) Verify the area and volume of each UASB reactor

- $A_r = W_{UASB} \times L_{UASB} = 11.6\ m \times 30.0\ m = 348\ m^2$
- $V_r = A_r \times H_u = 348\ m^2 \times 4.7\ m = 1{,}636\ m^3$

(7.8) Verify the hydraulic retention time (HRT)

- $HRT = V_r \times N_r/Q_{s\text{-}avg} = 1{,}636\ m^3 \times 8/1{,}483\ m^3 h^{-1} \approx 8.8\ h$ (The resulting HRT is practically the same of the previously adopted HRT)
  Note: the variables H_u, L_{UASB} (*as long as it is a multiple of 3.0 m*), $W_{aperture\text{-}wall}$ (usually between 0.30 and 0.40 m) and the design HRT can be managed to obtain different UASB reactor configurations.

(7.9) Verify the upflow velocities

- For $Q_{s\text{-}avg}$: $v = Q_{s\text{-}avg}/(A_r \times N_r) = 1{,}483\ m^3 h^{-1}/(348\ m^2 \times 8) = 0.5\ m\ h^{-1}$
- For $Q_{s\text{-}max}$: $v = Q_{s\text{-}max}/(A_r \times N_r) = 2{,}583\ m^3 h^{-1}/(348\ m^2 \times 8) = 0.9\ m\ h^{-1}$

(7.10) Verify the velocities through the apertures to the settler compartment

- Number of GLS separators (N_{GLS}): 80 (ten in each module)
- Number of apertures between GLS separators ($N_{apertureGLS\text{-}GLS}$): 72 (nine in each module)
- Number of apertures between the GLS separator and the reactor wall ($N_{apertureGLS\text{-}wall}$): 16 (two in each module)
- Adopted width for the aperture between two contiguous GLS separators ($GLS_{aperture}$): 0.70 m (see example 4.2)
- Adopted width for the aperture between the GLS separator and the UASB reactor wall ($W_{aperture\text{-}wall}$): 0.35 m
- Length of each aperture: $L_{aperture} = 11.6\ m$ (along the width of the reactor)
- Total area of the apertures: $A_{apertures} = L_{aperture} \times (N_{apertureGLS\text{-}GLS} \times GLS_{aperture} + N_{apertureGLS\text{-}wall} \times W_{aperture\text{-}wall}) = 11.6\ m \times (72 \times 0.70\ m + 16 \times 0.35\ m) = 649.6\ m^2$

- Verification of the velocities through the apertures:
  - For $Q_{s\text{-}avg}$: $v = Q_{s\text{-}avg}/A_{apertures} = 1{,}483\ m^3\ h^{-1}/649.6\ m^2 = 2.3\ m\ h^{-1}$
  - For $Q_{s\text{-}max}$: $v = Q_{s\text{-}max}/A_{apertures} = 2{,}583\ m^3\ h^{-1}/649.6\ m^2 = 3.9\ m\ h^{-1}$

(7.11) Verify the surface loading rates on the settler compartment

- Number of settler compartments ($N_{settler}$): 80 (ten in each reactor)
- Length of each settler (the same of $L_{GLS\text{-}lower}$) = 10.9 m
- Total length of the settlers ($L_{tsettlers}$) = 80 × 10.9 m = 872.0 m
- Width of each GLS separator gas hood ($W_{gashood}$): 0.40 m (see example 4.2). Note: the thickness (e) of the gas hood wall can be disregarded, due to the use of glas fibre GLS separator (e = 0.005 m)
- Useful width of each settler compartment: $W_{settler} = [L_{UASB} - (N_{GLS} \times W_{gashood})]/N_{settler} = [30.0\ m - (10 \times 0.40\ m)]/10 = 2.6\ m$
- Total area of the settlers: $A_{settler} = L_{tsettlers} \times W_{settler} = 872\ m \times 2.6\ m = 2{,}267.2\ m^2$

  - Surface loading rates on the settler compartment
    - For $Q_{s\text{-}avg}$: $v = Q_{s\text{-}avg}/A_{settler} = 1{,}483\ m^3\ h^{-1}/2{,}267.2\ m^2 = 0.7\ m\ h^{-1}$
    - For $Q_{s\text{-}max}$: $v = Q_{s\text{-}max}/A_{settler} = 2{,}583\ m^3\ h^{-1}/2{,}267.2\ m^2 = 1.1\ m\ h^{-1}$

(7.12) Verify the HRT on the settler compartment ($HRT_{settler}$)

- Adopted depth for the inclined GLS separator wall (GLS_H): 1.3 m (see example 4.2)
- Adopted depth for the vertical GLS separator wall (GLS_V): 0.40 m (see example 4.2)
- Adopted slope for the inclined GLS separator wall (GLS_S): 53.8° (see example 4.2)
- Width of the inclined GLS separator wall (GLS_W): 0.95 m (see example 4.2)
- Adopted width for the aperture between two contiguous GLS separators ($GLS_{aperture}$): 0.70 m (see example 4.2)

  - Determine the volume of the settler compartment ($V_{settler}$)
    – Triangular area; between the inclined walls of the GLS settler (see Figure 4.14): $A_{settler1} = (GLS_W \times GLS_H)/2 = (0.95\ m \times 1.3\ m)/2 = 0.62\ m^2$
    – Rectangular area; between the triangular areas (see Figure 4.14): $A_{settler2} = GLS_H \times GLS_{aperture} = (1.3\ m \times 0.70\ m) = 0.91\ m^2$
    – Rectangular area; between the vertical walls of the GLS settler (see Figure 4.14): $A_{settler3} = GLS_V \times W_{settler} = 0.40\ m \times 2.6\ m = 1.04\ m^2$
    – Total area along the settler compartment depth ($A_{settler}$): $A_{settler} = (A_{settler1} \times 2) + A_{settler2} + A_{settler3} = (0.62\ m^2 \times 2) + 0.91\ m^2 + 1.04\ m^2 = 3.19\ m^2$
    – Total volume of the settler compartment ($V_{settler}$): $V_{settler} = N_{settler} \times (L_{GLS\text{-}lower} + GLS_{aperture}) \times A_{settler} = 80 \times (10.9\ m + 0.70\ m) \times 3.19\ m^2 = 2960.3\ m^3$
  - Determine the HRT of the settler compartment ($HRT_{settler}$):
    – For $Q_{s\text{-}avg}$: $HRT_{settler} = V_{settler}/Q_{s\text{-}avg} = 2{,}960.3\ m^3/1{,}483\ m^3\ h^{-1} = 2.0\ h$
    – For $Q_{s\text{-}max}$: $HRT_{settler} = V_{settler}/Q_{s\text{-}max} = 2{,}960.3\ m^3/2{,}583\ m^3\ h^{-1} = 1.1\ h$

(7.13) Determine the influent distribution system

- Adopting an influence area (A_i) of 2.25 m² per distribution pipe (according to Table 4.4), the number of distribution pipes is:
- $N_d = A_r/A_i = 355\ m^2/2.25\ m^2 = 158$ distributors. Due to symmetry, adopt 160 distributors, as follows:

- Along the length of each module (30.0 m): 16 pipes
- Along the width of each module (11.6 m): 10 pipes

- Therefore, each module will have 160 (16 × 10) distribution pipes, each one with an influence area equivalent to: $A_i = 355\ m^2/160 = 2.22\ m^2$.

(8) Estimate the COD removal efficiency (E_{COD})

- According to Section 4.5: $E_{COD} = 48 \times (HRT^{0.14})$
- $E_{COD} = 48 \times (8.8^{0.14}) = 65\%$

  Note: a COD removal efficiency of 80% (in terms of filtered COD) will be adopted for the methane production estimation (see Example 1, Chapter 8)

(9) Estimate the BOD removal efficiency (E_{BOD})

- According to Section 4.5: $E_{BOD} = 49.2 \times (HRT^{0.18})$
- $E_{BOD} = 49.2 \times (8.8^{0.18}) = 73\%$

(10) Estimate the effluent COD and BOD concentrations (C_{effl})

- $C_{efflCOD} = C_{COD} - (E_{COD} \times C_{COD})/100 = 562\ mgCOD\ L^{-1} - (65\% \times 562\ mgCOD\ L^{-1})/100 = 197\ mgCOD\ L^{-1}$
- $C_{efflBOD} = C_{BOD} - (E_{BOD} \times C_{BOD})/100 = 309\ mgBOD\ L^{-1} - (73\% \times 309\ mgBOD\ L^{-1})/100 = 83\ mgBOD\ L^{-1}$

(11) Estimate the biogas flow rate (Q_g)

- $Q_{CH4} = 2080 m^3 CH_4\ d^{-1}$ (see Example 1, Chapter 8)
- Adopting a methane content of 75% (see Chapter 8): $Q_g = Q_{CH4}/0.75 = 2{,}773\ m^3_{biogas}\ d^{-1}$

(12) Verify the biogas release rate (K_g) in the GLS gas hood

- Total length of the gas collector ($L_{gascollector}$) $= N_{GLS} \times L_{GLS\text{-}upper} \times N_r = 10 \times 9.0\ m \times 8 = 720\ m$
- Adopted free space between the scum removal channel and the GLS separator wall (W_{free}): 0.25 m (see example 4.2)
- Total area of the gas collectors ($A_{gascollector}$) $= L_{gascollector} \times W_{free} = 720\ m \times 0.25\ m = 180.0\ m^2$
- Biogas release rate in the gas collectors $(K_g) = Q_g/A_{gascollector} = (2{,}773\ m^3 biogas\ d^{-1}/24\ h\ d^{-1})/180\ m^2 = 0.64\ m^3 m^{-2}\ h^{-1}$

  Note: For the biogas production estimation, the methane yield considers the losses of methane, such as leaks and dissolution in the liquid effluent (see Chapter 8). Therefore, it is difficult to meet the minimum biogas release rate (see Table 4.11).

(13) Determine the biogas piping diameter (D_{gas})

- Adopted velocity: $2.0\ m\ s^{-1}$

$D_{gas} = ((4 \times Q_g/86{,}400)/v_{gas} \times Ï)^{0.5} \times 1{,}000)/N_r = ((4 \times 2{,}773\ m^3 biogas\ d^{-1}/86{,}400)/2\ m\ s^{-1} \times Ï)^{0.5} \times 1{,}000)/8 = 35\ mm$

Adopt a minimum diameter of 50 mm.

The sludge management system was addressed in example 4.3.

Note: as mentioned in the book foreword, a design spreadsheet can be found on the IWA Publishing website. Additionally, the conventional design method for UASB reactors treating sewage can be found in Chernicharo (2007).

4.7 ALTERNATIVE CONFIGURATIONS OF UASB REACTORS

A series of alternative configurations of UASB reactors for sewage treatment has been tested. However, none of the configurations, except for the hybrid anaerobic reactor (HAnR), which will also be discussed in Chapter 7, has been implemented effectively at full-scale. Therefore, sparse or even no information on adequate construction materials, maintenance, instrumentation, and control needs is available. Configurations using membranes are specifically addressed in Chapter 12.

4.7.1 Modified UASB reactor integrated to energy recovery

This configuration, called "Modified UASB reactor integrated to energy recovery" is the result of a cooperation between the Federal University of Minas Gerais and Engenho Nove (a Brazilian company), registered as a national patent (BR 10 2016 017395 7 A2). The modified UASB reactor comprises a flexible GLS separator structure, formed by a slope with a gas chamber hermetically sealed. The slope can vary in a range of 20°–60° according to the type of soil. The reactor comprises a GLS separator anchored to the slope by an anchor block and floating support structures. A schematic representation of the system is shown in Figure 4.22.

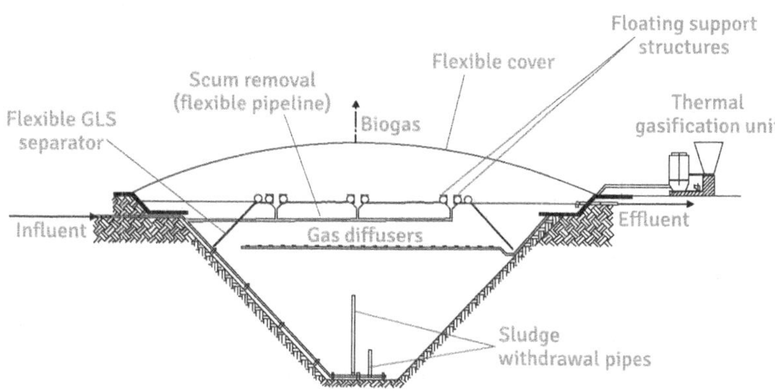

Figure 4.22 Schematic representation of the modified UASB reactor integrated to energy recovery. (Source: Patent deposit BR 10 2016 017395 7 A2)

The sewage inflow/sludge withdrawn systems are integrated, being flexible enough to perform these operations simultaneously. The inflow sewage, recirculation or sludge withdrawal can be managed by opening valves. Energy recovery from sludge and scum can be possible through an additional thermal gasification unit.

4.7.2 Y-shaped UASB reactor

The Y-shaped UASB reactor configuration differs from the classical UASB reactor configuration by incorporating a two-part settling device (van Haandel & Lettinga, 1994) instead of a GLS separator. The central part of the Y-shaped UASB reactor is comprised of a vertically positioned cylinder with a side arm inclined at 45° in the upper part (a schematic representation is shown in Figure 4.23). This device constitutes the liquid–solid separator, and also the discharge point of the treated effluent. Gas–liquid separation takes place in the central cylinder. This Y-shaped configuration could impose hurdles to the construction phase in large-scale systems; therefore, it would be an interesting option for small communities (rural sanitation).

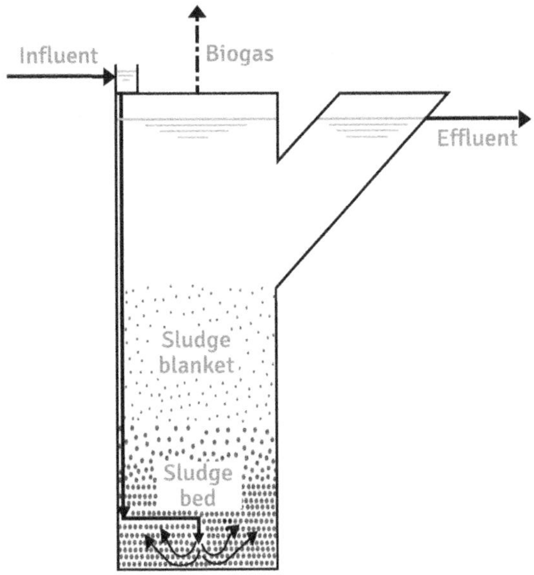

Figure 4.23 Schematic representation of the Y-shaped UASB reactor.

4.7.3 Two-stage UASB reactor

The two-stage UASB configuration emerged as an alternative for the treatment of high-strength wastewater, such as effluents from the agricultural sector (e.g. swine wastewater, wet-processed coffee wastewater, dairy wastewater). The main idea is to decouple the hydrolysis–acidification and acetogenesis–methanogenesis processes, taking into account that the volatile suspended solids (VSS) hydrolysis is the limiting step of the anaerobic digestion, particularly in cold climate regions (see Chapter 3).

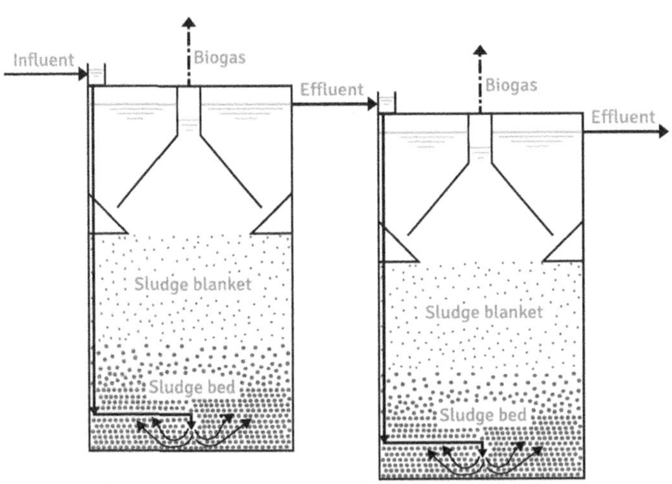

Figure 4.24 Schematic representation of the two-stage UASB reactor.

Therefore, the two-stage UASB system consists of two reactors in series, the first directed to the partial substrate hydrolysis and the second directed to the digestion of soluble compounds formed in the previous unit. A schematic representation is shown in Figure 4.24. Only a few experiences related to the application of the two-stage UASB system for sewage treatment have been reported in the literature (Halalsheh *et al.*, 2005; Sayed & Fergala, 1995). Halalsheh *et al.* (2005) pointed out a simultaneous decrease of biogas methane content and increase of suspended COD removal efficiencies during the winter season. Nevertheless, removal of colloidal and soluble COD was lower during winter than during the summer period. This condition was reported to be associated with the low gas production. The performance of the second stage was considered to be unsatisfactory. On the other hand, Sayed and Fergala (1995) reported that the performance of the second stage was satisfactory at moderate temperatures (18–20°C); nonetheless, they suggested that the second UASB reactor could be replaced by a post-treatment unit.

4.7.4 Combined UASB–digester system

The combined UASB-digester is of particular interest for low-temperature regions. As discussed further in Chapter 11, the concept is that the retained solids are conveyed to a digester for enhanced stabilisation at 30°C and the digested sludge is recirculated to the UASB reactor to improve the methanogenic capacity (Mahmoud *et al.*, 2004). A schematic representation of the system is shown in Figure 4.25. Anaerobic treatment of sewage is possible at low temperature when the system is operated at long SRTs (Seghezzo *et al.*, 2003; van der Last & Lettinga, 1992; van Haandel *et al.*, 2006). The latter can be attained easily by applying primary settling prior to the anaerobic reactor. However, as presented in Chapter 11, the application of anaerobic membrane (AnMBR) technology for sewage treatment at psychrophilic environments (<20°C) seems to be essential for wide-scale application.

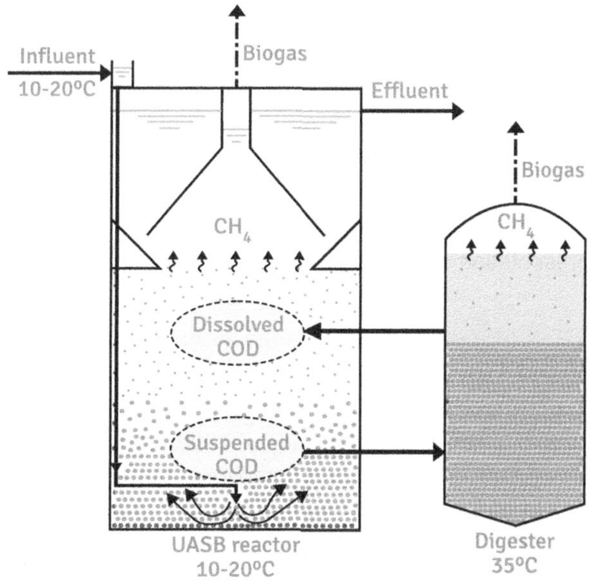

Figure 4.25 Schematic representation of the combined UASB-Digester system.

4.7.5 UASB reactor with enhanced solids separation

In classical UASB reactors, GLS separators have a major role in solid–liquid separation and subsequent solids return to the digestion compartment. Therefore, a more efficient solids separation will enhance sludge retention and contribute to a higher SRT, which is interesting for the reactor performance.

An enhanced solids separation can be achieved by applying high-rate settlers (lamella plates) in the settler compartment of UASB reactors, favouring sedimentation and retaining particles that would normally be washed out. According to Bressani-Ribeiro *et al.* (2017), a UASB reactor equipped with lamella plates achieved a decrease of 30% in TSS and 12% in COD concentration in the effluent compared with the conventional settler compartment. However, strict control over the solids content in the interface between the digestion and settler compartments should be implemented. Otherwise, solids can accumulate between lamellas and consequently be washed out during peak flows. A schematic representation of the system is shown in Figure 4.26.

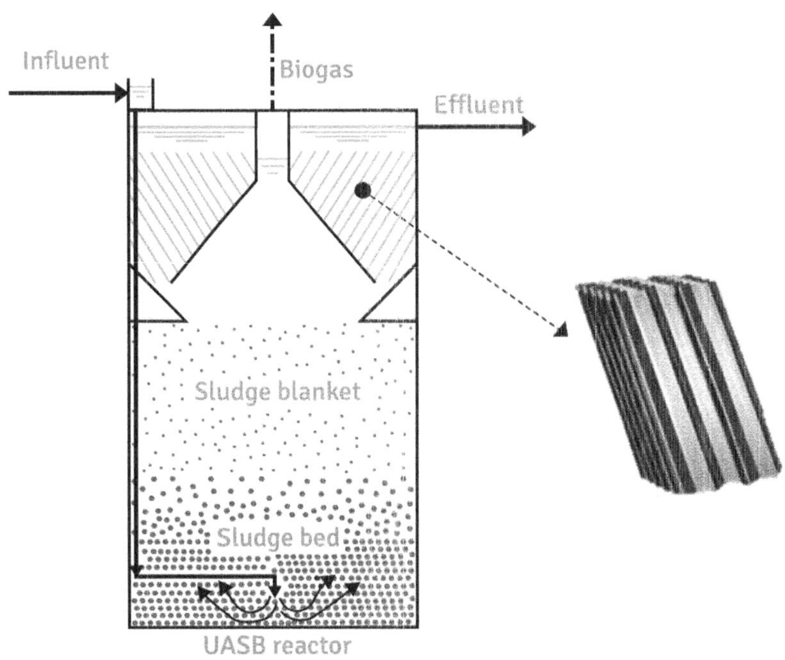

Figure 4.26 Schematic representation of the UASB reactor with enhanced solids separation.

4.7.6 Partitioned UASB reactor

This variation was conceived for the treatment of sewage from small villages and areas (PE < 2,000) with large variation of daily flow rate (Chernicharo & Cardoso, 1999). A schematic representation is shown in Figure 4.27. In such situations, conventional single compartment reactors are subjected to very extreme hydraulic conditions, which can disturb the behaviour of the treatment system. In the partitioned reactor, using two or three digestion chambers allows the establishment of more stable upflow velocities and less occurrence of dead zones. As a result, better contact between substrate and biomass can be achieved and improved performance of the system can be expected. In the partitioned reactor, the variation of upflow velocity in each digestion chamber is much smaller, which also contributes to minimising the solids washout to the settler compartment.

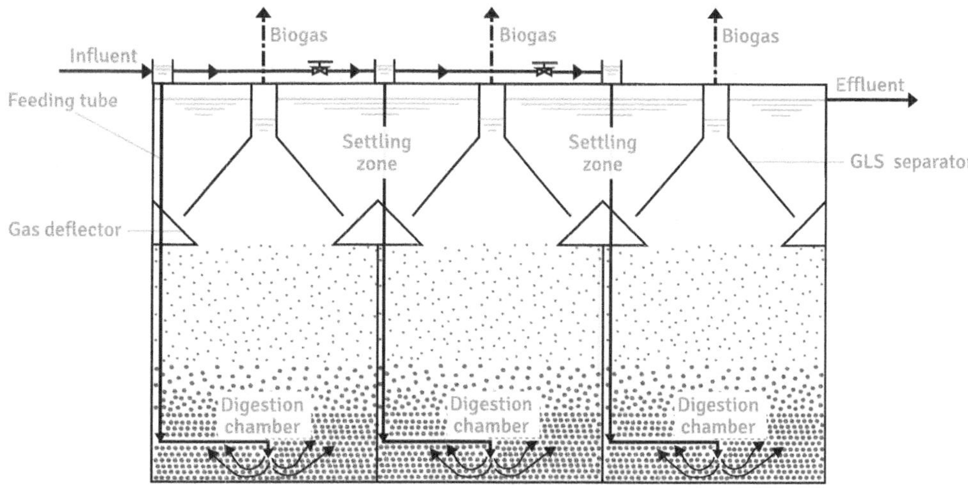

Figure 4.27 Schematic representation of the partitioned UASB reactor. (Source: Chernicharo & Cardoso, 1999).

The partitioning of a single digestion compartment (conventional reactor) into three chambers that operate in parallel (partitioned reactor) does not pose any significant extra costs to the reactor structure because the partitioning can be accomplished with cheap materials (e.g. cement board, thin brick wall and the like).

4.7.7 Hybrid UASB reactor

The hybrid UASB reactor configuration (UASB coupled with sponge media on top of the digester compartment) differs from the classic UASB reactor configuration due to the presence of a sponge-based support media on top of the digester compartment. A schematic representation of the system is shown in Figure 4.28. The purpose of the physical media is to enhance biomass immobilisation and prevent solids washout. Two types of support media have been reported for the biomass immobilisation: structured polyurethane sponge (Biobob) and vertical sheets of reticulated polyurethane foam (RPF). A full-scale experience of a hybrid UASB reactor is reported in Chapter 7.

4.7.8 UASB reactor with double stage biogas collection

The UASB reactor with double-stage biogas collection was conceived based on the following assumptions: (i) intense gas mixing in the upper part of the GSL separator, therefore allowing scum disaggregation and spontaneous exit to the settler compartment; and (ii) better contact between biomass and scum in the upper part of the GSL separator, therefore favouring scum degradation inside the reactor (Pereira *et al.*, 2009). A schematic representation is shown in Figure 4.29.

Experiments at demonstration scale have showed scum disaggregation and spontaneous exit to the settler compartment. However, the hypothesis that the double stage of biogas collection would promote a better contact between biomass and scum in the upper part of the GSL separator was not confirmed. This configuration seems to be promising but could impose significant hurdles at the construction stage.

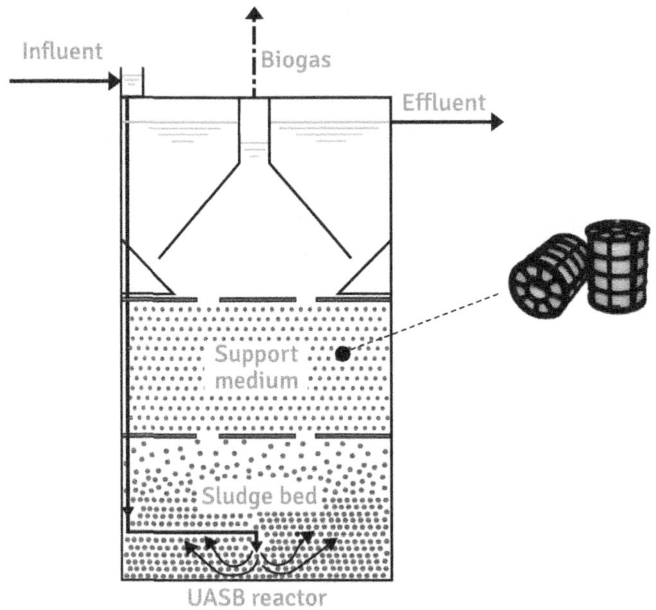

Figure 4.28 Schematic representation of the hybrid UASB reactor.

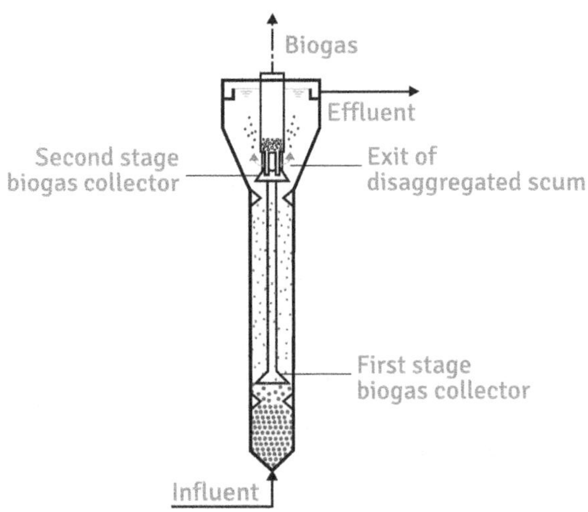

Figure 4.29 Schematic representation of the UASB reactor with double stage biogas collection. (Source: Pereira *et al.*, 2009).

4.8 OPERATIONAL FUNCTIONALITIES

Concern about ergonomic criteria as well as safety at workplace should be considered during the design and construction of STP units. For instance, the provision of guardrails and access ladders (corrosion resistant), service water points, minimum openings for the passage of operators and adequate tools for cleaning and maintenance operations (e.g. in the preliminary treatment units and effluent distribution boxes on top of UASB reactors) must be duly regarded. Provision of devices to empty the preliminary treatment units also makes the cleaning operations more agile and simple. The bucket for collecting the removed coarse material can be strategically placed near the screening at a level lower than the preliminary treatment unit. This mechanism contributes in the improvement of the ergonomic conditions associated with the manual cleaning of the bar rails.

Hatches for facilitating access to the reactor after emptying can be placed near the ground level. They should have a minimum diameter of 800 mm. Chapter 6 discusses the main operational aspects of UASB reactors treating sewage, which should be considered during the design stage in order to facilitate the work of operators.

4.9 FINAL REMARKS

The design of UASB reactors should be adapted to the different particularities of sewage treatment (e.g. low concentration of organic matter). Over the past 20 years, many important aspects of the process start-up and operational requirements have been identified (e.g. sludge withdrawal frequency and scum removal from the inner GLS separator). Important issues still need to be addressed. However, they exist at a level that goes beyond the design itself, such as better control of rainwater infiltration in the sewerage system. Compliance with the design criteria mentioned in this Chapter alongside the construction recommendations (Chapter 5) and operational protocols further discussed in Chapter 6 can certainly ensure that BOD and COD removal efficiencies are achieved within the typical reported ranges for UASB reactors treating sewage.

REFERENCES

Associação Brasileira de Normas Técnicas (ABNT) (2011). NBR 12209 – Elaboração de projetos hidráulico-sanitários de estações de tratamento de esgoto (Elaboration of hydraulic-sanitary designs of sewage treatment plants). ABNT, Rio de Janeiro, Brazil.

Baréa L. C. and Alem Sobrinho P. (2006). Behavior of polishing lagoon in metropolitan Curitiba and the possibility of using of duckweed for improvement of effluent quality. *Sanare Revista Técnica Sanepar Curitiba*, **24**(24), 46–60.

Batstone D. J., Keller J., Angelidaki I., Kalyuzhnyi S. V., Pavlostathis S. G., Rozzi A., Sanders W. T., Siegrist H. and Vavilin V. A. (2002). Anaerobic Digestion Model No. 1. IWA Publishing, London, UK.

Bressani-Ribeiro T., Brandt E. M. F., Almeida P. G. S., Flórez C. A. D. and Chernicharo C. A. L. (2017). Technological improvements in compact UASB/SBTF systems for decentralized sewage treatment in developing countries. *Desalination and Water Treatment*, **91**, 112–120.

Busato R. (2004). Desempenho de um filtro anaeróbio de fluxo ascendente como tratamento de efluente de reator UASB: estudo de caso da ETE de Imbituva. (Performance of an anaerobic filter for the treatment of UASB reactor effluent: case study). PhD Thesis, Universidade Federal do Paraná, Curitiba/PR, Brazil.

Carraro A. (2006). Evaluation of a sewage treatment system with intermittent receipt of water treatment plant sludge following RALF and conventional aerobic biological filtration. *Sanare Revista Técnica Sanepar Curitiba*, **24**(24), 46–60.

Chernicharo C. A. L. (2007) Anaerobic Reactors. IWA Publishing, London, UK.

Chernicharo C. A. L. and Cardoso M. R. (1999). Development and evaluation of a partitioned upflow anaerobic sludge blanket (UASB) reactor for the treatment of domestic sewage from small villages. *Water Science and Technology*, **40**(8), 107–113.

Chernicharo C. A. L., Almeida P. G. S., Lobato L. C. S, Couto T. C. and Borges J. M. (2009). Experience with the design and start up of two full-scale UASB plants in Brazil: enhancements and drawbacks. *Water Science and Technology*, **60**(2), 507–515.

Chernicharo C. A. L., van Lier J. B., Noyola A. and Bressani-Ribeiro T. (2015). Anaerobic sewage treatment: state of the art, constraints and challenges. *Environmental Science Biotechnology*, **14**, 649–679.

Das S., Sarkar S. and Chaudhari S. (2018). Modification of UASB reactor by using CFD simulations for enhanced treatment of municipal sewage. *Water Science and Technology*, **77**, 766–776.

Florencio L., Kato M. T. and Morais J. C. (2001). Domestic sewage treatment in full-scale UASB plant at Mangueira, Recife, Pernambuco. *Water Science and Technology*, **44**(4), 71–77.

Flórez C. A. D. (2016). Avaliação de protocolos de descarte e desidratação da escuma acumulada no interior de separadores trifásicos de reatores UASB. (Evaluation of protocols of withdrawal and dewatering of scum accumulated inside GLS separators of UASB reactors). MSc thesis, Federal University of Minas Gerais, Belo Horizonte, Brazil.

Franco P. L. P. (2010). Análise da Potencialidade do Reuso Indireto Potável: Estudo de Caso da ETE Atuba Sul, Região Metropolitana de Curitiba (Analysis of the potential of indirect drinking reuse: Case study of Atuba Sul WWTP, Curitiba's Metropolitan Region). MSc thesis. Universidade Federal do Paraná, Curitiba/PR, Brazil.

Garcia G. P. P., Souza C. L., Gloria R. M., Silva S. Q. and Chernicharo C. A. L. (2012). Biological oxidation of sulphides by microorganisms present in the scum layer of UASB reactors treating domestic wastewater. *Water Science and Technology*, **66** (9), 1871–1878.

Halalsheh M., Sawajneh Z., Zu'bi M., Zeeman G., van Lier J. B., Fayyad M. and Lettinga G. (2005). Treatment of strong domestic sewage in a 96 m^3 UASB reactor operated at ambient temperatures: two-stage versus single-stage reactor. *Bioresource Technology*, **96**, 577–585.

Khalil N., Mittal A. K., Raghav A. K. and Rajeev S. (2006). UASB technology for sewage treatment in India: 20 years experience. *Environmental Engineering and Management Journal*, **5**, 1059–1069.

Khan A. A., Gaur R. Z., Tyiagi V. K., Khursheed A., Lew B., Mehrotra I. and Kazmi A. A. (2011). Sustainable options of post treatment of UASB effluent treating sewage: a review. Resources, *Conservation and Recycling*, **55**, 1232–1251.

Mahmoud N., Zeeman G., Gijzen H. and Lettinga G. (2004). Anaerobic sewage treatment in a one-stage UASB reactor and a combined UASB-digester system. *Water Research*, **38**, 2348–2358.

Monroy O., Famá G., Meraz M., Montoya L. and Macarie H. (2000). Anaerobic digestion for wastewater treatment in Mexico: state of the technology. *Water Research*, **34**, 1803–1816.

Mungray A. K. and Patel K. (2011). Coliforms removal in two UASB/ASP based systems. *International Biodeterioration & Biodegradation*, **65**, 23–28.

Nada T., Ali H. I., Farid M. N., El-Gohary F. A. and Moawad A. (2006). A full-scale application for anaerobic wastewater treatment technology (up-flow anaerobic sludge blanket UASB in Egypt). Conference National Research Centre, March 13–15, Cairo, Egypt.

Oliveira S. C. and von Sperling M. (2011). Performance evaluation of different wastewater treatment technologies operating in a developing country. *Journal of Water, Sanitation and Hygiene for Development*, **1**, 37–56.

Pandey N. and Dubey S. K. (2014). Up-flow anaerobic sludge bed (UASB) based sewage treatment plant (STP) at Mirzapur: a review. *International Research Journal of Environmental Sciences*, **3**(8), 67–71.

Peña M. R., Rodriguéz J., Mara D. D. and Sepulveda M. (2000). UASBs or anaerobic ponds in warm climates? A preliminary answer from Colombia use of biogas for cogeneration of heat and electricity for local application: performance evaluation of an engine power generator and a sludge thermal dryer. *Water Science and Technology*, **42**(10), 59–65.

Peña M. R., Mara D. D. and Avella G. P. (2006). Dispersion and treatment performance analysis of an UASB reactor under different hydraulic loading rates. *Water Research*, **40**, 445–452.

Pereira J. (2012). Desenvolvimento de reator UASB com duplo estagio de coleta de biogás (DECB) aplicado ao tratamento de esgoto sanitário objetivando o controle da escuma e a granulação do lodo. (Development of a double-stage biogas collection UASB reactor (DEBC) applied to the treatment of domestic wastewater aiming at scum control and sludge granulation), PhD thesis, Federal University of Minas Gerais, Belo Horizonte, Brazil.

Pereira J. O., Celani J. S. S. and Chernicharo C. A. L. (2009). Control of scum accumulation in a double stage biogas collection (DSBC) UASB reactor treating domestic wastewater. *Water Science and Technology*, **59**(6), 1077–1083.

Petropoulos E., Dolfing J., Davenport R. J., Bowen E. J. and Curtis T. P. (2017). Developing cold-adapted biomass for the anaerobic treatment of domestic wastewater at low temperatures (4, 8 and 15°C) with inocula from cold environments. *Water Research*, **112**, 100–109.

Rosa A. P., Lobato L. C. S., Chernicharo C. A. L., Martins D. C. R. B., Maciel F. M. and Borges J. M. (2012). Improving performance and operational control of UASB reactors via proper sludge and scum discharge routines. *Water Practice and Technology*, **7**(3), 11.

Ross B. Z. L. (2015). Escuma de reatores anaerÃ³bios tratando esgotos domÃ©sticos em escala real: ProduÃ§Ã£o, caracterizaÃ§Ã£o e proposiÃ§Ã£o de parÃ¢metros para seu gerenciamento (Scum from full-scale anaerobic reactors treating sewage: production, characterization and management proposal). PhD Thesis, Universidade Federal do ParanÃ¡, Curitiba/PR, Brazil.

Sato N., Okubo T., Onodera T., Ohashi A. and Harada H. (2006). Prospects for a self-sustainable sewage treatment system: a case study on full-scale UASB system in India's Yamuna River Basin. *Journal of Environmental Management*, **80**, 198–207.

Sayed S. K. I. and Fergala M. A. A. (1995). Two-stage UASB concept for treatment of domestic sewage including sludge stabilization process. *Water Science and Technology*, **32**(11), 55–63.

Scalize P. S. (2003). Disposição de resíduos gerados em estações de tratamento de água em estações de tratamento de esgoto (Discharge of water treatment plant residues in sewage treatment plants). PhD thesis, Universidade de São Paulo, São Paulo, Brazil.

Seghezzo L., Trupiano A. P., Liberal V., Todd P. G., Figueroa M. E., Gutiérrez M. A., Da Silva Wilches A. C., Iribarnegaray M., Guerra R. G, Arena A., Cuevas C. M., Zeeman G. and Lettinga G. (2003). Two-step upflow anaerobic sludge bed system for sewage treatment under subtropical conditions with posttreatment in waste stabilization ponds. *Applied Biochemical Biotechnology*, **109**(1–3), 167–180.

Silva F. J. A., Lima M. G. S., Mendonca L. A. R. and Gomes M. J. T. L. (2013). Septic Tank combined with anaerobic filter and conventional UASB-results from full scale plants. *Brazilian Journal of Chemical Engineering*, **30**(1), 133–140.

Souza C. L., Silva S. Q., Aquino S. F. and Chernicharo C. A. L. (2006). Production and characterization of scum and its role in odour control in UASB reactors treating domestic wastewater. *Water Science and Technology*, **54**(9), 201–208.

Tachini M., Filho P. B. and Pinheiro A. (2006). Evaluation of integrated treatment of sanitary sewage and septic tank sludge in a RALF: case study. *Sanare Revista Técnica Sanepar Curitiba*, **24**, 46–60.

van Haandel A. C. and Lettinga G. (1994). Anaerobic Sewage Treatment: A Practical Guide for Regions with a Hot Climate, John Wiley, Chichester, UK, 198pp.

van Haandel A., Kato M. T., Cavalcanti P. F. F. and Florencio L. (2006). Anaerobic reactor design concepts for the treatment of domestic wastewater. *Reviews in Environmental Science and Bio/Technology*, **5**(1), 21–38.

van der Last A. R. M. and Lettinga G. (1992). Anaerobic treatment of domestic sewage under moderate climatic (Dutch) conditions using upflow reactors at increased superficial velocities. *Water Science and Technology*, **25**(7), 167–178.

van Lier J. B., Vashi A., van der Lubbe J. and Heffernan B. (2010). Anaerobic sewage treatment using UASB reactors: engineering and operational aspects. In: Environmental Anaerobic Technology, H. H. Fang (ed.), Imperial College Press, London, UK, pp. 59–89.

von Sperling M. and Chernicharo C. A. L. (2005). Biological Wastewater Treatment in Warm Climate Regions. Two volumes. IWA Publishing. Available at http://www.iwapublishing.com/open-access-ebooks/3567.

Walia R., Kumar P. and Mehrotra I. (2011). Performance of UASB based sewage treatment plant in India: polishing by diffusers an alternative. *Water Science and Technology*, **63**(4), 680–688.

WERF (2010). Energy efficiency in wastewater treatment in North America: a compendium of best practices and case studies of novel approaches. Final report. Water Environment Research Foundation, Alexandria, VA, USA.

Chapter 5

Construction of UASB reactors for sewage treatment

Carlos A. L. Chernicharo, Thiago Bressani-Ribeiro and Lívia C. S. Lobato

Department of Sanitary and Environmental Engineering (DESA), Universidade Federal de Minas Gerais (UFMG), Av. Antonio Carlos, 6.627, Campus Pampulha, Belo Horizonte 31270-901, Brazil. Corresponding author: calemos@desa.ufmg.br

5.1 INTRODUCTION

As presented in the previous chapters, one of the advantages associated with the treatment of sewage in UASB reactors relates to the constructional and operational simplicity of the process. The simplicity of the construction is ascribed to the extremely low (or non-existent) mechanisation levels and the absence of packing media inside the reactor. However, in practice, recurring problems have been observed that are inherent to the construction of these reactors, which is possibly because the application of the technology for sewage treatment is relatively recent. The first full-scale reactors date from the early 1980s, but it is believed that the problems and construction failures related to the implementation of the technology have not served as learning opportunities, and little has been done to improve relevant features, such as materials, methods, and construction supervision. These factors alone or in combination can be identified as the main causes of most problems that have been observed in several full-scale reactors, as discussed in the following sections.

Generally, most key aspects inherent to the construction of UASB reactors for sewage treatment are related directly or indirectly to the gaseous phase, and specifically to hydrogen sulphide (H_2S) (because of its corrosive properties), as detailed in the following sections. However, several additional conditions must be observed during the construction of UASB reactors, which are also presented and discussed in this chapter.

Editor(s): Carlos Augusto de Lemos Chernicharo and Thiago Bressani-Ribeiro
doi: 10.2166/9781780409238_0112

5.2 KEY ASPECTS IN THE CONSTRUCTION OF UASB REACTORS

5.2.1 Corrosion of concrete and metallic structures

5.2.1.1 Preliminaries

The corrosion problems in UASB-based STPs are associated mainly with the H_2S generated during the anaerobic treatment process (see Chapters 3 and 8). A large fraction of this H_2S remains dissolved in the bulk liquid, while another fraction is released into the gaseous atmosphere. The H_2S emissions from the liquid phase can occur either punctually (into the gas–liquid–solid (GLS) separator, thereby becoming one of the constituents of the biogas) or diffusely (along the course of the liquid effluent). Accordingly, the origin of the corrosion problems can be related to both the *biogas collection and treatment lines* and the *diffuse emissions from the liquid phase.*

When emitted to the atmosphere, H_2S can cause severe corrosion in structures made of reinforced concrete and in materials and equipment made from carbon steel and cast iron, because in the presence of moisture and small amounts of oxygen, the sulphide can be oxidised to sulphuric acid, which has corrosive properties. Splitting boxes, spillways, and other points that cause turbulence in the bulk liquid favour the release of H_2S, which consequently leads to the corrosion of concrete and metallic structures in the vicinity of these release points. Examples of materials and structures most prone to corrosion are splitting boxes, walls, slabs, and inspection caps located on the top of the reactors; stairways, walkways, guard rails, metal fences, and pipes in general; and materials and equipment used in the biogas collection and transport line (Figure 5.1(a–f)). In addition, corrosion (and odour) problems can occur in sewage pumping stations, preliminary treatment units, and sludge handling units, among others.

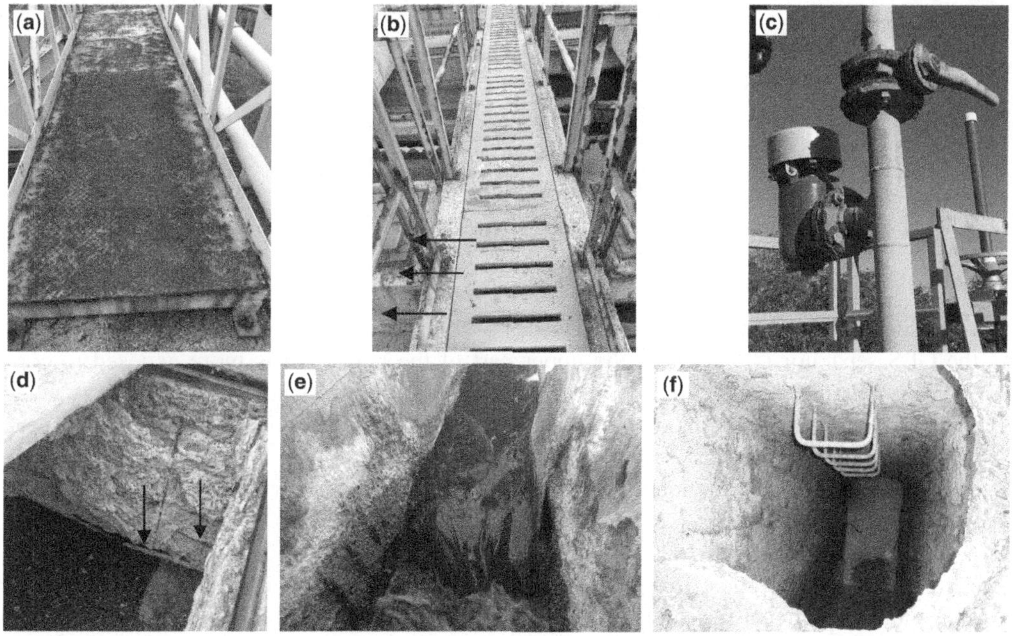

Figure 5.1 View of corroded structures of UASB reactors: (a) metallic walkway, (b) metallic guard rail, (c) biogas pipeline in carbon steel, (d) inspection hatch of a GLS separator in concrete (exposed steel bars), (e) main effluent collection channel (exposed steel bar), and (f) manhole of the effluent collection channel (corroded walls).

Accordingly, corrosion problems (and odour emissions) are related to four main factors, namely (i) sites that cause turbulence along the hydraulic profile of the STP and enable the release of H_2S dissolved in the liquid phase, (ii) materials and coatings that are not resistant to corrosion, (iii) poor gas chamber tightness and inadequate handling of reactor inspection hatches, and (iv) low combustion efficiency of some types of biogas burners.

5.2.1.2 Main construction and design concerns

(a) **Biogas line**
The main recommendations to minimise corrosion problems in the biogas collection lines were addressed in Chapter 4. In regard to the use of high-density polyethylene (HDPE), owing to the low structural stability of the plastic, it is important to install a sufficient number of supports to prevent undesirable curvatures and points of condensate accumulation (Silveira *et al.*, 2015). Furthermore, local standards must be observed, such as the Brazilian standard NBR 12209 (ABNT, 2011). Figure 5.2(a–c) shows biogas pipes made of stainless steel and HDPE.

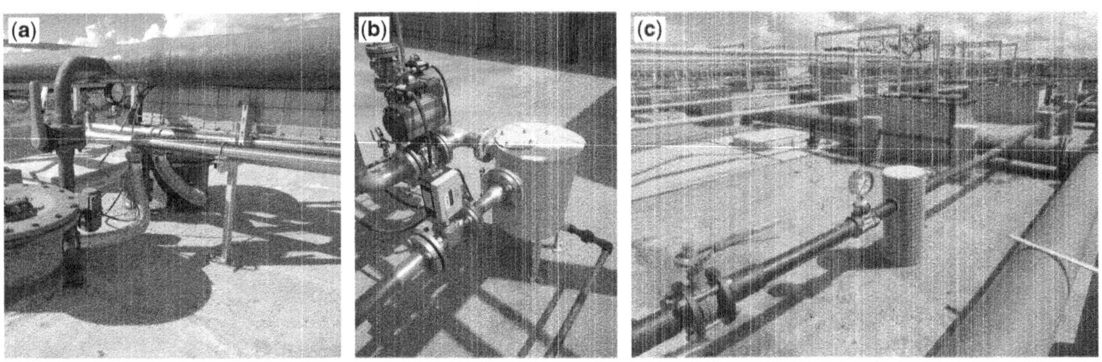

Figure 5.2 View of biogas lines manufactured from stainless steel (a, b) and HDPE (c).

(b) **Gas–liquid–solid separators (GLS separators)**
As the GLS separator is located in the upper part of the reactor, it is more exposed to corrosion; therefore, it must be manufactured from material that is more resistant or appropriately coated. As GLS separators made from concrete are not only heavy and bulky, but also prone to gas leaks and corrosion, alternative materials that are corrosion-resistant and have smaller volumes and weights, such as glass fibre, polypropylene, and HDPE (see Chapter 4), have been gaining importance in recent years (Figure 5.3(a–c)).

(c) **Reinforced concrete structures**
Concrete is the material most frequently applied in the construction of UASB reactors (Figure 5.4 (a–f)). Suitable chemical-resistant concrete must be used and anticorrosion protection of the structure must be applied from the start of the construction phase. Several aspects must be considered carefully in an effort to achieve low rates of absorption and permeability, namely using concrete with a low water/cement ratio, rigorous vibration of the concrete, applying appropriate healing processes, and choosing a suitable cement type, such as Portland Pozzolanic.

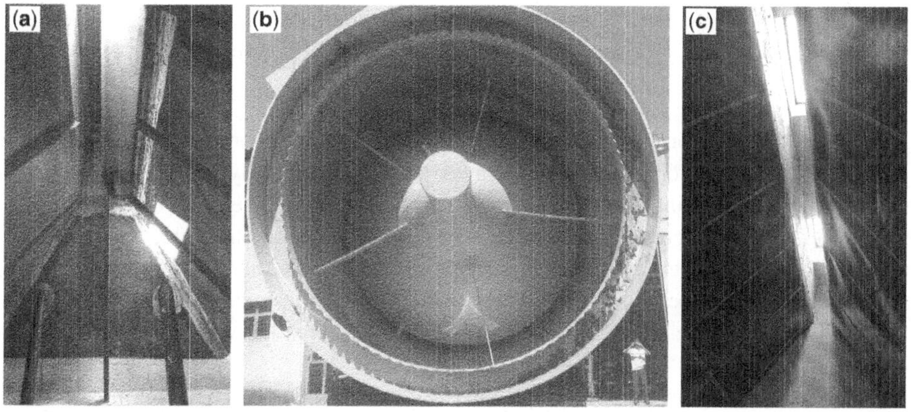

Figure 5.3 View of GLS separators manufactured from (a) glass fibre (bottom view), (b) polypropylene (top view) and (c) HDPE (bottom view).

Figure 5.4 View of a UASB reactor under construction: (a) steel framework of the bottom slab; (b) concrete floor slab; (c) sidewall steel framework; (d) steel framework of the side walls and gas deflectors; (e) completed main reactor structure; and (f) details of the general effluent collection channel and inserts for receiving the feeding distribution pipes.

Failure to observe these aspects in terms of the specifications of the materials and the application techniques can lead to a significant reduction in the useful life of the concrete structure, as explained below:

- A high water/cement factor would result in a structure with greater porosity/capillarity because of the exit of water that is not chemically associated (water added only to provide workability to the material), thereby facilitating the entry of aggressive agents into the concrete.
- Poor vibration of the concrete can result in a non-homogeneous structure, which can facilitate the penetration of water and gases inside this material, and consequently the attack and oxidation of the steel framework. In the concrete, the products of corrosion occupy volumes 3–10 times that of the original volume of the steel, thereby causing internal tensions with values higher than 40 MPa (Cánovas, 1988).
- An inappropriate healing process would produce concrete with cracks because of hydraulic retraction. These would enable the penetration of water and gases and the attack and oxidation of the steel framework.
- Using an inappropriate type of cement would reduce the resistance of the concrete to chemical attacks from the aggressive agents present in sewage (i.e. H_2S).

Additionally, applying acid-resistant coatings to concrete structures, especially in areas above the liquid level (where oxygen can enter), improves or inhibits the effects of corrosion. Coating the reactor with chlorinated rubber (minimum thickness of 4 mm) or bituminous epoxy (minimum thickness of 400 μm) are possible solutions. Although chlorinated rubber is cheaper, it has low acid resistance, whereas epoxy paint is more expensive but more resistant to acids and fewer coats are required (Fortunato *et al.*, 1998). More recently, elastomeric polyurethane coatings have been used successfully, sometimes in conjunction with geotextile blankets (Figure 5.5(a)). Glass fibre has also been applied successfully as a coating for concrete (Figure 5.5(b)) and steel structures.

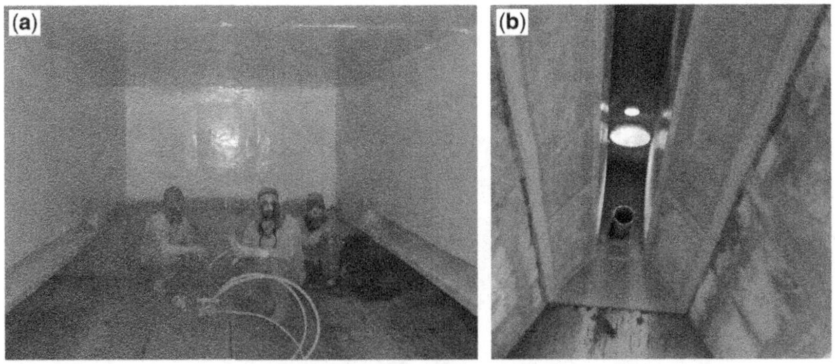

Figure 5.5 Internal coatings applied in UASB reactors: (a) application of elastomeric polyurethane in conjunction with geotextile blankets; and (b) glass fibre.

(d) Steel structures

In reactors manufactured from steel (Figure 5.6(a–c)), coatings are widely used as an anticorrosion protection method, but with the necessary strict control over the type of coating and type of steel. In this regard, a thick epoxy coating (>400 μm) applied after abrasive surface treatment is indicated, and stainless steel containing molybdenum (AISI 316 L) is more corrosion-resistant.

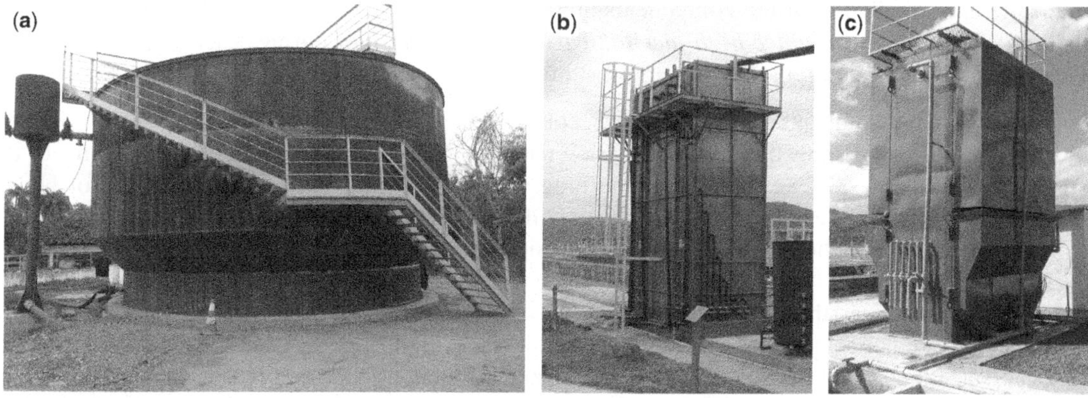

Figure 5.6 View of UASB reactors constructed from carbon steel with a cylindrical shape (a) and rectangular shape (b, c).

5.2.2 Collapse of GLS separators

5.2.2.1 Preliminaries

The thickening and solidification of the scum layer inside the GLS separator can impede the natural passage of biogas (see Chapter 6) and lead to an increase in the pressure under the thick layer of scum. This causes structural stress to the walls of the GLS separator. Depending on the magnitude of such stresses and the structural quality of the walls, unlevelling, twisting, and rupture of the structure can occur (Figure 5.7(a, b)). The propensity of the GLS separator to rupture is associated strongly with the types of material used, particularly when the separators are made from materials with lower resistance, such as plastic canvas, asbestos sheets, and glass fibre of inferior quality.

Figure 5.7 Effects of the accumulation of solidified scum inside gas–liquid–solid (GLS) separators: (a) twisting of a glass fibre GLS separator; and (b) structural collapse of a glass fibre GLS separator.

5.2.2.2 Main construction and design concerns

The main design and construction safety measures to prevent damage to the structure of GLS separators are choosing appropriate construction materials and complying with the strict technical specifications for the manufacture and installation of the components and accessories of the separators. The chosen materials and specifications must ***guarantee high mechanical/structural strength*** during reactor operations when the GLS separators will be exposed to adverse conditions, such as the following:

- A corrosive environment, as discussed in Section 5.2.1.
- Internal biogas pressures that can reach a few tens of millibars, especially in case of failures of the pressure-breaking devices installed in the biogas collection line. The internal pressure of biogas acts in an upward direction and can cause the GLS separators to fluctuate, twist, or rupture if they are not properly anchored or do not have enough resistance to withstand the working pressures (see Section 5.2.3).
- Mechanical stresses that can be caused by the accumulation of scum, especially when the GLS separators are not equipped with scum removal devices (see Chapters 4 and Chapters 6). However, even when the reactor does have a scum removal device, additional care is still required during the design and construction phases to ensure the strength and integrity of the GLS separators to withstand the accumulation of dense and solidified scum.

5.2.3 Sealing of GLS separators

5.2.3.1 Preliminaries

In addition to the concerns about structural resistance, particular attention to the design and construction of GLS separators is needed to ensure the complete tightness of the separators, as recurrent problems of biogas leakage have been verified in several full-scale anaerobic reactors (Chernicharo *et al.*, 2018). Biogas leaks can cause odour problems (owing to H_2S) and greenhouse gas emissions (owing to CH_4), as discussed in Chapter 9, as well as the loss of energy potential, as discussed in Chapter 8. In this regard, special attention is required not only to the manufacture and installation of the GLS separators, but also to the entire gas line.

5.2.3.2 Main construction and design concerns

To guarantee both resistance and perfect gas tightness throughout the operational life of the reactors, the construction materials have to be chosen carefully and the manufacture and installation of the parts and accessories of GLS separators have to comply with the technical specifications. The GLS separators must be resistant to solar radiation, corrosive environments, and pressures within the reactor gas chambers. Although the working pressures within GLS separators under normal UASB reactor operating conditions should not exceed 15–20 mbar (15–20 cm w.c), it is crucial that pressure and sealing tests are conducted for values at least 50% higher than those predicted in the design. Accordingly, the construction specifications should clearly establish and stipulate the requirement for reactor pressurisation tests to determine the leakage conditions of the GLS separators, inspection hatches, and the entire gas line.

The results of sealing tests performed on four full-scale UASB reactors are shown in Figure 5.8. The reactors were filled with water up to the level of the effluent collection channel, and then the GLS separators were pressurised with the aid of a compressor until the desired test pressure was reached (in this instance, 25 mbar). Pressure monitoring was conducted for 60 min and measurements were taken every 15 min using a manometer with a scale of 0 mbar to 100 mbar.

The tests indicated high pressure losses in the GLS separators and gas lines of reactors 1 and 3 (4.5 and 5.5 mbar h^{-1}, respectively), whereas reactors 2 and 4 showed pressure drops of only 0.6 and 1.0 mbar h^{-1}, respectively. From these results, further work was conducted to identify the locations where gas losses were occurring by applying foam along the area most prone to leaks. The most critical leakage points were the connections between the GLS separators and the biogas collection pipeline. However, other points that have been proven to be critical for leaks are the inspection compartment covers located on the top of the GLS separators.

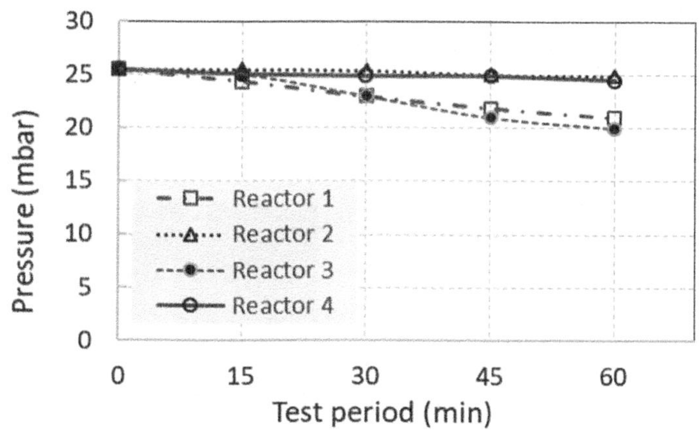

Figure 5.8 Results of the sealing tests performed in GLS separators and gas lines of four full-scale UASB reactors.

5.2.4 Installation of influent distribution pipes

5.2.4.1 Preliminaries

As discussed in Chapter 4 (Section 4.4.3), the influent distribution systems of UASB reactors treating sewage are crucial to improving the mixing conditions within the reactor, as well as the performance of the reactor in general. The crucial factors are the material, size (diameter), shape (avoidance of sharp bends), correct positioning, and robust anchoring (fastening) of the distribution pipes. Problems associated with the material, size, and shape of the distribution pipes have not been reported very often, but the opposite has been noticed with the positioning, alignment, and fastening of such pipes in some UASB reactors. They become apparent either during construction or after a few years of operation when the reactor is emptied for inspection and maintenance, as shown in Figure 5.9(a–d).

5.2.4.2 Main construction and design concerns

The design guidelines proposed in Chapter 4 (Section 4.4.3) should be followed strictly during the construction stage to prevent problems related to the installation of the distribution pipes. Close attention should be paid not only to the correct alignment and positioning of the end part of the pipe, but also to the anchoring (fastening) of the pipeline both on the top and bottom of the UASB reactor. This is particularly important, as the distribution pipes will be exposed to stresses and movements during clearing (un-blocking) operations. Depending on the extent of the blockage, the pipes can be exposed to severe stresses and can become loose if not properly anchored. An efficient influent distribution system can be constructed in various ways. An example is shown in Figure 5.10(a, b).

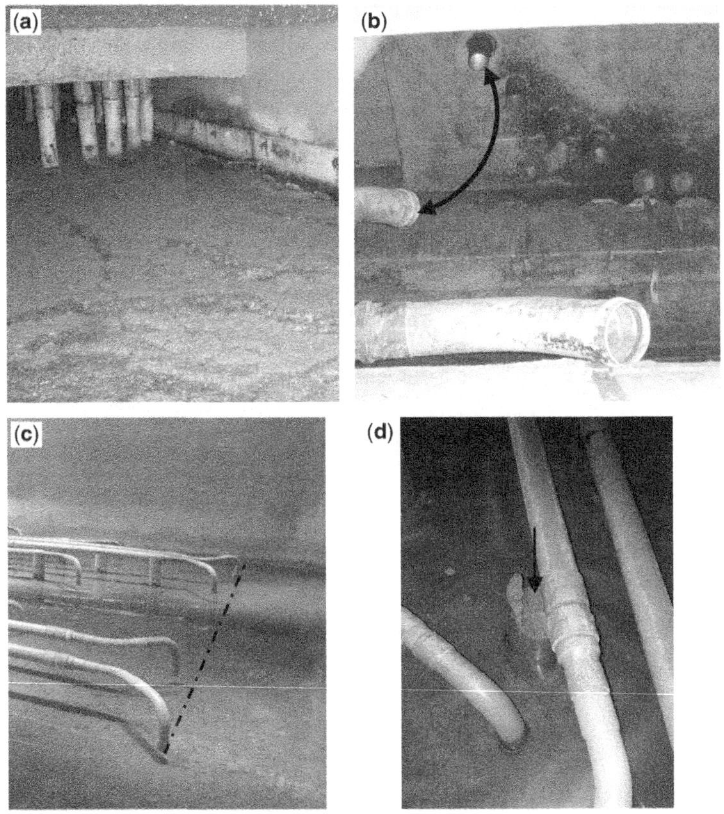

Figure 5.9 Incorrect installation of distribution pipes on the top (a, b) and bottom (c, d) of a UASB reactor: (a) inadequate fixing between the insert tube from the splitting box (above the reactor) and the polyvinyl chloride (PVC) feeding pipe; (b) a similar problem to that shown in (a) after a few years of operation; (c) incorrect positioning and alignment of PVC feeding pipes; and (d) flimsy anchoring (loose pipe).

Figure 5.10 Proper installation of distribution pipes on the bottom of a UASB reactor under construction: (a) correct positioning and alignment and (b) robust anchoring (fastening).

5.2.5 Installation of sludge sampling points

5.2.5.1 Preliminaries

The correct positioning of the sludge sampling points is crucial to the proper operation of UASB reactors treating sewage. This is because operators evaluate the solids concentration and amount of biomass present along the height of the reactor by analysing the sludge collected via the sampling points. The positioning (installation) of the ***uppermost sampling point*** (the highest point of the digestion compartment) is of particular interest, as it must be placed at the correct height (***immediately before the gas deflector***) for the operator to monitor whether there is excess sludge inside the reactor (see Chapter 6 Section 6.3.3). Accordingly, it is essential that the number, material, size (diameter), and positioning of the sampling points comply with the design, as discussed in Chapter 4 (Section 4.4.6). Although problems with the material and size of the sampling points have not been reported often, incorrect positioning is common, as shown in Figure 5.11(a, b).

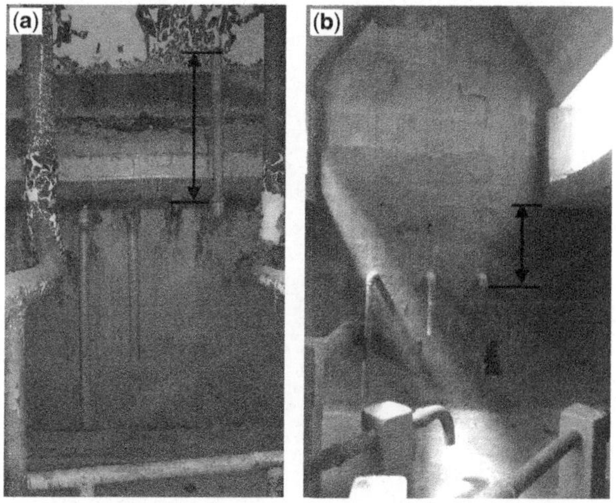

Figure 5.11 Incorrect installation of the uppermost sampling point along the height of UASB reactors: (a) positioning well above the gas deflector and (b) positioning well below the gas deflector.

5.2.5.2 Main construction and design concerns

As discussed previously, because of the importance of the correct installation of the sludge sampling points, the design guidelines presented in Chapter 4 (Section 4.4.6) should be followed strictly during the construction stage. Therefore, the operator will be able to determine the proper time for sludge withdrawals. An example of proper sludge sampling points is shown in Figure 5.12.

5.2.6 Installation of effluent collection channels

5.2.6.1 Preliminaries

There are two concerns related to the installation of the effluent collection channels, one being the fixing of the device and the other its correct levelling, both of which cause scum accumulation on the surface of settler compartments of UASB reactors, as detailed below:

- Incorrect levelling of the effluent collection channel (Figure 5.13(a)). In this instance, the channel was installed a few centimetres (2–3 cm) above the level of the other collection channel of the same reactor; therefore, the floating particle materials in the effluent were prevented from flowing into the effluent collection channel and caused the formation and accumulation of scum.

- Incorrect positioning and fixing of the glass fibre spillway (Figure 5.13(b)). In this instance, the scum level formed below the 'v' notch of the spillway, thereby indicating that the liquid fraction was flowing between the glass fibre plate and concrete channel. In addition to probable incorrect levelling, this problem was mainly caused by the inadequate fixing of the pouring plate.

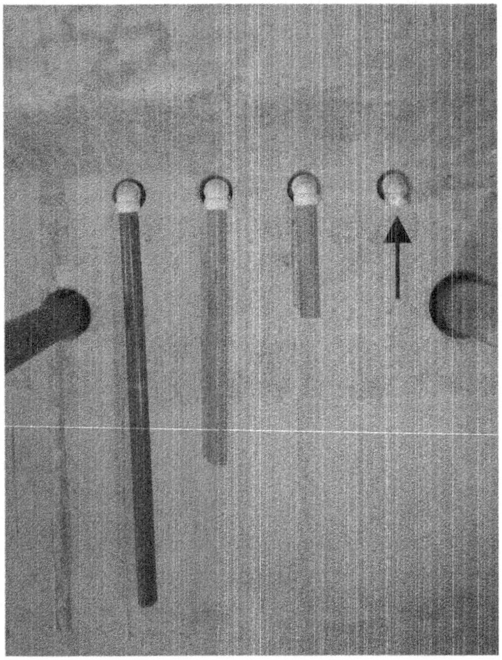

Figure 5.12 Proper installation of sludge sampling points along the height of the digestion compartment of a UASB reactor (uppermost sampling point immediately below the lower part of the gas deflector).

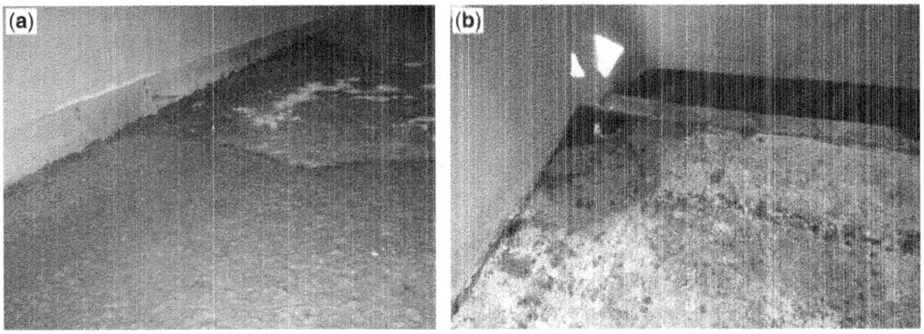

Figure 5.13 Incorrect installation of effluent collection channels in UASB reactors: (a) collection channel installed 2–3 cm above the water level of the reactor and (b) inadequate sealing of the glass fibre spillway.

5.2.6.2 Main construction and design concerns

Problems related to the installation of effluent collection channels are reasonably easy to avoid, but the construction supervisor must be extremely attentive to the correct positioning, fixing, and levelling of the effluent collection channels. The following corrective measures can be employed if constructions problems are noticed:

- Using elastomers (or silicone fillers) between the spillways and the concrete walls to prevent liquid from entering between the two structures (Figure 5.14(a)).
- Using supports of adequate strength to prevent twisting of collection channels made of thinner and less resistant materials, such as aluminium, glass fibre, and polypropylene (Figure 5.14(b)).
- Ensuring effective and accurate levelling of collection channels along their length and between the channels of the same reactor.

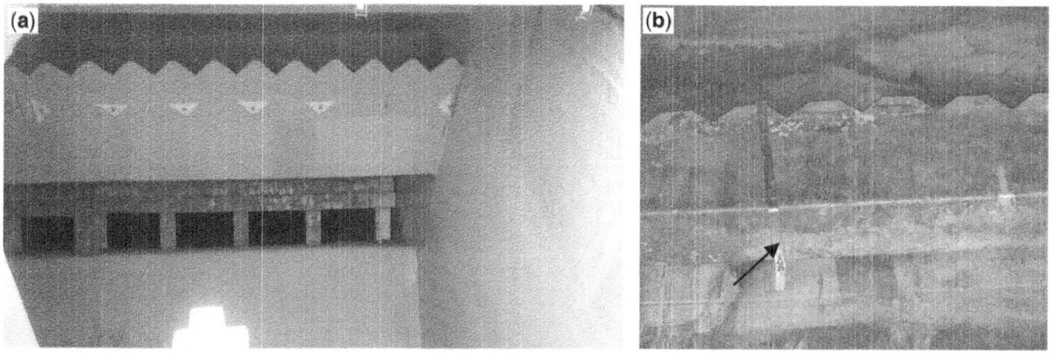

Figure 5.14 Proper installation of effluent collection channels in UASB reactors: (a) use of a properly sealed spillway and (b) use of a stainless-steel support to prevent twisting of the aluminium collection channel.

The effectiveness of the sealing and levelling procedures needs to be checked carefully when the reactors are filled with water for leak testing.

5.3 OTHER CONCERNS

5.3.1 Water accumulation on top of the reactor slab

Flat slabs can promote the accumulation of rainwater (Figure 5.15(a)), which can occasionally penetrate the concrete and cause the detachment of any existing protective coating underneath the slab. Proper and adequate measures should be taken to prevent this, one of which is constructing the slab with an adequate slope to drain the water to a side gutter (Figure 5.15(b)).

5.3.2 Leakages in general

Failure to observe the concrete specifications highlighted in Section 5.2.1.2(c) can lead to liquid and gas leakages, especially in areas of pipe inserts, such as the ones for biogas collection (Figure 5.16(a)) and sludge withdrawal (Figure 5.16(b)). Proper and adequate measures should be taken to prevent such liquid and gas leakages, such as using elastomeric sealing systems (Figure 5.16(c)).

Figure 5.15 Accumulation of rainwater on the top of a flat slab of a UASB reactor (a) and rainwater collection gutter of a sloped slab (b).

Figure 5.16 Leakages due to defective inserts on concrete structures (a, b) and use of an elastomeric sealing system (c): (a) repaired area of a biogas inspection hatch on a concrete slab; (b) water leakage next to a sludge pipe; and (c) use of Link-seal® modular seal in a reactor wall.

5.4 FINAL REMARKS

The problems reported, discussed, and illustrated in this chapter indicate that such challenges could be more common than anticipated. The lesson to be learned is that employing skilled construction supervisors who are experts on structural building aspects is not enough if they do not have at least a minimum understanding of UASB reactor operations, particularly when applied to sewage treatment. Even minimal deviations from the design without the prior approval of the engineer responsible have been shown to seriously compromise the operation and performance of the entire treatment system.

REFERENCES

ABNT (Associação Brasileira de Normas Técnicas) (2011). NBR 12209 – Elaboração de projetos hidráulico-sanitários de estações de tratamento de esgoto (Elaboration of hydraulic-sanitary designs of sewage treatment plants). ABNT, Rio de Janeiro, Brazil.

Cánovas M. F. (1988). Patologia e Terapia do Concreto Armado (Pathology and Therapy of Reinforced Concrete). Pini, São Paulo, Brazil, 522pp.

Chernicharo C. A. L., Bressani-Ribeiro T., Pegorini E., Possetti G. R. C., Miki M. K. and Nonato S. (2018). Contribuição para o aprimoramento de projeto, construção e operação de reatores UASB aplicados ao tratamento de esgoto sanitário – Parte 1: Tópicos de Interesse (Contribution to the improvement of the design, construction and operation of UASB reactors applied to sewage treatment – Part 1: Topics of interest). *Revista DAE*, **66**(214), 5–15.

Fortunato C. M., Gomes C. S., Andreoli F. N. and Aisse M. M. (1998). Monitoramento de reatores anaeróbios tipo RALF- Relatório n.5: Agressividade de sulfetos ao concreto dos RALFs (Monitoring of Anaerobic reactor type RALF – Assessment n.5: Sulphide attack on concrete). Instituto de Saneamento Ambiental, Pontifícia Universidade Católica do Paraná, Brazil.

Silveira B., Chernicharo C. A. L., Cabral C., Suidedos C., Platzer C., Silva G., Possetti G. R. C., Hoffmann H., Moreira H. C., Adamatti H. B., Miki M., Silva M., Takahashi R., Miki R., Rosenfeldt S., Araújo V. S. F., Bustani V. V. and Villani W. (2015). Guia técnico de aproveitamento energético de biogás em estações de tratamento de esgoto (Technical guide on biogas recovery from sewage treatment plants. Deutsche Gesellschaft für Internationale Zusammenarbeit GmbH & Ministério das Cidades, Brasília, Brazil, 183pp.

Chapter 6

Operation of UASB reactors for sewage treatment

Carlos A. L. Chernicharo, Thiago Bressani-Ribeiro and Lívia C. S. Lobato

Department of Sanitary and Environmental Engineering (DESA), Universidade Federal de Minas Gerais (UFMG), Av. Antonio Carlos, 6.627, Campus Pampulha, 31270-901, Belo Horizonte, Brazil
Corresponding author: calemos@desa.ufmg.br

6.1 INTRODUCTION

The operational simplicity of anaerobic treatment systems is one of its main advantages; however, the presence of operation and maintenance personnel is crucial to ensure that the systems function adequately. In this regard, three main control activities are necessary (Chernicharo, 2007):

- Operation: refers to the daily or periodic activities needed to assure the adequate and stable performance of the treatment system.
- Maintenance: refers to the activities to maintain the structures in the treatment plant in good condition.
- Information: refers to the communication, preferably in writing, between the members of the operation and maintenance teams, which, at the same time, creates records of the performance of the treatment system.

Adequate operational control includes aspects related to the *optimisation* of operational conditions and the *health and safety* of workers, as discussed in Chernicharo (2007).

This chapter focuses only on the operational aspects of anaerobic-based sewage treatment plants (STPs) and, particularly, on the activities related to the monitoring and management of the preliminary treatment unit, upflow anaerobic sludge blanket (UASB) reactor, and sludge dewatering unit.

Editor(s): Carlos Augusto de Lemos Chernicharo and Thiago Bressani-Ribeiro
doi: 10.2166/9781780409238_0126

6.2 MONITORING OF ANAEROBIC-BASED SEWAGE TREATMENT PLANTS

6.2.1 Preliminaries

The effective operational control of any anaerobic-based STP would be achieved only if a proper monitoring programme were implemented. This would enable both the verification of the operational parameters and the optimisation of the operational routine. An adequate monitoring programme should be broad enough to include all the aspects relevant to the operation of the treatment system, without disregarding the local realities and the availability of human resources and material. The operational activities of anaerobic-based STPs are related to the different parts of the treatment system, and can be divided into four groups (adapted from Chernicharo, 2007):

- Activities to ensure the appropriate operation of the pre-treatment units, usually consisting of: (i) screens; (ii) grit chamber; and (iii) flow measuring device (usually a Parshall flume coupled to the grit chamber in small-scale and medium-scale STPs).
- Activities to evaluate the efficiency of the anaerobic reactor. The anaerobic stage is applied primarily for the removal of organic matter, but marginally reduces pathogenic organisms, as well.
- Activities to evaluate the operational stability of the anaerobic reactor, that is, to establish whether there is any risk of pH being reduced to a value lower than the minimum for optimum methanogenesis ($pH_{min} = 6.5$). However, as addressed in Chapter 3, sewage alkalinity is usually sufficient to buffer the pH at values close to 7.
- Activities to determine the amount and quality of the sludge in the reactor and in the sludge dewatering unit. The amount of sludge in the reactor is important in determining the excess sludge withdrawal moment.

Generally, the different groups of activities are conducted by different people, and the activities related to the pre-treatment system require the frequent presence of personnel to verify whether there are any blockages. Usually, in small-scale STPs, the coarse solids and grit collected in the pre-treatment units and the dewatered sludge from the drying beds are removed manually by unskilled workers. However, qualified personnel are required for sampling of the biological treatment system and conducting analyses to verify the treatment efficiency, operational stability, and the sludge mass in the reactor.

6.2.2 Monitoring programme

Figure 6.1 shows a typical flowsheet of an anaerobic-based STP, comprising preliminary treatment, anaerobic reactor, post-treatment, and sludge dewatering unit. The monitoring and operational procedures should cover the entire plant, as is discussed in the following sections. Chernicharo (2007) provides a detailed discussion of the monitoring points and monitoring frequency.

6.2.2.1 Monitoring and operation of preliminary treatment

The satisfactory operation of the anaerobic reactor depends fundamentally on the adequate and accurate operation of the preliminary treatment units. This factor is even more important in developing countries, where the influent to the STP usually includes various improper materials that normally should not be part of the sewage or are present in amounts exceeding the accepted limits. In particular, these undesirable materials include various types of debris (Figure 6.2(a)) and grit (Figure 6.2(b, c)). Consequently, it is crucial to establish an operational routine of regular cleaning of the screens and grit chambers to ensure the effective removal of the coarse solids and grit present in the raw sewage. The cleaning frequency depends on the amount of coarse solids and grit contained in the raw sewage, which, in turn, is related closely to the habits of the population and the characteristics of the contribution basin.

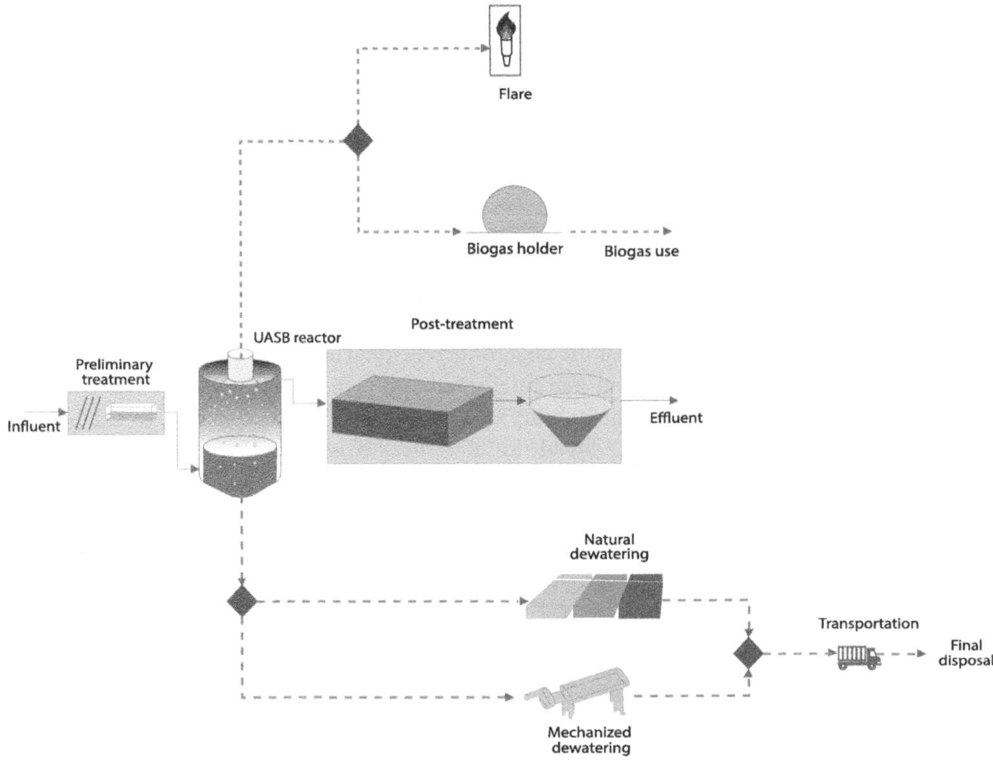

Figure 6.1 Typical flowsheet of an anaerobic-based STP with preliminary treatment, anaerobic reactor, post-treatment, and sludge dewatering unit.

Figure 6.2 Presence of debris and grit in the influent of a large-scale anaerobic-based STP: (a) debris at the inlet tank, prior to the coarse screen, (b) accumulation of grit in a grit chamber (in maintenance), and (c) accumulation of grit in a pumping station (in maintenance).

Suggestions on the points, parameters, and frequency of monitoring of the preliminary treatment units are available in Chernicharo (2007).

6.2.2.2 Monitoring of the anaerobic reactor

The successful operation of an anaerobic reactor depends on the systematisation and implementation of appropriate operational procedures during the start-up phase and over the steady-state operation. Three types of monitoring of the UASB reactor can be emphasised (adapted from Chernicharo, 2007):

(1) *Monitoring of the efficiency of the reactor*: The historical behaviour of the unit and whether its performance is in accordance with the design specifications are identified by monitoring the liquid phase at the inlet and outlet of the anaerobic reactor. As only marginal removal of pathogens and nutrients (nitrogen and phosphorus) is expected, the monitoring programme is focused mainly on the removal efficiency of solids and organic matter by determining their concentrations in the influent and effluent of the UASB reactor. The main parameters to be monitored are (i) *solids:* total suspended solids (TSS), volatile suspended solids (VSS), and settleable solids; and (ii) *organic matter*: chemical oxygen demand (COD) and biochemical oxygen demand (BOD). In addition, the biogas (methane) production is a useful parameter to monitor the efficiency of the reactor.

(2) *Monitoring of the stability of the reactor*: Monitoring of the operational stability of the treatment system aims at evaluating whether there are signs that the acid fermentation can prevail over the methanogenic fermentation, with the consequent acidification of the reactor, particularly in instances of industrial wastewater contribution. Therefore, it is important to determine pH, alkalinity, and the concentration of volatile acids in the effluent, and to compare these values with those in the influent. In addition, a sudden variation in the biogas composition and, particularly, an increased percentage of carbon dioxide can be an indication of operational instability.

(3) *Monitoring of the sludge quantity and quality*: In addition to monitoring the efficiency and stability of the reactor, tests should be performed to establish the quantitative and qualitative development of the sludge in the UASB reactor. As the concentration of sludge will not be uniform along the height of the reactor (vertical solids profile), samples should be taken at different heights to facilitate determining the solids profile from the bottom to the upper part of the digestion compartment.

Therefore, sludge samples collected at different heights in the reactor (Figure 6.3(a, b)) are analysed gravimetrically, and the results are expressed in terms of gram of TS and TVS per litre. These concentration values (for each of the sludge sampling points along the reactor height) multiplied by the volumes corresponding to each sampled zone provide the mass of the biomass (TVS) and sludge (TS) along the reactor profile. The sum of the biomass and sludge quantities in each zone is equal to the total mass of TVS and TS in the reactor. A detailed description of the solids profile determination is presented in Chernicharo (2007).

As regards the qualitative aspects of the sludge, the most important factors are (adapted from Chernicharo, 2007):

- *Specific methanogenic activity (SMA)*: reflects the capacity of the sludge to produce methane from an acetate substrate under optimised conditions. The test is performed in the laboratory according to standard procedures (Chernicharo, 2007). Once the SMA and the quantity of biomass present in the reactor are known, it is possible to estimate the maximum organic load that can be digested in the reactor. This load is equal to the product of the SMA value ($gCOD_{CH4}$ $gTVS^{-1}$ d^{-1}) multiplied by the total amount of biomass (gTVS).

- *Stability*: aims at establishing which fraction of the sludge mass consists of still undigested biodegradable organic matter. A large fraction of biodegradable material in the sludge is not only an indication of an overloaded system but also can cause significant problems in the separation of solids and liquid in the excess sludge withdrawn.
- *Settleability*: can be determined by employing a specific test described by Catunda and Van Haandel (1987). This test is time consuming and complicated, and the application of a simpler, although less accurate method, namely, the determination of the Sludge Volume Index (SVI) or the Diluted Sludge Volume Index (DSVI) is preferable in the operational routine.

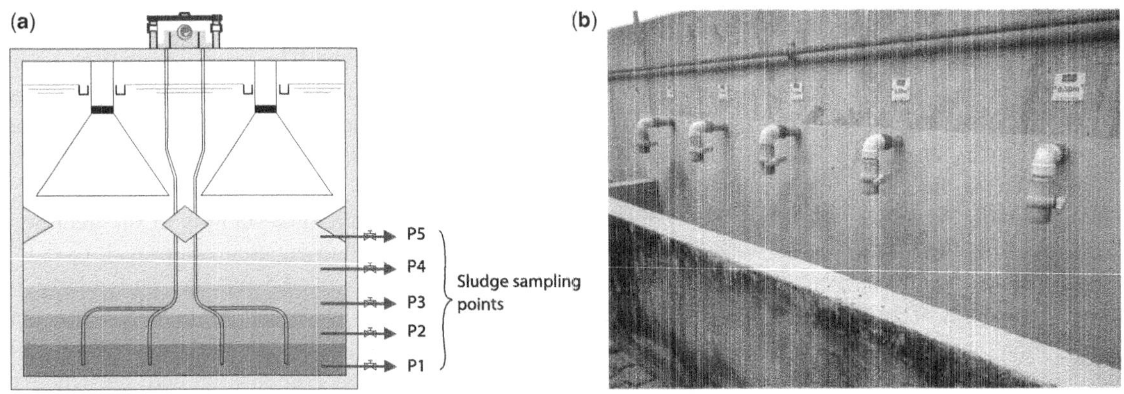

Figure 6.3 Sludge sampling points along the height of UASB reactors: (a) schematic configuration (and distribution); and (b) external view.

Information regarding the points, parameters, and frequency of monitoring of UASB reactors is available in Chernicharo (2007).

6.2.2.3 Monitoring of the dewatering unit

Adequate monitoring of the dewatering unit is essential to achieving optimised operational conditions for the entire treatment plant. This is because the function of the sludge dewatering unit is associated not only with reducing the volume of dry sludge to be transported to the final disposal site, but also with the quality of the effluent leaving the anaerobic reactor. Therefore, to reduce the drying cycles (for sludge drying beds) or the dewatering functioning hours (for mechanical units), continuous monitoring of the solids should be conducted inside the reactor (prior to sludge withdrawal) and in the drying/dewatering unit (after sludge withdrawal). Such monitoring is vital in defining the optimum sludge withdrawal and dewatering routine to contribute ultimately to reducing the drying cycles and producing dry sludge with low water content.

The purpose of sludge dewatering is to reduce the percentage of water in the sludge maximally and, simultaneously, to improve its hygienic quality (especially when using drying beds). This should be done with the aim of maintaining the organic matter and the nutrients (nitrogen and phosphorus) in the most suitable form to turn the sludge into a biosolid for soil application.

Information on the points, parameters, and frequency of the monitoring of drying beds is available in Chernicharo (2007).

6.3 MANAGEMENT OF SLUDGE

6.3.1 Preliminaries

Sewage treatment in UASB reactors generates solid by-products (sludge and scum), which must be managed effectively to guarantee the efficiency and advantages of this type of technology. However, in most STPs in Brazil, such management is often neglected, giving rise to several problems in the treatment system. During the start-up of the STP, the accumulation of sludge (biomass + inert solids) in UASB reactors occurs after a few months of continuous operation, and the rate of accumulation depends on both biomass growth (microbial cell synthesis) and on the presence of inert or non-biodegradable solids in suspension. The latter, although not part of the biomass itself, can contribute significantly to the increase in the sludge yield coefficient and must, therefore, be included in the calculations of sludge production in anaerobic reactors.

One of the main characteristics of UASB reactors, when properly operated, is their high solids retention capacity, resulting in high sludge ages and a high degree of sludge stabilisation. The high sludge retention capacity is ascribed to the following: (i) the gas–liquid–solid (GLS) separator, (ii) the upflow velocities imposed on the reactor, and (iii) the high settleability of the sludge that develops in the system (Leitão *et al.*, 2011). However, because of its compact volume, the reactor has a maximum capacity for the retention and storage of sludge, dictated mainly by the volumes of the digestion and settler compartments of the reactor. Therefore, the excess sludge must be withdrawn systematically to prevent exceeding the maximum reactor capacity. A lack of systematic sludge withdrawal can cause (Chernicharo *et al.*, 2018):

- Excessive passage of sludge from the digestion compartment to the settler compartment, with a consequent increase in solids washout with the reactor effluent.
- Solids washout with the effluent, leading to an immediate decrease in the physical efficiency of the removal of organic matter from the reactor because of the increase in the amounts of particulate COD and BOD present in the effluent.
- Solids washout with the effluent can change the kinetic parameters previously defined in the design of the post-treatment units.
- Increase in the formation of scum in the settler compartment of the reactor.
- Overloading and eventual impairment of the post-treatment unit, particularly if it operates with a packing medium (e.g. biofilters, trickling filters, and the like).
- Difficulties in managing the aerobic sludge generated in the post-treatment unit when it is sent for thickening and digestion in the UASB reactor.
- Excessive accumulation of grit on the bottom of the reactor.
- Decrease in the useful volume of the reactor.

Accordingly, it is mandatory to remove the excess sludge periodically and appropriately. The absence of and/or ineffective sludge management can compromise the efficiency of the entire treatment system.

Therefore, it is imperative to improve the design and construction of UASB reactors, as well as to establish adequate operational routines for sludge management that can be implemented effectively in these reactors, as discussed below.

6.3.2 Problems usually associated with excess sludge

Several UASB reactors are operated with sludge quantities that exceed their retention and storage capacities leading to the reactors generally having an excessive loss of solids (solids washout). This negatively affects the quality of the effluent and the physical efficiency of removing particulate matter owing to the depletion of the storage/retention capacity of sludge inside the reactor.

As an example, Figure 6.4(a) shows the sludge mass and settleable solids concentrations in the effluent of four UASB reactors of the same volume that integrate a full-scale STP. Figure 6.4(b) shows the concentrations of sludge (TS, %) along the height of the digestion compartment of the four units. Note that all the reactors have a sludge mass above the recommended maximum, defined specifically for these four reactors in the order of 9,000 kgTS (Figure 6.4(a)). Moreover, the sludge concentrations at the highest sampling point, 2.0 m from the bottom of the reactors, are also high, in the order of 2.3% (Figure 6.4(b)).

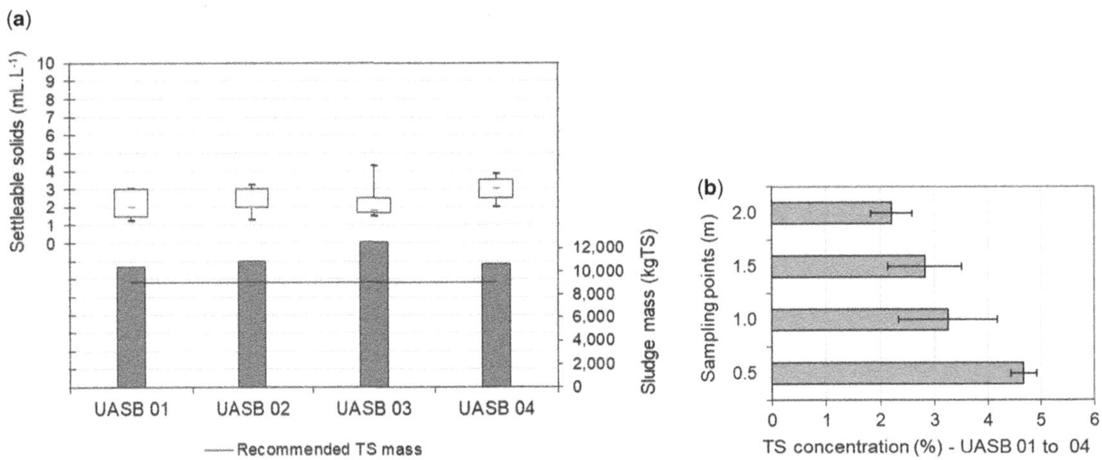

Figure 6.4 (a) Relationship between sludge mass (kgTS) in the digestion compartment and concentration of settleable solids (mL L^{-1}) in the effluent of four UASB reactors of a full-scale STP; and (b) solids profile along the height of the digestion compartment of the same four UASB reactors.

Therefore, all four UASB reactors present high concentrations of settleable solids in the effluent (medians varying between 2 and 3 mL L^{-1} (see Figure 6.4(a)), confirming that these reactors have masses and sludge concentrations exceeding their storage capacity. It should be noted that adequately operated UASB reactors (without excess sludge) could potentially produce effluents with settleable solids concentrations below 1 mL L^{-1}.

The origin of the problem of excess sludge is related possibly to the following: (i) a lack of capacity of the dewatering unit to receive/process the amount of sludge that should be removed from the reactor, and/or (ii) lack of systematic withdrawal protocols.

Additionally, the solids washout with the effluent of UASB reactors can lead to clogging or flooding in the post-treatment units, as in the instance of attached growth processes (e.g. trickling filters (TFs)), as shown in Figure 6.5(a, b).

(a) (b)

Figure 6.5 Effect of excessive solids washout on the post-treatment unit: (a) clogging of the packing media, and (b) flooding of trickling filters after UASB reactor.

In addition, the return of aerobic sludge from the post-treatment unit (e.g. trickling filter, activated sludge) for thickening and digestion in the UASB reactors has been pointed out as one of the causes of solids washout with the effluent in these reactors. However, practical experience has indicated that the establishment and implementation of operational protocols for the return of aerobic sludge and for the withdrawal of excess sludge from the UASB reactors systematically and with adequate frequency, can ameliorate the difficulties reported in the management of secondary aerobic sludge generated in the post-treatment unit, as addressed in Chapter 7 and by Floripes *et al.* (2018).

Another factor for consideration that affects the production and management of sludge is receiving contributions in the STP that were not considered during the design stage of the plant. These contributions include, e.g. sludge from septic tanks, leachate from landfills, chemical sludge from water treatment plants (WTPs), non-domestic effluents (NDE), and the like. Figure 6.6(a–c) shows the appearance of chemical sludge from the flocculators and settlers of a WTP when wet (Figure 6.6(a)) and after dewatering (Figure 6.6(b)). The 'muddy' aspect of the sludge, basically composed of silt and clay, is clearly visible. Figure 6.6(c) shows the effect of the WTP sludge on the sewerage system, showing that the settleable solids content in the sewage that reaches the STP is increased approximately15 times.

Studies on sludge production in a full-scale STP, in which the UASB reactors were designed originally to treat only sewage and secondary aerobic sludge generated in the post-treatment unit, have shown that the real sludge coefficient yield (Y_{sludge}) was approximately the double than the value considered in the design. Upon investigation, the reason for the discrepancy between the design and the field Y_{sludge} was identified as the STP receiving contributions not considered in the design stage, such as landfill leachate, sludge from septic tanks, and chemical sludge from a WTP, as addressed in Chapter 7 and Chernicharo *et al.* (2014). This implies that the decisions at management level had not considered whether the dewatering unit of the STP would have the capacity to process the additional amount of sludge generated in the UASB reactors.

Usually, the lack of an operational routine for systematic sludge withdrawals from UASB reactors, or even their being ineffective, is associated with one or more of the factors described in Table 6.1. Note that the factors can be related to the design, construction, and operation of the reactors.

(a) (b) (c)
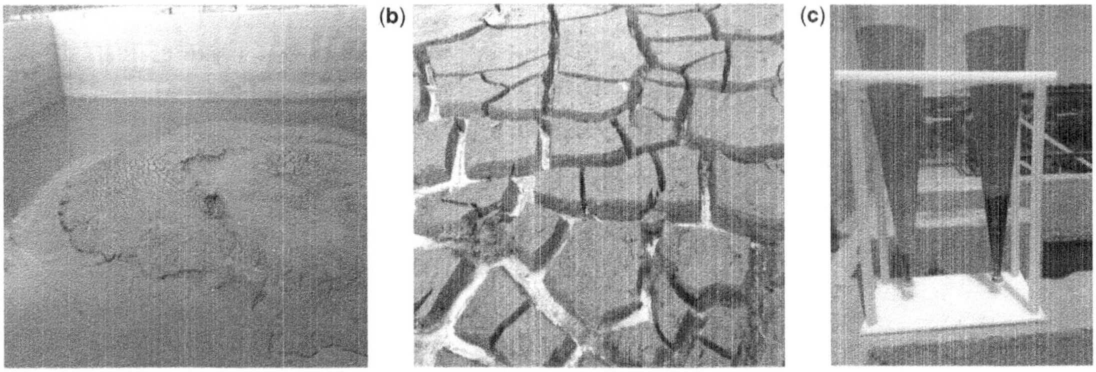

Figure 6.6 Chemical sludge from a WTP: (a) retained in the balance tank after being removed from flocculators and settlers; (b) after the dewatering step; and (c) effect on the settleable solids content of the influent sewage of a STP (left cone: only raw sewage, approximately 5 mL L^{-1}; right cone: raw sewage + WTP chemical sludge discarded in the sewerage system, approximately 80 mL L^{-1}).

Table 6.1 Main factors that can affect the management of sludge from UASB reactors treating sewage.

Factors that Can Affect the Operational Routine of Sludge Management	Non-compliance Source
Sludge sampling points, used to monitor the growth and concentration of the sludge along the height of the reactor digestion compartment, are insufficient in number and/or installed in the wrong position, i.e. non-compliant with the design (see Figure 5.11(a, b) in Chapter 5).	Design and construction
Sludge withdrawal pipes in insufficient numbers or poorly distributed along the height of the digestion compartment (Figure 6.7).	Design and construction
Manuals or operating guidelines for UASB reactors and dewatering systems that do not provide adequate detail on the procedures to be followed by the operators.	Design
Lack of systematic removal of excess sludge from UASB reactors and drying beds because of logistical and administrative problems (e.g. service contract for sludge transport incompatible with the required frequency, preventing sludge withdrawal from the reactor).	Operational management
Failures in the design of drying beds or mechanised dewatering systems, with capacity below that needed to process the sludge produced in the STP or incompatible with the characteristics of the sludge to be dewatered, as important factors were not considered, e.g. local specificities, climatic and operational conditions, such as the timescale of the operators, transport of sludge, availability of chemical products and spare parts, need for preventive maintenance, possibility of shutdowns because of corrective maintenance, and the like.	Design and operational management
Equipment of the mechanised dewatering system (e.g. sludge pumps, polymer dosers, centrifuges, and the like) not in operation owing to difficulties with preventive and corrective maintenance.	Operational management

Figure 6.7 Sludge withdrawal piping positioned extremely close to the gas deflector, i.e. non-compliant with the design.

In addition to the problems related to the management of the sludge itself, as discussed previously, other problems are caused by the inadequate functioning of the units comprising the preliminary treatment, with the passage of high amounts of debris (litter) being a common factor. Debris with higher density and, in some instances, grit, are deposited in the bottom of the reactors and become part of the sludge. The aggregation of debris and grit in the sludge can result in a decrease in the useful volume of the reactors and, when removed, can cause clogging in the sludge pipes, and impair the operation of the dewatering unit. Furthermore, this will result in the generation of a solid by-product (dewatered sludge) with objectionable aesthetic quality. Figure 6.8(a, b) shows images of debris retained in a 6 mm sieve (Figure 6.8(a)) and debris that entered the reactor and became part of the sludge and scum dewatered in the drying beds (Figure 6.8(b)).

(a) (b)

Figure 6.8 (a) Debris retained in a 6 mm sieving unit; (b) presence of debris in sludge and scum dewatered in drying beds.

6.3.3 Possible improvements related to sludge management

In view of the above, managing the sludge produced in the STP is of fundamental importance. The adoption of the measures discussed in the following items can contribute significantly to improved management of the sludge and can solve various problems that occur currently in UASB-based STPs.

6.3.3.1 Improvements related to the design of the reactor

6.3.3.1.1 Accurate determination of sludge production

As mentioned before, the production of sludge in a STP can exceed the value considered during the design phase, which can ***impose unforeseeable difficulties on the operational team***, as the dewatering unit might not have the capacity to process extra quantities of sludge. Therefore, it is imperative that the design defines the value of the sludge yield coefficient (Y_{sludge}), with the maximum possible safety, as it will affect the following directly: (i) the production of sludge from the reactor, (ii) dimensioning of the dewatering unit, and (iii) the establishment of sludge withdrawal routines.

Recommendations for more precise determination of sludge production in anaerobic reactors treating sewage are presented in Chapter 4.

6.3.3.1.2 Proper sludge withdrawal system

The reactors must have two sets of sludge withdrawal pipes to provide more operational flexibility. These are a lower set of pipes, which allows the withdrawal of the more concentrated sludge that has better settling conditions, and an upper set of pipes, which allows the withdrawal of the less concentrated sludge, with worse settling conditions, as shown in Figure 6.9.

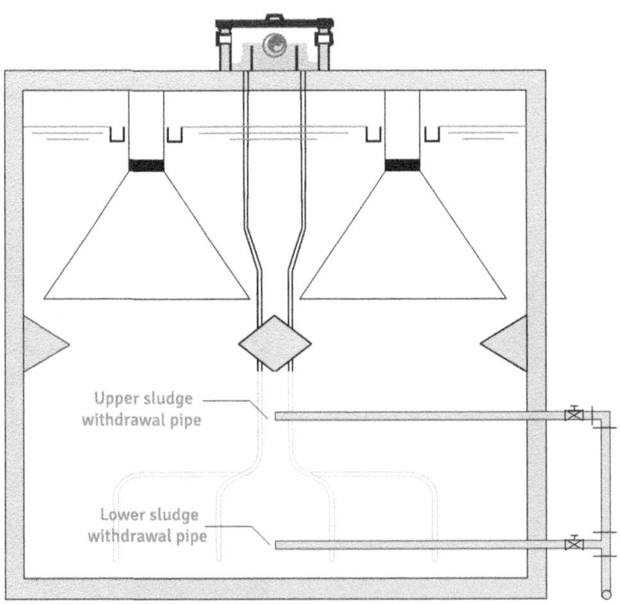

Figure 6.9 Schematic representation of the positioning of the lower and upper sludge withdrawal pipes of UASB reactors.

This configuration of an upper and lower sludge withdrawal system facilitates the scheduled ***removal of the worst quality sludge*** (upper, less concentrated) and the ***preservation of the best quality sludge*** (lower, more concentrated), which provides:

- Reaching a higher sludge age and, consequently, the withdrawal of more stabilized (digested) sludge.
- Maintaining a dense sludge bed near the bottom of the reactor, which can act as a 'filter' that assists in the retention of suspended solids long enough for them to be hydrolysed.
- Lower volume of solids loss to the settler compartment, directly reflecting the improvement of the reactor effluent quality. In addition, the amount of sludge discarded by the bottom allows the partial removal of inert solids (grit) that enter the reactor.

Recommendations and design details on the sludge withdrawal system are presented in Chapter 4.

6.3.3.2 Improvements in operational procedures

6.3.3.2.1 Amount of sludge to be maintained and removed in the reactor

To prevent the undesired solids washout with the final effluent, the sludge mass within the reactor must be maintained ***between a minimum value*** that should match the amount of biomass sufficient to digest the influent organic load, ***and a maximum value*** that is dependent on the sludge retention capacity in the UASB reactor. The minimum mass is calculated from the organic load applied to the reactor and the specific methanogenic activity of the anaerobic sludge, as shown in Example 6.1.

EXAMPLE 6.1

Calculate the minimum amount of biomass to be maintained in a UASB reactor treating sewage, considering the following design and operational data:

Design data (see Example 4.3 in Chapter 4):

- Population equivalent: PE = 20,000 inhabitants
- *Per capita* COD contribution: $QPC_{COD} = 0.10$ kgCOD PE^{-1} d^{-1}
- COD removal efficiency: $E_{DQO} = 70\%$
- Specific methanogenic activity: SME = 0.20 kg COD_{CH4} $kgTVS^{-1}$ d^{-1}

Operational data:

- Population equivalent: PE = 15,000 inhabitants (*the STP operates below its design capacity*)
- *Per capita* COD contribution: $QPC_{COD} = 0.10$ kgCOD PE^{-1} d^{-1}
- COD removal efficiency: $E_{DQO} = 70\%$
- Specific methanogenic activity: SME = 0.20 $kgCOD_{CH4}$ $kgTVS^{-1}$ d^{-1}

Solution

(1) Estimate the applied organic load (OL_{app}):

$$OL_{app} = PE \times QPC_{COD} = 15{,}000 \text{ inhab.} \times 0.10 \text{ kgCOD } PE^{-1} d^{-1} = 1{,}500 \text{ kgCOD}_{app} d^{-1}$$

(2) Estimate the removed organic load (OL_{rem}):

$$OL_{rem} = E_{DQO} \times OL_{app} = 0.70 \times 1{,}500 \text{ kgCOD}_{app} d^{-1} = 1{,}050 \text{ kgCOD}_{rem} d^{-1}$$

(3) Estimate the minimum mass necessary to treat the removed load (M_{min}):

$$M_{min} = OL_{rem}/SME = 1{,}050 \text{ kgCOD}_{rem} d^{-1}/0.20 \text{ kgCOD}_{CH4} \text{ kg TVS}^{-1} d^{-1} = 5{,}250 \text{ kgTVS}$$

On the other hand, to define the maximum mass of the sludge the reactor can store, without sludge passing to the settler compartment and the excessive solids washout with the effluent, close observation is necessary of the results of TSS and settleable solids in the effluent of the reactor, and the TS concentrations along the entire height of the reactor digestion compartment (sludge profile). Such monitoring information facilitates inferring which sludge mass in the reactor initiates a greater loss of solids and deterioration of the effluent quality. When there is no historical data series that allows the definition of the maximum mass, another possibility is referring to the maintenance of the TS concentration of the sampled sludge at the highest point of the digestion compartment, located immediately before the gas deflector (point P5 in Figure 6.3), ***always below 0.5%.***

Definition of the withdrawal strategy: The corresponding volume of sludge to be discarded according to the adopted withdrawal strategy is calculated from the estimation of the sludge production in the reactor and the mass of sludge to be discarded. One recommended strategy is to discard ***at least 50% of the excess sludge mass from the upper pipeline*** (1.0–1.5 m above the bottom). To compensate for the higher volumes of sludge extraction resulting from this practice, new STPs should consider the ***inclusion of a balance and homogenisation tank*** before the dewatering unit (see Sections 4.4.6 and 4.4.7 in Chapter 4). Partial thickening of the sludge allows the supernatant to be returned to the UASB reactor and the remaining more concentrated fraction to be homogenised suitably to reach the ideal range of TS content required by the dewatering unit. The removal of the upper sludge also favours the maintenance of low TS concentrations at the interface of the digestion and settler compartments, which should always remain below 0.5% TS, as mentioned previously.

Pre-programmed and continued removal of the sludge is recommended to enable the reactor to function as steadily as possible. The definition of the percentage of sludge to be discarded from the upper and lower points, as well as the frequencies of the pre-programmed discards should be stipulated according to the capacity and operational routine of the dewatering system (drying beds or mechanised), as shown in the calculations presented in Chapter 4 and in Examples 6.2 and 6.3.

Return of secondary (aerobic) sludge from the post-treatment unit to the UASB reactors: The transfer of excess aerobic sludge for thickening and digestion in UASB reactors has been pointed out as one of the causes of the solids washout with the effluent of these reactors. There are several reports of STP operators that associate this loss of solids with the characteristics of the aerobic sludge, which supposedly would have a different composition from the anaerobic sludge and, therefore, would tend to float and be lost with the UASB reactor effluent. On the other hand, careful studies on two full-scale STPs showed good settleability characteristics of the aerobic sludge sent to the UASB reactors (Chernicharo *et al.*, 2014; Floripes *et al.*, 2018) and indicated that the cause of the solids washout with the anaerobic effluent would be associated more with the absence of an adequate sludge management protocol in the STP. This should include both the withdrawal of the excess sludge from the UASB reactors and sending the aerobic sludge from the post-treatment units for thickening and digestion in the anaerobic reactors.

This perception is confirmed by several STPs presenting significant difficulties in managing sludge, particularly, with substantial amounts of excess sludge occurring both in the UASB reactors and in the secondary settlers of the post-treatment units. As regards the post-treatment units, exceptionally high TS concentrations are found commonly in the secondary settlers (TS~4%), well above the assumed design concentrations (<1.0%). Although such a condition would appear unlikely, in reality it has been recurring often. This can be attributed, quite possibly, to the significant contribution of the amount of sludge lost with the anaerobic effluent that eventually reaches the secondary settler, particularly in the instance of settlers following trickling filters. The accumulation of anaerobic sludge in the secondary settlers can be clearly noticed through the gas bubbles escaping from the surface of the settlers. Figure 6.10(a, b) shows the appearance of excessive accumulation of anaerobic sludge in a secondary settler after a trickling filter.

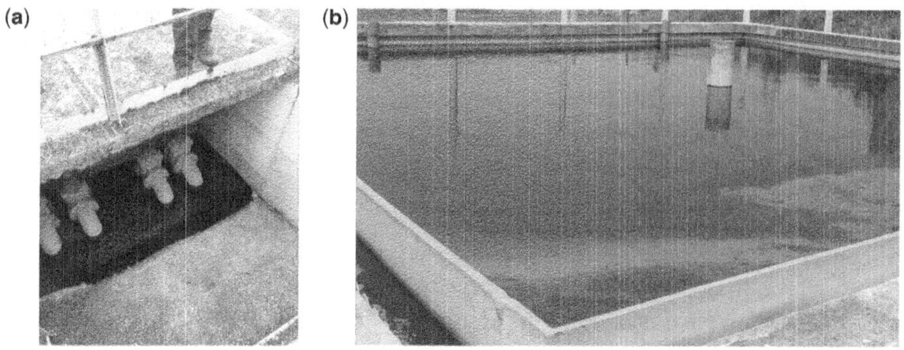

Figure 6.10 (a) Black (anaerobic) sludge discarded from a secondary settler after a trickling filter; (b) floating sludge and gas bubbles on the surface of a secondary settler.

Possible reasons for the excessive accumulation of anaerobic sludge in secondary settlers are: (i) a lack of installed capacity of the dewatering units (either mechanised or drying beds); (ii) long maintenance periods of the mechanised dewatering unit and/or cleaning of the drying beds, which makes it impossible to remove the excess sludge; and (iii) logistic difficulties associated with the routing of the sludge to the final disposal unit, as pointed out in Table 6.1.

In view of the above, to eliminate the excess sludge accumulated in the secondary settlers, the sludge flows sent to the UASB reactors must be significantly (about four times) larger compared with the flows that would be practiced if the sludge in the secondary settler had the expected design concentration (lower than 1.0% TS). Additionally, these sludge flow rates are often practiced in typically short batch times (in the order of a few hours), often coinciding with times of higher sewage flows, which tend to cause: (i) an increase in the upflow velocities in the UASB reactors, which, in turn, can cause higher solids washout with the effluent; (ii) significant reduction of the sludge storage capacity of the UASB reactors because of the overload of received solids. Accordingly, with the non-simultaneous removal of excess sludge from the secondary settlers and from the UASB reactors, this sludge will eventually return, in a cyclic way, to the trickling filters, to the secondary settlers and, finally, again to the UASB reactors.

It can be concluded, therefore, that the removal of excess sludge from UASB reactors in smaller amounts than the effective production is a potential cause that contributes to the retention of sludge with lower settleability characteristics in UASB reactors. In addition, inadequate criteria for sending the sludge from secondary settlers to the UASB reactors can contribute to the increase in the solids load and the deterioration in the quality of the anaerobic effluent. Therefore, in addition to the prerequisite to match the estimates of the effective sludge production with the processing capacity of the sludge dewatering unit, the recommendations for improving the effluent quality in systems that adopt the return of secondary (aerobic) sludge for thickening and digestion in UASB reactors are:

- *Preservation of the best quality sludge in the bottom of the UASB reactors.* As discussed in detail previously, the protocol of sludge removal from UASB reactors should allow the preservation of a significant portion of the bottom sludge (more concentrated) so that it can act as a 'filter' that assists in the retention and hydrolysis of suspended solids contained in the aerobic sludge returned from the secondary settlers of the post-treatment unit.

- *Maintenance of low sludge concentrations in the upper part of the digestion compartment of the UASB reactor*. As discussed previously, it should be ensured that the reactor does not contain excess sludge. The simplest way to verify that there is no excess sludge in the reactor is by systematically checking the TS concentration at the highest point of the digestion compartment (just before the gas deflector). *The concentration of sludge at this point should always be below 0.5% TS.*
- *Maintenance of low concentrations of sludge in secondary settlers*. Matching the effective sludge yield in the UASB reactors with the design of the sludge dewatering system will reduce the demand on the secondary settling stage because of the lower concentrations of TSS in the anaerobic effluent and after the post-treatment by trickling filters or activated sludge. Consequently, the secondary (aerobic) sludge load sent to the UASB reactors will be lower, as the sludge concentration in the secondary settlers will also be lower. *It is recommended that the sludge concentration in the secondary settling step does not exceed 1.0%.*

  Two possible alternatives to solving the problem of sludge build-up in the secondary settlers are: (i) to increase the removal of sludge from the UASB reactors so as not to exceed the maximum mass or the recommended maximum concentration of sludge in the upper part of the digestion compartment; and (ii) to remove the excess sludge from the secondary settler without considering the transfer to the UASB reactors (e.g. by means of cesspool cleaning trucks).
- *Establishment of the period and flow rate of the aerobic sludge to be returned to the UASB reactors*. It is recommended that the return of aerobic sludge (from the secondary settlers to the UASB reactors) during periods when the influent sewage flowrate is below the average flow (as a rule, the period of the lower flow occurs during dawn). In such instance, the sludge pumps must be operative to route the aerobic sludge *over the entire period of low flow*. This operational strategy aims at maintaining the sludge with less settleability in the regions of the digestion compartment where a higher TS concentration is observed.

6.4 MANAGEMENT OF SCUM

6.4.1 Preliminaries

The formation, accumulation, and type of scum that occurs inside the UASB reactors depend essentially on the composition of the raw sewage, particularly in relation to the presence of debris (e.g. fat, oil and grease, soaps, food debris, fruit and vegetable peels, hair, cotton swabs, dental floss, paper and cotton, cigarette butts, plastics, and similar materials). The accumulation and formation of a solid and thick scum layer is difficult to remove by the STP operators, which is one of the main operational problems in full-scale UASB reactors. The scum layer tends to prevent the release of the biogas produced in the anaerobic reactor and, in turn (Chernicharo *et al.*, 2018):

- Causes the passage of the biogas to the settler compartment of the UASB reactor, with direct effects on the quality of the effluent (solids washout) and on the emission levels of malodorous and greenhouse gases.
- Reduces the recovery (capture) of biogas for purposes of burning and/or energy generation.
- Reduces the useful volume of the gas chamber and changes its operating pressure.

Furthermore, the thickening of the scum layer can impose high structural stresses on the walls of the gas–liquid–solid separator (GLS separator). This, in turn, can cause damage to and even the rupture of the GLS separator, particularly when manufactured from less resistant materials, such as plastic canvas, metal sheets, asbestos, and the like (see Figure 5.7 in Chapter 5).

Accordingly, similar to the management of sludge, it is mandatory that the scum be removed periodically and adequately. The absence of and/or inefficient scum management procedures can compromise the efficiency of the entire treatment system. Therefore, it is imperative to improve the design and construction of UASB reactors, and to establish appropriate operational routines for scum management that can be implemented effectively in such reactors, as discussed in the following items.

6.4.2 Problems associated with scum

In UASB reactors, scum can form in two distinct locations: (i) on the surface of the settler compartment, and (ii) inside the GLS separator at the interface where the gases formed during the anaerobic digestion process are released (Figure 6.11).

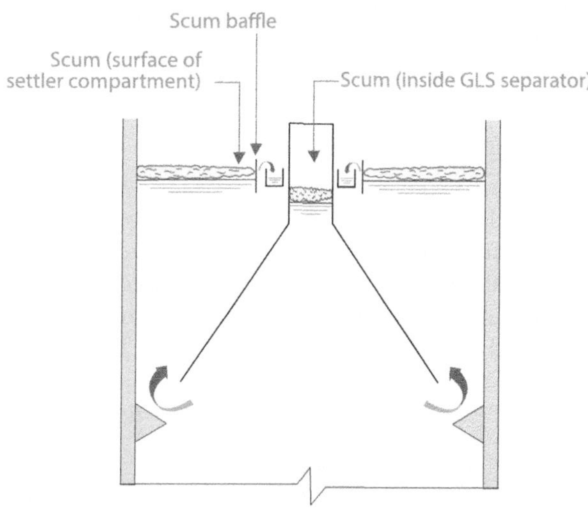

Figure 6.11 Identification of scum accumulation zones in UASB reactors treating sewage.

6.4.2.1 Scum accumulated on the surface of the settler compartment

The scum that accumulates on the surface of the settler, although it is characterised as extremely concentrated residue, has relatively low accumulation coefficients. It is possible to design the settler compartments of the UASB reactors without scum baffles to allow this residue to be removed along with the effluent, without causing a significant deterioration of its quality. However, it is essential to ***ensure adequate levelling, fixation, and sealing of the effluent collection gutters*** (as discussed in Chapter 5) to avoid preferential flows and the accumulation of scum in the upper part of the settler compartments, as shown in Figure 6.12(a, b).

It is important to note that the rate of scum accumulation on the surface of the settlers depends also on the adequate management of the excess sludge. The absence or removal of excess sludge at a frequency not compatible with the effective sludge production in the system can result in a more significant loss of solids to the settler compartment. Once in the settler, this sludge is entrapped in the scum layer and has the appearance of 'floating sludge' along with the scum (Figure 6.12(c)). This situation tends to worsen when the reactor functions with hydraulic overload and upflow velocities higher than those considered in the design. If not managed effectively, excessive accumulation of scum in the settler compartment could contribute to the deterioration of the final effluent quality and pose severe difficulties to the operators.

Figure 6.12 Accumulation of scum on the surface of settler compartments because of: (a) non-levelled effluent collection gutters; (b) lack of sealing during the installation of the pouring plates (the effluent passes between the glass fibre pouring plate and the concrete collecting channel); and (c) lack of sludge removal and hydraulic overload in the reactor.

6.4.2.2 Scum accumulation in the GLS separator

The absence of or inefficient scum removal devices, associated with poor access conditions, lack of occupational safety conditions, and the absence of a systematic protocol for the removal of this material can lead to thickening and solidification of the scum layer (Figure 6.13(a, b)). In turn, the solidified scum layer can prevent the natural passage of the biogas generated in the reactor. This can cause the biogas to pass to the settler compartment, compromising the solids retention and leading to deterioration of the effluent quality (Lettinga & Hulshoff Pol, 1991; Souza *et al.*, 2006).

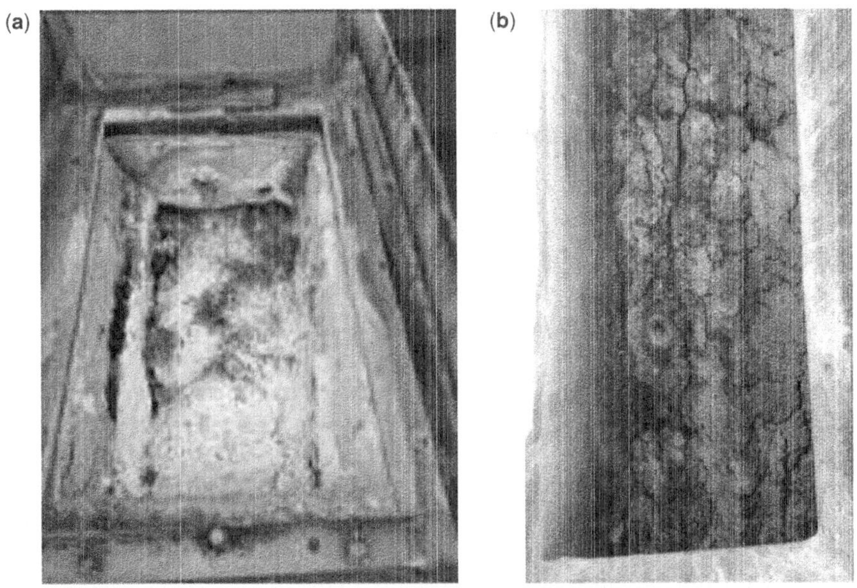

Figure 6.13 Thick and solidified scum layer inside GLS separators (a, b).

The passage of the biogas to the settler compartment also has a direct effect on the increase in emission levels of odorous and greenhouse gases. In addition, it reduces the recovery of biogas for purposes of burning and/or energy generation.

The same considerations as for 'floating sludge' that is entrapped in the scum layer of the settler compartment are valid for the scum formed inside the GLS separator, which is caused by excess sludge in the digestion compartment and hydraulic overload in the reactor. In addition, valid concerns exist about debris (litter) not retained in the preliminary treatment that enter the reactor. A considerable portion of the lighter debris will float on and become entrapped in the scum layer in the GLS separator or on the surface of the settler compartment.

6.4.3 Possible improvements related to scum management

Similar to the management of the sludge, the correct management of the scum produced in the UASB reactors is of fundamental importance. The adoption of the measures described in the following items can contribute significantly to improve the management of the scum and solve several problems currently occurring in UASB-based STPs.

6.4.3.1 User awareness

Several of the current operational problems in Brazilian STPs arise from the inadequate use of residential sanitary facilities (and also of commercial establishments, e.g. restaurants). This results in a significant alteration of the generated sewage, in quantitative and qualitative terms, e.g. excessive rainwater, detergent (surfactant), debris (e.g. hair, cotton swab, absorbent, dental floss, cotton, wet wipe, cigarette butt, plastic materials, and the like), and fat, oil, and grease. Consequently, it is essential to raise the awareness of users of the sewerage services, which implies the creation of opportunities and spaces for dialogue, enabling understanding of the knowledge, perception, and perspectives of the population in relation to the importance of sanitation in the community. In particular, this relates to the sewage generation and solutions to its treatment and final disposal. In addition, in such spaces, it is important to hold debates on the direct responsibility of every user for the origin of the process, namely what is understood as reasonable to dispose of in the household sanitary facilities.

6.4.3.2 Improvement of the preliminary treatment

To reduce the entry of debris into UASB reactors, particular care must be taken in the definition of the screens and sieves of the preliminary treatment unit. If the pretreatment units could not retain such materials, a sizable volume of debris would be eliminated from the sludge (and scum) composition. Among the improvements recommended are using ultra-thin screens and sieves with small apertures (3–4 mm). In addition, possibly, oil and grease could be removed from the upstream of UASB reactors. An alternative that has been adopted in some anaerobic-based STPs is aerated grit chambers (spiral flow), as reported in Chapter 4. However, the scant Brazilian experiences reported in the literature indicate low efficiency of such grit chambers associated with the removal of oil and grease.

6.4.3.3 Control of the reception of non-domestic effluents

Some types of non-domestic effluents, such as sludge from septic tanks and industrial effluents with high oil and grease content (e.g. dairies, slaughterhouses, and the like) can contribute to an increase in the formation and accumulation of scum in UASB reactors and, consequently, can worsen the problems mentioned previously. It is recommended, therefore, that strict control be exercised over receiving these and other contributions that can lead to the worsening in the formation of scum in UASB reactors.

6.4.3.4 Improved control over the formation of scum in the settler compartment

To improve the control over the formation of scum on the surface of the settler compartment, the guidelines presented in Table 6.2 should be considered. It should be noted that the guidelines are related to the design, construction, and operation of reactors.

Table 6.2 Main guidelines related to the control over the formation of scum on the surface of the settler compartment of UASB reactors treating sewage.

Guidelines to Minimise the Formation of Scum	Non-compliance Source
Improvement of the design, manufacture, and installation of the effluent-collecting weirs (guarantee a minimum of 15 mm pouring water level and an installation/fixation that secures levelling and resistance against torsion).	Design and construction
Correct design and use of the sewage overflow structure and/or the pumping station that feeds the reactors to avoid hydraulic overloads and excessive upflow velocities in the digestion and settler compartments of the reactor.	Design and operation
Correct management of the sludge to avoid excessive passage and 'flotation' of sludge in the settler compartment.	Design and operation

6.4.3.5 Use of GLS separators equipped with scum removal device

In the development of alternatives for improving UASB reactors, the most recent designs incorporate scum collection channels inside the GLS separator to facilitate the hydrostatic removal of the scum. This removal procedure is based on the change in the water level inside the GLS separator, which allows the scum to flow to a channel, internal to the separator, and to be directed to the withdrawal pipe, outside the reactor. The change in the water level inside the GLS separator is achieved by increasing or reducing the pressure in the gas line between the GLS separator and the water seal located on top of each UASB reactor, as discussed in Chapter 4.

Different configurations of hydrostatic scum removal devices have been tested, with satisfactory results in demonstration- and full-scale UASB reactors, as addressed in Chapter 7. Scum removal efficiencies between 75 and 90% have been observed by Rosa *et al.* (2012), Santos (2014) and Diaz Flórez (2016). It should be noted that such devices only function adequately for the removal of scum in its fluid state (not excessively thickened or solidified). Table 6.3 shows the main design, construction, and operational guidelines for the adequate functioning of a GLS separator equipped with hydrostatic scum removal devices.

Table 6.3 Guidelines for design, construction, and operation of hydrostatic scum removal devices.

Guidelines for Adequate Functioning of Hydrostatic Scum Removal Devices	Responsibility
The amount of scum accumulated inside the GLS separator should be checked for each specific condition, considering the characteristics of the raw sewage, the efficiency of the preliminary treatment, and the constructive characteristics of the reactor, when the scum removal device starts operations.	Operation
The GLS separator must be integrated to a water seal to allow the control of the internal pressures and, consequently, the adjustment of the water levels in relation to the scum collecting channel (Figures 6.14(a) and 6.15 (a, b)).	Design and construction
The scum collection channels should preferably be positioned adjacent to one of the walls of the GLS separator and not at the centre. The positioning of the channel next to one of the sides allows a larger water level and superior efficiency in the removal of scum. A scum collection channel in triangular section is shown in Figure 6.14b.	Design and construction
Correct positioning of the scum collection channel at the time of construction, as installation at levels other than those specified in the design can prevent proper GLS pressurising and operation of the device.	Construction
Provision of inspection points and valves on the scum outlet pipes of each GLS separator, aiming to identify clogging points and avoiding uneven hydraulic head loss (Figure 6.14(c)).	Design
Provision of extraction nozzles along the length of the scum collection channel to allow more adequate hydraulic flow conditions (see Figure 4.13 in Chapter 4).	Design
Ensuring gas compartment tightness to allow the establishment of small pressures in the biogas line (5–10 cm w.c (490–980 Pa)).	Design, construction, and operation
Establishment of the optimum frequency of scum removal from the interior of GLS separators will be based on visual inspection, which can be performed from the opening of the hermetic inspection hatches (or through transparent displays when available), before and after the withdrawal operation trials during the start-up period. The operational practice will enable establishing the optimal removal routine (frequency and time of withdrawal, degree of valve opening, volumes removed, and the like).	Operation
Whenever possible, provisioning of motorised valves to facilitate the scum removal operation.	Design

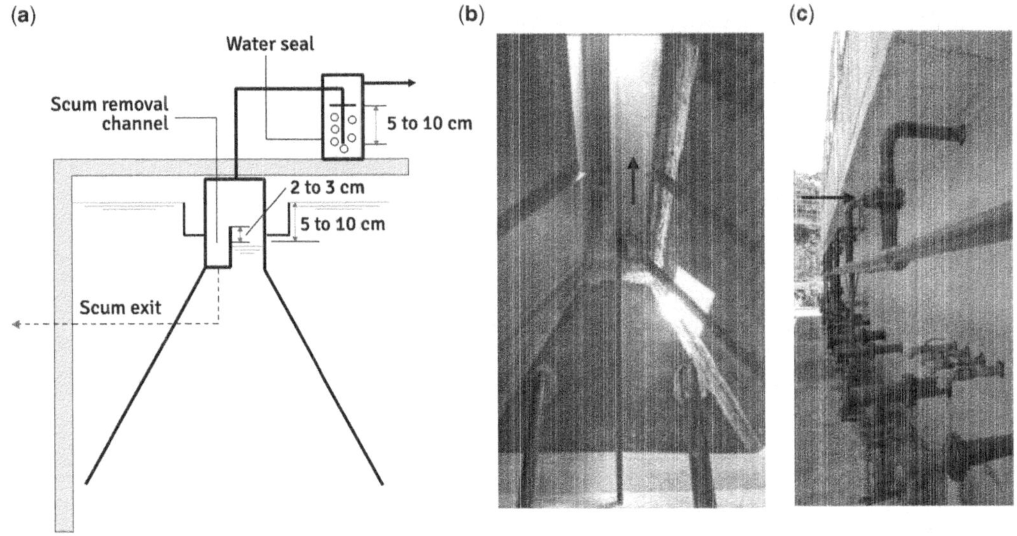

Figure 6.14 Hydrostatic scum removal device: (a) schematic illustration of pressure control and level of scum inside the GLS separator; (b) view of scum collecting channel in triangular section; (c) valve in the scum outlet pipes of each GLS separator.

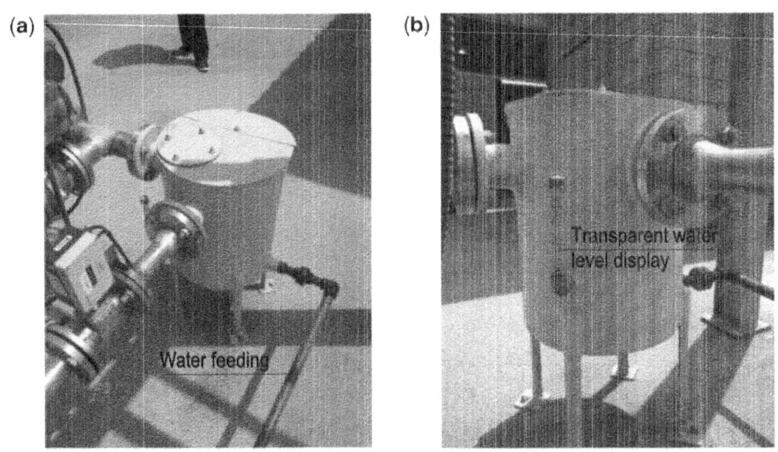

Figure 6.15 Water seal for controlling the pressure and level of scum inside the GLS separator: (a) general view of the installation; and (b) detail of the water level display.

6.5 ROUTINE OPERATIONAL ACTIVITIES

6.5.1 Preliminary treatment

The routine operational activities of the preliminary treatment depend on the degree of mechanisation of screens, sieves, and grit chambers. In more mechanised systems, the routine activities will mainly be the inspection of equipment and good housekeeping of the area, in addition to managing the correct destination of all the solid material removed in the units that comprise the preliminary treatment.

As regards STPs that employ mechanised screens, the adjustments of the rakes must be programmed during the start-up period as a function of the inflow and the amount of solids retained in the screens. In this way, the mechanised cleaning system will be operated intermittently during the 24 hours of the day, depending on the amount of debris to be removed by the screen.

As regards STPs that employ non-mechanised systems, the removal of the solids retained in the screens will be carried out manually by the plant operators, equipped with appropriate rakes. Removal occurs periodically when necessary, i.e. when there is a significant difference between the level of the sewage upstream and downstream of the screen (head loss) or when there is excessive accumulation of solids in the screen. The removal of the retained material is usually done at least every two hours.

To evaluate the performance of the screening unit and to analyse the composition of the sewage in relation to the presence of coarse solids, the following procedures are useful:

- *Observation of presence of coarse solids in the units downstream of the screen*: coarse solids in the units downstream of the screen, such as in the grit chamber channel or causing clogging in the pipes and hydraulic devices is a useful indication of the poor performance of the screen, implying that specific analysis of the operation of this unit is necessary.
- *Evaluation of the amount of material removed*: this can be visual, but will be performed based on previous measurements that will serve to estimate the amount of solids removed in the screening unit. The procedures must be recorded in a separate file. The volume of screened material must be recorded daily, and a performance evaluation must be performed on a monthly basis.
- *Determination of the quality of the material*: this is done in the laboratory, with samples collected before the final disposal of the material, ignoring high volume materials such as toys, pieces of wood, and the like. The occurrence of these materials should also be recorded, as their presence indicates negligent use of the sewerage system by the users, implying measures to raise the awareness of users of the sewerage services (see section 6.4.3.1).

6.5.2 Verification and unblocking of the influent distribution and feeding system

The influent distribution chamber and feeding devices should be checked continuously to detect obstruction. This measure is particularly important in UASB reactors, as the correct sewage distribution to the bottom of the reactors is essential for appropriate operation. It is recommended that the sewage distribution tubes are verified (and, if necessary, obstructions cleared) daily. Figure 6.16(a–c) shows one unblocking device used in one large-scale STP in Brazil.

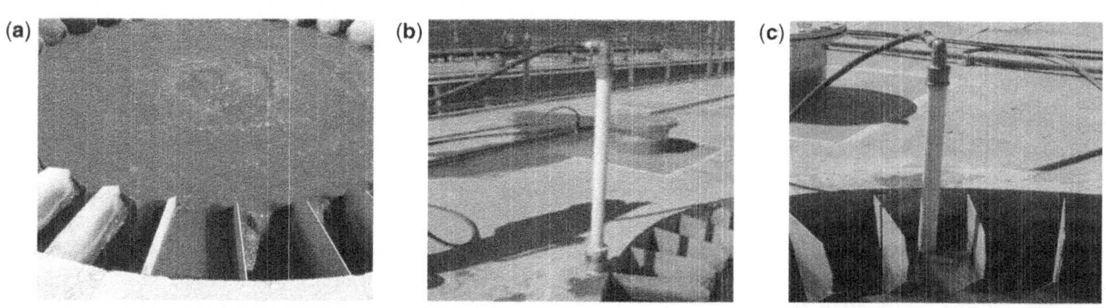

Figure 6.16 Distribution chamber on top of a UASB reactor: (a) blocked compartment; (b) view of an unblocking device fed on pressurised water (silicone tip at the far end); and (c) unblocking device in operation.

6.5.3 Withdrawal of excess sludge from the reactor

Excess sludge withdrawal from the UASB reactors should be done preferably from the bottom and the middle-height of the digestion compartment, as discussed previously in section 6.3.3. The following benefits can be achieved:

- Withdrawal from the middle-height allows the removal of more dispersed sludge, usually with less activity and settleability.
- To compensate for the higher volumes originating from the withdrawal of less dense sludge, a smaller parcel of the more concentrated sludge from the bottom can be removed.
- Depending on the quality of the preliminary treatment, inert solids (grit) can accumulate in the bottom sludge that should be discarded periodically from the reactor.

6.5.3.1 Withdrawal of sludge to drying beds

As regards the dewatering of excess sludge in the drying beds, the frequency of sludge withdrawal usually varies between one and three weeks, according to the sludge storage capacity of the reactor and the mean dewatering time in the beds. Pre-programmed and continuous withdrawal of the sludge is recommended to enable the reactor to operate as steadily as possible. Example 6.2 shows the sequence of calculations that should be carried out by the operator of the plant to estimate the mass and volume of sludge to be removed, considering the strategy of weekly removals and the cyclic use of each drying bed equal to three weeks.

EXAMPLE 6.2

Determine the mass and volume of sludge to be removed from the UASB reactors of Example 4.3 (Chapter 4), considering the following design and operational data:

Design data (see Example 4.3 in Chapter 4):

- Population equivalent: PE = 20,000 inhabitants
- *Per capita* COD contribution: $QPC_{COD} = 0.10$ kgCOD PE^{-1} d^{-1}
- Area of drying beds ($A_{drying-bed-total}$): 1,050 m^2
- Number of drying beds in use: 4 units (for this situation there is one spare bed)
- Area of each drying bed ($A_{drying-bed-unit}$): 210 m^2

Operational data:

- Applied organic load: $OL_{app} = 1{,}500$ $kgCOD_{app}$ d^{-1} (note that the actual organic load is lower than the one considered in Example 4.3)
- Sludge yield: $Y_{sludge} = 0.20$ kgTS $kgCOD^{-1}{}_{app}$ (the same as considered in Example 4.3)
- Sludge specific mass: $\gamma = 1{,}020$ kgTS m^{-3} (the same as considered in Example 4.3)

Important note: The example only considers the sludge production coefficient of the UASB reactor. If the STP receives other types of contributions (e.g. sludge from septic tanks, landfill leachate, non-domestic effluents, and the like) and/or secondary (aerobic) sludge from the post-treatment unit (e.g. trickling filters, activated sludge) for thickening and digestion in the UASB reactors, the increase in the value of the sludge production coefficient (Y_{sludge}) from these contributions should be included in the calculation of the dewatering system.

Solution

(1) Estimate the daily sludge production (P_{sludge})

- $P_{sludge} = Y_{sludge} \times OL_{app} = 0.20$ kgTS $kgCOD_{app}^{-1} \times 1{,}500$ $kgCOD_{app}$ $d^{-1} = 300$ kgTS d^{-1}

(2) Define the sludge frequency of withdrawal and cycle time (C_t) of the drying beds

- Frequency of withdrawals: $F_w = 7$ d (every 7 days)
- Cycle time: $C_t = 21$ d (should be defined based on local climatic conditions and operational aspects of the STP)

(3) Sludge concentrations at the different reactor heights (obtained from the solids profile – see section 6.2.2.2)

- Upper withdrawal pipe (see Figure 6.9): $C_{sludge-upper} = 2.2\%$ (22,000 mgTS L^{-1})
- Lower withdrawal pipe (see Figure 6.9): $C_{sludge-lower} = 4.5\%$ (45,000 mgTS L^{-1})

(4) Percentages of sludge mass to be removed from the different reactor heights (adopted)

- Upper withdrawal pipe: 60%
- Lower withdrawal pipe: 40%

(5) Mass of sludge to be removed from the different reactor heights (upper and lower pipes – see Figure 6.9)

- $Mass_{sludge\text{-}upper} = P_{sludge} \times F_w \times 60\% = 300$ kgTS $d^{-1} \times 7$ d $\times 0.60 = 1{,}260$ kgTS every 7 d
- $Mass_{sludge\text{-}lower} = P_{sludge} \times F_w \times 40\% = 300$ kgTS $d^{-1} \times 7$ d $\times 0.40 = 840$ kgTS every 7 d

(6) Total mass of sludge to be removed from the reactors ($Mass_{sludge\text{-}total}$)

- $Mass_{sludge\text{-}total} = Mass_{sludge\text{-}upper} + Mass_{sludge\text{-}lower} = 1{,}260$ kgTS $+ 840$ kgTS $= 2{,}100$ kgTS every 7 d

(7) Volume of sludge to be removed from the different reactor heights (upper and lower pipes – see Figure 6.9)

- $V_{sludge\text{-}upper} = Mass_{sludge\text{-}upper}/(\gamma \times C_{sludge\text{-}upper}) = 1{,}260$ kgTS$/(1{,}020$ kgTS $m^{-3} \times 0.022) =$ 56.1 m^3 every 7 d
- $V_{sludge\text{-}lower} = Mass_{sludge\text{-}lower}/(\gamma \times C_{sludge\text{-}lower}) = 840$ kgTS$/(1{,}020$ kgTS $m^{-3} \times 0.045) =$ 18.3 m^3 every 7 d

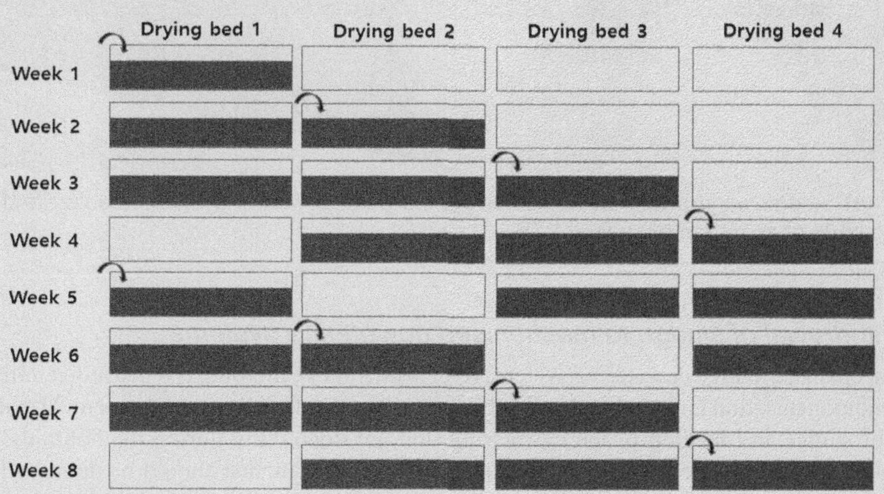

Figure 6.17 Schematic representation of the routines of sludge discharge and drying beds' cleaning.

(8) Total volume of sludge to be removed from the reactors ($V_{sludge\text{-}total}$)
- $V_{sludge\text{-}total} = V_{sludge\text{-}upper} + V_{sludge\text{-}lower} = 56.1\ m^3 + 18.3\ m^3 = 74.4\ m^3$ every 7 d

(9) Resulting sludge height on each drying bed ($H_{drying\text{-}bed}$)
- $H_{drying\text{-}bed} = V_{sludge\text{-}total}/A_{drying\text{-}bed\text{-}unit} = 74.4\ m^3/210\ m^2 = 0.35\ m$ (note that only four out of five drying beds are in use in this example)

(10) Resulting solids load on each drying bed (SL_{app})
- $SL_{app} = Mass_{sludge\text{-}total}/A_{drying\text{-}bed\text{-}unit} = 2{,}100\ kgTS/210\ m^2 = 10\ kgTS\ m^{-2}$ (expected to be lower than $15\ kgTS\ m^{-2}$ in each operation cycle).

A schematic of the sludge withdrawal routine and drying periods is presented in Figure 6.17.

Once the volumes of sludge to be discarded from each point (upper and lower withdrawal pipes) are determined, according to the calculations in Example 6.2, the volumes can be measured automatically with flow meters or by measuring the sludge level height and the corresponding volume in the drying bed.

The sludge withdrawal operation consists of opening the valves located on the outside wall of the reactor (Figure 6.18(a)), which enables draining the sludge to the drying bed (Figure 6.18(b)). ***The individualised opening of each valve (one at a time), corresponding to each sludge withdrawal point (level), is imperative***, otherwise the sludge removal will not be uniform inside the reactor.

(a) (b)

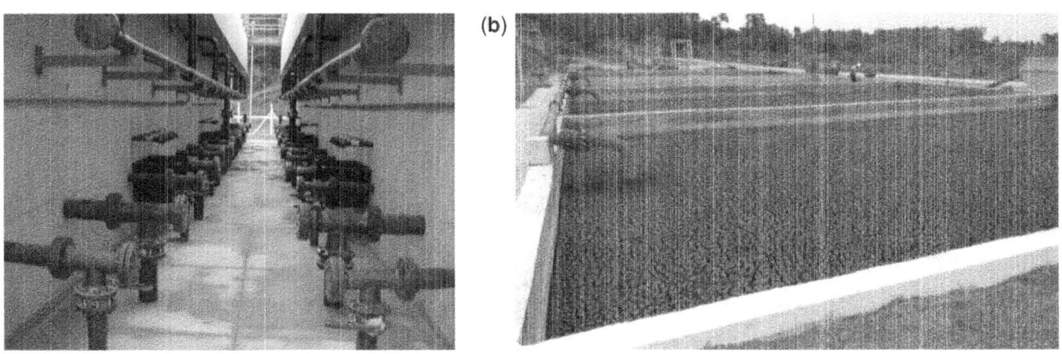

Figure 6.18 (a) Sludge withdrawal valves located on the outside wall of a UASB reactor; and (b) view of sludge drying beds of an anaerobic-based STP.

6.5.3.2 Withdrawal of sludge to mechanised dewatering systems

As regards mechanised sludge dewatering systems (e.g. centrifuges), the sludge is discarded initially into the balance and homogenisation tank and pumped subsequently to the dewatering equipment. Usually, both the withdrawal of sludge and feeding of the dewatering unit are done daily during the hours the operator is present at the plant. Example 6.3 shows the sequence of calculations that should be done by the operator to estimate the mass and volume of sludge that should be discarded, considering the strategy of daily removals, five days per week. From the equipment processing capacity (defined during the design stage – Example 4.4, Chapter 4) the daily number of operational hours can be estimated.

EXAMPLE 6.3

Determine the mass and volume of sludge to be removed from the UASB reactors of Example 4.4 (Chapter 4) and the number of hours of operation of the dewatering unit, considering the following design and operational data:

Design data (see Example 4.4, Chapter 4):

- Population equivalent: PE = 200,000 inhabitants
- *Per capita* COD contribution: $QPC_{COD} = 0.10$ kgCOD PE^{-1} d^{-1}
- Processing capacity of the dewatering unit: $V_{cap\text{-}dew.unit}$: 17.4 m^3 h^{-1} or 461 kgTS h^{-1}
- Number of hours of operation of the dewatering unit: $N\text{-}hours_{design} = 12$ h d^{-1}

Operational data:

- Applied organic load: $OL_{app} = 15{,}000$ $kgCOD_{app}$ d^{-1} (note that the actual organic load is lower than the one considered in Example 4.4)
- Sludge yield: $Y_{sludge} = 0.20$ kgTS $kgCOD^{-1}{}_{app}$ (the same as considered in Example 4.4)
- Sludge specific mass: $\gamma = 1{,}020$ kgTS m^{-3} (the same as considered in Example 4.4)

Important note: The example only considers the sludge production coefficient of the UASB reactor. If the STP receives other types of contributions (e.g. sludge from septic tanks, landfill leachate, non-domestic effluents, and the like.) and/or the secondary (aerobic) sludge from the post-treatment unit (e.g. trickling filters, activated sludge) for thickening and digestion in the UASB reactors, the increase in the value of the sludge production coefficient because of these contributions must be considered in the calculation of the dewatering system.

Solution

(1) Estimate the daily sludge production (P_{sludge})
- $P_{sludge} = Y_{sludge} \times OL_{app} = 0.20$ kgTS $kgCOD^{-1}{}_{app} \times 15{,}000$ $kgCOD_{app}$ $d^{-1} = 3{,}000$ kgTS d^{-1}

(2) Define the sludge frequency of withdrawal (F_w)
- Frequency of withdrawals: $F_w = 5$ d wk^{-1} (5 days per week)

(3) Sludge concentrations at the different reactor heights (obtained from the solids profile – see section 6.2.2.2)
- Upper withdrawal pipe (see Figure 6.9): $C_{sludge\text{-}upper} = 2.2\%$ (22,000 mgTS L^{-1})
- Lower withdrawal pipe (see Figure 6.9): $C_{sludge\text{-}lower} = 4.5\%$ (45,000 mgTS L^{-1})

(4) Percentage of sludge mass to be removed from the different reactor heights (adopted)
- Upper withdrawal pipe: 60%
- Lower withdrawal pipe: 40%

(5) Mass of sludge to be removed from the different reactor heights (upper and lower pipes – see Figure 6.9)
- $Mass_{sludge\text{-}upper} = [(P_{sludge} \times 7\ d\ wk^{-1})/F_w] \times 60\% = [(3{,}000\ kgTS\ d^{-1} \times 7\ d\ wk^{-1})/5\ d\ wk^{-1}] \times 0.60 = 2{,}520$ kgTS d^{-1} (5 d wk^{-1} Monday to Friday)
- $Mass_{sludge\text{-}lower} = [(P_{sludge} \times 7\ d\ wk^{-1})/F_w] \times 40\% = [(3{,}000\ kgTS\ d^{-1} \times 7\ d\ wk^{-1})/5\ d\ wk^{-1}] \times 0.40 = 1{,}680$ kgTS d^{-1} (5 d wk^{-1} Monday to Friday)

(6) Total mass of sludge to be removed from the reactors ($Mass_{sludge\text{-}total}$)
- $Mass_{sludge\text{-}total} = Mass_{sludge\text{-}upper} + Mass_{sludge\text{-}lower} = 2{,}520\ kgTS\ d^{-1} + 1{,}680\ kgTS\ d^{-1} = 4{,}200$ kgTS d^{-1} (5 d wk^{-1} Monday to Friday)

(7) Volume of sludge to be removed from the different reactor heights (upper and lower pipes — see Figure 6.9)

- $V_{sludge\text{-}upper} = Mass_{sludge\text{-}upper}/(\gamma \times C_{sludge\text{-}upper}) = 2{,}520\ kgTS/(1{,}020\ kgTS\ m^{-3} \times 0.022) = 112.3\ m^3\ d^{-1}$ (5 d wk^{-1} Monday to Friday)
- $V_{sludge\text{-}lower} = Mass_{sludge\text{-}lower}/(\gamma \times C_{sludge\text{-}lower}) = 1{,}680\ kgTS/(1{,}020\ kgTS\ m^{-3} \times 0.045) = 36.6\ m^3\ d^{-1}$ (5 d wk^{-1} Monday to Friday)

(8) Total volume of sludge to be removed from the reactors ($V_{sludge\text{-}total}$)

- $V_{sludge\text{-}total} = V_{sludge\text{-}upper} + V_{sludge\text{-}lower} = 112.3\ m^3\ d^{-1} + 36.6\ m^3\ d^{-1} = 148.9\ m^3\ d^{-1}$ (5 d wk^{-1} Monday to Friday)

(9) Average concentration of the sludge pumped to the centrifuge ($C_{sludge\text{-}avg}$)

- $C_{sludge\text{-}avg} = [(C_{sludge\text{-}upper} \times V_{sludge\text{-}upper}) + (C_{sludge\text{-}lower} \times V_{sludge\text{-}lower})]/V_{sludge\text{-}total}$
- $C_{sludge\text{-}avg} = [(0.022 \times 112.3\ m^3\ d^{-1}) + (0.045 \times 36.6\ m^3\ d^{-1})]/148.9\ m^3\ d^{-1} = 0.028$ (2.8%)

(10) Sludge load processed by the dewatering unit ($Solids_{load\text{-}dew\text{-}unit}$)

- $Solids_{load\text{-}dew\text{-}unit} = V_{sludge\text{-}total} \times \gamma \times C_{sludge\text{-}avg} = 148.9\ m^3\ d^{-1} \times 1{,}020\ kgTS\ m^{-3} \times 0.028 = 4{,}252.5\ kgTS\ d^{-1}$ (5 d wk^{-1} Monday to Friday)

(11) Number of hours of operation of the dewatering unit ($N\text{-}hours_{field}$)

- $N\text{-}hours_{field} = Solids_{load\text{-}dew\text{-}unit}/V_{cap\text{-}dew\text{-}unit} = 4{,}252.5\ kgTS\ d^{-1}/461\ kgTS\ h^{-1} = 9.2\ h\ d^{-1}$

As regards the centrifuges, the sludge removal operation is done from the opening of the valves located on the outside wall of the reactors, which allows the sludge to be directed to the balance and homogenisation tank (Figure 6.19(a)) and pumped (Figure 6.19(b)) subsequently to the centrifuge (Figure 6.19(c)). In addition, it is essential to open each sludge withdrawal valve individually (one at a time), corresponding with each sludge withdrawal point (level).

(a) (b) (c)

Figure 6.19 Mechanised dewatering system: (a) view of a sludge balance and homogenisation tank (receives sludge from the UASB reactors and feeds the dewatering equipment); (b) positive displacement pumps used to feed the dewatering equipment; and (c) view of a centrifuge.

From the balance and homogenisation tank, the operator can obtain samples to measure the concentration of sludge and define the dosage of polymer to be applied in the feeding line of the centrifuge.

6.5.4 Removal of scum from the inside of GLS separators

6.5.4.1 GLS separators equipped with hydrostatic scum removal devices

Once the ideal scum removal protocol is established (see Table 6.3), the routine operation of the hydrostatic removal device will consist simply of opening the external withdrawal valve, according to the time and degree of valve opening defined during the start-up period, and allow the *fluid scum* to flow to the drying bed (Figure 6.20 (a, b)) or to the sludge balance and homogenisation tank for processing in the mechanised dewatering unit.

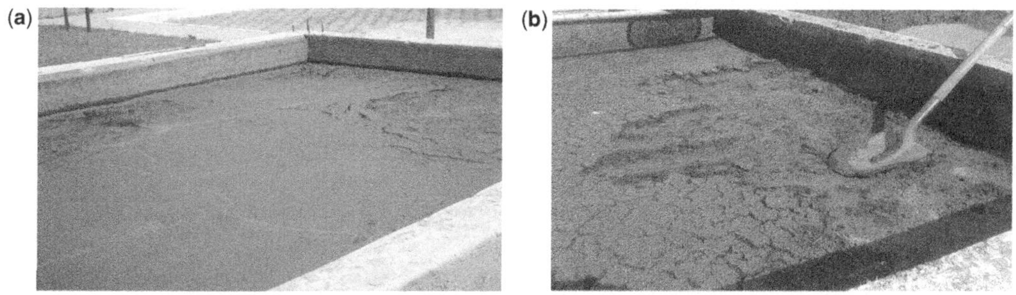

Figure 6.20 Drying bed containing scum removed from the inside of a GLS separator equipped with a hydrostatic removal device: (a) photo taken immediately after disposal; (b) photo taken three days after disposal.

6.5.4.2 GLS separators not equipped with hydrostatic scum removal devices

In UASB reactors lacking proper scum removal devices (old generation reactors), the removal operation tends to be less frequent and the scum layer usually becomes thicker and more solidified, posing difficulties to the operators. In such instances, the scum is typically removed through the hermetic inspection hatches located on top of the GLS separator. The scum can be removed manually (Figure 6.21(a)), with a 'skimmer' or a suction hose coupled to a cesspool cleaning truck (Figure 6.21(b, c)). The manual removal option can be employed only when the accumulated quantities are relatively small and can be assimilated into the operational routine of the STP. The solidified scum can be disposed of directly into landfills.

Figure 6.21 Removal of scum from the inside of GLS separators: (a) manual removal, (b) and (c) removal with cesspool cleaning trucks and suction hose.

Important note: as this scum removal procedure requires opening the inspection hatches of the GLS separator, biogas will escape to the atmosphere, posing health and security risks to the operators close to the inspection hatches. The risks are related to the emission of hydrogen sulphide (a malodorous and toxic compound) and methane (an explosive compound), as discussed in Chapters 8 and 9. Therefore, all essential health and safety rules must be followed before opening these inspection hatches.

6.5.5 Inspection of equipment and protection of structures

In addition to the activities mentioned previously, the operational routine of anaerobic-based STPs should include other equally important aspects (adapted from Chernicharo, 2007):

- Inspection of all equipment comprising the STP regularly, especially those without a back-up unit.
- Verification of the occurrence of corrosion in the structure of the anaerobic reactor, particularly in the steel and concrete parts, such as effluent collection channels, gas collectors, guard rails, and the like; in case of corrosion, the affected structures should be repaired quickly, aiming at both the integrity of the treatment unit and the safety of the system operators.
- Correct destination of all solid material removed in the preliminary treatment unit (screens, sieves, and grit chambers) and sludge wasted from the anaerobic reactor.
- Cleaning and weeding of the STP area (access roads, drying beds, and the like).

6.6 OPERATIONAL TROUBLESHOOTING

The following tables (Tables 6.4 to 6.7) present information for detecting and correcting operational problems in anaerobic-based STPs.

Table 6.4 Troubleshooting list related to the preliminary treatment unit.

Observation	Probable Cause	Aspect to be Verified	Solution
Odour or insects at the screen	• Long cleaning intervals	• Cleaning interval	• Increase the cleaning frequency
Sudden increase of coarse solids retained	• Illegal dumping of solid wastes	• Existence of illegal sources	• Identify and take action on the sources to correct the problem
Sudden decrease in coarse solids retention	• Retention failure at the screen	• Condition of the screen	• Repair the screen
Sudden increase in grit retained	• Discharge of storm water into the sewerage system	• Sewage flow	• Identify and disconnect illegal connection
Sudden decrease in grit retention	• Grit dragged from the grit chamber	• Sewage flow and hydraulic loading rate	• Reduce flow or expand the capacity of the grit chamber
Malodorous emissions in the grit chamber	• Sedimentation of organic matter	• Sewage flow and hydraulic loading rate	• Increase flow by reducing the number of grit chambers in operation
Grit retained is grey, has unpleasant odour, and contains grease	• Sedimentation of organic matter	• Sewage flow and hydraulic loading rate	• Increase flow by reducing the number of grit chambers in operation
Corrosion in the preliminary treatment units	• Release of H_2S from the incoming sewage	• Concentration of H_2S in the nearby atmosphere	• Use of corrosion-proof material. Confine the channels, extract, and treat the gases

Source: Adapted from Chernicharo (2007).

Table 6.5 Troubleshooting list related to the performance of the UASB reactor.

Observation	Probable Cause	Aspect to be Verified	Solution
Unequal influent distribution	• Distribution structure unlevelled	• Level of the distribution structure	• Level the distribution structure
Feeding pipe does not receive sewage	• Blocking • Undesirable solids retention in the preliminary treatment	• Water level of the distribution chambers • Opening screens	• Unblock feeding pipes • Improve solids removal in the preliminary treatment units
Non-uniform effluent collection	• Collection structure unlevelled • Surface layer obstructs collection points	• Level of the collection structure • Flow conditions	• Level the collection structure • Remove obstruction
High level of settleable solids in the effluent	• Excessive hydraulic load • Excess sludge in the reactor	• Flow and upflow velocities • Sludge concentration in sampling point 5 (Figure 6.3(a))	• Reduce flow by expanding the number of reactors in operation • Withdraw the excess sludge
Gas production lower than normal	• Biogas leakage • Defective gas meter • Small sewage flow • Influent dilution because of rainwater • Toxic material in the influent • Excessive organic load	• Gas collection system • Gas meter • Influent flow • Influent flow and concentration • SMA test • SMA and stability test	• Eliminate leakage • Repair or replace • Unblock/repair sewers • Identify and disconnect illegal connections • Identify and take action on sources of toxic material • Reduce organic load
Sludge production higher than normal	• Overloaded sludge • Coarse and/or inorganic solids entering the reactor	• Sludge stability • Operation of the preliminary treatment	• Reduce applied load • Re-establish operation of the preliminary treatment units
Sludge production lower than normal	• Small sewage flow • Influent dilution because of rainwater • Deficient sludge retention	• Influent sewage flow • Influent flow and influent concentration • GLS separator; settleable solids in the effluent	• Unblock/repair sewers • Identify and disconnect illegal connections • Repair GLS separator
Sludge with high fraction of inorganic solids	• Defective grit chamber • Hydraulic overload of grit chamber • Influent with excess of grit	• Operation of grit chamber • Hydraulic loading rate and velocity • Amount of grit removal in the preliminary treatment unit	• Repair grit chamber • Reduce flow • Identify and act on sources of grit intrusion
Floating sludge accumulates rapidly	• Excessive hydraulic load	• Hydraulic retention time and upflow velocities	• Reduce influent flow by expanding the number of reactors in operation
Low organic matter removal efficiency	• Hydraulic overload • Deficient influent distribution	• Hydraulic retention time • Influent distribution system (tracer studies)	• Reduce influent flow by expanding the number of reactors in operation • Repair distribution system

Source: Adapted from Chernicharo (2007).

Table 6.6 Troubleshooting list related to the sludge characteristics.

Observation	Probable Cause	Aspect to be Verified	Solution
SMA lower than expected	• Entrance of inert solids	• Settleable solids in influent • Sludge stability and organic matter removal efficiency	• Reduce source or revise preliminary treatment • Reduce load of inert material
	• Presence of toxic material	• Test stored sludge	• Identify and take action on sources of toxic materials
Poor stability	• Sludge overload	• Specific organic load	• Reduce specific load by increasing the amount of biomass and/or expanding the number of reactors in operation
High sludge volume index	• Biodegradable organic matter	• Stability of the process	• Reduce organic load by expanding the number of reactors in operation
	• Low hydraulic load	• Upflow velocity	• Increase dragging temporarily
Poor settleability	• Dispersed flocs from excessive organic load	• Sludge stability (test)	• Reduce organic load by expanding the number of reactors in operation
	• Presence of toxic material	• SMA of the sludge	• Identify and take action on sources of toxic materials
Increased specific sludge production	• Flocculation without metabolism	• Sludge stability (test)	• Reduce specific load by increasing the amount of biomass and/or expanding the number of reactors in operation
Increased inorganic fraction	• Entrance of silt and grit	• Velocity in the grit chamber	• Reduce flow or expand the capacity of the grit chamber
	• Low upflow velocity	• Upflow velocity	• Increase hydraulic load by reducing the number of reactors in operation

Source: Adapted from Chernicharo (2007).

Table 6.7 Troubleshooting list related to the drying beds.

Observation	Probable cause	Verify	Solution
Unpleasant odour during sludge withdrawal	• Sludge instability	• Sludge stability (test)	• Reduce organic load by expanding the number of reactors in operation
Blockage of sludge pipeline	• Accumulation of solids and grit	• Occurrence of blocked pipes	• Clean sludge pipeline after use
Long percolation time	• Excessive TS load applied • Inadequate bed cleaning • 'Blind' sand	• Applied load • Verify permeability • Upflow washing with water	• Reduce load of TS applied • Improve maintenance • Replace sand
	• *High rainfall* • Drainage system blocked • Air trapped in the bed preventing passage of water		• Cover drying bed • Apply upflow washing • Apply water in upward direction
Long evaporation time	• Excessive load applied • High rainfall, low temperatures, high air humidity	• Load applied	• Reduce load • Reduce load, cover drying bed
Extremely diluted excess sludge	• Sludge removal from an extremely high level in the reactor	• Solids concentration profile	• Remove the sludge from a lower level (closer to the bottom of the reactor)
Mosquito reproduction	• Semi-permanent water layer	• Drainage system	• Reduce load, improve permeability

Source: Adapted from Chernicharo (2007).

REFERENCES

Catunda P. F. C. and van Haandel A. C. (1987). Activated sludge settlers: Design and optimization. *Water Science and Technology*, **19**(3–4), 613–623.

Chernicharo C. A. L. (2007). Anaerobic Reactors. IWA Publishing, London, UK, 184pp.

Chernicharo C. A. L., Lobato L. C. S., Paiva Júnior C. A., Barbosa C. A., Almeida P. G. S., Gonçalves E. V., Souza J. R., Silva L. R., Judice M. A. M. and Moraes O. J. (2014). Return of secondary sludge for thickening and digestion in UASB reactors: case study of Onça STP – Brazil. In: Proc. 11 Simposio Latinoamericano de Digestión Anaerobia, International Water Association, Havana, Cuba.

Chernicharo C. A. L., Bressani-Ribeiro T., Pegorini E., Possetti G. R. C., Miki M. K. and Nonato S. (2018). Contribuição para o aprimoramento de projeto, construção e operação de reatores UASB aplicados ao tratamento de esgoto sanitário – Parte 1: Tópicos de Interesse (Contribution to the improvement of the design, construction and operation of UASB reactors applied to sewage treatment – Part 1: Topics of Interest). *Revista DAE*, **66**(214), 5–15.

Diaz Flórez C. A. (2016). Avaliação de protocolos de descarte e desidratação da escuma acumulada no interior de separadores trifásicos de reatores UASB (Evaluation of protocols for the disposal and dewatering of accumulated scum inside GLS separators of UASB reactors). MSc thesis, Graduate Programme on Sanitation, Environmental and Water Resources, Federal University of Minas Gerais, Belo Horizonte, Minas Gerais.

Floripes T. C., Chernicharo C. A. L. and Mota-Filho C. R. (2018). Avaliação do descarte de excesso de lodo secundário de FBP sobre o desempenho de reatores UASB em escala plena: estudo de caso da ETE Laboreaux – Itabira-MG (Assessment of excess sludge withdrawal from trickling filters on the performance of full-scale UASB reactors: case study of ETE Laboreaux – Itabira-MG). *Revista DAE*, **66**(211), 89–104.

Lettinga G. and Hulshoff Pol L. W. (1991). UASB-process design for various types of wastewaters. *Water Science & Technology*, **24**(8), 87–107.

Leitão R. C., Santaellla S. T., van Haandel A. C., Zeeman G. and Lettinga G. (2011). The effect of operational conditions on the hydrodynamic characteristics of the sludge bed in UASB reactors. *Water Science & Technology*, **64**(9), 1935–1941.

Rosa A. P., Lobato L. C. S., Chernicharo C. A. L., Martins D. C. R. B., Maciel F. M. and Borges J. M. (2012). Improving performance and operational control of UASB reactors via proper sludge and scum discharge routines. *Water Practice & Technology*, **7**, 1–11.

Santos A. (2014). Remoção, tratamento e valoração de escuma proveniente de reatores UASB aplicados ao tratamento de esgoto doméstico (Scum removal, treatment and evaluation from UASB reactors applied to domestic sewage treatment). PhD thesis, Graduate Programme on Sanitation, Environmental and Water Resources, Federal University of Minas Gerais, Belo Horizonte, Minas Gerais, Brazil.

Souza C. L., Silva S. Q., Aquino S. F. and Chernicharo C. A. L. (2006). Production and characterization of scum and its role in odour control in UASB reactors treating domestic wastewater. *Water Science & Technology*, **54**(9), 201–208.

Chapter 7

Experience with full-scale UASB reactors treating sewage

Carlos A. L. Chernicharo[1], Lívia C. S. Lobato[1], Thiago Bressani-Ribeiro[1], Rodrigo A. S. Pereira[2], André Lermontov[2] and Leonardo Luis Francisco[3]

[1]Departament of Sanitary and Environmental Engineering (DESA), School of Engineering, Universidade Federal de Minas Gerais (UFMG), Av. Antônio Carlos 6.627, Belo Horizonte/MG, 31270-901, Brazil

[2]Grupo Águas do Brasil, Rua Luiz Leopoldo Fernandes Pinheiro, 534 – Centro, Niterói/RJ, 24030-121, Brazil

[3]Paques Brasil Sistemas para Tratamento de Efluentes Ltda, Av. Prof. Benedito de Andrade, 511, Piracicaba/SP, 13422-000, Brazil

**Corresponding author: E-mail: calemos@desa.ufmg.br*

7.1 INTRODUCTION

Anaerobic-based technology for mainstream sewage treatment in warm climate regions has expanded significantly (Chernicharo *et al.*, 2018; Noyola *et al.*, 2012), and the disclosure of the performance data of full-scale UASB-based sewage treatment plants (STPs) is an important issue for discussion. This is particularly relevant in view of the different STP sizes, technologies for effluent post-treatment, recovery of byproducts (e.g. biogas), innovative configurations available in the market (e.g. hybrid anaerobic reactor), operational features (e.g. return of aerobic sludge for thickening and digestion), and the associated operational costs.

The aim of this chapter is to discuss these aspects in relation to experience with six different full-scale UASB-based STPs currently in operation in Brazil, as follows:

- *STP 1*: UASB reactors (without post-treatment step).
- *STP 2*: UASB reactors followed by trickling filters (TFs) – medium-scale STPs.

Editor(s): Carlos Augusto de Lemos Chernicharo and Thiago Bressani-Ribeiro
doi: 10.2166/9781780409238_0159

- *STP 3*: UASB reactors followed by trickling filters (TFs) – large-scale STPs.
- *STP 4*: UASB reactors followed by trickling filters (TFs) – large-scale STPs (comprising energy recovery from biogas).
- *STP 5*: Innovative UASB reactor configuration (Biopaq® Ubox reactor).
- *STP 6*: Innovative UASB reactor configuration (hybrid anaerobic reactor – HAnR) followed by integrated fixed-film activated sludge (IFAS).

Overall, the operational performance of each STP was assessed based on the final effluent quality and removal efficiency in terms of biochemical oxygen demand (BOD) and chemical oxygen demand (COD). The total suspended solids (TSS) and ammonium ($NH_4^+ - N$) concentrations/removal efficiencies were also examined when these data were made available. As regards STP 2, specifically, the performance evaluation also comprised: (i) the effect of the implementation of a protocol for sludge withdrawal on the effluent quality, and (ii) the results of the implementation of a protocol for systematic scum removal inside the gas–liquid–solid (GLS) separator. The assessment of the performance of STP 3 considered the effect of receiving sludge contributions not foreseen at the design stage. Additionally, the evaluation comprised the monthly operating costs per cubic metre of treated sewage. As regards STP 4, the biogas recovery facility of the STP was also assessed. STP 5 and STP 6 are examples of process innovations that have been implemented recently and are available in the worldwide market. Table 7.1 summarises the main characteristics of each evaluated STP.

Table 7.1 Main characteristics of each assessed STP.

Characteristic	Sewage Treatment Plant					
	1	2	3	4	5	6
Configuration	UASB only	UASB + TF	UASB + TF	UASB + TF + ER	Biopaq® Ubox	HAnR + IFAS
Population equivalent (PE)	79,000	70,000	1,000,000	392,000	30,000	79,000
Design flow rate ($L\ s^{-1}$)	200	170	2,050	990	66	164
Current flow rate ($L\ s^{-1}$)	99	80	1,850	320	45	90

Note: TF: trickling filter; ER: energy recovery; HAnR: hybrid anaerobic reactor; IFAS: integrated fixed-film activated sludge.

7.2 SEWAGE TREATMENT PLANT 1

7.2.1 Overall plant description

STP 1 was designed for a sewage flow rate of 200 $L\ s^{-1}$, corresponding to a treatment capacity of approximately 79,000 PE. The flowsheet comprises preliminary treatment units (screens and grit chamber), UASB reactors, drying beds for sludge dewatering, and a flare for biogas burning (there is no energy recovery), as shown in Figures 7.1 and 7.2. The post-treatment step of the anaerobic effluents was not implemented. The main characteristics of the preliminary treatment units and the design and operational parameters of the UASB reactors are presented in Tables 7.2 and 7.3.

7.2.2 System performance

The operational performance of STP 1 was assessed based on data provided by the sanitation company responsible for the plant, accounting for an operational period of 36 months. The results comprise the historical series of the influent and effluent BOD and COD concentrations (Figure 7.3(a, c)). These data are also presented as boxplot graphs (Figure 7.3(b, d)). Figure 7.4 shows the overall BOD and COD removal efficiencies for the analysis period.

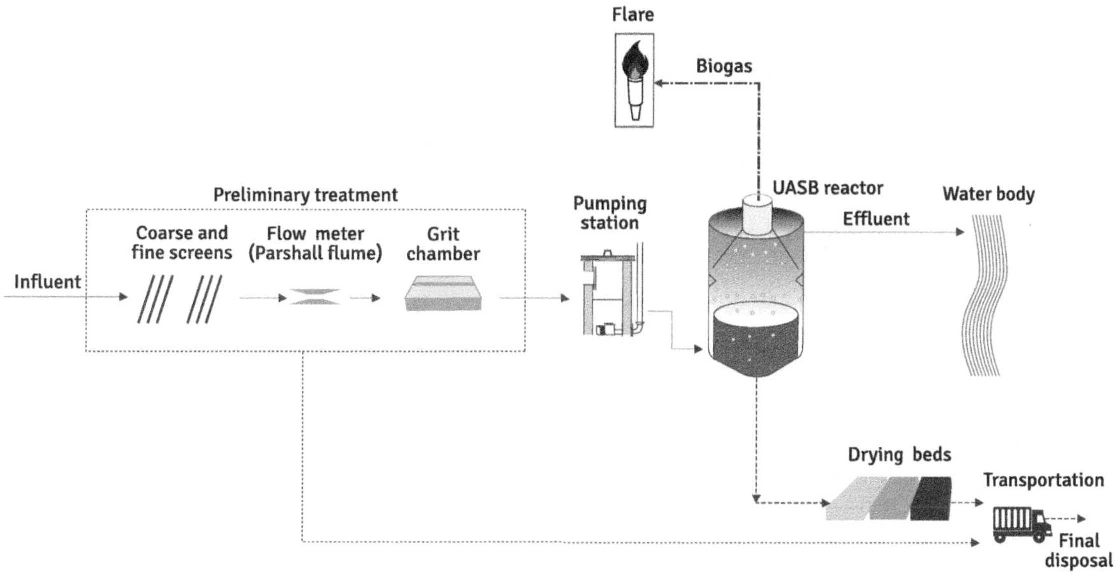

Figure 7.1 Flowsheet of STP 1.

Figure 7.2 General views of STP 1: (a) Top view of semi-buried UASP reactors, and (b) view of the sludge drying beds.

Table 7.2 Main characteristics of the preliminary treatment units of STP 1.

Design Parameters	Coarse Screen	Fine Screen	Grit Chamber
Type	**Manual**	**Mechanised**	**Mechanised**
Number of units	1	1	1
Aperture (mm)	100	15	–
Hydraulic loading rate (HLR) ($m^3\ m^{-2}\ d^{-1}$)	–	–	1,600

Table 7.3 Main design and current operational parameters of the UASB reactor in STP1.

Parameters	UASB Reactors
Design	
Population equivalent (PE)	79,000
Flow rate (L s^{-1})	200
Hydraulic retention time (HRT) (h)	10.7
Upflow velocity (m h^{-1})	0.51
Number of units[a]	4
Type	Rectangular
Area (m^2)	353.8
Useful depth (m)	5.80
Useful volume of each tank (m^3)	1,940
Current operation	
Flow rate (L s^{-1})	99
HRT (h)	16.4
Upflow velocity (m h^{-1})	0.33

[a]Only three UASB reactors were in operation during the assessment period.

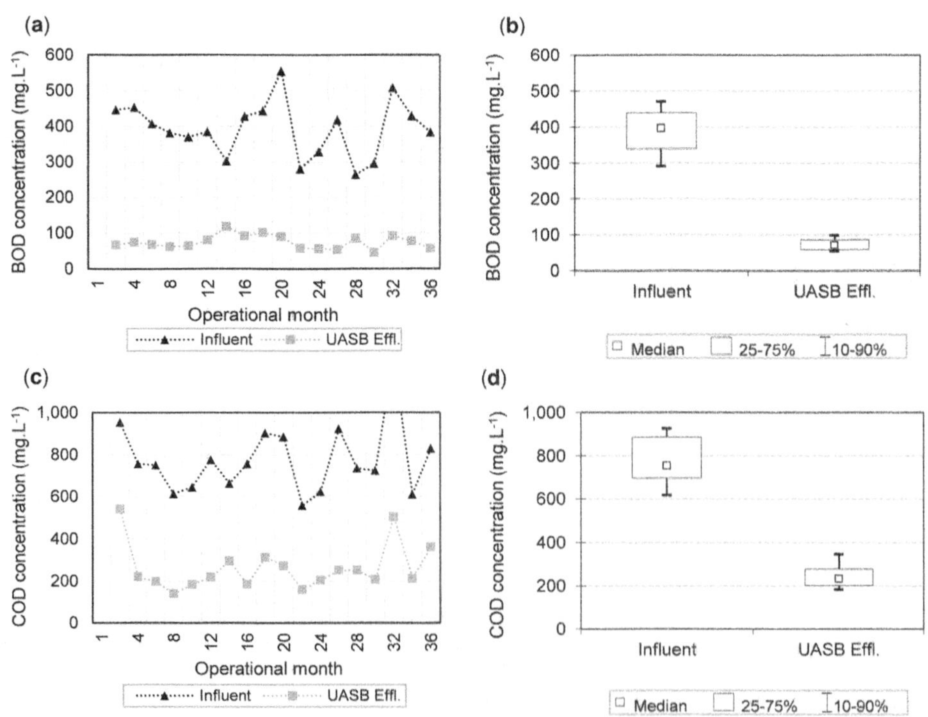

Figure 7.3 Historical series (a, c) and boxplots (b, d) of influent and effluent BOD and COD concentrations of STP 1.

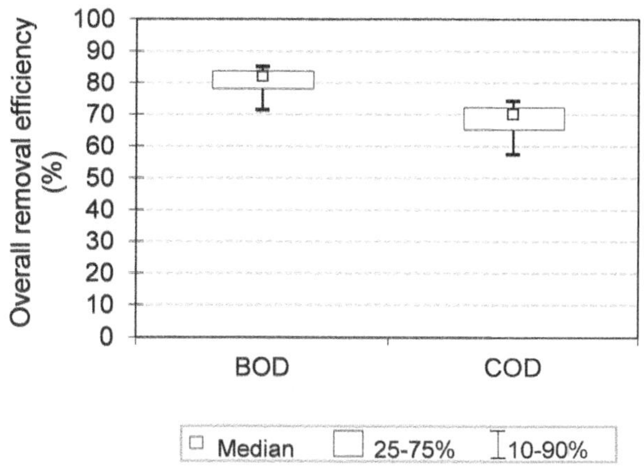

Figure 7.4 Boxplot of the overall BOD and COD removal efficiencies of STP 1.

The following observations can be made from the operational results of STP 1:

- The median influent (raw sewage) BOD and COD concentrations (396 and 754 mg L^{-1}, respectively) resemble sewage that is more concentrated, as these values are higher than are the typical concentrations reported by von Sperling and Chernicharo (2005) for sewage (300 mgBOD L^{-1} and 600 mgCOD L^{-1}).
- The median effluent BOD and COD concentrations (71 and 232 mg L^{-1}, respectively) are, to some extent, typical of the effluents of UASB reactors treating sewage, considering the reports on full-scale UASB reactor performances by Chernicharo *et al.* (2015).
- The median overall BOD and COD removal efficiencies (82 and 70%, respectively) are higher than the typical ranges reported in the literature for UASB reactors treating sewage (60 to 75% and 55 to 70% for BOD and COD, respectively) according to von Sperling and Chernicharo (2005). These high efficiencies are possibly associated with the current, excessive, hydraulic retention time of the UASB reactors (16.4 h).

7.3 SEWAGE TREATMENT PLANT 2

7.3.1 Overall plant description

STP 2 was designed for a sewage flow rate of 170 L s^{-1}, corresponding to a treatment capacity of approximately 70,000 PE. The flowsheet comprises preliminary treatment units (screens and grit chambers), UASB reactors followed by trickling filters (TFs) and secondary settlers, a sludge dewatering unit, and a biogas flare (there is no energy recovery), as shown in Figures 7.5 and 7.6. The excess aerobic sludge withdrawn from the secondary settler is returned to the UASB reactors for thickening and digestion. The excess sludge withdrawn from the UASB reactors is routed to a filter press and, subsequently, to final disposal in a landfill. This STP features spare drying beds should an operational breakdown occur, or preventive/corrective maintenance of the filter press unit were needed. The main characteristics of the preliminary treatment units and the design and operational parameters of the UASB + TF system are presented in Tables 7.4 and 7.5.

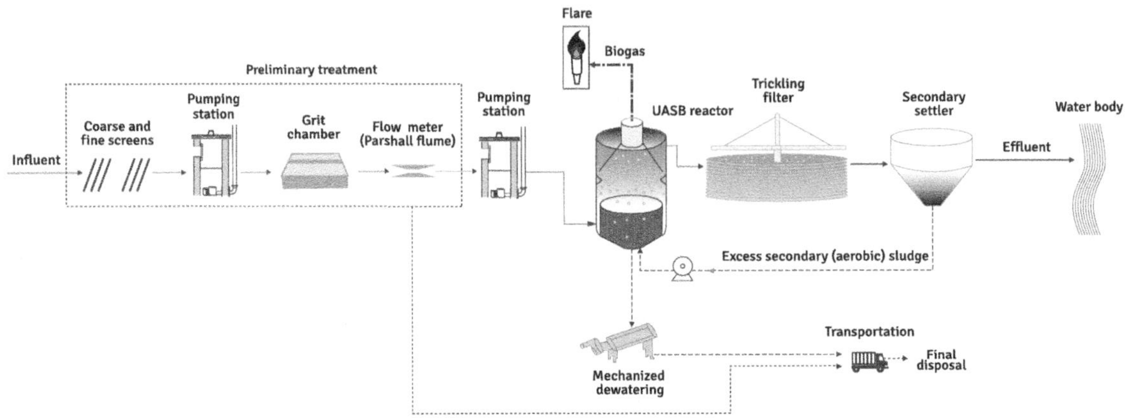

Figure 7.5 Flowsheet of STP 2.

Figure 7.6 Aerial view of STP 2.

Table 7.4 Main characteristics of the preliminary treatment units of STP 2.

Design Parameters	Coarse Screen	Fine Screen	Grit Chambers
Type	**Manual**	**Mechanised**	**Mechanised**
Number of units	1	1	2
Aperture (mm)	50	15	–
Hydraulic loading rate (HLR) ($m^3\ m^{-2}\ d^{-1}$)	–	–	900

Table 7.5 Main design and current operational parameters of the UASB + TF system in STP 2.

Parameters	UASB Reactors	Trickling Filters[a]	Secondary Settlers
Design			
Population equivalent (PE)		70,000	
Flow rate (L s^{-1})		170	
Hydraulic retention time (HRT) (h)	7.9	–	3.1
Upflow velocity (m h^{-1})	0.57	–	–
Organic loading rate (OLR) (kgBOD m^{-3} d^{-1})	–	<1.00	–
Hydraulic loading rate (HLR) (m^3 m^{-2} d^{-1})	–	18.5	23.4
Number of units	8	2	2
Type	Rectangular	Circular	Circular
Dimensions (m)	21.70 × 6.20	$D = 22.50$	$D = 20.00$
Area (m^2)	134.5	397.6	314.0
Useful depth (m)	4.50	2.50	3.00
Useful volume of each unit (m^3)	605.25	994	942
Current operation			
Flow rate (L s^{-1})		80	
HRT (h)	8.4[b]	–	6.5
Upflow velocity (m h^{-1})	0.54[b]	–	–
HLR (m^3 m^{-2} d^{-1})	–	8.7	11.0

[a]The trickling filters are filled with rock.
[b]Only four out of the eight UASB reactors were operating during the assessment period.
Source: Adapted from Rosa *et al.* (2012) and Floripes *et al.* (2018).

7.3.2 System performance

The operational performance of STP 2 was assessed based on data provided by the sanitation company responsible for the plant, and accounts for an operational period of 18 months. It should be noted that the return of the aerobic sludge from the trickling filters (secondary settler units) for thickening and digestion in the UASB reactors was in operation throughout the analysis period. The results comprise the historical series of the influent and effluent COD and TSS concentrations (Figure 7.7(a, c)). These data are also presented as boxplot graphs (Figure 7.7(b, d)). Figures 7.8(a, b) shows the UASB reactor efficiency and the overall STP efficiency (UASB + TF) for COD and TSS removal.

The following observations can be made from the operational results of STP 2:

- The median influent COD and TSS concentrations were 654 and 380 mg L^{-1}, respectively, which match the typical concentrations reported by von Sperling and Chernicharo (2005) for sewage (600 mgCOD L^{-1} and 350 mgTSS L^{-1}).
- The median effluent COD concentration and removal efficiency of the UASB reactors (270 mg L^{-1} and 56%, respectively) matches the typical range reported by von Sperling and Chernicharo (2005) for UASB reactors (180–270 mgCOD L^{-1} and 55–70%).
- The median effluent TSS concentration of the UASB reactors (196 mg L^{-1}) was higher than is the typical range reported by von Sperling and Chernicharo (2005) for UASB reactors (60–100 mgTSS L^{-1}). The median TSS removal efficiency (59%) of the UASB reactors was lower than is

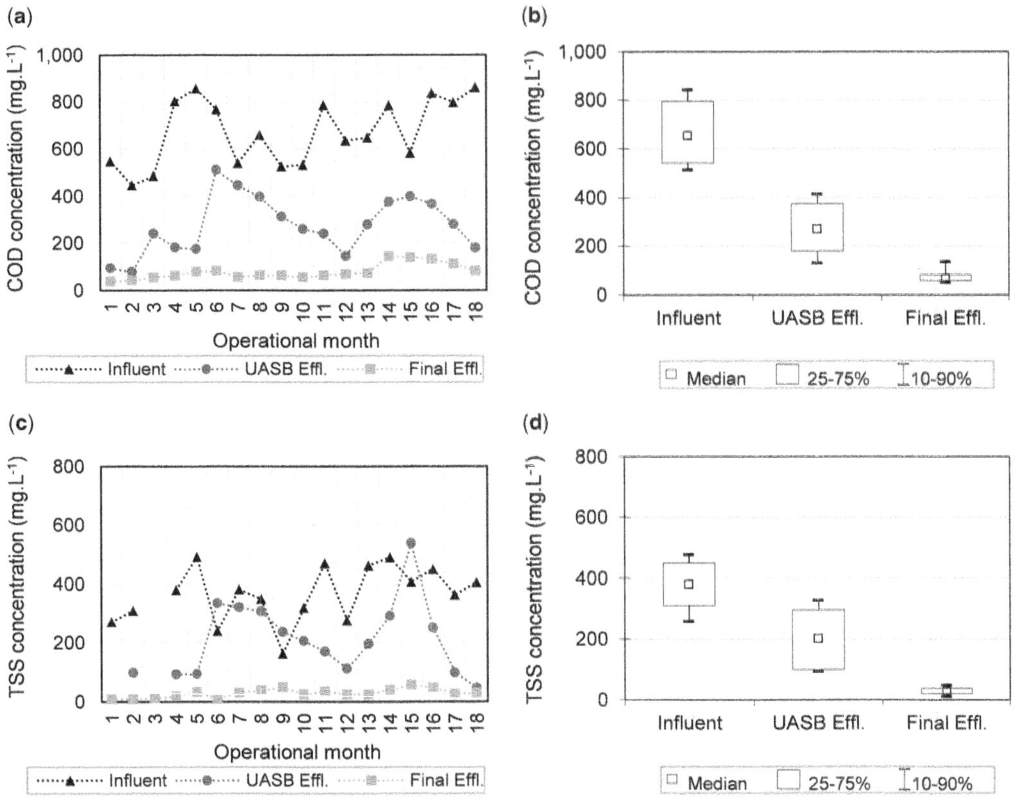

Figure 7.7 Historical series and boxplots of COD (a, b) and TSS (c, d) concentrations of STP 2. (Source: adapted from Floripes *et al.*, 2018).

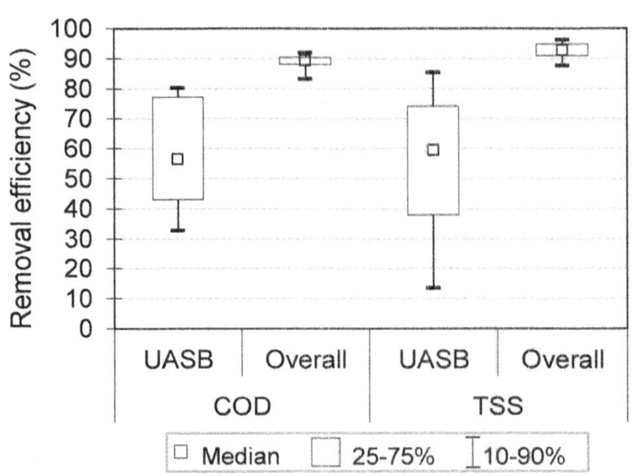

Figure 7.8 Boxplot of COD and TSS removal efficiencies of STP 2. (Source: adapted from Floripes *et al.*, 2018).

the typical range reported by von Sperling and Chernicharo (2005) for anaerobic sewage treatment (65–80%). The higher effluent TSS concentrations and lower removal efficiencies can be ascribed to the presence of the excess sludge stored in the UASB reactors, leading to the washout of solids with the effluent.

- As regards the final effluent of STP 2, the median COD and TSS concentrations (67 and 28 mg L^{-1}, respectively) are within the typical ranges reported by von Sperling and Chernicharo (2005) for UASB + TF systems treating sewage (30–120 mgCOD L^{-1} and 20–40 mgTSS L^{-1}).
- The median overall COD and TSS removal efficiencies (89 and 93%, respectively) match the typical ranges reported by von Sperling and Chernicharo (2005) for UASB + TF systems treating sewage (80–90% and 87–93%, for COD and TSS, respectively).

7.3.3 Return of aerobic sludge for thickening and digestion

As discussed in Chapter 6, the return of aerobic sludge from post-treatment units for thickening and digestion in UASB reactors has been determined as one of the causes of solids washout with the effluent from UASB reactors. However, practical experience has shown that the implementation of proper operational protocols for the return of aerobic sludge and withdrawal of excess sludge from UASB reactors facilitates a reliable STP performance. Figure 7.9(a–d) shows the results of the COD and TSS concentrations and the removal efficiencies for two operational periods, namely: (i) period 1 (two months) – UASB reactors without excess sludge withdrawal (sludge accumulated above the maximum reactor storage capacity); and (ii) period 2 (two months) – UASB reactors with a proper sludge withdrawal protocol. The aim was to improve the assessment of the effect on the effluent quality by implementing protocols for the withdrawal of excess sludge.
The following observations can be made from Figure 7.9:

- The median effluent COD and TSS concentrations (296 and 243 mg L^{-1}, respectively) of the UASB reactors in period 1 were much higher than were those observed in period 2 (228 and 68 mg L^{-1}, respectively).
- The median COD and TSS removal efficiencies for the UASB reactors in period 1 (57 and 45%, respectively) were much lower than were those observed in period 2 (75 and 84%, respectively).
- As regards the final effluent (UASB + TF), the median COD and TSS concentrations in period 1 (76 and 38 mg L^{-1}, respectively) and period 2 (97 and 29 mg L^{-1}, respectively) were closer.
- The median overall (UASB + TF) COD and TSS removal efficiencies in period 1 (88 and 90%) and period 2 (89 and 92%) were also closer, corresponding to the typical ranges reported by von Sperling and Chernicharo (2005) for UASB + TF systems treating sewage (80–90% and 87–93%, for COD and TSS, respectively).

In period 1, the low removal efficiencies of TSS (and, consequently, high TSS concentrations) in the UASB reactor effluent were related to constraints in the dewatering system, namely the filter press was undergoing maintenance repair, there was a lack of chemical products and, consequently, overloading of the spare drying beds. Such problems prevented the proper withdrawal of excess sludge and, therefore, led to the significant accumulation of sludge inside the UASB reactors. However, in period 2, with the implementation of a proper excess sludge withdrawal protocol, comprising a systematic sludge discharge frequency, the excess sludge was withdrawn effectively from the UASB reactors, which improved the effluent quality. Despite the high losses of solids in the UASB reactors in period 1, which have overloaded the trickling filters and secondary clarifiers, the overall performance of the plant did not change.

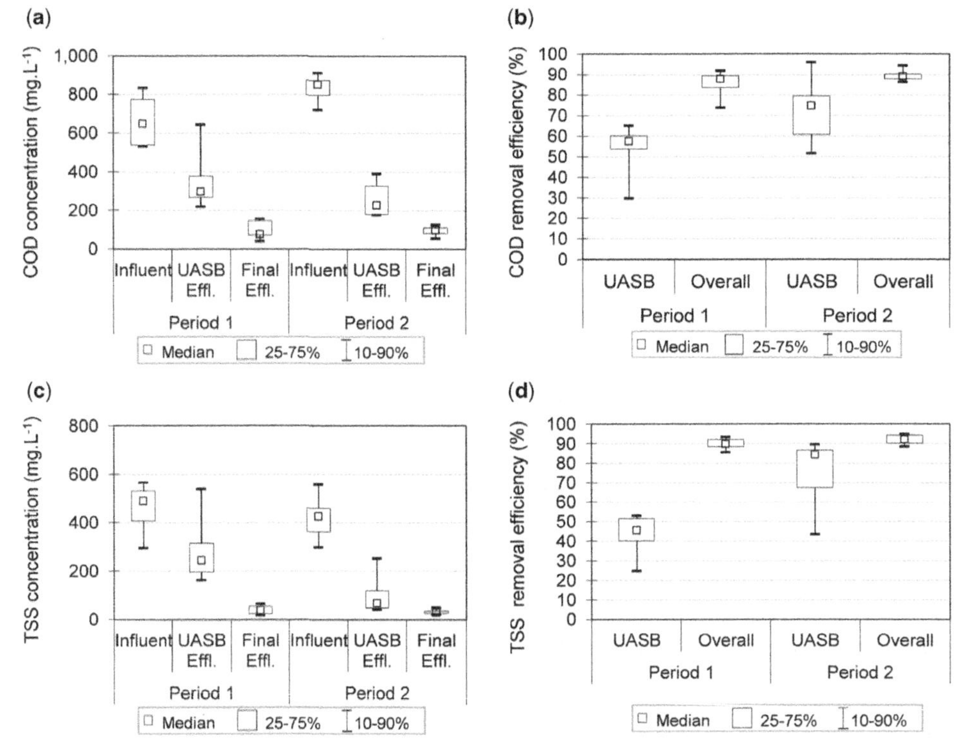

Figure 7.9 Boxplot of concentrations and removal efficiencies for period 1 (UASB reactors operated with excess sludge above the maximum storage capacity) and period 2 (UASB reactors operated with proper sludge withdrawal protocol) for COD (a, b) and TSS (c, d). (Source: adapted from Floripes *et al.*, 2018).

Additionally, Figure 7.10(a–b) shows the sludge concentration (TS, %) along the height of the digestion compartments of the UASB reactors (solids profile) for the two analysis periods.

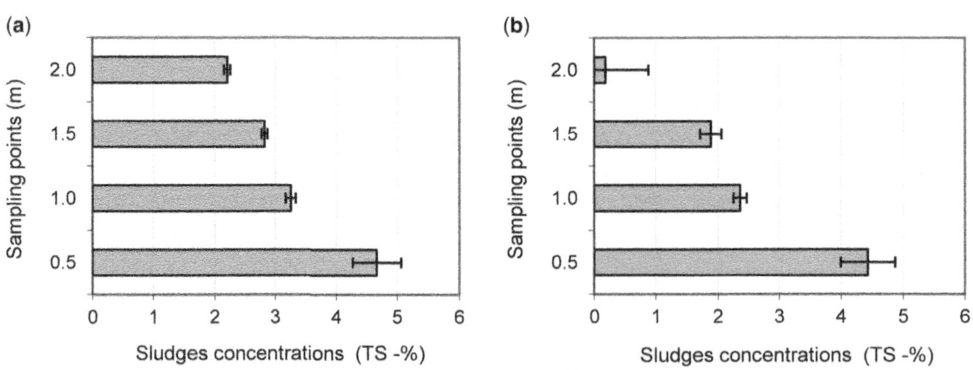

Figure 7.10 Concentrations of sludge (TS, %) along the height of the digestion compartment of the UASB reactors for period 1 (a) and period 2 (b). (Source: adapted from Floripes *et al.*, 2018).

The following observations can be made from Figure 7.10:

- In period 1, the sludge concentrations at the digestion and settling compartments interface of the UASB reactors (sampling point at 2.0 m) remained above 2% TS, indicating the probability of solids washout with the effluent and, indeed, was observed (see Figure 7.9(c, d)).
- In period 2, with the implementation of a proper sludge withdrawal protocol, the TS concentrations at the highest point of the digestion compartment (2.0 m height) were below 0.5%, as was recommended in Chapter 6. This prevented the storage of sludge over the maximum capacity of the UASB reactors and, therefore, prevented the deterioration of the effluent quality.

During both analysis periods, the aerobic sludge was returned for thickening and digestion in the UASB reactors. Accordingly, it can be concluded that the success of this operational practice depends intrinsically on a rigorous sludge management control to prevent high concentrations of sludge (TS) at the upper part of the digestion compartment, near the apertures to the settling compartment.

7.3.4 Scum removal from inner GLS separators

As discussed in Chapter 6, one of the major operational constraints reported in full-scale STPs is the removal of scum that accumulates inside the GLS separators. The accumulation and irregular removal of scum lead to a blockage of the natural passage of gas, which, in turn, impairs its collection and imposes hurdles for energy recovery.

As discussed in Chapter 4, advances have been made in the removal of scum from UASB reactors through the design of hydrostatic removal devices, which are based on the control of the water level within the GLS separator. The operational protocols tested in STP 2 indicated that an appropriate pressure arrangement in the gas line (5–10 cm.w.c (490–980 Pa)) and the resulting level of scum within the gas chambers enabled the effective removal of the scum at approximately 75–90% of the initial volume recorded (Figure 7.11(a–b)) (Rosa *et al.*, 2012; Santos, 2014). The level of scum inside the GLS separator was maintained at 3 cm below the upper edge of the internal collection channel. Therefore, opening the outside valves for the removal of scum produced pressure relief in the gas line, allowing the elevation of the scum level and its entrance in the internal GLS separator weir.

(a) (b)

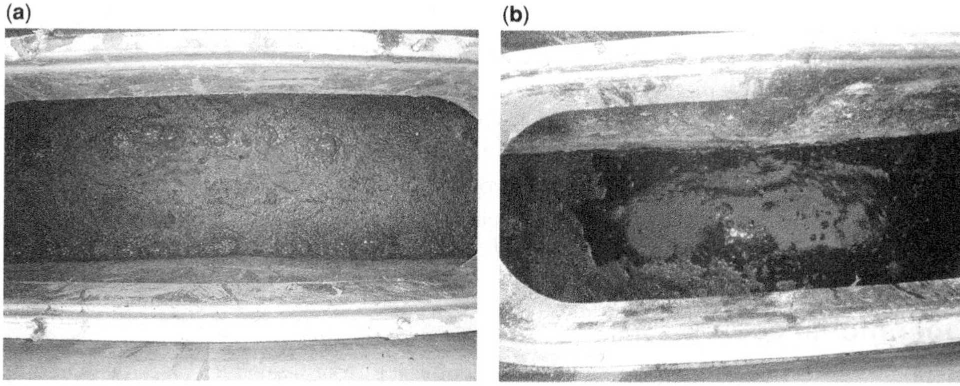

Figure 7.11 Visual inspection of the scum layer under the hatch of a GLS separator: (a) before the removal (thick scum layer) and (b) after (virtually no scum). (Source: Chernicharo *et al.*, 2013).

The management and final disposal of scum followed the recommendation proposed in Chapter 4. In STP 2, using static sieves and drying beds (Figure 7.12(a–d)) facilitated the separation of the coarse material present in the scum. The liquid fraction separated from the coarse material showed a satisfactory infiltration rate in the drying bed, resulting in virtually complete dewatering in less than three days (25% of its original volume), without any noticeable detrimental clogging effect on the drying beds.

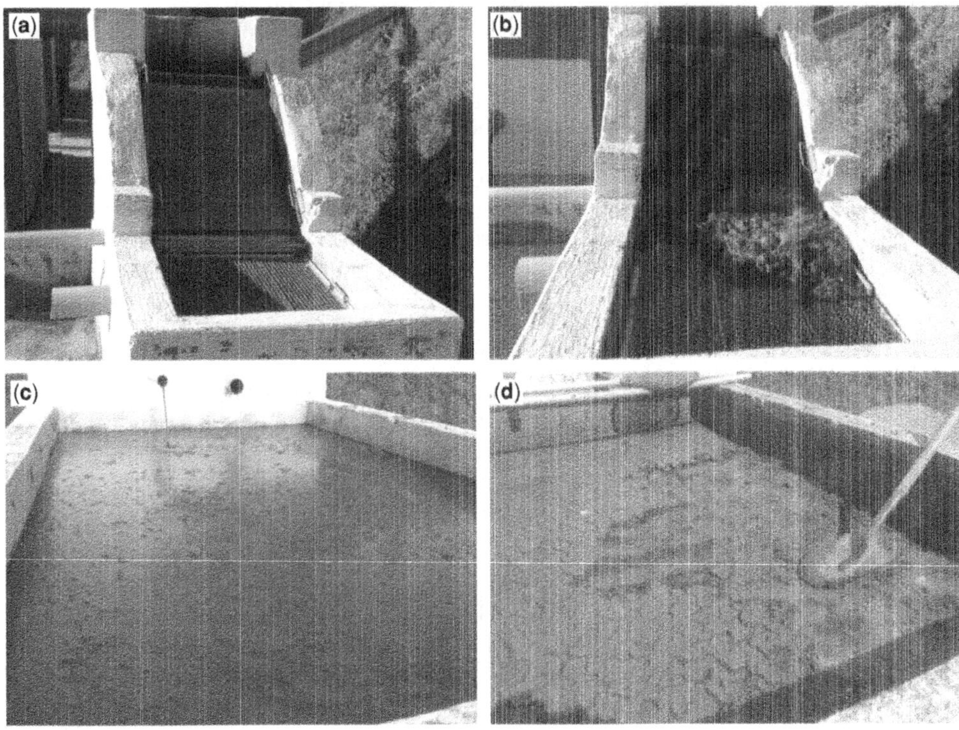

Figure 7.12 Scum management system tested in STP 2: (a) and (b) static sieve before and after scum withdrawal; (c) and (d) scum in the drying bed shortly after the removal operation and after the drying period (72 hours). (*Source*: Santos, 2014; Chernicharo *et al.*, 2015).

The clarified fraction (liquid passing from the drying beds) showed median concentrations of COD (180 mg L^{-1}) and oils and grease (20 mg L^{-1}) (Santos, 2014) within the usual ranges for raw sewage. Therefore, this fraction could be returned to the start of the STP without any difficulties to the following units. In this instance, as discussed in Chapter 4, the proposal of a tank for thickening the sieved fraction of the scum is very important, as it allows the separation of a large volume of the clarified fraction that would be routed to the start of the STP. In this way, only the supernatant and settled fractions would be sent for dewatering in the drying beds, together or separately, thereby reducing the area required for such units.

7.3.5 Operational costs

Figure 7.13(a) shows the monthly operating costs per cubic metre of treated sewage for the components, namely electricity, personnel, chemicals (laboratory consumables, polymers, and deodorants products), system maintenance (civil and electromechanical infrastructure) and transport and final disposal of the solids byproducts (material retained in the preliminary treatment and dewatered sludge). To facilitate interpreting these indicators, each component is plotted in a boxplot (Figure 7.13(b)).

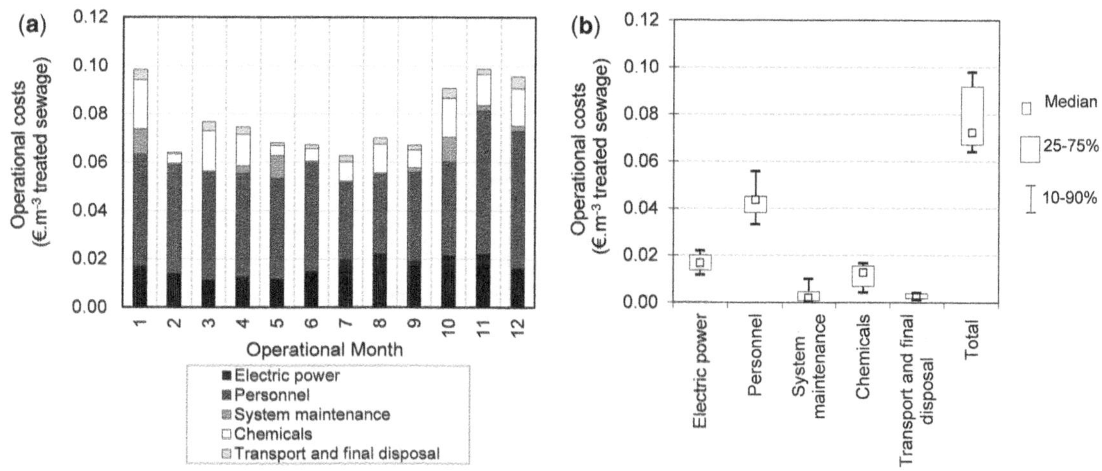

Figure 7.13 Operational costs of STP 2 per cubic metre of treated sewage per component: (a) monthly values; (b) boxplot.

The total average operating cost of STP 2 was € 0.077 per cubic metre of treated sewage, comprising the following main categories: (i) 21% for electricity (€ 0.017 per cubic metre of treated sewage); (ii) 56% for personnel (€ 0.044 per cubic metre of treated sewage); (iii) 16% for chemicals (€ 0.013 per cubic metre of treated sewage); (iv) 3% for system maintenance (€ 0.002 per cubic metre of treated sewage); and (v) 3% for transport and final disposal of the solids byproducts (€ 0.002 per cubic metre of treated sewage). Considering that STP 2 serves 70,000 PE, the operational cost was € 2.53 per population equivalent per year.

7.4 SEWAGE TREATMENT PLANT 3

7.4.1 Overall plant description

STP 3 was designed for a sewage flow rate of 2,050 L s^{-1}, corresponding to a treatment capacity of approximately 1 million PE. The flowsheet comprises preliminary treatment units (screens, grit chambers, and sieves), UASB reactors followed by trickling filters and secondary settlers, as well as a dewatering unit (centrifuges), as shown in Figures 7.14 and 7.15. Biogas generated in the UASB reactors is currently burned in an open flare, but new studies consider heat recovery for the thermal drying of the sludge. The plant was designed to function with the secondary (aerobic) sludge returned for thickening and digestion in the UASB reactors, with the withdrawn excess sludge forwarded to the centrifuges and to final disposal in a landfill. The main characteristics of the preliminary treatment units and the design and operational parameters of the UASB+TF system are shown in Tables 7.6 and 7.7.

7.4.2 System performance

The operational performance of STP 3 was assessed based on data provided by the sanitation company responsible for the plant, and accounts for an operational period of 36 months. The results comprise the historical series of the influent and effluent BOD, COD, and TSS concentrations (Figure 7.16(a, c, e)). These data are also presented as boxplot graphs (Figure 7.16(b, d, f)). Figure 7.17 shows the UASB reactor efficiency and the overall STP efficiency (UASB+TF) for BOD, COD, and TSS removal.

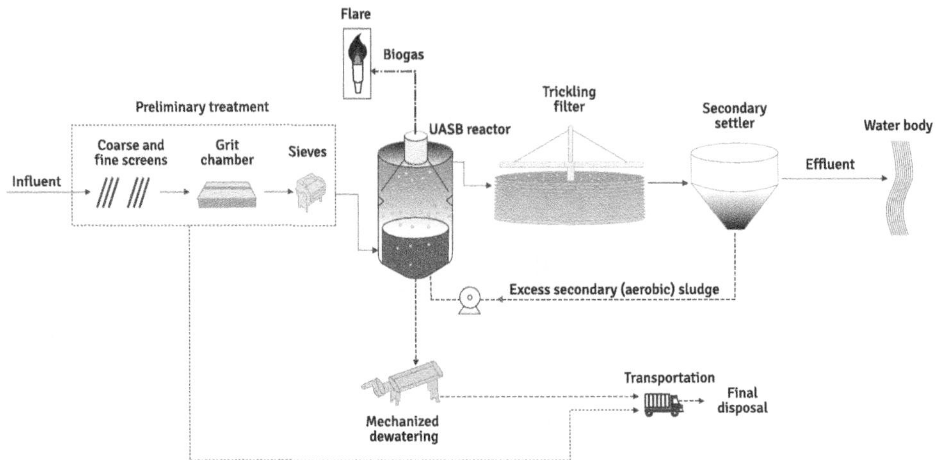

Figure 7.14 Flowsheet of STP 3.

Figure 7.15 Aerial view of STP 3.

Table 7.6 Main characteristics of the preliminary treatment units of STP 3.

Design Parameters	Coarse Screens	Fine Screens	Grit Chambers	Sieves
Type	**Manual**	**Mechanised**	**Mechanised**	**Mechanised**
Number of units	2	4	4	3
Aperture (mm)	100	13	–	5
Hydraulic loading rate (HLR) ($m^3\ m^{-2}\ d^{-1}$)	–	–	734	–

Source: adapted from Chernicharo *et al.* (2014).

Table 7.7 Main design and current operational parameters of the UASB+TF system in STP 3.

Parameters	UASB Reactors	Trickling Filters[a]	Secondary Settlers
Design			
Population equivalent (PE)		1,000,000	
Flow rate (L s^{-1})		2,050	
Hydraulic retention time (HRT) (h)	7.2	–	4.4
Upflow velocity (m h^{-1})	0.63	–	–
Organic loading rate (OLR) (kgBOD m^{-3} d^{-1})	–	<1.00	–
Hydraulic loading rate (HLR) (m^3 m^{-2} d^{-1})	–	16.8	21.8
Number of units	24	8	8
Type	Rectangular	Circular	Circular
Dimensions (m)	38.40 × 12.80	$D=41.00$	$D=36.00$
Area (m^2)	491.5	1,319.6	1,017.4
Useful depth (m)	4.65	2.50	4.00
Useful volume of each unit (m^3)	2,212	3,299	4,069
Current operation			
Flow rate (L s^{-1})		1,850	
HRT (h)	7.6	–	4.4
Upflow velocity (m h^{-1})	0.59	–	–
HLR (m^3 m^{-2} d^{-1})		15.2	19.7

[a]The trickling filters are filled with rock.
Source: Adapted from Chernicharo *et al.* (2014).

The following observations can be made from the operational results of STP 3:

- The median influent BOD, COD, and TSS concentrations (267, 578, and 295 mg L^{-1}, respectively) match the typical concentrations reported by von Sperling and Chernicharo (2005) for sewage (300 mgBOD L^{-1}, 600 mg COD L^{-1}, and 350 mgTSS L^{-1}).
- The median anaerobic effluent BOD, COD, and TSS concentrations (87, 248, and 120 mg L^{-1}, respectively) are, to some extent, typical of the effluents of UASB reactors treating sewage, considering the reported performances of full-scale UASB reactors by Chernicharo *et al.* (2015).
- The median BOD and COD removal efficiencies of the UASB reactors (66% and 58%, respectively) match the typical ranges reported by von Sperling and Chernicharo (2005) for anaerobic sewage treatment (BOD 60–75% and COD 55–70%).
- The median TSS removal efficiency (59%) of the UASB reactors is lower than the typical range reported by von Sperling and Chernicharo (2005) for anaerobic sewage treatment (65–80%). The lower TSS removal efficiencies can be ascribed to the presence of excess sludge stored in the UASB reactors, leading to solids washout with the effluent.
- As regards the final effluent of STP 3, the median BOD, COD, and TSS concentrations (33, 93, and 34 mg L^{-1}, respectively) match the typical ranges reported by von Sperling and Chernicharo (2005)

for UASB + TF systems treating sewage (15–40 mgBOD L^{-1}, 30–120 mgCOD L^{-1}, and 20–40 mgTSS L^{-1}).

- The median overall BOD, COD, and TSS removal efficiencies (87, 83, and 89%, respectively) also match the typical ranges reported by von Sperling and Chernicharo (2005) for UASB + TF systems treating sewage (87–93%, 80–90%, and 87–93% for BOD, COD, and TSS, respectively).

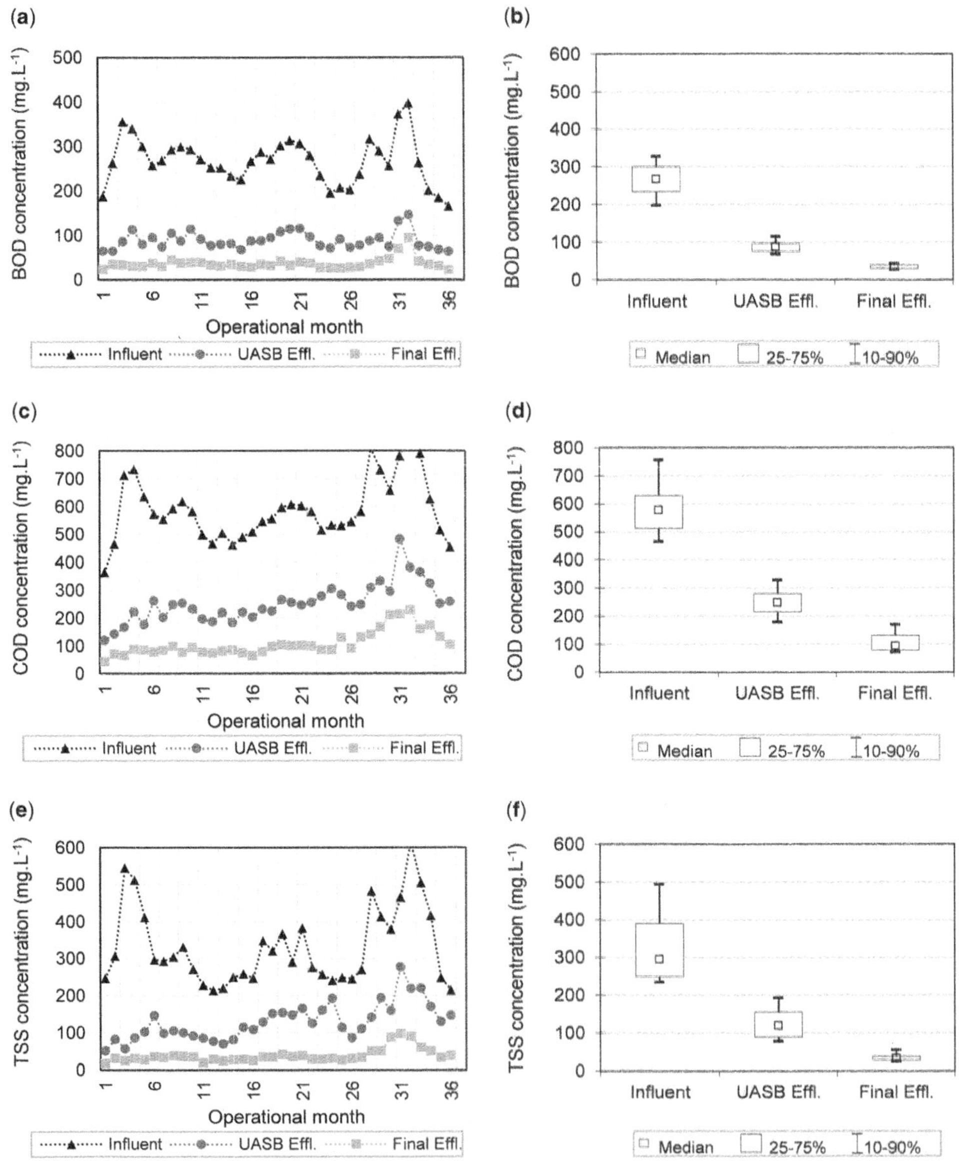

Figure 7.16 Historical series and boxplot of the BOD (a, b), COD (c, d), and TSS (e, f) concentrations of STP 3.

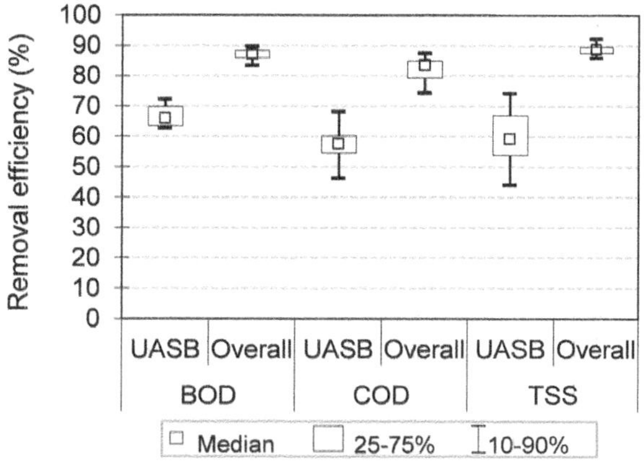

Figure 7.17 Boxplot of BOD, COD, and TSS removal efficiencies of STP 3.

The COD and TSS concentrations reported in Figure 7.16(c, e) show the gradual deterioration of the UASB effluent quality from the 17th month onward. As mentioned previously, this can be ascribed to the excess sludge stored in the UASB reactors over their storage capacity. This is the main operational drawback of this STP, caused by a lack of dewatering capacity, as the centrifuges are interrupted frequently for corrective maintenance and there is no spare dewatering equipment.

7.4.3 Sludge yield coefficient (Y_{sludge})

Studies on the production of sludge in STP 3, of which the UASB reactors were designed to treat only sewage and sludge generated in the aerobic post-treatment unit, showed that the actual sludge yield coefficient (Y_{sludge}) was double the amount considered originally in the design. This can be ascribed to the contributions not considered at the design stage, such as landfill leachate, sludge from septic tanks, and chemical sludge from water treatment plants.

As discussed in Chapter 6, operational problems in the dewatering unit (e.g. centrifuges) could generate a 'vicious circle' of sludge in the STP, as the sludge that should be withdrawn from the UASB reactors escapes to the following trickling filters and secondary clarifiers, and, subsequently, back to the anaerobic reactors, restarting the cycle. Accordingly, the sludge concentration in the secondary settlers remains extremely high, between 2.5 and 4.0%, clearly overloading the operation of the UASB reactors. The excessive concentration of the return sludge from the secondary clarifiers (2.5–4.0 times higher than predicted in the design ~1.0% TS) is undoubtedly one of the main constraints to the operation of UASB reactors, as discussed in Chapter 6.

Simulations considering all the incoming contributions to STP 3 (domestic sewage, landfill leachate, sludge from septic tanks, and return sludge from the secondary settlers) showed Y_{sludge} for the UASB reactors of approximately 0.36 kgTS $kgCOD_{applied}^{-1}$ (considering the design concentration of return sludge equal to 1.0% TS) and 0.43 kgTS $kgCOD_{applied}^{-1}$ (considering the concentration of return sludge equal to 4.0% TS – present data). Therefore, the operation of the STP with excess sludge stored in the UASB reactors and the high concentrations of TS in the secondary settlers resulted in Y_{sludge} even higher than the design criteria.

These data clearly show the importance of carefully defining the value for the Y_{sludge} coefficient at the design stage, considering all possible contributions, as this affects the sludge production in the system, as well as the design of the dewatering system and the establishment of a proper sludge withdrawal routine, as discussed in Chapters 4 and 6.

7.4.4 Operational costs

Figure 7.18(a) shows the monthly operating costs per cubic metre of treated sewage for the components, namely electricity, personnel, chemicals, system maintenance and transport and final disposal of the solids byproducts. To facilitate interpreting these indicators, each component is plotted in a boxplot (Figure 7.18(b)).

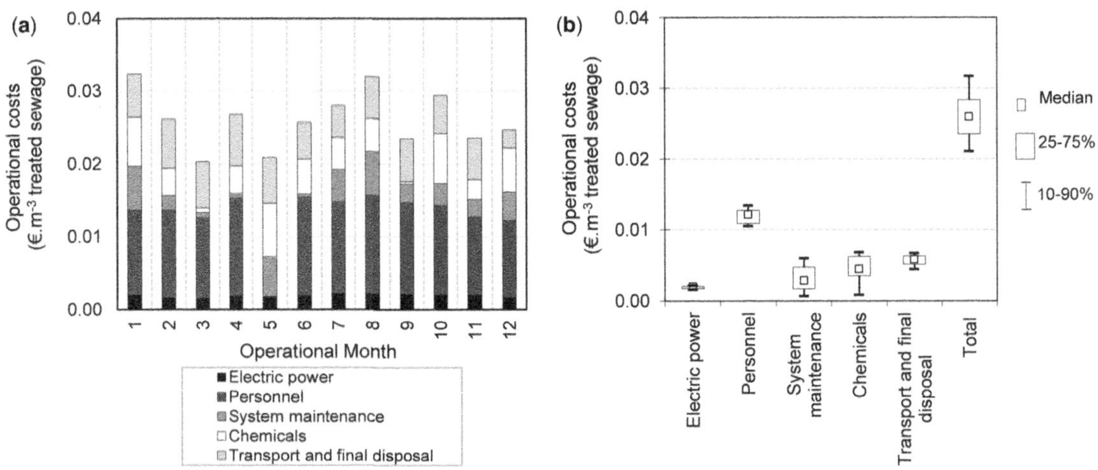

Figure 7.18 Operational costs of STP 3 per cubic metre of treated sewage per component: (a) monthly values; (b) boxplot.

The total average operating cost of STP 3 was € 0.027 per cubic metre of treated sewage, comprising the following main categories: (i) 7% for electricity (€ 0.002 per cubic metre of treated sewage); (ii) 45% for personnel (€ 0.012 per cubic metre of treated sewage); (iii) 16% for chemicals (€ 0.004 per cubic metre of treated sewage); (iv) 10% for system maintenance (€ 0.003 per cubic metre of treated sewage); and (v) 21% transport and final disposal of the solids byproducts (€ 0.006 per cubic metre of treated sewage). Considering that STP 3 serves 1 million PE, the operational cost was € 1.56 per population equivalent per year.

7.5 SEWAGE TREATMENT PLANT 4

7.5.1 Overall plant description

STP 4 was designed for a sewage flow rate of 990 L s^{-1}, corresponding to a treatment capacity of approximately 392,000 PE. The flowsheet comprises preliminary treatment units (screens and grit chambers), UASB reactors followed by trickling filters and secondary settlers, dewatering units (centrifuges and a rotary sludge thermal dryer unit), and gasholder and flare, as depicted in Figures 7.19 and 7.20. This STP features spare drying beds should an operational breakdown occur, or preventive/corrective maintenance of the centrifuges were needed. The main characteristics of the preliminary treatment units and the design and operational parameters of the UASB + TF system are presented in Tables 7.8 and 7.9.

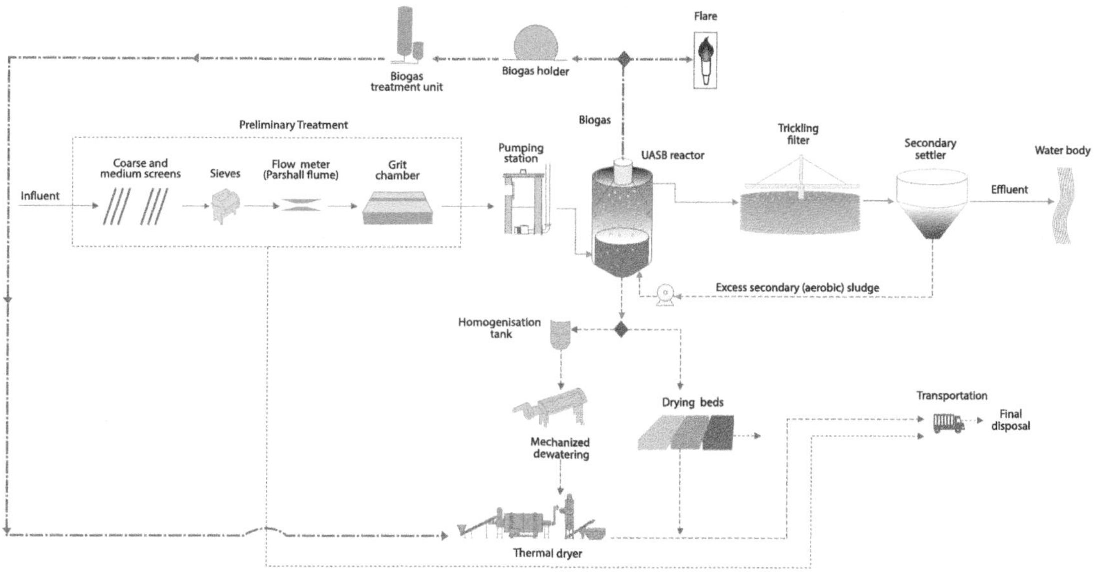

Figure 7.19 Flowsheet of STP 4.

Figure 7.20 Aerial view of STP 4.

Table 7.8 Main characteristics of the preliminary treatment units of STP 4.

Design Parameters	Coarse Screens	Medium Screens	Sieves	Grit Chambers
Type	**Manual**	**Mechanised**	**Mechanised**	**Mechanised**
Number of units	3	3	3	4
Aperture (mm)	75	25	6	–
Hydraulic loading rate (HLR) ($m^3\ m^{-2}\ d^{-1}$)	–	–	–	1,024

Table 7.9 Main design and current operational parameters of the UASB + TF system in STP 4.

Design Parameters	UASB Reactors[a]	Trickling Filter[b]	Secondary Settlers
Design			
Population equivalent (PE)		392,000	
Flow rate ($L\ s^{-1}$)		990	
Hydraulic retention time (HRT) (h)	6.4	–	3.1
Upflow velocity ($m\ h^{-1}$)	0.73	–	–
Organic loading rate (OLR) (kgBOD $m^{-3}\ d^{-1}$)		$<$1.00	
Hydraulic loading rate (HLR) ($m^3\ m^{-2}\ d^{-1}$)	–	23.18	31.54
Number of units	12	6	6
Type	Rectangular	Circular	Circular
Dimensions (m)	32.00 × 12.80	$D = 28{,}00$	$D = 24{,}00$
Area (m^2)	409.6	615	452
Useful depth (m)	4.65	2.50	4.00
Useful volume of each unit (m^3)	1,905	1,539	1,809
Current operation			
Flow rate ($m^3 \cdot s^{-1}$)		320	
HRT (h)	13.2	–	7.9
Upflow velocity ($m\ h^{-1}$)	0.35	–	–
HLR ($m^3\ m^{-2}\ d^{-1}$)	–	9.0	12.2

[a]Only eight UASB reactors were in operation.
[b]The trickling filters are filled with rock.

7.5.2 System performance

The operational performance of STP 4 was assessed based on data provided by the sanitation company responsible for the plant, and accounts for an operational period of 36 months. The results comprise the historical series of the influent and final effluent BOD and COD concentrations (Figure 7.21(a, c)). These data are presented also as boxplot graphs (Figure 7.21(b, d)). Figure 7.22 shows the overall STP efficiency (UASB + TF) for BOD and COD removal.

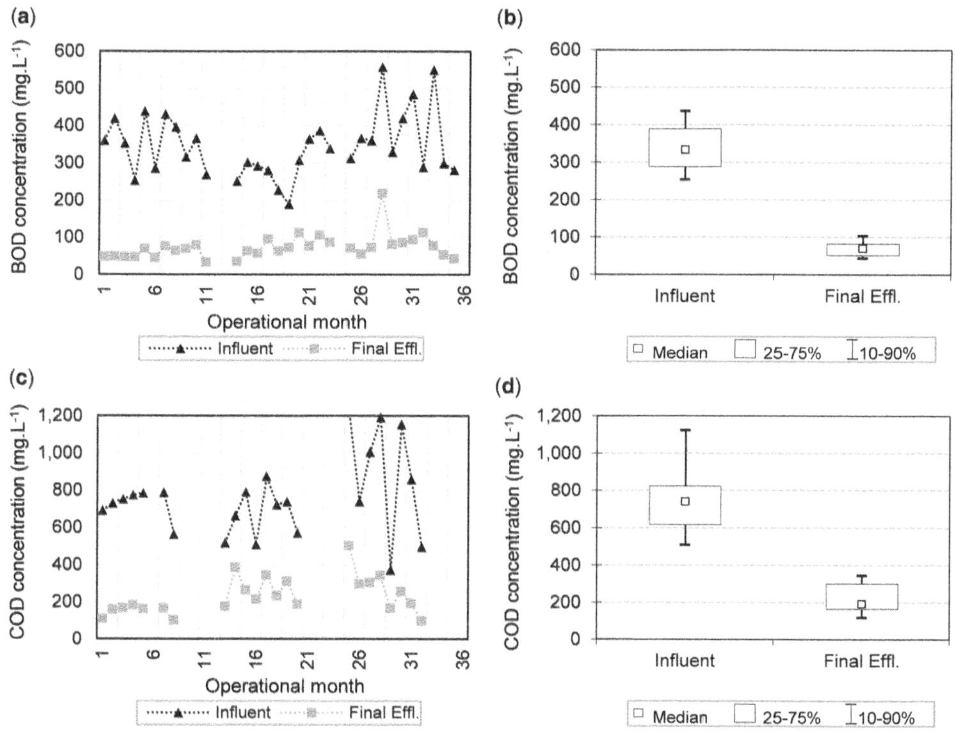

Figure 7.21 Historical series and boxplot of BOD (a, b) and COD (c, d) of STP 4.

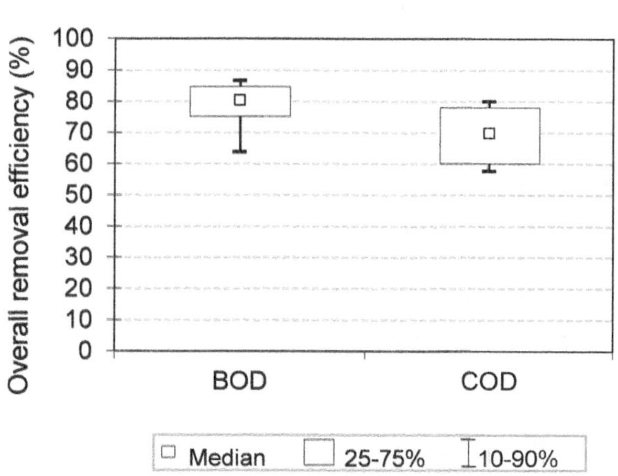

Figure 7.22 Boxplot of BOD and COD overall removal efficiencies of STP 4.

The following observations can be made from the operational results of STP 4:

- As regards STP 4, the median influent (raw sewage) BOD and COD concentrations (334 and 738 mg L^{-1}, respectively) resemble sewage that is more concentrated, with higher values than the typical concentrations reported by von Sperling and Chernicharo (2005) for domestic sewage (300 mgBOD L^{-1} and 600 mgCOD L^{-1}).

- The median final effluent BOD and COD concentrations (70 and 192 mg L^{-1}, respectively) were higher than that expected from UASB + TF systems (15–40 and 30–120 mg L^{-1}, respectively), as reported by von Sperling and Chernicharo (2005).
- The median overall BOD and COD removal efficiencies (80 and 70%, respectively) were below the typical ranges reported by von Sperling and Chernicharo (2005) for UASB + TF systems treating sewage (85–93 and 80–90%, for BOD and COD, respectively).
- The final effluent BOD and COD concentrations were above the usual values for well operated UASB + TF systems because of the excessive solids washout with the effluent of the UASB reactors, which could be ascribed to the frequent breakdowns of the dewatering unit (centrifuges). As discussed in Chapter 6, excessive solids washout from UASB reactors could disturb the functioning of the TFs and its secondary settlers, resulting in the loss of quality of the final effluent. Accordingly, the continuous and proper operation of the dewatering unit is of utmost importance.

7.5.3 Energy recovery from biogas

STP 4 was equipped with a sludge thermal dryer system (Figure 7.23(a)) that used the biogas generated in the UASB reactors as a thermal energy source. The biogas is stored in a low-pressure 510 m^3 volume double-membrane gas holder, inflated as counter pressure with compressed air (Figure 7.23(b)).

The operation of the rotary sludge thermal dryer is a combination of direct (convection) and indirect (conduction) functioning, processing close to 8 ton d^{-1} of sludge and consuming approximately 150–210 $Nm^3\ h^{-1}$ of biogas, with 75% methane content. The dewatered sludge from the centrifuge enters the thermal dryer with a moisture content in the range 75–78%, whereas the dried sludge produced by the system has a moisture content of only 5–10% (Figure 7.23(c)). The drying system has an evaporation capacity of approximately 3,000 $LH_2O\ h^{-1}$ and produces close to 1,200 kg h^{-1} of dry pelletised sludge (Figure 7.23(c)).

Figure 7.23 Thermal sludge drying system of STP 4 fuelled by biogas: (a) rotary thermal dryer; (b) double membrane biogas holder; and (c) dry sludge (pellets) produced by the system.

7.5.4 Operational costs

Figure 7.24(a) shows the monthly operating costs per cubic metre of treated sewage for the components, namely electricity, personnel, chemicals, system maintenance and transport and final disposal of the solids byproducts. To facilitate interpreting these indicators, each component is plotted in a boxplot (Figure 7.24(b)).

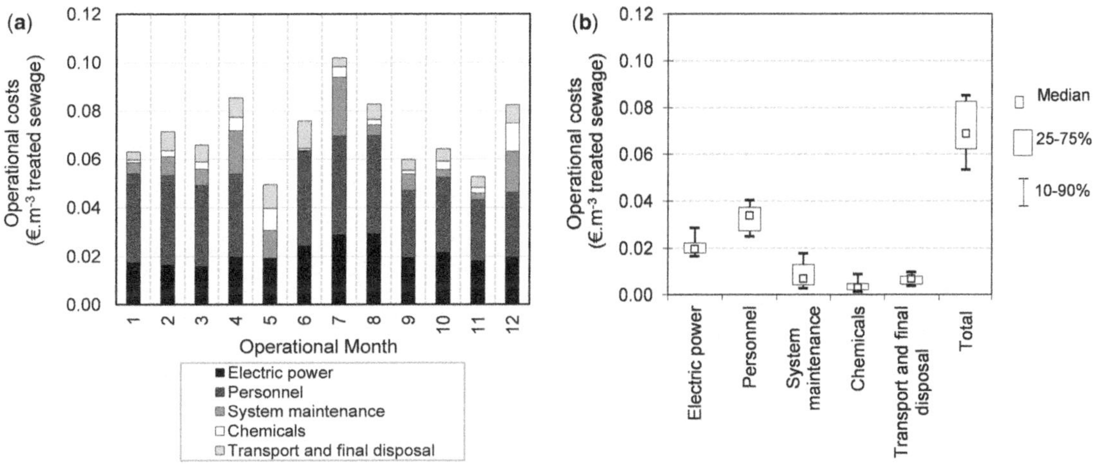

Figure 7.24 Operational costs of STP 4 per cubic metre of treated sewage per component: (a) monthly values; (b) boxplot.

The total average operating cost of STP 4 was € 0.069 per cubic metre of treated sewage, comprising the following main categories: (i) 28% for electricity (€ 0.019 per cubic metre of treated sewage); (ii) 48% for personnel (€ 0.034 per cubic metre of treated sewage); (iii) 4% for chemicals (€ 0.003 per cubic metre of treated sewage); (iv) 10% for system maintenance (€ 0.007 per cubic metre of treated sewage); and (v) 10% transport and final disposal of the solids byproducts (€ 0.007 per cubic metre of treated sewage). Considering that STP 3 serves 392,000 PE, the operational cost was € 5.96 per population equivalent per year.

7.6 SEWAGE TREATMENT PLANT 5

7.6.1 Overall plant description

STP 5 comprises an innovative patented process called a Biopaq® Ubox reactor (the Paques company, the Netherlands) designed for a sewage flow rate of $66\ L\ s^{-1}$, corresponding to a treatment capacity of approximately 30,000 PE. The flowsheet of this STP comprises preliminary treatment units (roto-sieves, and grit chamber), the Biopaq® Ubox reactors, and the dewatering unit (centrifuge), as shown in Figures 7.25 and 7.26. The main characteristics of the preliminary treatment units and the design and operational parameters of the Biopaq® Ubox system are presented in Tables 7.10 and 7.11.

The Biopaq® Ubox reactor couples the anaerobic and aerobic process (including the typical secondary settling stage) but uses only one tank. Therefore, the area requirement is smaller ($0.09\ m^2\ PE^{-1}$) than that typically required for an STP flowsheet comprising UASB reactors + activated sludge (0.08–$0.20\ m^2\ PE^{-1}$) The influent raw sewage is distributed equally in the anaerobic digestion zone (pressurised feeding system). Biogas generated in the anaerobic section is collected in the GLS separators and delivered into the gas scrubber for desulphurisation, without the addition of chemicals. The treated sewage in the anaerobic section flows to the aerobic zone by an anti-return passage. The effluent undergoes a high-rate settling step (lamella settlers). The excess sludge from the aerobic zone of the Biopaq® Ubox is returned to its anaerobic section for thickening and digestion. The excess sludge withdrawn from the anaerobic compartment is routed to a dewatering unit (centrifuge) and, subsequently, to final disposal in a landfill.

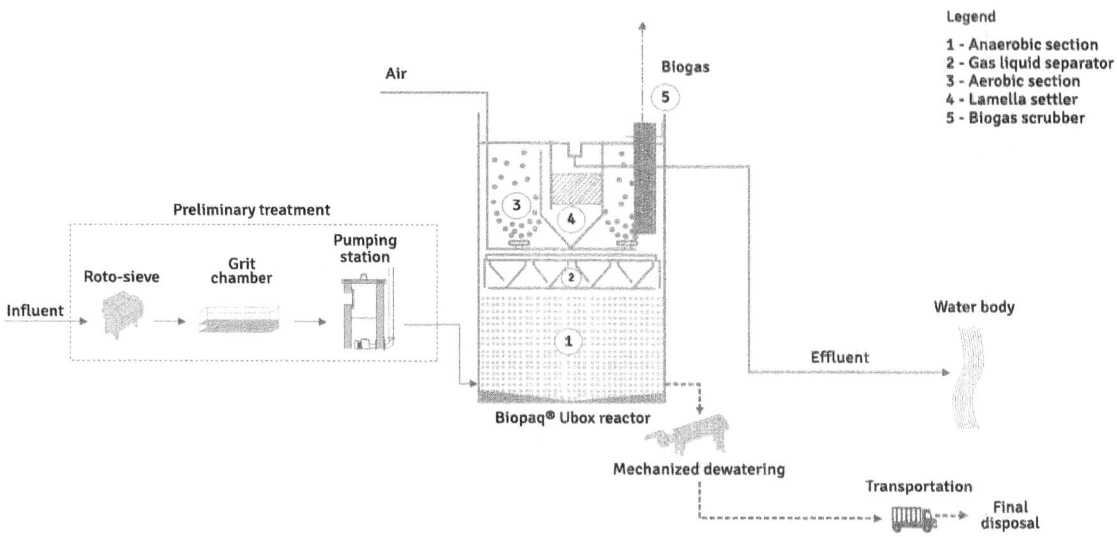

Figure 7.25 Flowsheet of STP 5.

Figure 7.26 Biopaq® Ubox reactor in STP 5.

Table 7.10 Main characteristics of the preliminary treatment units of STP 5.

Design Parameters	Coarse Screens[a]	Roto-sieves	Grit Chamber
Type	**Manual**	**Mechanised**	**Manual**
Number of units	–	2	1 with 2 channels
Aperture (mm)	40	1	–
Hydraulic loading rate (HLR) ($m^3\ m^{-2}\ d^{-1}$)	–	–	≤1,300

[a]The coarse screens are located in an external pumping station.

Table 7.11 Main design and current operational parameters of the Biopaq® Ubox system in STP 5.

Parameters	Anaerobic Compartment	Aerobic Compartment	Settlers
Design			
Population equivalent (PE)		30,000	
Flow rate ($L\ s^{-1}$)		70	
Hydraulic retention time (HRT) (h)	5.5	5.5	1.00
Upflow velocity ($m\ h^{-1}$)	0.75	0.75	–
Organic loading rate (OLR) ($kgBOD\ m^{-3}\ d^{-1}$)	–	<0.50	–
Hydraulic loading rate (HLR) ($m^3\ m^{-2}\ h^{-1}$)	–	–	12.8
Number of reactors	3	3	3
Type	Rectangular	Rectangular	Lamellar
Dimensions (m)	14.50 × 7.50	14.50 × 7.50	–
Area (m^2)	108.8	108.8	149
Useful depth (m)	4.00	4.00	4.00
Useful volume of each tank (m^3)	435	435	84.2
Current operation			
Flow rate ($L\ s^{-1}$)		45	
HRT (h)	8.0	8.0	1.50
Upflow velocity ($m\ h^{-1}$)	0.50	0.50	–
HLR ($m^3\ m^{-2}\ d^{-1}$)	–	–	8.7

7.6.2 System performance

The operational performance of STP 5 was assessed based on data provided by the Paques Brazil company, accounting for an operational period of 10 months. The results comprise the historical series of the influent and effluent BOD, COD, TSS, and $NH_4^+ - N$ concentrations (Figure 7.27(a, c, e, g)). These data are also presented as boxplot graphs (Figure 7.27(b, d, f, h)). Figure 7.28 shows the overall STP efficiency for BOD, COD, TSS, and $NH_4^+ - N$ removal.

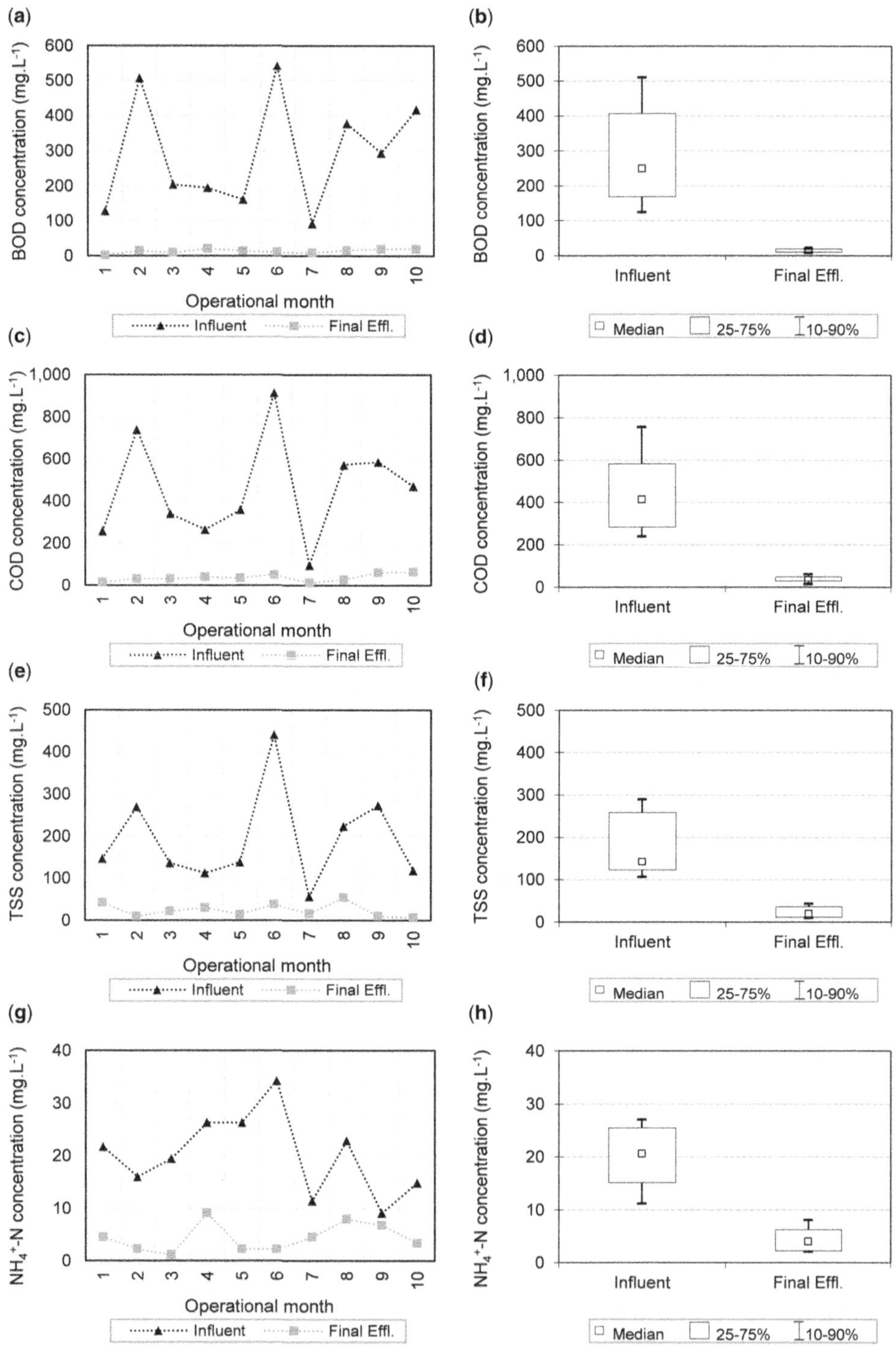

Figure 7.27 Historical series and boxplot of BOD (a, b), COD (c, d), TSS (e, f), and $NH_4^+ - N$ (g, h) of STP 5.

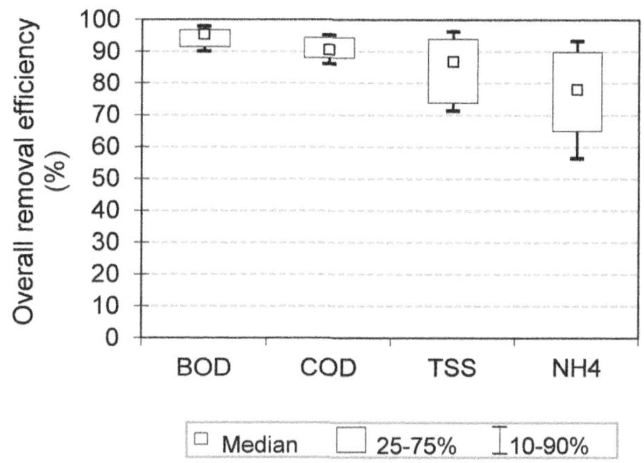

Figure 7.28 Boxplot of the overall BOD, COD, TSS, and $NH_4^+ - N$ removal efficiencies of STP 5.

The following observations can be made from the operational results of STP 5:

- The median influent (raw sewage) BOD, COD, TSS, and $NH_4^+ - N$ concentrations (249, 414, 142, and 21 mg L^{-1}, respectively) were normal for diluted sewage, of which the values are lower than the typical concentrations reported by von Sperling and Chernicharo (2005) for domestic sewage (300 mgBOD L^{-1}, 600 mgCOD L^{-1}, 350 mgTSS L^{-1}, and 25 mg NH_4^+ L^{-1}).
- The median final effluent BOD, COD, TSS, and $NH_4^+ - N$ concentrations (15, 34, 19, and 4 mg L^{-1}, respectively) were below the typical ranges reported by von Sperling and Chernicharo (2005) for a UASB + activated sludge system (20–50 mgBOD L^{-1}, 60–150 mgCOD L^{-1}, 20–40 mgTSS L^{-1}, and 5 to 15 mg$NH_4^+ - N L^{-1}$).
- The median overall BOD, COD, TSS, and NH_4 removal efficiencies for the period evaluated were high (95, 91, 87, and 78%, respectively), corresponding to the typical ranges reported by von Sperling and Chernicharo (2005) for a UASB + activated sludge system (83–93%, 75–88%, 87–93%, and 50–85% for BOD, COD, TSS, and $NH_4^+ - N$, respectively).

7.7 SEWAGE TREATMENT PLANT 6

7.7.1 Overall plant description

STP 6 was designed for a sewage flow rate of 164 L s^{-1}, corresponding to the treatment capacity of approximately 79,000 PE. As the effluent is discharged into a lagoon, the ability to reach stringent levels of nitrogen and phosphorus removal was a design prerequisite. The STP flowsheet comprises a compact preliminary treatment unit (roto-sieve and aerated grit chambers equipped with grease trap), UASB reactors filled with a sponge-based support material, so-called hybrid anaerobic reactor (HAnR), and activated sludge tanks (with aerobic and anoxic compartments) also filled with a sponge-based support material, so-called integrated fixed-film activated sludge (IFAS). The flowsheet comprises secondary settlers and a flotation unit especially designed for phosphorus removal, as well as a dewatering unit (screw press module), as shown in Figures 7.29 and 7.30. The main characteristics of the preliminary treatment units and the design and operational parameters of the HAnR+IFAS system are presented in Tables 7.12 and 7.13.

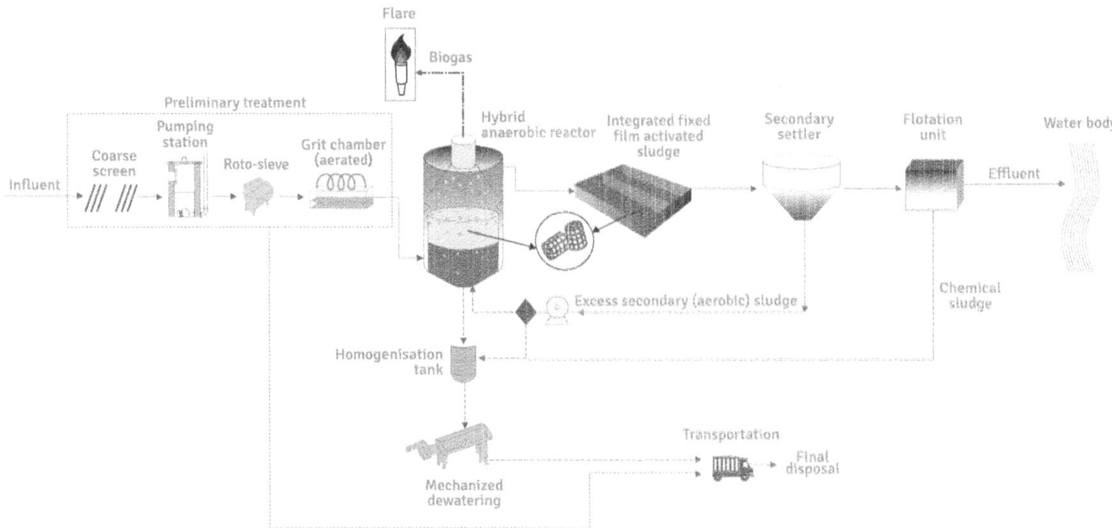

Figure 7.29 Flowsheet of STP 6.

Figure 7.30 Aerial view of STP 6.

Table 7.12 Main characteristics of the preliminary treatment units of SPT 6.

Design Parameters	Roto-sieves	Grit Chambers	Grease Trap
Type	**Mechanised**	**Aerated**	**Mechanised**
Number of units	2	2	2
Aperture (mm)	3	–	–

Table 7.13 Main design and current operational parameters of the HAnR + IFAS system in STP 6.

Parameters	HAnR	Anoxic Reactor	IFAS	Secondary Settlers	Flotation Unit
Design					
Population equivalent (PE)			79,000		
Flow rate (L s^{-1})			164		
Hydraulic retention time (HRT) (h)	4.3	Unit 1: 1.5 Unit 2: 0.8	Unit 1: 2.4 Unit 2/3: 1.1	1.5	–
Upflow velocity (m h^{-1})	1.3	–	–	–	–
Organic loading rate (OLR) (kgBOD m^{-3} d^{-1})	–	9.0	Unit 1: 1.8 Unit 2/3: 2.7	–	–
Hydraulic loading rate (HLR) (m^3 m^{-2} d^{-1})	–	–	–	70.8	121.2
Number of reactors	1	2	3	1	1
Type	Rectangular	Rectangular	Rectangular	Rectangular	Circular
Dimensions (m)	18.40 × 25.00	Unit 1: 4.5 × 25.0 Unit 2: 5.4 × 12.4	Unit 1: 5.4 × 25.0 Unit 2/3: 5.4 × 12.4	8.0 × 25.0	$D = 12.2$
Area (m^2)	460	Unit 1: 113 Unit 2: 67	Unit 1: 135 Unit 2/3: 67	200	116.9
Useful depth (m)	5.54	Unit 1: 4.0 Unit 2: 3.4	Unit 1: 3.5 Unit 2/3: 3.4	4.54	1.0
Useful volume of each tank (m^3)	2,548	Unit 1: 452 Unit 2: 224	Unit 1: 472 Unit 2/3: 224	908	116.9
Current operation					
Flow rate (L s^{-1})			90		
HRT (h)	7.9	Unit 1: 2.8 Unit 2: 1.4	Unit 1: 4.4 Unit 2/3: 2.1	2.9	–
Upflow velocity (m h^{-1})	0.7	–	–	–	–
HLR (m^3 m^{-2} d^{-1})	–	–	–	37.1	63.5

The flowsheet of the STP 6 was designed originally as a UASB reactor followed by activated sludge (anoxic and aerobic compartments). Support media was inserted purposely for biomass growth (hybridisation) in both reactors (HAnR and IFAS) to increase the treatment capacity while maintaining the same operational area.

The main characteristics of the support media (Biobob®) (Figure 7.31) are presented in Table 7.14, according to information from the manufacturer. Using 3 mm sieves in the preliminary treatment unit is necessary to retain the maximum amount of solids to prevent clogging of the media carrier.

The excess sludge from the secondary settlers, UASB reactor, and the flotation unit is routed to a sludge tank and, subsequently, to a dewatering unit (screw press module). Subsequently, the dewatered sludge is sent to final disposal in a landfill.

Figure 7.31 Schematic representation of the Biobob® media.

Table 7.14 Main characteristics of the Biobob® media.

External structure	Polypropylene
Internal structure	Polyurethane
Diameter (mm)	45 mm
Height (mm)	60 mm
Specific surface area (m^2 m^{-3})	94,000
Media porosity (%)	70

7.7.2 System performance

The operational performance of STP 6 was assessed based on data provided by the sanitation company responsible for the operation of the plant, accounting for an operational period of 36 months. The results comprise the historical series of the influent and effluent COD, TSS, BOD, and $NH_4^+ - N$ concentrations (Figure 7.32(a, c, e, g)). These data are also presented as boxplot graphs (Figure 7.32(b, d, f, h)). Figure 7.33 shows the HAnR efficiency and the overall STP (HAnR + IFAS) efficiency for COD, TSS, BOD, and $NH_4^+ - N$ removal.

The following observations can be made from the operational results of STP 6:

- The median influent (raw sewage) COD, TSS, BOD, and mg $NH_4^+ - N$ concentrations (511, 190, 292, and 38 mg L^{-1}, respectively) were normal for diluted sewage, and the values were lower than the typical concentrations reported by von Sperling and Chernicharo (2005) for domestic sewage (600 mgCOD L^{-1}, 350 mgTSS L^{-1}, 300 mgBOD L^{-1}, and 25 mg NH_4^+ L^{-1}).
- The median HAnR effluent COD and TSS concentrations were equal to 183 and 66 mg L^{-1}, being the median COD and TSS removal efficiencies in this unit, equal to 62% and 67%, respectively. The effluent concentrations and removal efficiencies of HAnR match the typical ranges reported by von Sperling and Chernicharo (2005) for UASB reactors treating sewage (COD: 180 to 270 mg L^{-1} and 55 to 70%; TSS: 60 to 100 mg L^{-1} and 65 to 80%).
- As regards the final effluent of STP 6, the median COD, TSS, BOD, and mg $NH_4^+ - N$ concentrations (25, 15, 8, and 3 mg L^{-1}, respectively) were below the typical ranges reported by von Sperling and Chernicharo (2005) for a UASB + activated sludge system (60–150 mgCOD L^{-1}, 20–40 mgTSS L^{-1}, 20–50 mgBOD L^{-1}, and 5 − 15 mg NH_4^+ L^{-1}). The results were consistent with the discharge standards. Furthermore, STP 6 consistently reached concentrations of nitrogen

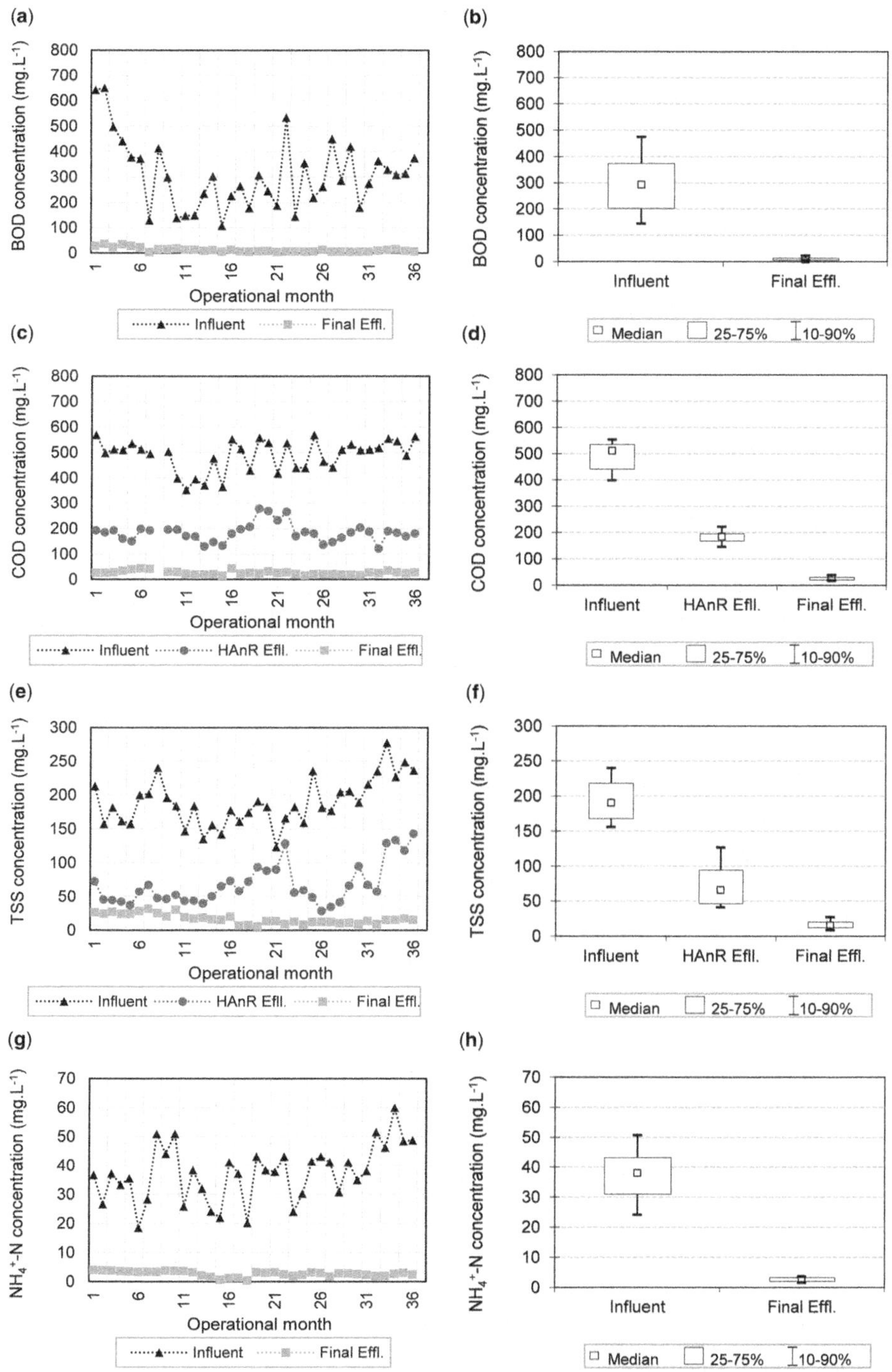

Figure 7.32 Historical series and boxplot of COD (a, b), TSS (c, d), BOD (e, f), and $NH_4^+ - N$ (g, h) of SPT 6. Note: the BOD and $NH_4^+ - N$ concentrations of the HAnR effluent were not available.

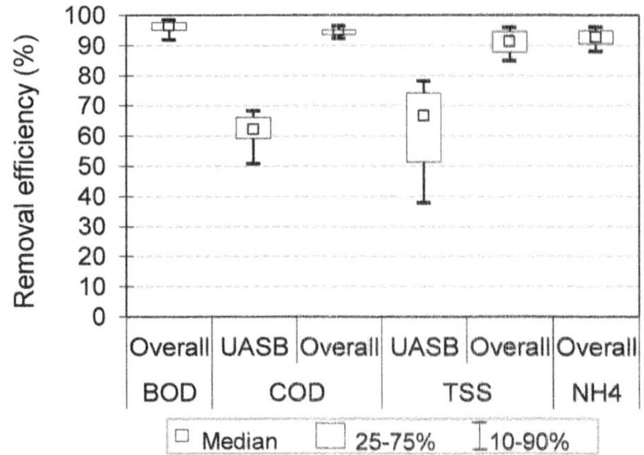

Figure 7.33 Boxplot of BOD, COD, TSS, and NH_4^+ – N removal efficiencies of STP 6.

(N) and phosphorus (P) (10.0 ± 1.2 mg TN L^{-1} and 0.98 ± 0.33 mg TP L^{-1}) that were in accordance with the stringent limits for discharge into lentic water bodies.

- The median overall COD, TSS, BOD, and NH_4^+ – N removal efficiencies (95, 91, 97, and 93%, respectively) were higher than are the typical ranges reported by von Sperling and Chernicharo (2005) for a UASB + activated sludge system (75–88%, 87–93%, 83–93%, and 50–85% for COD, TSS, BOD, and NH_4^+ – N, respectively).

7.8 OVERVIEW OF THE STPs ANALYSED

This section presents a brief overview of the performance and operational indicators of the analysed STPs. Figure 7.34 shows the COD removal efficiencies of the anaerobic treatment step (STPs 1, 2, 3 and 6) and the overall efficiencies of the STPs, including the post-treatment stage.

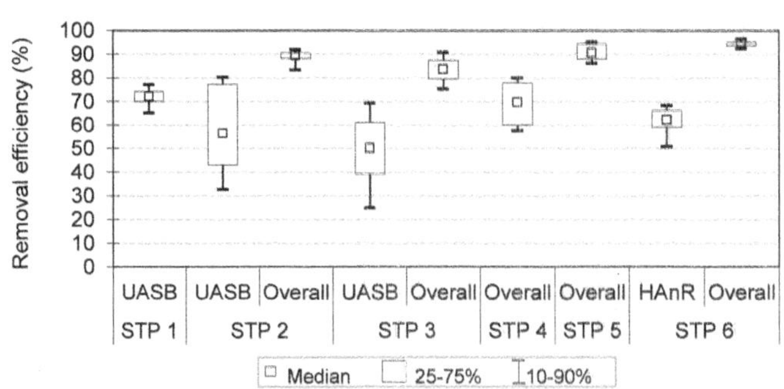

Figure 7.34 COD removal efficiencies in the assessed STPs.

The following observations can be made from Figure 7.34:

- The median overall COD removal efficiencies in STPs 2, 3, 5 and 6, based on anaerobic–aerobic combined systems, were higher than 80%, matching the values reported by von Sperling and Chernicharo (2005).

- Although the median COD removal efficiency of STP 1 was approximately 70%, this system is comprised only by UASB reactors without a post-treatment step. This COD removal efficiency matches the typical range (55–70%) reported by von Sperling and Chernicharo (2005).
- As regards STP 4, the median overall COD removal efficiency was lower than that usually expected for anaerobic–aerobic combined systems (UASB + TF: 80–90%), because of the excessive solids washout with the effluent of the UASB reactors, as was observed also in STP 3. This can be ascribed to a lack of sludge dewatering capacity, because of the frequent interruptions of the centrifuges to undertake corrective maintenance.

As regards the operation of the STPs, Table 7.15 shows a comparison between the indicators for the preliminary treatment units, the sludge production and the operational costs. These data are also presented as graphs (Figure 7.35).

Table 7.15 Operational indicators of the STPs.

Indicators	Unit	Sewage Treatment Plant					
		1	2	3	4	5	6
Coarse material and grit retained	ton m^{-3} of treated sewage	11	29	22	9	53	27
Production of dewatered sludge	kg TS m^{-3} of treated sewage	0.08	0.26	0.37	0.23	*	0.66
	gTS PE^{-1} d^{-1}	18	55	66	49	*	120
Operational costs	€. m^{-3} treated sewage	*	0.07	0.03	0.07	*	*

*Data not available.

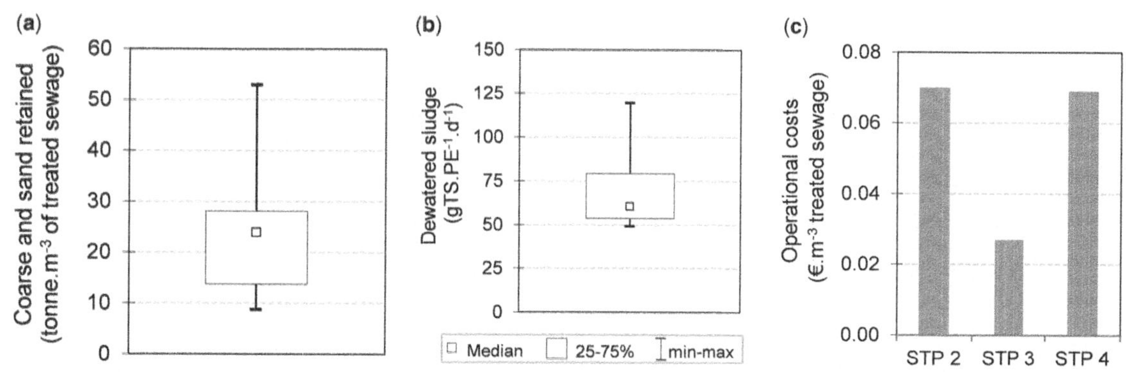

Figure 7.35 STPs operational indicators: (a) boxplot of the data on coarse and grit retained in the preliminary treatment for STPs 1 to 6; (b) boxplot of the data on production of dewatered sludge for STPs 2, 3, 4 and 6; (c) average operational costs of STPs 2, 3 and 4.

The following observations can be made from Table 7.15 and Figure 7.35:

- The volume of coarse material and grit retained in the preliminary treatment step is quite variable, depending on the characteristics of the screens (spacing between rods) and grit chambers, as well as the characteristics of the sewage contribution basin and the habits of the contributing population.

- The dewatered sludge production in STP 1 (18 gTS PE^{-1} d^{-1}) matches the typical range reported in the literature (12–18 gTS PE^{-1} d^{-1}) for UASB reactors. For STPs 2, 3 and 4, the sludge production (55, 66 and 49 gTS PE^{-1} d^{-1}, respectively) was higher than the typical range reported for UASB + TF systems treating sewage (18–30 gTS PE^{-1} d^{-1}). This is possibly because those two plants treat other effluents rather than sewage (e.g. industrial effluents, sludge from septic tanks etc.). For STP 6, the recorded value (120 gTS PE^{-1} d^{-1}) was much higher than the typical range reported by von Sperling and Chernicharo (2005) for UASB + activated sludge systems (20–32 gTS PE^{-1} d^{-1}). This is because this plant operates with a flotation post-treatment step to remove phosphorus, which significantly increases the sludge production.
- The unitary operational costs of STPs 2, 3 and 4 (0.02–0.07 €. m^{-3} treated sewage), all of them based on UASB + TF process, shown to be significantly lower than that reported for the activated sludge process (0.10–0.15 €. m^{-3} of treated sewage) (adapted from von Sperling and Chernicharo, 2005). The low operational costs constitute one of the major advantages of the anaerobic-based STPs and trickling filters.

7.9 FINAL REMARKS

To consolidate knowledge for future projects, it is most important to disclose the full-scale operational performance of anaerobic-based STPs (e.g. effluent concentrations and removal efficiencies), and information on practical experience of establishing operational protocols to manage sludge and scum, as well as biogas recovery.

Additionally, as discussed in this Chapter, the performance of an STP is dependent strictly on the implementation of adequate operational routines, particularly those related to the preliminary treatment unit and the management of the solid byproducts (sludge and scum). The excessive solids washout with the effluent of UASB reactors, because of frequent interruptions of the sludge dewatering units for maintenance, leads to the over accumulation of sludge in the anaerobic reactors, disturbing the functioning of the entire STP, and resulting in the loss of the final effluent quality. Therefore, it is imperative to implement adequate operational routines for sludge and scum management.

REFERENCES

Chernicharo C. A. L., Rosa A. P., Schmidt A. D. U. C., Borges J. M., Fonseca W. F. and Monteiro A. D. (2013). Current Limitations and The Necessary Improvements in the Anaerobic Technology for Domestic Sewage Treatment. 13 IWA World Congress on Anaerobic Digestion, International Water Association, Santiago de Compostela, Spain.

Chernicharo C. A. L., Lobato L. C. S., Paiva Júnior C. A., Barbosa C. A., Almeida P. G. S., Gonçalves E. V., Souza J. R., Silva L. R., Judice M. A. M. and Moraes O. J. (2014). Return of Secondary Sludge for Thickening and Digestion in UASB Reactors: Case Study of Onça STP – Brazil. 11 Simposio Latinoamericano de Digestión Anaerobia, International Water Association, Havana, Cuba.

Chernicharo C. A. L., van Lier J. B., Noyola A. and Bressani-Ribeiro T. (2015). Anaerobic sewage treatment: state of the art, constraints and challenges. *Environmental Science Biotechnology*, **14**, 649–679.

Chernicharo A. A. L., Bressani-Ribeiro T., Garcia G. B., Lermontov A., Pereira C. B., Platzer C. J., Possetti G. R. C., Leites M. A. L. and Rosseto R. (2018). Panorama do tratamento de esgoto sanitário nas regiões Sul, Sudeste e Centro-Oeste do Brasil: tecnologias mais empregadas (Overview of sewage treatment in the south, south-east and center-west regions of Brazil: most employed technologies). *Revista DAE*, **66**(213), 5–19.

Floripes T. C., Chernicharo C. A. L. and Mota-Filho C. R. (2018). Avaliação do descarte de excesso de lodo secundário de FBP sobre o desempenho de reatores UASB em escala plena: estudo de caso da ETE Laboreaux – Itabira-MG. (Assessment of excess sludge withdrawal from trickling filters on the performance of full-scale UASB reactors: case study of ETE Laboreaux – Itabira-MG). *Revista DAE*, **66**(211), 89–104.

Noyola A., Padilla-Rivera A., Morgan-Sagastume J. M., Güereca L. P. and Hernández-Padilla F. (2012). Typology of municipal wastewater treatment technologies in Latin America. *Clean – Soil, Air, Water*, **40**, 926–932.

Rosa A. P., Lobato L. C. S., Chernicharo C. A. L., Martins D. C. R. B., Maciel F. M. and Borges J. M. (2012). Improving performance and operational control of UASB reactors via proper sludge and scum discharge routines. *Water Practice & Technology*, **7**, 1–11.

Santos A. (2014). Remoção, tratamento e valoração de escuma proveniente de reatores UASB aplicados ao tratamento de esgoto doméstico (Scum Removal, Treatment and Evaluation from UASB Reactors Applied to Domestic Sewage Treatment). PhD Thesis, Graduate Programme on Sanitation, Environmental and Water Resources, Federal University of Minas Gerais, Belo Horizonte, Minas Gerais, Brazil.

von Sperling M. and Chernicharo C. A. L. (2005). Biological Wastewater Treatment in Warm Climate Regions (2 vols). IWA Publishing, London, UK. http://www.iwapublishing.com/open-access-ebooks/3567.

Chapter 8

Energy recovery from biogas in UASB reactors treating sewage

Gustavo R. C. Possetti[1,2], Julio C. Rietow[3], Carolina B. G. Cabral[4], Hélinah C. Moreira[5], Christoph Platzer[4], Thiago Bressani-Ribeiro[6] and Carlos A. L. Chernicharo[6]*

[1]Water and Sanitation Company of Paraná State (Sanepar), Rua Eng. Antônio Batista Ribas 151, Tarumã, Curitiba, 82.800-130, Paraná-Brazil
[2]Higher Institute of Administration and Economy of Mercosur (ISAE), Av. Visconde de Guarapuava, 2943, Curitiba, 80.010-100, Paraná, Brazil
[3]Department of Hydraulics and Sanitation, Universidade Federal do Paraná, Av. Cel. Francisco H. dos Santos, 100, Centro Politécnico, Curitiba, 81530-000, Paraná, Brazil
[4]Rotária do Brasil, Rua Teodoro Manoel Dias, 421, Florianópolis, 88050-540, Santa Catarina, Brazil
[5]Deutsche Gesellschaft für Internationale Zusammenarbeit (GIZ) GmbH, SCN Quadra 01, Bloco C, Sala 1501, Ed. Brasília Trade Center, Brasília, 70.711-902, Distrito Federal, Brazil
[6]Department of Sanitary and Environmental Engineering (DESA), Universidade Federal de Minas Gerais (UFMG), Av. Antonio Carlos, 6.627, Campus Pampulha, Belo Horizonte, 31.270-901, Belo Horizonte, Brazil
**Corresponding author: Gustavo R. C. Possetti (E-mail: grcpossetti@gmail.com)*

8.1 INTRODUCTION

Anaerobic systems, mainly those based on upflow anaerobic sludge blanket (UASB) reactors, have been used widely in Latin America and the Caribbean and, particularly, in Brazil (Chernicharo *et al.*, 2018a; Noyola *et al.*, 2012). The cost for their construction and operation is typically less than is that for aerobic systems (Chernicharo, 2007). In addition, UASB reactors enable the generation of by-products such as biogas, a compound that can be used for energy purposes in the sewage treatment system.

Editor(s): Carlos Augusto de Lemos Chernicharo and Thiago Bressani-Ribeiro
doi: 10.2166/9781780409238-8_0194

The biogas from the UASB reactors comprises mainly methane, nitrogen, and carbon dioxide, whereas hydrogen sulphide, oxygen, hydrogen, carbon monoxide, ammonia, siloxanes, water, and particulate matter are present in smaller proportions (Noyola *et al.*, 2006). Methane has a lower calorific value of 9.9 kWh Nm^{-3} and its concentration defines the potential of recovering energy from the biogas. In addition, methane has a global warming potential 28 times higher than that of carbon dioxide over a 100-year horizon (IPCC, 2014).

Currently most of the sanitation companies in Brazil using UASB reactors in their sewage treatment systems do not store and/or use the biogas produced. Generally, the captured biogas is conveyed to an open flare, where it is destroyed thermally to reduce the emission of methane and hydrogen sulphide to the atmosphere. Hydrogen sulphide is harmful to human health at higher levels and is perceptible to humans from 0.47 to 100 ppb, having an unpleasant smell (Tchobanoglous *et al.*, 2013; Park *et al.*, 2014). In addition, hydrogen sulphide can contribute to the corrosion of materials (see Chapter 5). Obviously, gaseous emissions need to be mitigated and controlled, as addressed in Chapter 9.

However, the practice of destroying biogas in open flare leads to the loss of its chemical energy and, therefore, tests and studies have emerged in Brazil involving the use of biogas energy, particularly for thermal utilization and electricity generation (Possetti *et al.*, 2018). Heat recovery from biogas combustion has been studied as an alternative for the drying and sanitization of the sewage sludge. One of the advantages of this system is the non-use of the chemicals usually necessary to treat sludge, such as lime. In addition, the heat recovery from biogas in these systems ensures a high rate of water removal in the sludge, as well as the removal of pathogens, reducing the costs inherent to its handling, transportation, and final destination (Chernicharo *et al.*, 2015). Electricity generation from biogas has also been studied, particularly for distributed generation, including net metering (Gomes *et al.*, 2017). In addition, the cogeneration of electricity and heat has been considered as on-site utilization of the residual heat produced in the thermodynamic processes of electricity generation. Aiming at the energy efficiency of the process, residual heat can be used also in thermal sludge drying systems (Rosa *et al.*, 2016).

For these reasons, the utilization of biogas energy seems to offer powerful leverage to increase the historically lagging coverage of sanitation infrastructure in the developing world, as addressed further in Chapter 12. In addition to reducing methane and hydrogen sulphide emissions to the atmosphere, using biogas in sewage treatment plants (STP) is an energy alternative that could lead the sanitation sector toward development that is more sustainable, socially just, environmentally correct, safe, and economically viable.

The objective of this chapter is to present the latest developments in energy recovery from biogas in UASB reactors treating domestic sewage, the lessons learned, and the conditions necessary for improving the designs. The chapter reports the biogas production patterns and typical quality, as well as some parameters for the design of energy recovery facilities, including biogas treatment and its final use. Application examples are presented and some full-scale instances of energy recovery from biogas are reported.

8.2 BIOGAS: QUANTITATIVE AND QUALITATIVE CHARACTERIZATION

The key to successful biogas recovery design is the accurate quantification of the biogas available. Each system presents daily and monthly variations in biogas production at a smaller or larger scale. The influent sewage characteristics, applied organic load, reactor structure tightness to prevent leaks, scum removal devices, operational control, and significant rainfall events all influence the net biogas production, as reported by Possetti *et al.* (2013), Silva (2015), Cabral *et al.* (2017a), and Possetti *et al.* (2018). These publications showed that measuring biogas was recommended highly before investing in biogas recovery facilities, as, otherwise, the investment might be quite conservative. This section is intended to provide guidance on this matter.

8.2.1 How much biogas is produced in UASB reactors treating sewage?

Reliable data on the effective usable (net) biogas production in UASB reactors provide the most important base for dimensioning energy recovery facilities; therefore, the quantification of the available biogas is a crucial factor. A comparison between biogas production estimates and measurement results was performed recently by Possetti *et al.* (2013), Silva *et al.* (2014), Cabral *et al.* (2016) and Cabral *et al.* (2017a). These studies reported that the amount of energy estimated by existing methods was on average much higher than was the amount actually available for recovery. The model proposed by Lobato *et al.* (2012) showed less deviation in relation to the values measured, but still showed values exceeding those found in most STPs.

The model proposed by Lobato *et al.* (2012) includes all the routes of conversion and losses in UASB reactors, including the portion of COD used for the reduction of sulphates and the loss of methane in the residual gas and dissolved in the effluent. Therefore, it probably expresses the physical, chemical, and biological phenomena inherent in UASB reactors.

This model considers three scenarios that lead to various levels of methane recovery potential, namely, the (i) worst scenario, (ii) typical scenario, and (iii) best scenario. The worst scenario, in which the energy potential is lower, involves systems operating with more diluted sewage, higher sulphate concentrations, lower COD removal efficiencies, and higher rates of methane loss. The best scenario involves systems operating with more concentrated sewage, lower sulphate concentrations, higher COD removal efficiencies, and lower rates of methane loss. In the typical scenario, the input data have intermediate values. Different unitary relationships have been obtained from these scenario simulations. The results are summarized in Table 8.1.

Table 8.1 Unitary relationships of methane yield in UASB reactors treating sewage.

Unitary Relationship	Unit	Worst Scenario			Typical Scenario			Best Scenario		
		Max	Min	Mean	Max	Min	Mean	Max	Min	Mean
Unitary methane yield	$NLCH_4 PE^{-1} d^{-1}$	9.9	3.6	6.8	13.3	7.4	10.2	16.7	11.1	13.7
	$NLCH_4\ m_{sewage}^{-3}$	81.7	16.7	42.2	103.7	34.8	64.2	134.6	51.8	81.3
	$NLCH_4\ kgCOD_{remov}^{-1}$	154.1	66.0	113.4	185.8	124.2	158.3	219.1	173.9	196.0
Unitary energy potential	$MJ\ m_{sewage}^{-3}$	2.9	0.6	1.5	3.7	1.2	2.3	4.8	1.9	2.9
	$MJ\ kgCOD_{removed}^{-1}$	5.5	2.4	4.1	6.7	4.5	5.7	7.9	6.2	7.0
	$MJ\ PE^{-1}\ year^{-1}$	129.5	47.7	89.7	173.8	96.8	133.8	218.4	145.7	179.3

Source: Lobato *et al.* (2012).

The unitary relationships presented serve as a first reference for methane yield in UASB reactors treating sewage. For more accurate estimates, we suggest using the software ProBio (Program of Estimation of Biogas Production in UASB Reactors), developed in cooperation between two Brazilian institutions, namely, the Water and Sanitation Company of Paraná State (Sanepar) and the Federal University of Minas Gerais (UFMG). Version 1.0 of this software is free of charge and is available for download at site.sanepar.com.br/probio/probio.zip.

Biogas yields in UASB reactors treating sewage have been measured recently in various full-scale STPs (Cabral, 2016; Cabral *et al.*, 2016, 2017a; Possetti *et al.*, 2013), over a prolonged period (more than 1 year). Table 8.2 presents a synthesis of the methane yields calculated for several investigated STPs, considering various intervening factors (sewage flow, influent COD, COD removal efficiency, and leakage status). Considerable variation was indicated between the medians of the monitored STPs, with the values ranging between approximately 42.7 and 138.3 $NLCH_4\ kgCOD_{remov}^{-1}$. The significant variation of more than 100% shows that the measuring of biogas production for existing plants is indispensable when designing energy recovery facilities. Such unitary relationships allow comparison between the treatment plants that is more reliable, as it comprises the variables of COD concentration and reactor efficiency (Duarte *et al.*, 2018).

Table 8.2 Comparative table of the methane yields for full-scale STPs.

Sewage Flow ($L\ s^{-1}$)	Influent COD ($mg\ L^{-1}$)	COD Removal Efficiency (%)*	Leakage Status**	Methane Yield ($NLCH_4\ kgCOD_{remov}^{-1}$)	Comparison with Theoretical Scenarios***
431	684	87	low	138.3	typical–worst
97	1146	65	low	125.3	typical–worst
223	714	90	high	118.7	worst
91	670	75	medium	112.1	worst
79	431	87	medium	110.3	worst
395	476	72	high	84.8	worst
31	656	93	high	72.5	worst
127	938	75	high	56.4	lower than worst
247	450	73	high	42.7	lower than worst

*Efficiency calculated with total influent COD and filtered effluent COD.
**Classification based on on-site conditions.
***Comparison with scenarios reported by Lobato *et al.* (2012) in Table 8.1.
Source: Adapted from Cabral *et al.* (2016) and Cabral *et al.* (2017b).

The methane yields were found to approach the typical scenario proposed by Lobato *et al.* (2012) only in STPs with a low level of leakage in the gas-liquid-solid (GLS) separator of the UASB reactors and in the gas line. In most of the STPs, methane yields were found to be close to or even lower than those of the worst scenario predicted by the model. Notably, biogas leaks were observed in anaerobic reactors through cracks and poorly sealed pipes, causing greater gas losses than those estimated by the model.

Additionally, it is important to highlight that a considerable amount of the biogas produced in UASB reactors is dissolved in the liquid phase (Noyola *et al.*, 2006). This was not considered in the measurements reported above. In a typical scenario, the dissolved methane could represent up to 40% of the total produced methane (Glória *et al.*, 2016; Lobato *et al.*, 2012; Nelting *et al.*, 2017; Souza, 2010). Therefore, the net biogas production in UASB reactors treating sewage could be increased significantly when methane is recovered from the liquid phase, as is discussed in Chapter 9.

For existing STPs to implement energy recovery from biogas, it is advisable to assume that biogas production would approach the predicted worst scenario (113 $NLCH_4\ kgCOD_{remov}^{-1}$) or even lower (80 $NLCH_4\ kgCOD_{remov}^{-1}$), if a high number of leaks were detected. For new STPs that have already incorporated design improvements, including gas-tight GLS separators and adequate sealing of the gas pipeline, it is reasonable to adopt predicted methane yields from the typical scenario (approximately ($158\,NLCH_4\ kgCOD_{remov}^{-1}$) or even the best scenario (approximately $196\,NLCH_4\ kgCOD_{remov}^{-1}$).

Typical biogas production patterns in UASB reactors treating sewage present time-variable, periodic, and non-stationary behaviour. The measurements show a daily pattern of biogas production and release, as first reported by Possetti *et al.* (2013). Subsequently, other real-scale studies have shown that the largest biogas production during the day was typically in the afternoon (Cabral, 2016; Cabral *et al.*, 2016; GIZ, 2017).

In general, it was noted that the biogas flow rate followed the sewage flow rate behaviour, as shown in Figure 8.1, however, with different lag times between the peaks of each of these parameters. Additionally, rainfall events were observed to decrease the biogas production probably because of the extreme dilution of the incoming sewage (Cabral, 2016; Possetti *et al.*, 2013; Silva, 2015). An example of this decrease can be seen in Figure 8.2, considering a significant rainfall event (138 mm during 2 days).

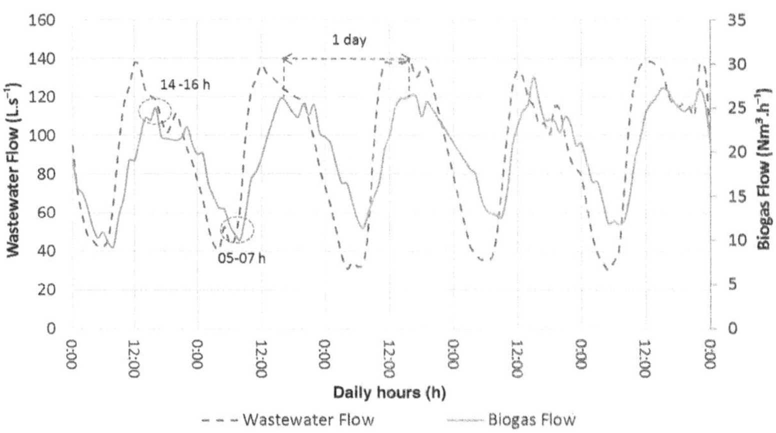

Figure 8.1 Typical biogas flow pattern for UASB reactors treating sewage in the absence of rainfall events. Source: Adapted from Cabral (2016).

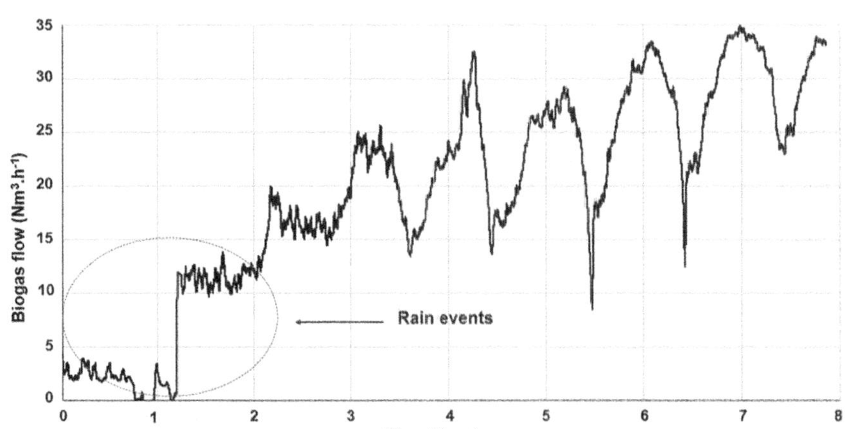

Figure 8.2 Biogas flow pattern for UASB reactors treating sewage, including a significant rainfall event. Source: Adapted from Possetti *et al.* (2013).

Temperature in the range 20–25°C was reported a minor variable relative to the biogas production behaviour (van Haandel & Lettinga, 1994), as was also observed in the monitored full-scale UASB reactors (20–29°C) (Cabral *et al.*, 2016). The concomitant influence of other variables on biogas production, such as the operating conditions, could mask the possible effect of the temperature on the efficiency of the UASB reactor. Although the correlation between biogas production and the reactor operational conditions is still under investigation, various trends have emerged, as shown in Table 8.3.

Table 8.3 Influence of the different variables in biogas production for full-scale UASB reactors, with temperatures from 20 to 29°C.

Parameter	Influence on Biogas Production	
Sewage flow	⬆ Positive during drought or low rainfall periods	⬇ Negative in extreme rainfall periods
Rainfall	⬇ Negative	
Influent COD concentration	⬆ Positive	
Removed organic load	⬆ Positive	
Temperature	▭ Not significant	

Source: Adapted from Cabral *et al.* (2016).

8.2.2 Biogas composition

Biogas composition depends on the physical and chemical conditions inside the UASB reactors, such as pH, alkalinity, temperature (particularly below 15°C), and the presence of anions (sulphate and nitrate). The biogas generated in UASB reactors comprises mostly methane, nitrogen, and carbon dioxide. In addition, oxygen, hydrogen, carbon monoxide, ammonia, siloxanes, hydrogen sulphide, water, and particulate matter are present in smaller proportions. Table 8.4 summarizes the typical composition of the biogas from UASB reactors treating domestic sewage.

Table 8.4 Typical composition of biogas from UASB reactors treating domestic sewage.

Component	Unit	Typical Concentration Range
Methane (CH_4)	% v/v	70–85
Nitrogen (N_2)	% v/v	2–25
Carbon dioxide (CO_2)	% v/v	5–15
Oxygen (O_2)	% v/v	0–2
Hydrogen (H_2)	% v/v	0–3
Carbon monoxide (CO)	% v/v	0–0.3
Hydrogen sulphide (H_2S)	ppmv	233–5,000

Source: Adapted from Silveira *et al.* (2015) and Cabral *et al.* (2016, 2017a).

The most important finding from measuring biogas composition online over extended periods (longer than 1 year) was its constant high methane content. The median methane concentrations were approximately 78% v/v, indicating 15–20% higher energy content in the biogas from UASB reactors in comparison with the biogas from a typical sludge digester (DWA-M 363, 2010).

According to Noyola *et al.* (2006), the high concentration of nitrogen in biogas originates from the dissolved nitrogen in the domestic sewage at the entrance to the UASB reactors. The lower concentration of carbon dioxide in the biogas from UASB reactors compared with that from sludge digesters can be attributed to the greater solubility of this gas in the liquid. Therefore, much of the carbon dioxide produced during the organic matter degradation process is solubilized within the treated effluent (van Haandel & Lettinga, 1994), remaining in the bulk liquid in the form of bicarbonate ion (Noyola *et al.*, 2006).

Biogas can still contain small fractions of siloxanes. This compound, produced by the anaerobic degradation of substances found in cosmetic products, deodorants, soaps, and food additives can bring about the formation of mineral compounds during the combustion of biogas, causing problems related to pipe incrustations (Chernicharo & Stuetz, 2008). However, siloxanes concentration in biogas from UASB reactors treating sewage are sparsely reported in the literature and, consequently, there are no typical range reported.

The presence of other components, some at trace level, requires the biogas to be conditioned and pretreated prior to conversion. Hydrogen sulphide is one of the substances that requires special attention because it can cause corrosion in components, as well as unpleasant odours, as addressed in Chapters 5 and 9.

Recent data obtained from the monitoring of nine full-scale STPs, during more than one year, showed that methane and hydrogen sulphide concentrations in biogas are above 70% v/v and 500 ppmv, respectively, for at least 90% of the data (Cabral *et al.*, 2015, 2016), as depicted in Figure 8.3(a–b).

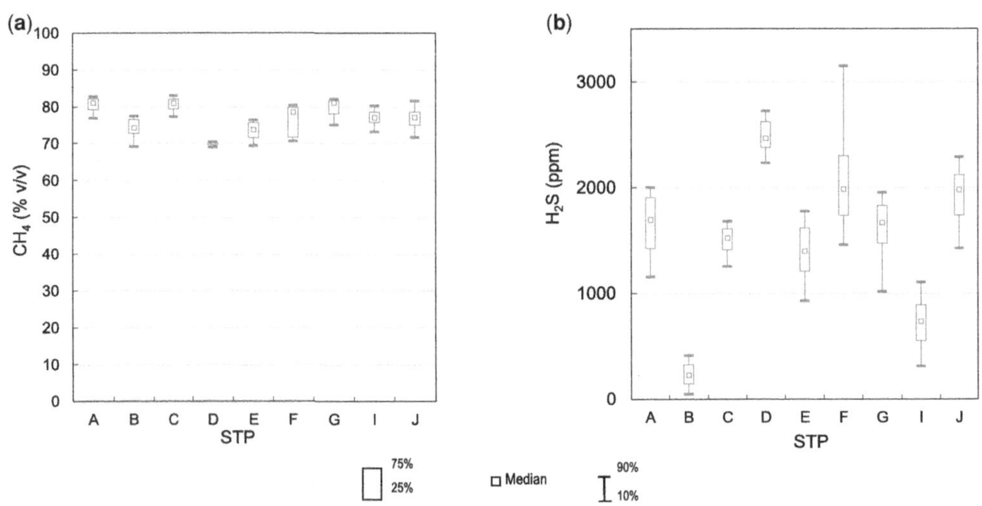

Figure 8.3 CH_4 (a) and H_2S (b) concentrations in biogas from UASB reactors. (Source: Cabral *et al.*, 2016).

8.2.3 Main design, construction, and operational aspects influencing biogas flow and composition

Table 8.5 summarizes the main aspects that can influence biogas flow and composition in UASB reactors treating sewage. These aspects, discussed further in different chapters of this book (e.g., Chapters 4, 5, and 6), should be considered in the design, construction, and operation of STPs to maximize the energy recovery from biogas.

Table 8.5 Main aspects that influence biogas flow and composition from UASB reactors treating sewage.

Aspect	Description
Preliminary treatment	It is essential to remove coarse solids effectively before the sewage is sent to the UASB reactors. The operational problems arising from the non-removal (or inadequate removal) of coarse solids could jeopardize the entire operation of the treatment system, as these materials can adversely affect both the distribution of the influent sewage in the bottom of the reactor and the generation and accumulation of sludge with low activity and difficulty to withdraw. The implementation of efficient preliminary treatment measures is essential, as is described further in Chapter 4.
Influent distribution system	Sewage distribution boxes have to be designed to prevent preferential flows and sedimentation of solid materials (see also Chapter 5), which can compromise the UASB reactor efficiency for organic carbon removal.
Volumetric hydraulic load	The volumetric hydraulic load applied to UASB reactors has considerable influence on the treatment efficiency. High hydraulic loads can cause solid washout and loss of biomass, which lead to a reduction in treatment efficiency (effluent quality) and the degree of sludge stabilization. In extreme cases, the minimum solid retention time can be compromised.
Gas-liquid-solid separator	One of the most critical parts of UASB reactors is the GLS separator. The construction of GLS separators with materials that have high resistance can minimize maintenance stops, reducing the risk of accidents, as well as the loss of biogas via leakage. Such structures should be tight to avoid biogas escaping through the sedimentation compartment and directly to the atmosphere, which enhances fugitive emissions of methane and hydrogen sulphide. GLS separator design and construction are discussed further in Chapters 4, 5, and 6.
Scum removal	Scum accumulation owing to improper operational routines or inefficient removal devices can contribute to the obstruction of the liquid–gas phase interface inside GLS separators, leading, in some instances, to the rupture of such structures. GLS separators equipped with hydraulic scum removal devices are an important alternative, as discussed in Chapters 4, 5, and 6.
Cross-connections between sewerage and drainage systems	Biogas production is associated directly with the influent organic load. Both the infiltration rate and the irregular cross-connections of rainwater into sewerage systems reduce the influent organic matter concentration, leading to lower biogas production. Using appropriate materials and techniques during the construction stage of the sewerage system, as well as the inspection of critical points along the network, can contribute significantly to reducing the infiltration rates, as well as the contributions of rainwater during rainy periods.
Sealing and corrosion of reactors	It is essential to perform pressure and leakage tests on the biogas collection lines, GLS separators, and inspection points prior to the start-up of the system. This could prevent future problems, such as gas leakage, explosion risks, significant losses of energy potential, and diffuse methane and hydrogen sulphide emissions. Simple tests, such as the injection of smoke into GLS separators, or more elaborate tests, such as reactor pressurization can contribute significantly to the detection of leakages and the prevention of the problems mentioned above (more details on pressurization in Chapter 5). In addition, the upper part of the gas manifolds features a number of openings for inspection and cleaning. Adequate sealing of these openings is essential to preventing biogas losses. The closure system material must be resistant to solar radiation, a corrosive environment, and pressures within the reactor gas chambers, generally approximately 1,500 Pa.

Source: Adapted from Silveira *et al.* (2015), Chernicharo *et al.* (2018b), Bressani-Ribeiro *et al.*, 2018 and Possetti *et al.* (2018).

8.3 BIOGAS RECOVERY INFRASTRUCTURE

8.3.1 General layout of a biogas recovery facility

An energy recovery facility based on a UASB reactor comprises basically the following biogas subsystems, namely, capture and piping, pre-conditioning, measurement and control, storage, burn, treatment, and final use. Figures 8.4 and 8.5 show the application of these general layouts to electricity production and thermal sludge drying, respectively. The description of each subsystem is reported below, based on technical norms and engineering best practices, which are described further in the *Technical Guide for the Use of Biogas in STPs* (GIZ, 2017; Silveira *et al.*, 2015) and in the German standard DWA-M 363 (210).

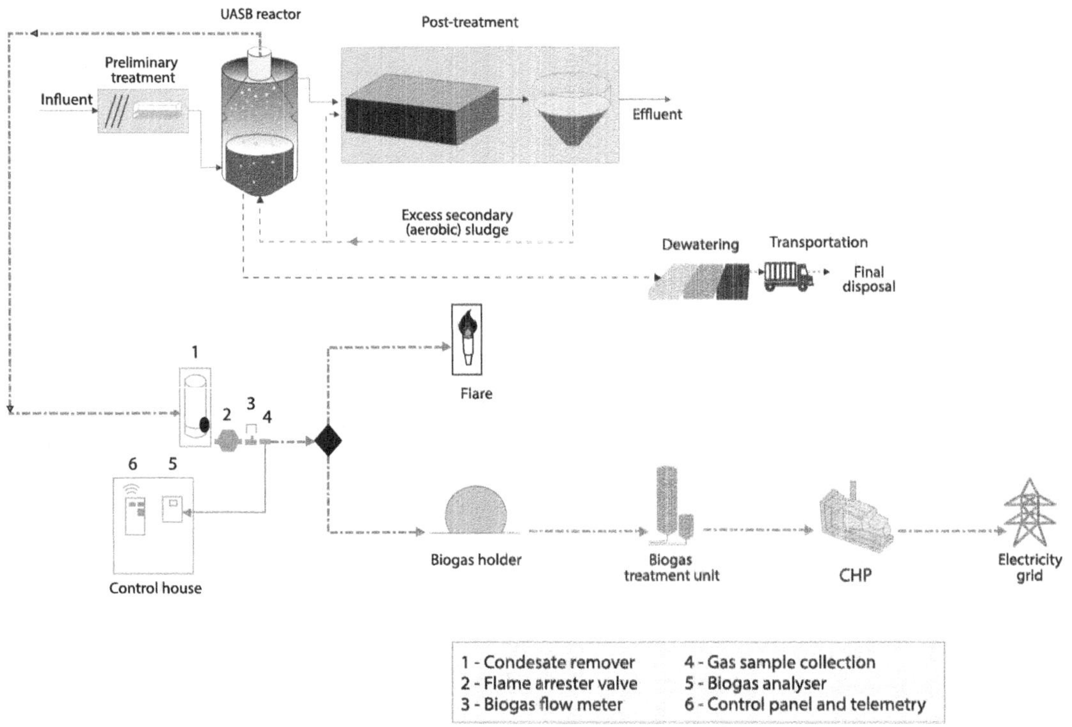

Figure 8.4 General layout of an energy recovery facility based on UASB reactors, with electricity production from biogas conversion.

8.3.2 Biogas collection, conveyance, and pre-treatment

The design and operation of UASB reactors for sewage treatment should ensure that: (i) the reactor is sealed tightly to prevent biogas leakage during operation; (ii) scum is removed frequently from the inner part of the GLS separator; and (iii) the biogas collection system is equipped with a condensate removal device, flame trap, and a biogas flare.

Effective sealing is crucial for efficient biogas capture, as it prevents biogas leakages that cause explosion risks, loss of the energy potential, emissions of greenhouse gases (methane), and odorant gases (hydrogen sulphide). Proper design and construction guidelines for the entire biogas system are discussed in Chapters 4 and 5.

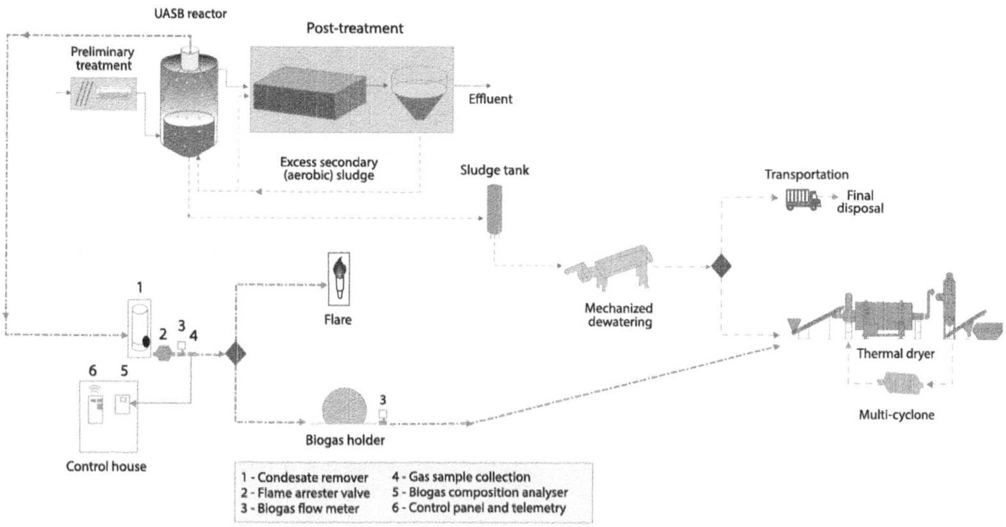

Figure 8.5 General layout of an energy recovery facility based on UASB reactors, with thermal sludge drying from biogas conversion.

Condensate removal is a crucial step in any biogas collection system. The condensate is produced when the temperature of the gas is higher than is the temperature inside the collection system. Cooling down the gas leads to condensation along the transport system. If the condensate is not removed, it can block the collection system at a lower point by building up a water trap. The control of condensate should be implemented at the beginning of the biogas collection system, using appropriate slopes for the pipes, to allow the drainage of the condensate, either back into the UASB reactor or to the drainage points (USEPA, 1996).

Problems related to high pressure (overpressure) or vacuum (underpressure) inside the GLS separators can interfere with the correct management of the biogas. Therefore, when burning or using biogas, a minimum pressure of 1,500 Pa (15 cm of water column) must be maintained to ensure the correct operation of the system (ABNT, 2011). To ensure the desired minimum pressure, pressure and vacuum relief valves must be installed in the biogas pipeline. Currently, many STPs have adopted water seals as a mitigating measure related to overpressure and vacuum. As a hydraulic device, a water seal offers high safety at a low construction cost (Silveira *et al.*, 2015).

In low-pressure gas chambers, biogas collection can be conducted with the aid of blowers or radial compressors driven by electric motors, provided that the process is controlled and safe. Blowers must be explosion proof (ATEX certified) and facilitate the controlled extraction of biogas according to consumption and, particularly, from automatic and simultaneous readings of oxygen and methane contents. This is because the active extraction of the biogas must be carried out in a manner that prevents contamination by air and, consequently, the formation of explosive mixtures inside pipes and gas holders. Further information on the operational and safety accessories is available in Silveira *et al.* (2015), GIZ (2017) or DWA-M 212 (2008).

8.3.3 Measuring

To improve the understanding of the behaviour of biogas production, the parameters shown in Table 8.6 should be monitored. The suggested monitoring frequencies and the importance of determining each parameter are also shown in Table 8.6.

Table 8.6 Parameters for monitoring and control of biogas recovery facilities associated with UASB reactors treating sewage.

Parameter	Frequency	Comments
Sewage flow	Continuously, real time	The continuous monitoring of the sewage flow is fundamentally important, as the production of biogas is directly proportional to it.
Biogas flow	Continuously, real time	Fundamental for the design of the biogas utilization plant and for the calculation of accurate economic viability. It can indicate the instability of the anaerobic process and how much biogas is actually available for use.
Biogas quality (CH_4)	Real-time measurement is recommended in the instance of electricity production. During the design phase, grab samples are sufficient.	Knowledge of the CH_4 concentration can help to detect operational problems, which could influence energy production. The results of continuous measuring in real-scale plants show low variation in biogas quality; therefore, punctual analysis is sufficient during scoping.
Biogas quality (H_2S, and O_2)	In general, only grab samples. O_2 in real time only if vacuum is applied to transport the biogas. Intensive measurement of H_2S is recommended during the design phase.	As H_2S concentrations vary significantly over the day and other operational conditions (Cabral, 2016), an intensive measurement is recommended during the design phase. For reasons of cost, the continuous measurement of H_2S can be dismissed during the operation of a biogas plant that has an activated carbon filter as a final polishing step. In the instance of negative pressure, the O_2 measurement in real time is indispensable to prevent explosive situations. In other instances, it can be used to determine leakages.
Biogas temperature	Real time	Parameter of easy control, and necessary to convert the biogas production into normal temperature and pressure conditions (NTP), when the meter itself does not provide the normalized biogas flow.
Biogas pressure	Real time	Parameter of easy control, and extremely helpful to detect leakages.
COD load removed	In instances of the biogas flow being lower than expected	Important to evaluate the COD load removed in the anaerobic process and determine the rate of methane production per kilogram of COD removed. If the biogas flow is lower than expected, the COD removal can indicate whether this is attributable to a decrease in the reactor efficiency or to other reasons (leaks, etc.). UV/VIS spectrometry-based sensors can be used for real time and in situ COD measurements (Gomes *et al.*, 2017; Duarte *et al.*, 2018).

Source: Adapted from Drosg (2013) and authors' personal experience.

The biogas flow pattern is the most important aspect in designing the equipment for energy recovery and, if necessary, the biogas storage unit. Future scenarios, such as an increase in the STP capacity (influent sewage flow) should be considered adequately during such design. Biogas flow can be measured using various technologies, as shown in Table 8.7. Furthermore, although the accuracy is equivalent, the criteria for measuring clean/dry gas or humid/not cleaned gas differ and have to be taken into consideration (Silveira *et al.*, 2015).

Table 8.7 Main biogas flow measurement technologies.

Parameter	Advantages	Disadvantages
Ultrasonic meter	• Satisfactory results at low pressure • No moving parts	• Long linear distance of measurement required • High cost
Vortex meter	• No moving parts • High accuracy • Corrosion resistant • Low head loss	• Long linear distance of measurement required • Temperature and pressure measurements are required for normalized flows
Measurement by means of dynamic pressure	• High durability • Little influence of dirty gases • Pressure variations have no negative influence on accuracy	• Works better at high pressure • Difficult calibration • Error of 1.5–5% • Long measuring distance
Fluidistor	• No moving parts • High accuracy • Low cost • Easy to clean, handle, and replace	• Complex flow calculation • Error of 1.5% • Sensitive to vibrations
Turbine meter	• Low investment cost • Simple measurement technology	• Impurities cause problems • Burdensome and routine maintenance • Low resolution
Mass meter by thermal dispersion	• Easy installation • Low investment cost • Low head loss • Recommended for process control of survey of potentials • Accurate measurements even at pressure fluctuations	• Sensitive to moisture and gas particles • Long linear distance of measurement required • Calibration requires standard gas • If the fluid has a composition variable, the flow must be corrected

Source: Adapted from Silveira *et al.* (2015).

The amount of available methane should be determined as it is represents the energy content of the biogas. The determination of other compounds present in the biogas is important in the design of the biogas treatment system, particularly in relation to the removal of hydrogen sulphide intended to generate electricity, and to prevent the corrosion of the components. When real-time monitoring is not possible, using a portable in-loco meter is recommended. It is also possible to take biogas samples and analyse them in specialized laboratories. In both instances, samples should be collected during a representative period of at least one week and at various times during the day to capture temporal variations. We recommend that this one-week campaign be repeated at least three times to detect the variation scale of the main biogas compounds, as the results can vary significantly throughout the day and year depending on the climatic conditions. If biogas composition measurements are not possible, it is reasonable to assume the concentrations reported in Table 8.4. In this instance, it is suggested to use the lowest typical methane concentration (70% v/v) and the highest typical hydrogen sulphide content (5,000 ppmv) to prevent the overestimation of the energy recovery and to assure the biogas cleaning.

For the installation of measurement and sampling equipment, it is important to have an efficient biogas collection and transport system. If this system does not exist, we recommend that it be installed before developing the biogas project.

8.3.4 Storage

Storage is indispensable when biogas is used for energy, and an operation lacking biogas storage is recommended only in the instance of direct flaring. The type and size of the gas storage depend on the intended energy use.

There are three main storage forms, namely, low-pressure storage, with a double membrane inflated as counter pressure with compressed air (up to 10 mbar); low-pressure storage, with a counter weight (mainly concrete); and, no-pressure storage in gas bags. For instance, currently in Brazil, double membrane has been designed as storage, as it offers the best trade-off and operational security.

Where the continuous use of the biogas is required, the storage serves to equalize the variation in the biogas production rate. Therefore, the minimum storage volume has been determined as 25% of the daily volume of gas used. Generally, 30% of the daily gas use is assumed as a design volume for storage. For biogas use for electricity generation in peak-hour mode (energy is generated during certain hours of the day with higher energy tariffs), we recommend drawing biogas production and consumption curves (volume/time). The largest distance between the two curves determines the necessary gas storage volume. The resulting storage will be significantly larger than the indicated 30%, which is considered a minimum storage size (Silveira *et al.*, 2015). Calculations have shown that compensation for several days is not economically feasible, as the storage size increases significantly.

8.3.5 Flares

When the biogas is not used after collection, the implementation of an efficient burning system is recommended. Direct combustion without heat recovery is performed by devices that enable the ignition and burning of gases by means of equipment called a flare. This equipment must always be present in biogas generation sites, as it is used in the prevention of accidents (avoiding possible explosions), control of odorous emissions, and control of greenhouse gas emissions. The installation of flares is useful also for situations where the energy recovery equipment is not in full operation or when the methane content is too low, making energy recovery unfeasible. In addition, flares are important in promoting the burning of surplus biogas from energy recovery equipment. Flares must be designed to maintain a capacity corresponding to the maximum biogas volume plus a safety factor of 10% (RIRDC, 2008), and stainless steel is recommended as the construction material. There are generally two models of flares, namely, open and enclosed.

Open flares are used mostly for biogas combustion in STPs in developing countries (Noyola *et al.*, 2006). Open flares have no combustion chamber, if the flame is exposed to the atmosphere. However, because the flame is open (exposed to weather conditions), the intensity is low, and the ambient air quickly cools or even extinguish the flame. Low flame temperature and rapid heat exchange with atmospheric air result in undesirable by-products and non-oxidized compounds, leading to a reduction in the efficiency of the combustion process. Furthermore, the temperature and residence time in the combustion process are difficult to measure and, therefore, determining the conversion efficiency of the biogas is difficult. The methane destruction efficiency, for example, is typically approximately 50% (Baukal, 2013).

Because the flame is exposed to the environment, open flares are installed at heights and distances that prevent the radiation of the flame reaching people and equipment. Protection is often placed around the nozzle to prevent wind from making the ignition process difficult (Barton *et al.*, 2008).

Enclosed flares enable the burning of biogas inside a combustion chamber. Its interior is insulated thermally to avoid cooling from heat exchange with the metal walls of the burner, thereby providing significantly higher temperatures. In addition, thermal insulation guarantees the protection of people by preventing the external surface of the burner from reaching extreme temperatures (Barton *et al.*, 2008). The structure of enclosed flares contains monitoring equipment for temperature and gas flow and a facility for sample collection, thereby facilitating monitoring of the performance of the combustion process. Because the biogas combustion occurs under controlled conditions, the enclosed burner ensures near-complete conversion of the biogas into oxides and water vapor. In addition, the smoke, flame, and noise levels are practically imperceptible (IEA, 2000) at the output of this equipment. The efficiency of compound destruction (CH_4 and H_2S) in enclosed burners can reach rates exceeding 99% (Barton *et al.*, 2008; Wagner *et al.*, 2017). Nevertheless, to achieve higher efficiencies, it is necessary to regulate the flow rate (Wagner *et al.*, 2017), as biogas production in UASB reactors shows variable behaviour, as was discussed previously.

Table 8.8 summarizes a comparison between open and enclosed flares according to economic, social, environmental, and operational aspects. The advantage of open flares is that their correct operation can be confirmed externally simply by observing the visible flame. In addition, the cost is significantly lower than is that of an enclosed flare (GIZ, 2017) – the cost of investing in open flares is usually 25–70% lower than is that of investing in enclosed flares (IEA, 2000).

Table 8.8 Comparison between open and enclose flares.

Evaluation Aspects		**Flare**	
		Open	**Enclosed**
Economic	Acquisition cost	+++++	+
Social	Control of odorant compounds	+	++++
	Visual impact of flame	+++	+++++
Environmental	Destruction of methane	+	+++++
	Release of non-oxidized compounds	+	+++++
Operational	Dispersion of combustion gases	+++++	++
	Operational simplicity	+++++	+++
	Burning temperature	++	+++++
	Flame stability	+	+++++
	Residence time	+	+++++
	Risk of explosion	++	++
	Radiation emission	++	+++
	Possibility of monitoring and control of the process	+	+++++
	Sound emission	++	++

+++++: more favourable; +: less favourable; ++++, +++, ++: intermediate, in descending order.
Source: Adapted from Kaminski *et al.* (2018).

8.3.6 Treatment

Prior to energy recovery from biogas, the biogas must be treated to the level required according to its final use. Treated biogas, after removal of major impurities such as moisture, siloxanes, and mainly H_2S, can be used for the production of electricity, thermal energy, or for combined heat and power (CHP) generation. The requirements shown in Table 8.9 must be observed for the most common uses in STPs.

The quality requirements for biomethane are significantly higher than the values shown in Table 8.9 and, additionally, it is usually necessary to remove CO_2 and N_2 to values lower than 3% v/v (more than 97% v/v methane). Therefore, currently, the production of biomethane is not considered viable in the instance of UASB reactors treating domestic sewage.

Table 8.9 Quality requirements for biogas according to the final use.

Component	Burning for Heat Generation*	Combined Heat and Power (CHP)*
Total sulphur (S)	<1000 ppm	N.S.
Hydrogen sulphide (H_2S)	–	<200 ppm
Total organic compounds with silicon	–	<5–10 mg.Nm^{-3}
Relative humidity	<dew point temperature soil	50–80%
Oxygen content (O_2)	<3% v/v	N.S.
Methane (CH_4)	>50% v/v	>45% v/v
Ammonia (NH_3)	–	<50 mg.10 kWh^{-1}
Gas temperature	–	<40°C
Operating pressure	8–25 mbar	>20–80 mbar

(–): No requirements; N.S.: Not specified.
*Variable according to the requirements of manufacturers of energy conversion technologies. Values in the table are only indicative.
Source: Adapted from DWA-M361PT (2011) and Lobato (2011).

The main uses for biogas from UASB reactors treating sewage are heat (particularly for drying sludge) and electricity. As mentioned before, the main component to be removed from biogas is H_2S. The methods for H_2S elimination could be physical, chemical, or biochemical (Chernicharo *et al.*, 2011). Measurements in full-scale systems show that H_2S removal prior to the injection in CHP engines is essential. Typically, the H_2S concentrations are higher than 500 ppm (see Figure 8.3(b)). Therefore, physical methods, such as activated carbon do not appear to be economically feasible once higher concentrations (>500 ppm) lead to quick saturation and, therefore, to high operational costs. In this instance, using a biochemical method, such as biodesulphurization is recommended.

Biodesulphurization is based on the biological oxidation of H_2S to elemental sulphur (S) and sulphate SO_4^{2-}. As regards UASB reactors treating sewage, applying the desulphurization together with the storage to reduce investment costs appears an interesting solution (see Figure 8.6). In such instance, a support material should be provided for oxidizing bacterial growth, and air injection must be supplied (approximately 2% v/v of the biogas volume as air). Air feeding must be controlled to prevent a critical situation that could lead to an explosive atmosphere in the storage area.

A combination with the aforementioned techniques appears to be effective for H_2S removal from biogas prior to electricity and heat cogeneration. In such instance, the biogas stream follows a two-step process, namely, it is routed first to the gas storage with integrated biodesulphurization, which is followed by an activated carbon polishing step. Cabral *et al.* (2017b) showed that after biodesulphurization (without inoculation), the H_2S concentration in biogas dropped from 1,590 ppm to approximately 868 ppm. The activated carbon step further decreases the H_2S concentration to approximately 0.54 ppm. Other methods to reduce the H_2S concentration are shown in Table 8.10, but the above combination appears to be one of the most promising alternatives in economic terms.

Figure 8.6 Biogas storage with integrated biodesulphurization at Jacuípe II STP, Bahia State, Brazil. (Source: Rotaria of Brasil cited in Silveira *et al.* (2015)).

Siloxanes are another relevant component that should be removed prior to biogas injection in CHP engines. As reported in Section 8.2.2, measuring siloxane in the biogas of UASB reactors treating sewage is in the early stage and the analysis and sampling methodology is not yet well established. Therefore, there is a lack of knowledge on this parameter. Preliminary analysis in full-scale UASB reactors has indicated some siloxanes traces. Cooling and drying have been considered capable of eliminating most of this contamination and the activated carbon capable of polishing the gas stream. Nevertheless, there is a strong need for data that are more reliable on siloxanes in biogas from UASB reactors treating sewage.

8.3.7 Energy recovery from biogas

Using biogas for energy purposes implies the transformation of its chemical energy. Table 8.11 presents the main alternatives for the biogas energy use from UASB reactors treating sewage. The best alternative should

Table 8.10 Some desulphurization methods.

Method	Energy Consumption	Indirect Inputs*		Air Injection	H_2S Concentration After Treatment (ppm)	Problems
		Consumption	Disposal			
Biodesulphurization in the digestor or the gas storage facility (injection of air)	++	++	++	Yes	50–2,000	Lack of accuracy in process control
External biodesulphurization (trickling filter)	–	+	+	Yes	50–100	Lack of accuracy in process control
Gas washing	–	–	+	No	50–100	High complexity
Precipitation of sulphide	0	– –	0	No	50–500	Slow process
Internal chemical desulfurization	0	– –	– –	Yes	1–100	Decreasing purifying effect
Activated carbon	0	– –	–	Yes	<5	Large amount of activated carbon to be disposed

++: extremely recommendable; +: recommendable; 0: neutral; –: not recommendable; – –: extremely negative.
*Consumption and disposal of the additional inputs inherent to the operation of each method.
Source: Adapted from FNR (2012).

be decided in every instance by considering the local financial, social, and environmental aspects. The most relevant biogas uses are direct heat, direct electricity production, or the combined generation of heat and power. The transformation of biogas into biomethane is typically not feasible because of the amount of the produced biogas in UASB reactors applied at mainstream sewage treatment (not for sidestream processes, such as sludge digestion). In addition, biogas purification to biomethane requires the removal of the high concentrations of N_2 present in the biogas from the UASB reactor, which appears to be unfeasible economically.

Table 8.11 Main alternatives for the management of biogas in STPs with UASB reactors.

Alternative	Proposed Use
Direct burning for heat production	For sludge drying and/or sludge sanitization
Direct electricity production	Burning in a motor combustion chamber and production of electricity Burning in a chamber and use of the heat for a microturbine for electricity production
Cogeneration (electricity and heat)	The above described solutions combined to provide high energy recovery efficiency in microturbines and motor combustion

8.3.7.1 Direct heat used for sludge drying and/or sanitization

An interesting possibility is using biogas for thermal sludge drying, primarily aimed at reducing the sludge volume and the attendant costs of transport and final disposal. Secondly, thermal drying can sanitize the sludge.

According to Andreoli *et al.* (2001), under ideal conditions, 2,744 kJ of energy is needed to evaporate 1 kg of the water present in sludge. However, under normal operating conditions, this value could double. The total energy demand will depend on the efficiency of the chosen equipment and the type of sludge processed. Recent studies conducted by Possetti *et al.* (2015) in a pilot rotating drum thermal dryer, have shown that 3,600 kJ of energy is required to remove 1 kg of water present in the sludge.

Biogas can replace other types of fuel in boilers to generate the heat required for thermal drying. Any type of fuel can be combusted as a heat source in boilers, such as wood, coal, oil, natural gas, or dried sludge. In addition, exhaust gases from other processes, such as gas turbines, can be used as a source of thermal energy for boilers. In the USA, using biogas instead of conventional fuel is a well known practice, with a history of success spanning more than 25 years (USEPA, 2008).

To ensure the proper functioning of the system, various factors should be considered relevant to the direct combustion of biogas in boilers. Biogas combustion is similar to burning dilute natural gas. If the fuel is a mixture of natural gas and biogas, small adjustments are necessary in the boiler. On the other hand, if biogas is the predominant fuel or the only fuel, larger modifications are necessary to compensate for the lower energy content of biogas compared with natural gas.

As biogas has a lower calorific value than natural gas has, the volume of biogas used in the boiler will be larger than that of natural gas. Therefore, the fuel supply and fuel system in the boiler must be modified to allow the processing of the higher gas flow and to guarantee a calorific value equivalent to that of natural gas. However, such adaptations do not interfere significantly with the design and operation of the boiler components (USEPA, 2008).

Corrosion because of the moisture and H_2S present in biogas is an additional problem in boilers that use biogas as fuel. Coating the preheater and chimney with corrosion-resistant material or replacing with corrosion-resistant materials (for example, stainless steel), control of exhaust gas temperature, and adequate circulation of water can reduce the corrosive effects. The biogas moisture levels should be controlled with the installation of traps and condensate lines to prevent damage to equipment and operational problems in the boilers. Flame instability because of fluctuations in the biogas flow rate can be controlled by the installation of ultraviolet sensors that monitor the flame, preventing it from being extinguished, or by using an auxiliary fuel system, acting as a reserve when variations occur in the flame (USEPA, 2008).

Figure 8.7 shows the possible configurations for drying and sanitization of sludge from small- and medium-scale STPs (PE $>$ 2,000; PE $<$ 100,000) in which biogas boilers can be used, as follows:

(a) modified drying bed to promote drying and sanitization simultaneously (Andreoli *et al.*, 2002; Ferreira and Nishiyama, 2003);
(b) sanitizing tank followed by conventional drying bed (Borges *et al.*, 2009);
(c) conventional drying bed followed by heated bed for sanitization (Possetti *et al.*, 2012; Wagner *et al.*, 2015).

These solutions typically apply to small- and medium-scale STPs because of the lower operating and construction costs. In addition, the heat produced from the biogas can be combined with that of another energy source, such as solar energy. Wagner *et al.* (2015) designed a radiant floor thermal system capable of promoting the drying and sanitization of the previously dewatered sludge with the combined use of solar energy and biogas. In this instance, solar energy promotes the preheating of the system, providing a significant reduction in the consumption of biogas. In addition to sludge sanitization, the system can provide an average daily increase of approximately 9.3% in the total solids content. Notably, the volume of the methane available appears to be higher than the volume required in this combined solar-biogas system (Wagner *et al.*, 2017).

In Table 8.12, a summary is presented of a technical comparison of the three possible configurations shown in Figure 8.7 for simplified sludge drying and sanitizing systems. This table also includes the traditionally used alternative of alkaline treatment.

The implementation of commercial sludge thermal dryers and the possibility of thermal energy recovery from CHP engines should be considered for large-scale STPs (PE $>$ 100,000). A combination that has been implemented in Europe is using solar drying with the support of thermal energy from a CHP fed on biogas. This alternative is extremely attractive because of its energy sustainability and relatively low cost.

Using sludge thermal dryers has several benefits, particularly the possibility of using dry sludge as fuel, which can support heat for dryers after a thermal conversion (Possetti *et al.*, 2015; Rosa *et al.*, 2016). Several types of thermal dryers are currently available in the market, such as rotary direct and indirect dryers, direct belt dryers, and direct hoppers with internal mixers, among others. Strumillo and Kudra (1986), Herndon (2008a, b) and Prata *et al.* (2017) indicate the criteria that could help in selecting sludge dryers for STPs. In direct dryers, heat transfer is carried out by convection, by direct contact of the sludge with the drying medium (hot air, combustion gases, or heated steam), which serves as a source of heat for humidity removal. Because of the direct contact between the sludge and the drying medium, the heat transfer rate is higher. In indirect dryers, the heat transfer takes place through a metal wall separating the sludge from the drying medium (hot air, steam, or thermal oil) (Perry & Green, 1997). The design and operation of sludge dryers must meet safety requirements (Fernandez-Anez & Medic-Pejic, 2014; Herndon, 2008a, b; Manchester, 2001; Medic-Pejic *et al.*, 2014).

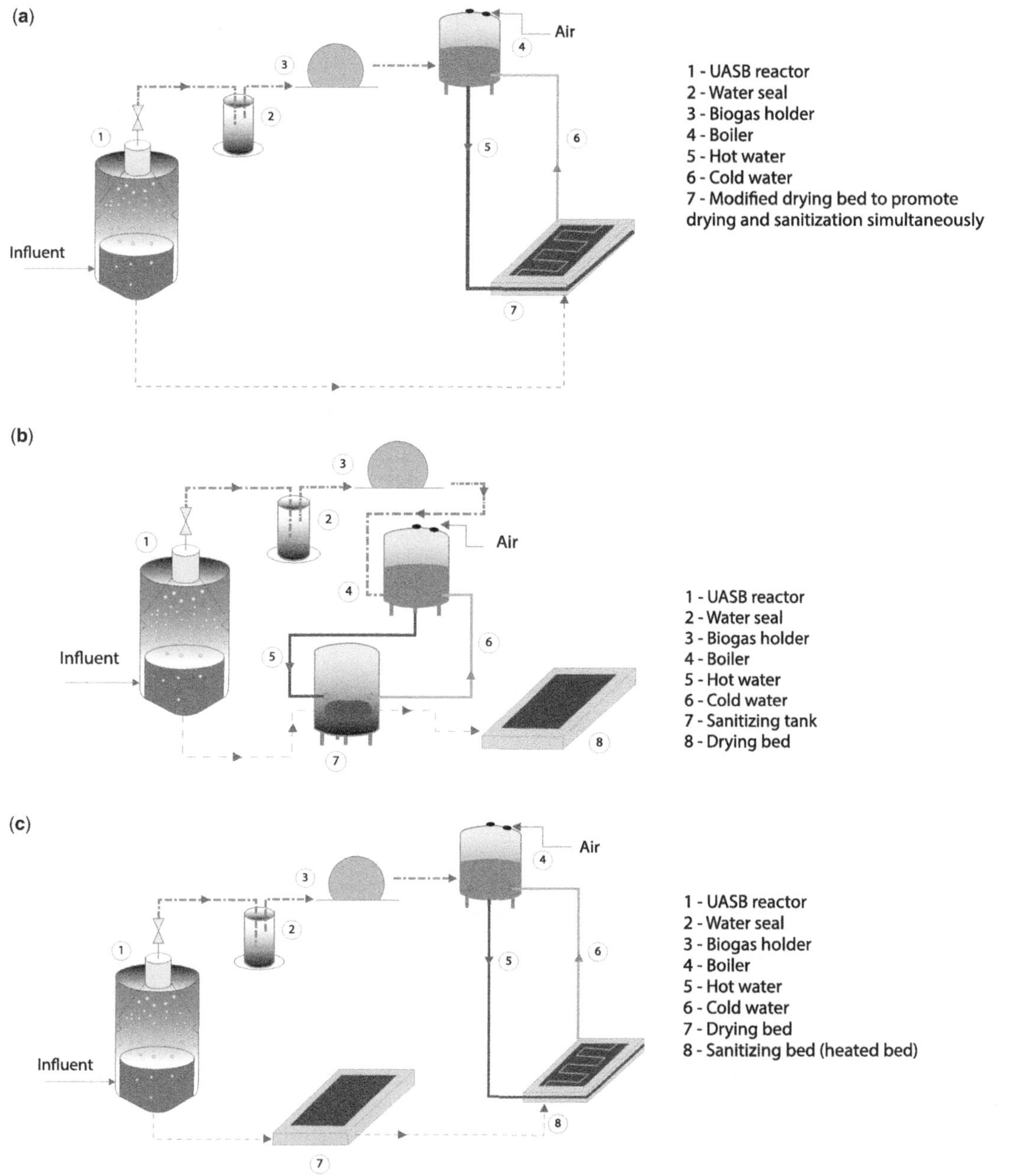

Figure 8.7 Alternative configurations for drying and sanitization of sludge in small- and medium-sized STPs: (a) modified drying bed to promote drying and sanitization simultaneously; (b) heated tank for sanitization followed by conventional drying bed; (c) conventional drying bed followed by heated bed for sanitization. (Source: Adapted from Rietow *et al.*, 2018).

Table 8.12 Summary of the technical aspects of the configurations shown in Figure 8.7 for simplified sludge drying and sanitization.

Aspects Evaluated		**Alternatives**			
		A	**B**	**C**	**D**
Economics	Electric power consumption	+	+++	++	+++
	Area demand	+++	++	+	++
	Energy self-sufficiency through the use of biogas	+	++	+++	Ø
Reliability	Sanitization safety	+	+++	++	+
	Drying safety	+++	++	+++	+++
Simplicity	Sludge handling	+++	+++	++	+
	Maintenance	+	+++	++	+++
	Operation	+	+++	++	++
	Robustness of installations	+	+++	++	+++
Social	Odour	++	+++	+	+
	Operator safety	++	+++	+	+
Environmental	Emission of greenhouse gases	++	+++	++	+

A: Modified drying bed to promote sludge drying and sanitization simultaneously; B: Sanitizing tank followed by conventional drying bed; C: Conventional drying bed followed by heated bed for sludge sanitization; D: Alkaline treatment.
+++: more favourable; ++: intermediate; +: less favourable; Ø: not applied.
Source: Adapted from Rietow *et al.* (2018).

8.3.7.2 Electricity production and cogeneration

The chemical energy of biogas can be converted into electricity. Some advantages of this process (Salomon & Lora, 2009; Gomes *et al.*, 2017) are:

- Decentralized generation, with energy generation close to the consumer
- Possibility of additional profit from the production of energy and commercialization/exchange with energy companies
- Reduction in energy consumption in the sewage treatment plant
- Potential use of both electricity and heat by cogeneration.

Electricity generation and consumption in the STP itself are options used worldwide. Some guidelines on distributed electricity from biogas are available in Gomes *et al.* (2017). In STPs with a significant demand for electricity, for example, when using aerated post-treatment units with electromechanical equipment, the electricity generated will probably be consumed completely in the unit. In such instances, electricity generation and self-consumption is an attractive option that must be considered. In exclusively anaerobic STPs, with a consequently low consumption of electricity, there will probably be surplus energy production. Various countries have relevant legislation for such instances, e.g., for the produced electricity injection into the public grid, which allows the surplus energy injected into the grid to be consumed in other plants/units owned by the same client (net metering).

In the continuous generation regime, the operation mode of the gas engines is guided by the average biogas production. In peak-hour generation mode, the energy is generated during certain hours of the day, which could bring about economic benefits. During this period of the day, electricity generation is increased, taking full advantage of biogas storage capacity and additional spare-machine capacities. Depending on the tariff difference between peak and off-peak hours, it could be more advantageous to store the biogas for longer hours and concentrate its use with maximum power at peak hours. This is because the tariff increase for peak hours is often greater than is the investment value increase for a larger gas holder and energy generation unit.

To increase energy efficiency and economic viability, we suggest taking advantage of the available residual heat, i.e., the combined heat and power (CHP) option. CHP is the simultaneous production of electricity and heat from a single fuel source, such as biogas. CHP is not a single technology, but an energy system that can be modified depending on the needs of the energy end user. CHP systems consist of a number of individual components configured into an integrated whole. These components include the prime mover, generator, heat recovery equipment, and electrical interconnection. The prime mover that drives the overall system typically identifies the CHP system. Prime movers for CHP systems include reciprocating engines, combustion turbines, steam turbines, microturbines, and fuel cells (USEPA, 2011).

Taking advantage of the combined use of heat and electricity allows the use of more than 70% of the available energy in biogas. Using the heat means roughly doubling the energy use efficiency.

The internal combustion engine is the most appropriate equipment for the conversion of biogas into electricity for STPs with UASB reactors. The operation mechanisms are based on air/fuel burning inside the combustion chamber, which generates an exothermic reaction with gases at high temperature and pressure. With the expansion of the gases the pistons are moved, converting the pressure into rotational movement (USEPA, 2003). The power of such equipment varies between 30 kW and 5 MW, being adequate for use in STPs with UASB reactors in the range of 30–800 kW. Some of the advantages of using internal combustion engines are rapid start-up, stable operation, high heat-recovery potential, and operational reliability. Additionally, such engines are usually priced more affordably.

The microturbine is another technology for generating electricity based on biogas from UASB reactors treating sewage. This equipment works with the air being pressed into the combustion chamber by the compressor, where it reacts with the fuel, thereby feeding the continuous combustion process. The expansion of the heated gases under high pressure causes the turbine blades to start rotating at high speed. The turbine provides the energy required to rotate the compressor and the electric generator, as it is mounted on the same axis as the compressor and the electric generator (Cortez *et al.*, 2008; Costa, 2006).

At a pressure of 4 to 6 atm (405–608 kPa), the fuel enters the combustion chamber, where it reacts with the air. With the release of heat from the combustion reaction, the gases expand, moving the generator and producing energy. The overall efficiency of the system is enhanced by using the exhaust gas, which promotes the preheating of the combustion air. The microturbines have equipment capable of recovering heat for the heating of water during the combined production of electricity and heat. Such equipment is available in the market, with typical power between 30 and 200 kW (USEPA, 2002).

Because of their small size, these systems have the potential for operational flexibility, which facilitates the generation of energy in small localities and increases the biogas recovery conditions. Compared with internal combustion engines, microturbines emit lower concentrations of mainly nitrogen oxides. Therefore, microturbine emissions are below 9 ppm, whereas those from gas turbines are 50 ppm, and for internal combustion engines up to 3,000 ppm. The main problem of the technology with small power supply relates to its cost related to the small scale of production (Pilavachi, 2002). In Brazil, the microturbines have a significant economic disadvantage because of their high maintenance costs. Tables 8.13 and 8.14 summarize the main characteristics of the solutions for generating electricity from biogas from UASB reactors treating sewage.

Table 8.13 Advantages and disadvantages of the alternatives to produce electricity based on biogas from UASB reactors treating sewage.

Alternative	Advantages	Disadvantages
Internal Combustion Engines	• Moderate acquisition cost • Mature and widely applied technology to generate energy from biogas • Wide range of stationary motors available on the market, suitable to a wide range of applications (30 kW up to 800 kW) and operating cycles • Quick start, stable operation, and usually quite reliable • Normally operated with natural gas, but can be adjusted to operate with propane or biogas • High overall potential for energy recovery (electric and thermal)	• More subject to corrosion, although the use of lubricating oils provides some protection against the acidic products of combustion
Microturbines	• Advantages over other alternative energy generation from biogas in the following instances: low biogas production, biogas with low methane content, concerns about NOx emission • Portable and easy to modulate equipment (one or multiple turbines) to suit biogas production and local needs • Compact equipment with few moving parts, requiring little attention to operate and maintain • Low NOx emissions, usually ten times lower than those of the best performing internal combustion engines, and also lower than those emitted by flares • High potential for heat recovery	• Lower efficiency compared with combustion engines and consume approximately 35% more fuel per kWh generated • Few compressors are currently available to meet microturbine requirements without the need for costly adaptations • Little information is available on the long-term reliability and costs of operation and maintenance of biogas microturbines

Source: Adapted from Lobato (2011) and Gomes *et al.* (2017).

Table 8.14 Comparison between the main alternatives of producing electricity based on biogas from UASB reactors treating sewage.

Parameter	Internal Combustion Engines	Microturbines
Usual application range (kW)	30–800	30–200
CAPEX cost (US$ kW^{-1})	900–1,000	4,000–5,000
OPEX costs (US$ kWh^{-1})	0.009–0.010	0.15–0.20
Rate of electricity and heat generation (kWh consumed per kWh generated)	2.8–3.0	4.1–4.7
Electrical efficiency (%)	33–36	22–33
Gas supply pressure (atm)	<0.2–3.0	3.7–5.8
Exhaust gas production (kg h^{-1})	6–7	23–37
Exhaust gas temperature (°C)	360–470	280–310
Total efficiency of combined electricity and heat generation (%)	75–76	60–73
NOx emissions[a] (ppm)	250–3,000	<9

[a]When comparing NOx emissions, the result of the concentrations of these emissions times the exhaust gas flow must be considered.
Source: Adapted from USEPA (2002; 2003); CENBIO (2005); Chambers and Potter (2000); Lobato (2011).

8.4 SAFETY ASPECTS ASSOCIATED WITH THE ENERGY RECOVERY OF BIOGAS

Biogas leaks contribute to the low potential of energy recovery and could cause various hazards in STPs. Studies by Moreno and Cozzan (2015) identified 169 documented accidents occurring between 1995 and 2014 in biogas plants worldwide. The results of the studies indicated that 82% of these accidents were caused by explosions and inhalation of toxic and asphyxiating substances.

UASB reactors have confined spaces, such as the interface between the liquid phase and the top cover of the reactor, and inspection channels, among others. Confined space is any area or environment not designed for continuous human occupation, having limited means of access and exit, and of which the existing ventilation is insufficient to remove contaminants. Therefore, the expulsion of air from these spaces can cause a lack of oxygen. In environments where the content of this gas is reduced to concentrations lower than 18%, suffocation and intoxication can occur (DWA-M 363, 2010; Silveira *et al.*, 2015).

Another relevant safety aspect in STPs with biogas production is the formation of explosive atmospheres. Characterized as a sudden and violent release of energy, an explosion could occur only when the environment presents an explosive air–fuel mixture and an ignition source with enough energy to start the process. All gas mixtures have a minimum ignition energy and, below this level, it is impossible to cause an explosion. A spark from an electrical circuit or even the heating of equipment installed in an explosive environment can be considered an ignition source and can, therefore, cause an explosion. Table 8.15 shows some of the explosive characteristics of the gases present in the biogas.

By adopting the explosion limits of CH_4, an explosive band can be established for the air–fuel ratios and the inert gases present in the biogas. The explosive range between CH_4 and atmospheric air is between 4.4% v/v (lower limit of explosion) and 15% v/v (upper limit of explosion) of CH_4 concentration (Deublein & Steinhauser, 2008). The presence of inert gases, such as N_2 and CO_2, for example, when at high concentrations in the medium, provides a narrowing of the explosive range, leading to the explosion limits being close to the atmospheric air concentration of 58% v/v (or approximately 11.6% v/v oxygen) (DWA-M 363, 2010; Silveira *et al.*, 2015).

Table 8.15 Characteristics of explosive gases present in biogas.

Characteristics	Unit	CH_4	H_2S
Lower explosion limit	% v/v	4.4	3.9
Upper explosion limit	% v/v	15.0	50.2
Autoignition temperature	°C	595	270
Minimum ignition energy	mJ	0.29	–

Source: Deublein and Steinhauser (2008); Umweltbundesamt (2018).

The classification of explosive atmospheres in anaerobic-based STPs is crucial. The main objectives of this classification are to eliminate the probability of an explosive environment occurring around an ignition source. In most practical situations, it is almost impossible to ensure that an explosive atmosphere will never form or that the installed equipment will never constitute a source of ignition. Therefore, in instances of a high probability for the development of an explosive atmosphere, equipment with a low probability of becoming a potential source of ignition must be used. General requirements for the installation of electrical equipment with different types of protection in classified areas are found in NBR IEC 60079-0 (ABNT, 2016) and Rietow *et al.* (2017).

Areas classified as explosive atmospheres should be divided into zones of risk. Such zones are based on the frequency of occurrence and duration of an explosive atmosphere. The zones can be divided into: (i) zone 0, an area in which an explosive atmosphere of gas is present continuously or for long periods of time; (ii) zone 1, an area in which an explosive gas atmosphere is likely to occur occasionally under normal operating conditions; and (iii) zone 2, an area in which an explosive gas atmosphere is not likely to occur under normal operating conditions, but if it does, it will persist for short periods of time.

The UASB reactor itself, the gas holder, and the flare under special operating conditions (such as maintenance, for example), can be classified as Zone 0. Equipment such as pressure relief valves, flame arresters, and water seals, which can release flammable gas during normal operation, can be classified as Zone 1. Flanges, connections, and pipe fittings can be classified as Zone 2 (Deublein & Steinhauser, 2008). Figure 8.8 shows a schematic map of the hazard zones identified in a UASB-based STP.

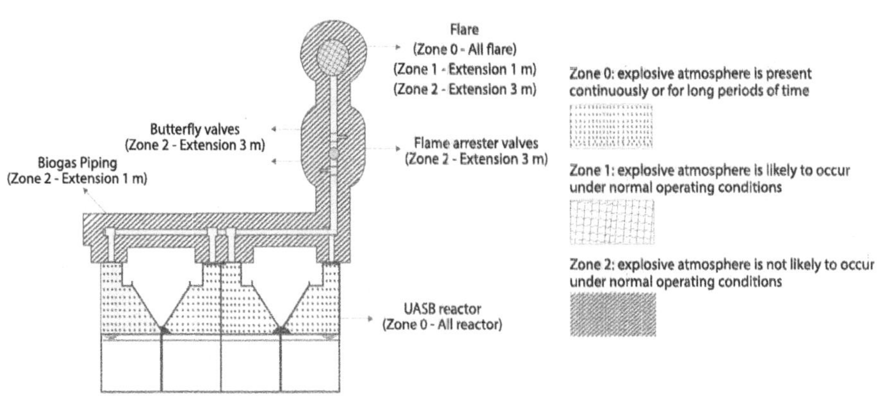

Figure 8.8 Schematic map of the zones of explosion risks in a STP with a UASB reactor. (Source: Possetti *et al.*, 2018).

In addition to mapping classified areas, it is necessary to identify, isolate, and provide signage to the confined spaces present in a STP. The identification of a confined space according to its characteristics and normative requirements allows controlling the risks of activities that must take place in such places. To reduce these risks, it is recommended that gas measurements be performed before and during operation.

8.5 FULL-SCALE EXPERIENCES

8.5.1 Vieira STP, Minas Gerais State, Brazil

Montes Claros STP in Minas Gerais (270,000 PE), of which the treatment flowsheet comprises UASB reactors followed by trickling filters, has a rotary sludge thermal dryer that operates in a combined manner directly (convection) and indirectly (conduction). This equipment operates with sludge from a centrifuge with approximately 80% moisture content and its evaporation capacity is approximately 3,000 $LH_2O\ h^{-1}$, resulting in a residual moisture content of 10% and production of approximately 1,200 $kg.h^{-1}$ of dried sludge. Figures 8.9(a) and (b) show the sludge thermal dryer and the dried material (solids content of the order of 80%), respectively.

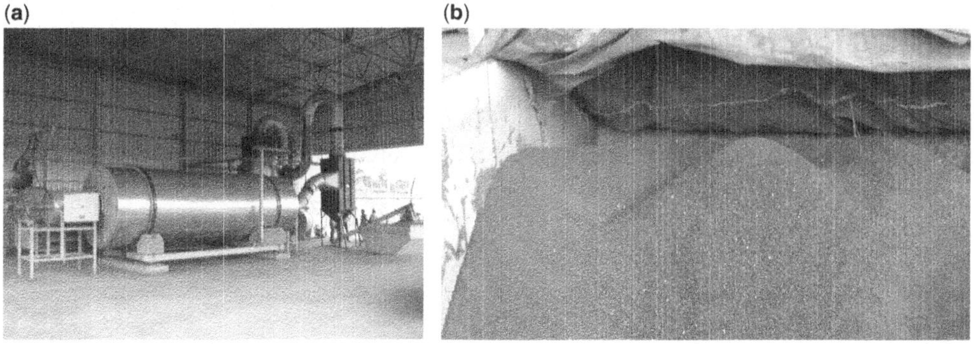

Figure 8.9 Thermal sludge drying system from biogas burning at the Montes Claros STP, Minas Gerais State: (a) sludge drying system, and (b) dried sludge. (Source: Copasa (2011)).

8.5.2 Ouro Verde STP, Paraná State, Brazil

Ouro Verde STP (35,000 PE) is located in Foz do Iguaçu, Paraná, and belongs to Sanepar. This STP is one of the first UASB reactor plants in Brazil to use biogas, and the first to receive a permit to inject electricity into the public grid according to the Brazilian standards ANEEL no. 482/2012 and 687/15 (ANEEL, 2012, 2015).

This STP produces approximately 20 ton of sludge per year that is used for agricultural purposes after sanitization. The biogas produced (approximately 50 $m^3\ d^{-1}$) is captured and guided through an iron oxide filter to remove H_2S. After treatment, the gas is stored in a 50 m^3 unpressurized gas bag (Figure 8.10(a)). When passed through the internal combustion engine (Figure 8.10(b)), the gas produces approximately 22 kW of energy.

The facility is equipped with instruments for the complete monitoring of the gas quality (CH_4, CO_2, H_2S, and O_2) and quantity (flow, pressure, and temperature). This guarantees a complete shutdown of the facility in cases when there is no energy in the grid, but the facility is generating energy. This is an indispensable safety feature to guarantee the safety of the operation on the public grid.

(a) (b)

Figure 8.10 Ouro Verde STP, Paraná State: (a) general picture of the energy recovery system, and (b) internal combustion engine. (Source: Sanepar cited by Gomes *et al.*, 2017).

The average energy consumption of the STP is approximately 812 kWh y^{-1}, and the estimated production is approximately 3,600 kWh y^{-1}. The net energy production to the grid is estimated at 2,788 kWh y^{-1}.

8.5.3 Jacuípe II STP, Bahia State, Brazil

STP Jaquipe II is a 100,000 PE facility. The solution is the straightforward use of the biogas for electricity production. The installed engine is a 200 kW gas motor. The design was done without prior biogas measurement and the UASB reactor was not sealed adequately during the start-up. Figures 8.11(a) and (b) are images of the system and the internal combustion engine, respectively.

(a) (b)

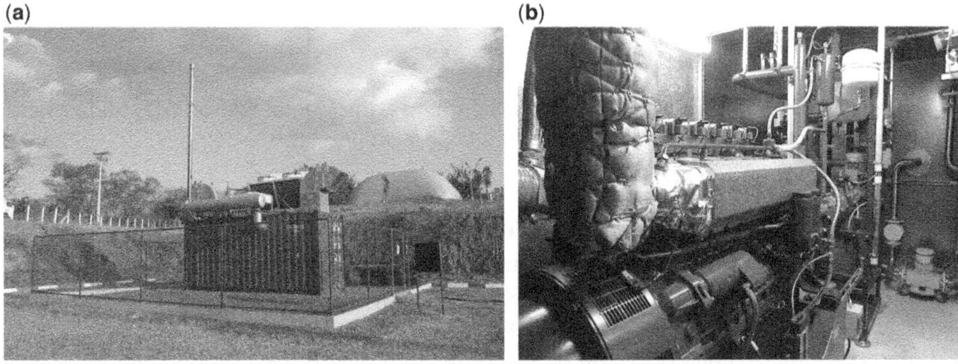

Figure 8.11 Distributed system of electricity generation based on biogas in Jacuípe II STP, Bahia State: (a) general picture of the energy recovery from biogas, and (b) internal combustion engine. (Source: Rotária do Brasil (2018)).

A retrofit of the existing UASB reactor was done in two steps. The first step was sealing the UASB reactor and restoring the maximum gas capture, measuring the gas quantity, and burning the gas, followed by a second step of designing the electricity production according to the measured biogas captured. However, the motor was oversized, leading to a change in the operation concept, from constant base load production (24 hours/day) to peak load production (3 hours/day, only week days). The originally calculated internal rate of return changed from 19.61% before the project to 6.31% in continuous operation and to 10.71% in peak operation, with an injection of surplus power into the electric grid (Cabral *et al.*, 2017b).

8.6 ECONOMIC FEASIBILITY ASSOCIATED WITH ENERGY RECOVERY FROM BIOGAS IN ANAEROBIC-BASED STPs

Although energy recovery from biogas in sidestream sewage treatment (e.g., sludge digestors) in Europe starts at approximately 30 kW (electricity production), the situation is completely different in anaerobic-based STPs treating sewage in developing countries. This is ascribed mainly to the following factors: (i) energy in developing countries is often cheaper than in Europe, (ii) there is a lack of special incentive programs for energy recovery from biogas, (iii) the need for thermal energy is often limited in warm climate countries, and (iv) the market is in the developing phase and, therefore, services are often more expensive and many spare parts have to be imported. However, there has been an increase in energy tariffs in many developing countries over the last few years, which has focused more attention on the possibility of using biogas from anaerobic-based STPs to produce electricity. In addition, more restrictive environmental requirements for final sludge disposal can stimulate the thermal use of the biogas for both drying and sanitization of the sludge.

From the Brazilian experience so far (Valente, 2015; Rosenfeldt *et al.*, 2017a, b), the following information has been considered in the new projects of energy recovery from biogas produced in UASB reactors treating sewage:

For small- and medium-scale STPs (PE $>$ 2,000; PE $<$ 100,000), biogas use for electricity production is generally not feasible. However, the use of thermal energy for sludge drying becomes feasible even in small STPs ($<$10,000 PE) when the cost of sludge disposal is high (see Figure 8.7). In this instance, biogas thermal energy is recommended as an additional support to other drying methods, as discussed previously in Section 8.3.7.

- For large-scale STPs ($>$100,000 PE), sludge thermal dryers using biogas as fuel are feasible in most instances, depending on the cost of sludge disposal. As regards electric energy generation, several studies conducted for 100,000–200,000 PE facilities showed a clear economic viability for using biogas from UASB reactors. The viability varies depending on the business model employed. The internal rate of return is approximately 8% per year, but mostly above 10%.
- For a 200 kW motor in continuous use, the internal rate of return was found to be approximately 19.6%. Generally, motor sizes above 150 kW should have reasonable viability for a minimum biogas production of approximately 1,750 $m^3\ d^{-1}$ for an average methane content of approximately 75%.
- The price of the peak hour tariff has to be considered in the economic feasibility analysis. Depending on the peak hour tariffs, it might be interesting to produce a surplus of energy during peak hours, inject this into the public grid, with this being paid as credit for those hours with a higher tariff by net metering. In this instance, as pointed out in Section 8.3.4, larger equipment has to be considered; however, storage capacity is not too expensive. Therefore, it is recommended to have at least the storage capacity for 100% energy production during peak hours.

To enhance the total energy efficiency, it is recommended to analysing the possibility of using thermal energy after electricity production. One clear opportunity is sludge drying, as discussed previously, as the avoided costs contribute largely to the viability. The heat required to dry 1 ton of dehydrated sludge, from 25–85% TS, is approximately 800 kWh. In large-scale STPs ($>$100,000 PE), this complementary energy can be obtained from the exhaust gases of the CHP engine or from thermal conversion of dried sludge. In small- and medium-scale STPs (PE > 2,000; PE < 100,000 PE), the energy for such a process can be complemented by the use of solar energy. In this instance, a combination of thermal heat with solar drying could be an interesting combination; however, so far there is no full-scale implementation of this option to clearly indicate its viability.

Determining the correct size of the energy recovery engines is a vital aspect in a design in the absence of data. Although the plant will be equipped with state-of-the-art scum removal and completely sealed to prevent biogas losses, it should be considered that, so far, no UASB reactors have been reported to produce biogas fitting the best scenario, mentioned in Section 8.2. For this reason, Figure 8.12 was elaborated to support the decision-making process for establishing the size of the biogas recovery facility.

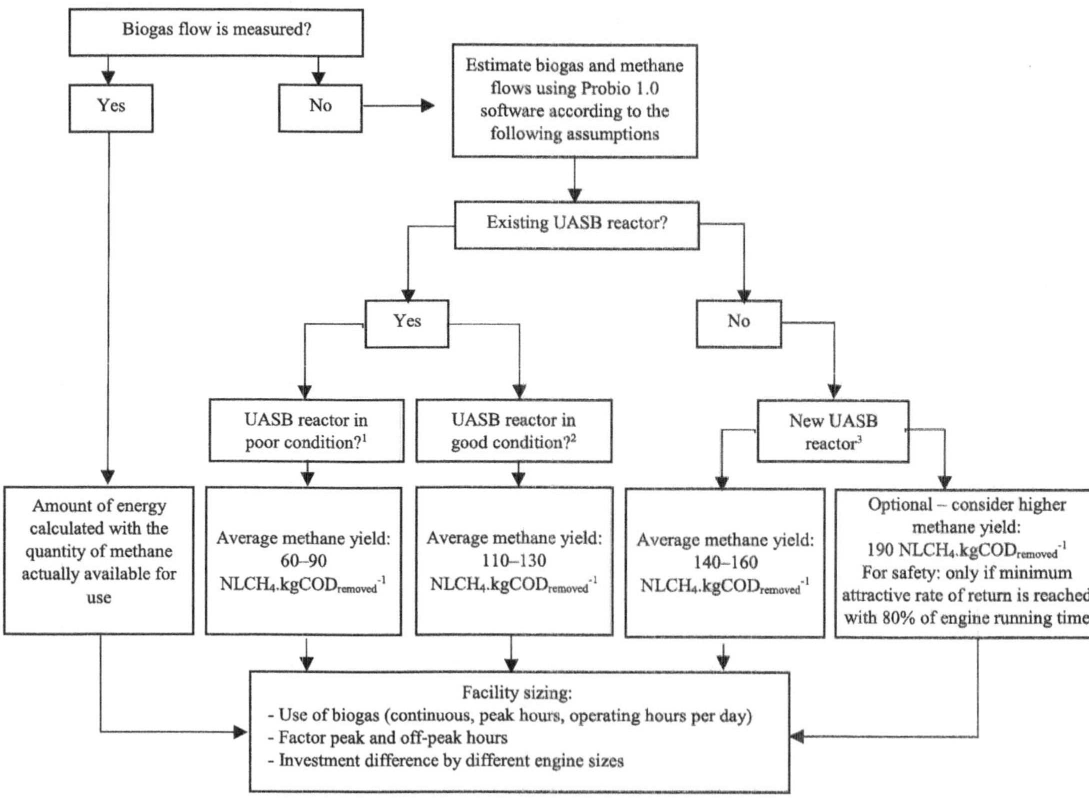

Comments		
[1]UASB reactor in poor condition	[2]UASB reactor in good condition	[3]New UASB reactor (optimized for biogas collection)
• Significant biogas leaks • Scum removed sporadically • Inadequate biogas collection and intermittent burning • Influent COD < 400 mg L^{-1} • COD removal below 65%	• Few biogas leaks • Scum removed frequently • Proper biogas collection and efficient burning • Influent COD > 400 mg L^{-1} • COD removal above 65%	• Proper sealing • Proper scum removal device • Pipes in stainless steel or high-density polyethylene (HDPE) • Influent COD > 400 mg L^{-1} • COD removal above 65%

Figure 8.12 Decision-making flowsheet to assess energy recovery from biogas in UASB-based STPs.

There are two options for a new facility, namely, using conservative methane yields or predicting higher methane yields (best scenario). However, when considering higher methane yields, it is recommended, for assurance, to invest only if the minimum attractive rate of return is met with 80% of the engine running time, e.g., from 4 a.m. to 11 p.m. (ensuring this includes the peak hours). When the internal rate of return is lower than the minimum attractive rate of return considering the 80% engine running time, it is recommended to use lower methane yields and a smaller internal combustion engine.

Additionally, a general rule for a well sealed, regularly functioning UASB reactor is 150 kW of internal combustion engine power (operation of 24 hour/day) for each 100,000 PE with typical organic influent concentrations (above 250 mg L^{-1}BOD).

For many countries, the dependency on import is critical and an abrupt change in exchange rates could significantly influence the viability of operations to the tipping point. This problem can be overcome only in time, by establishing a strong local biogas market. Also important is stressing the need for effective operations. In countries lacking a well established biogas market, public services will always have a problem in reacting in a timely manner to maintenance problems occurring during operations. Therefore, the business model is as important as the economic viability study. Particularly for a first implementation, it is important to choose a cost-saving solution and to avoid employing all the possible optimizations available. Larger safety margins are recommended for unusual decision-making processes.

8.7 MECHANISMS TO DEVELOP THE BIOGAS SUPPLY CHAIN IN EMERGING BIOGAS-USE ECONOMIES – EXAMPLE OF BRAZIL

As do most developing countries, Brazil has to contend with an enormous challenge in providing the entire population with sewage collection and treatment services; moreover, there is significant potential for these services to be more efficient. Using biogas for energy is not a consolidated practice in sewage treatment, as, in most instances, biogas is burned and emitted to the atmosphere, and sometimes vented directly without burning. In this context, biogas is considered a strategic by-product that could help to promote more efficient management in the sewage treatment sector, facilitating the following indirect benefits:

- Reduction of operational costs
- Improvement of process control and final effluent quality
- Increase in resilience against climate change
- Improvement of professionalization of the sector.

However, using biogas is not relevant only to the sanitation sector, as other sectors of the economy, such as agriculture, intensive animal husbandry, energy, and transport are involved as well. This means that the legal framework for biogas must be established between these sectors. As biogas is a developing alternative fuel, national regulations and directives are necessary to facilitate the development of a biogas market.

A crucial initiative in Brazil has allowed the consolidation of the biogas issue at the national level, namely, the Brazil–Germany cooperation project, titled Probiogás. This initiative was coordinated by the National Secretariat of Environmental Sanitation of the Ministry of Cities, and GIZ, the German cooperation agency for sustainable development. Starting in 2013 and lasting for four years, the project was aimed at promoting the use of biogas as energy in Brazil, with the goal of incorporating this alternative source of energy into the Brazilian energy market. As a side effect, this project promoted more efficient sanitation services and the reduction of the emission of greenhouse gases.

The initiative produced a wide range of publications and results, which served as a basis for this chapter. The list of publications is available at http://www.cidades.gov.br/saneamento-cidades/Probiogas/

publicacoes/publicacoes-do-Probiogas. The publications are all in Portuguese, although some of them have been translated, enhanced, and adapted to the Brazilian reality.

One important publication derived from a study that identified the main barriers and challenges of the biogas market in Brazil. The experience of authors in other developing countries has indicated clearly that the barriers and challenges were often quite similar (some major findings):

- The small number of referenced biogas projects is related directly to the weakness of the market in terms of service offers, skilled labour, and supply of equipment
- The necessary import of knowledge and equipment increases investment and operational costs, thereby reducing the attractiveness of the investment
- The lack of a biogas framework and incentives makes development of a market more difficult
- A general lack of knowledge exists in all related areas.

Therefore, the main actions and impacts of the Probiogás cooperation project in the sewage sector concentrated on three main themes: framework conditions; technical and scientific cooperation; and capacity building and training. The national biogas measurement program in UASB reactors has provided a substantial learning experience for all the sanitation utilities involved (see results described in Section 8.2).

Moreira (2017) compared the status of the biogas sector before and after the implementation of the cooperation project relevant to the mentioned aspects that directly influence the success of biogas projects. The results indicated significant development in all areas for the duration of the Probiogás project. These include the development of a set of national technical references, enhancement of the feasibility of biogas models, promotion of the integration of several key players, development of applied research, and capacity building of two thousand people, including the training of knowledge multipliers.

Additionally, during this project, there was a significant increase in regulations that strengthened the use of biogas, such as a directive on net metering (stimulating energy production in facilities with low energy consumption) and an internal regulation of the Ministry of Cities, determining that biogas use can be financed with government funds when viability is demonstrated. These indicate a paradigm shift in the basic assumptions from just increasing the number of conventional sewage treatment plants to implement more sustainable infrastructures.

The fragility of management strategies, which hinders the long-term vision and the development of a sustainable STP concept has been analysed and the challenges have been identified very clearly. For public service providers, the scenario is more challenging, as there is a need to go beyond the limits of politics to establish institutional arrangements that facilitate the management of the biogas system. Often, the procedures for public bidding complicate the maintenance and operation of biogas facilities and cause significant delays, e.g., in equipment purchase. In view of this, a combination of public and private operations could be the most feasible procedure to follow; however, a suitable operational model is still lacking.

The main challenge in this regard is based on the fundamental difference in how services are regarded. In developed countries, the focus is not only on the primary objective (i.e., sewage treatment) but also on all the means to reduce the life cycle costs of the treatment. In developing countries, the focus is often only on the sewage treatment and mainly on CAPEX, as there remains so much to achieve in terms of service coverage.

Despite the challenges, clear advances have been noted in Brazil. There are strong indications for more biogas projects in the sector, as, currently, several substantial biddings for the use of biogas in sewage treatment plants are expected. In addition, companies are contemplating their business strategies based on the reduction of greenhouse emissions associated with biogas recovery (Filippini *et al.*, 2017; Carvalho & Possetti, 2017). It is expected that other developing countries employing UASB reactors for treating sewage could take advantage of the experience in Brazil.

8.8 DESIGN EXAMPLES

EXAMPLE 8.1

Use of biogas generated in large-size anaerobic-based STPs for cogeneration of heat and electricity in CHP engines, the heat being used for drying and sanitization of the sludge

Input data

- Population equivalent: PE = 200,000 inhabitants
- Average influent sewage flowrate: $Q_{s\text{-}avg} = 411{,}9\ L\ s^{-1}$ ($35{,}596.8\ m^3\ d^{-1}$)
- *Per capita* COD contribution: $QPCP_{COD} = 0.10\ kg\ inhab^{-1}\ d^{-1}$
- Average influent COD concentration: $C_{COD} = 562\ mg\ L^{-1}$ ($0.562\ kg\ m^{-3}$)
- COD removal efficiency (in terms of filtered COD): $E_{COD} = 80\%$
- Methane yield: $Y_{CH4} = 0.13\ Nm^3CH_4\ kgCOD_{remov}{}^{-1}$
- Lower calorific value of methane: $LCV_{CH4} = 9.9\ kWh\ m^{-3}$
- Sludge yield: $Y_{sludge} = 0.15\ kgTS\ kgCOD_{app}{}^{-1}$
- Sludge specific mass: $\gamma = 1{,}020\ kgTS\ m^{-3}$
- TS concentration of the sludge removed from the UASB reactors: $C_{sludge\text{-}UASB} = 4.0\%$
- TS concentration of the sludge after centrifuge: $C_{sludge\text{-}centrif} = 20\%$
- TS concentration of the sludge after thermal drying: $C_{sludge\text{-}thdryer} = 80\%$
- Solids capture in the centrifuge ($SC_{centrif}$) = 90%
- Solids capture in the thermal dryer ($SC_{thdryer}$) = 100%
- Electric efficiency of CHP engines (η_{el}) = 35%
- Relation between the heat contained in the exhaust gases and the total energy of the fuel (biogas) consumed by the internal combustion engine ($R_{exhaust}$) = 25%
- Thermal efficiency of the thermal drier ($\eta_{th\text{-}dryer}$) = 80%
- Energy required to evaporate 1 kg water (E_{evap}) = 1.0 kWh (Possetti *et al.*, 2015)

Solution

(1) Estimate the organic load removed (OL_{remov})

$$OL_{rem} = Q_{s\text{-}avg} \times C_{COD} \times E_{COD} = 35{,}596.8\ m^3\ d^{-1} \times 0.562\ kg\ m^{-3} \times 0.80 = 16{,}000\ kgCOD_{removed}\ d^{-1}$$

(2) Estimate the daily methane production (Q_{CH4})

$$Q_{CH4} = OL_{remov} \times Y_{CH4} = 16{,}000\ kg\ COD_{remov}\ d^{-1} \times 0.130\ Nm^3\ CH_4\ kgCOD_{remov}{}^{-1} = 2,080\ m^3CH_4\ d^{-1}$$

(3) Estimate the amount of chemical energy in the biogas (E_{biogas})

$$E_{biogas} = Q_{CH4} \times LCV_{CH4} = 2{,}080\ m^3CH_4\ d^{-1} \times 9.9\ kWh\ m^{-3} = 20{,}590\ kWh\ d^{-1}$$

(4) Estimate the daily sludge mass production in the UASB reactors ($P_{sludge\text{-}UASB}$)

$$P_{sludge-UASB} = Q_{s\text{-}avg} \times C_{COD} \times Y_{sludge} = 35{,}596.8\ m^3\ d^{-1} \times 0.562\ kg\ m^{-3} \times 0.15\ kgTS.kgCOD_{app}{}^{-1} = 3{,}000\ kgTS\ d^{-1}$$

(5) Estimate the daily sludge volumetric production in the UASB reactors ($Q_{sludge\text{-}UASB}$)

$$Q_{sludge\text{-}UASB} = P_{sludge\text{-}UASB}/(C_{sludge\text{-}UASB} \times \gamma) = 3{,}000\ kgTS\ d^{-1}/(0.04 \times 1{,}020\ kgTS\ m^{-3}) = 73.5\ m^3\ d^{-1}\ \text{with 4\% TS}$$

(6) Estimate the daily sludge volumetric production after dewatering ($Q_{sludge\text{-}centrif}$)

$$Q_{sludge\text{-}centrif} = (P_{sludge\text{-}UASB} \times SC_{centrif})/(C_{sludge\text{-}centrif} \times \gamma)$$
$$= (3{,}000\ kgTS\ d^{-1} \times 0.90)/(0.20 \times 1{,}020\ kg\ m^{-3}) = 13.2\ m^3\ d^{-1}\ \text{with 20\% TS}$$

(7) Estimate the amount of electricity generation (E_{el})

$$E_{el} = E_{biogas} \times \eta_{el} = 20{,}590\ kWh\ d^{-1} \times 0.35 = 7{,}206\ kWh\ d^{-1}$$

(8) Estimate the amount of heat generation from the exhaust gases of the CHP ($E_{th\text{-}exh\text{-}gas}$)

$$E_{th\text{-}exh\text{-}gas} = E_{biogas} \times R_{exhaust} \times \eta_{th\text{-}dryer} = 20{,}590\ kWh\ d^{-1} \times 0.25 \times 0.80 = \mathbf{4{,}118\ kWh\ d^{-1}}$$

(9) Calculate the daily sludge volumetric production after thermal drying ($Q_{sludge\text{-}thdryer}$)

$$Q_{sludge\text{-}thdryer} = (P_{sludge\text{-}centrif} \times SC_{thdryer})/(C_{sludge\text{-}thdryer} \times \gamma)$$
$$= (2{,}700\ kgTS\ d^{-1} \times 0.9)/(0.80 \times 1{,}020\ kgm^{-3}) = 3.0\ m^3\ d^{-1}\ \text{with 80\% TS}$$

(10) Estimate the mass of water evaporated (M_{water})

$$M_{water} = Q_{sludge\text{-}centrif} - Q_{sludge\text{-}thdryer} = 13.2\ m^3\ d^{-1} - 3.0\ m^3\ d^{-1}$$
$$= 10.2\ m^3_{water}\ d^{-1}\ \text{or}\ 10{,}200\ kg_{water}\ d^{-1}$$

(11) Estimate the amount of thermal energy needed to meet demand ($E_{th\text{-}needed}$)

$$E_{th\text{-}needed} = M_{water} \times E_{evap} = 10{,}200\ kg_{water}\ d^{-1} \times 1.0\ kWh.kg_{water}^{-1} = \mathbf{10{,}200\ kWh\ d^{-1}}$$

(12) Thermal energy balance

It can be concluded that the amount of heat generated from the exhaust gases of the CHP engine ($E_{th\text{-}exh\text{-}gas} = 4{,}8\ kWh\ d^{-1}$) is enough to meet only about 40% of the thermal energy needed to dry all the sludge produced in the STP ($E_{th\text{-}needed} = 10{,}200\ kWh\ d^{-1}$).

Scenarios to manage the heat and the sludge generated in the STP

Scenario 1 (drying of all sludge generated in the centrifuges: 2,700 kgTS d^{-1})

One way to dry all this sludge is the combustion of this material and utilization of the heat gases of this process to also inject in the thermal dryer. Considering a lower calorific value of the dry sludge of 2.9 kWh $kgTS^{-1}$ (LCV_{sludge}, dry base) and a thermal efficiency of the combustor of 80% ($\eta_{combustor}$), it would be possible to obtain about 6,200 kWh d^{-1} of energy. Thus, from the complementation of the heat gases of the combustion process, the thermal drying process could be carried out in a sustainable way in the STP. Additionally, with the use of the combustor, the volume reduction of the sludge that enters the thermal dryer can reach up to 90%, generating ashes as final residue (Possetti *et al.*, 2015; Rosa *et al.*, 2016).

Figure 8.13 illustrates the overall energy balance in the STP considering the use of a CHP engine for cogeneration of heat and electricity. In this scenario, the heat of the exhaust gases of the CHP engine is complemented by the heat generated from the combustion of the dry sludge in a combustor with 80% efficiency.

Scenario 2 (drying of 40% of the sludge generated in the centrifuges: 1,080 kgTS d^{-1})

A second option would be to assume that 40% of the sludge from the centrifuge (5.3 m^3) will be routed to the thermal drier. The remainder of the sludge should be lime-sanitized (Prolonged Alkaline Stabilization), as illustrated in Figure 8.14.

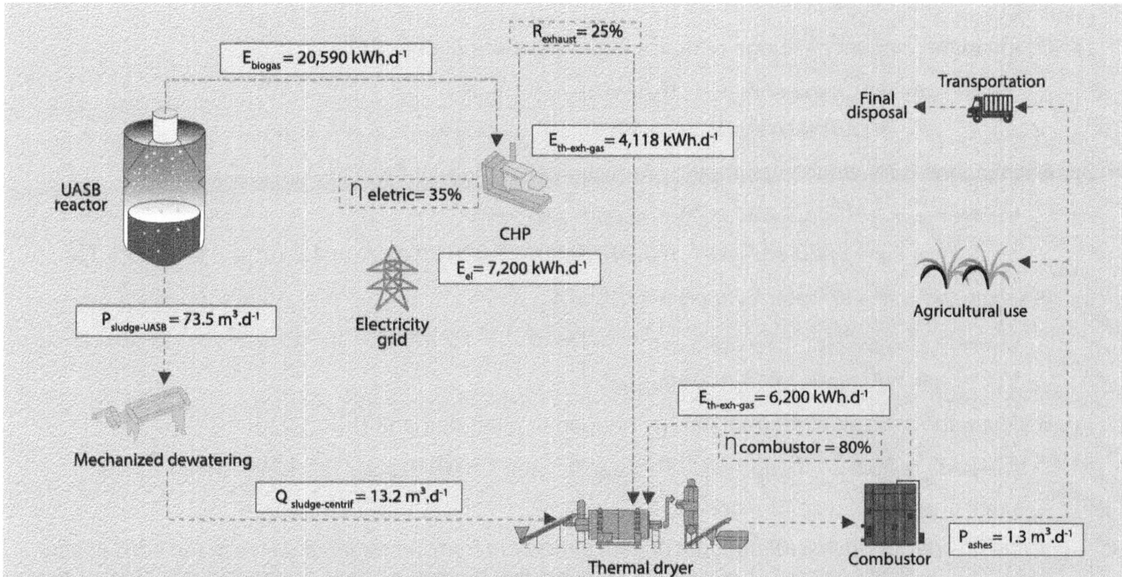

Figure 8.13 Overall energy and sludge mass balance in the STP considering the use of biogas for cogeneration of heat and electricity in a CHP engine and the dry sludge from the thermal dryer to generate more heat.

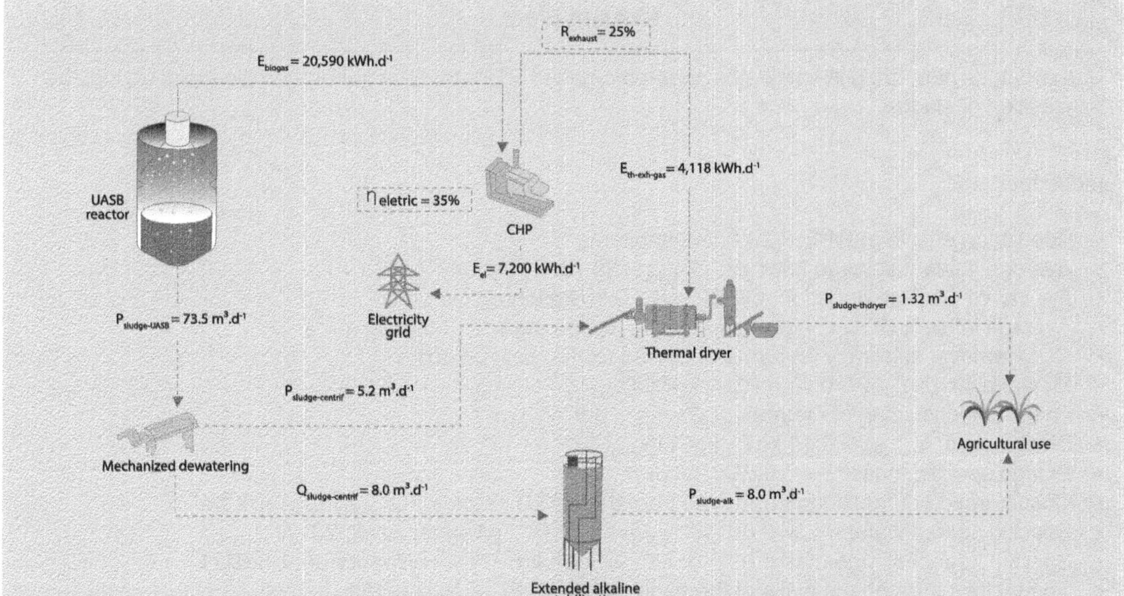

Figure 8.14 Overall energy and sludge mass balance in the STP considering the use of biogas for cogeneration of heat and electricity in a CHP engine. The sludge sanitization is reached by thermal treatment (40% of the sludge) and prolonged alkaline stabilization (60% of the sludge).

(1) Estimate the amount of sludge to be routed to the thermal dryer ($Q_{sludge\text{-}centrif}$)

$$Q_{sludge\text{-}centrif} = (P_{sludge\text{-}centrif} \times 0.40)/(C_{sludge\text{-}centrif} \times \gamma)$$
$$= (2{,}700\ \text{kgTS d}^{-1} \times 0.40)/(0.20 \times 1{,}020\ \text{kgm}^{-3}) = 5.3\ \text{m}^3\ \text{d}^{-1}\ \text{with 20\% TS}$$

(2) Estimate the daily sludge volumetric production after thermal drying ($Q_{sludge\text{-}thdryer}$)

$$Q_{sludge\text{-}thdryer} = (P_{sludge\text{-}centrif} \times SC_{thdryer})/(C_{sludge\text{-}thdryer} \times \gamma)$$
$$= (1{,}080\ \text{kgTS d}^{-1} \times 0.90)/(0.80 \times 1{,}020\ \text{kg m}^{-3}) = 1.2\ \text{m}^3\ \text{d}^{-1}\ \text{with 80\% TS}$$

(3) Estimate the mass of water evaporated (M_{water})

$$M_{water} = Q_{sludge\text{-}centrif} - Q_{sludge\text{-}thdryer} = 5.3\ \text{m}^3\ \text{d}^{-1} - 1.2\ \text{m}^3\ \text{d}^{-1}$$
$$= 4.1\ \text{m}^3_{water}\ \text{d}^{-1}\ \text{or}\ 4{,}100\ \text{kg}_{water}\ \text{d}^{-1}$$

(4) Estimate the amount of thermal energy needed to meet demand ($E_{th\text{-}needed}$)

$$E_{th\text{-}needed} = M_{water} \times E_{evap} = 4{,}100\ \text{kg}_{water}\ \text{d}^{-1} \times 1.0\ \text{kWh kg}_{water}^{-1} = \mathbf{4{,}100\ kWh\ d^{-1}}$$

(5) Thermal energy balance for scenario 2

It can be concluded that the amount of heat generated from the exhaust gases of the CHP engine ($E_{th\text{-}exh\text{-}gas} = 4{,}118$ kWh d^{-1}) is enough to meet the thermal energy needed to dry 40% of the sludge produced in the STP ($E_{th\text{-}needed} = 4{,}100$ kWh d^{-1}).

EXAMPLE 8.2

Use of biogas generated in small-size anaerobic-based STPs for heat generation aiming at drying and sanitization of sludge.

Input data

- Population equivalent: PE = 20,000 inhabitants
- Average influent sewage flowrate: $Q_{s\text{-}avg} = 38.7\ \text{L s}^{-1}$ ($3{,}343.7\ \text{m}^3\ \text{d}^{-1}$)
- *Per capita* COD contribution: $QPCP_{COD} = 0.10$ kg inhab^{-1} d^{-1}
- Average influent COD concentration: $C_{COD} = 598$ mg L^{-1} (0.598 kg m^{-3})
- COD removal efficiency (in terms of filtered COD): $E_{COD} = 80\%$
- Methane yield: $Y_{CH4} = 0.13\ \text{m}^3\text{CH}_4\ \text{kgCOD}_{remov}^{-1}$
- Lower calorific value of methane: $LCV_{CH4} = 9.9$ kWh m^{-3}
- Sludge yield: $Y_{sludge} = 0.15\ \text{kgTS kgCOD}_{app}^{-1}$
- Sludge specific mass: $\gamma = 1{,}020$ kgTS m^{-3}
- Concentration of the sludge removed from the UASB reactors: $C_{sludge\text{-}UASB} = 4.0\%$
- Specific heat of water: $H_w = 0.001162$ kWh kg^{-1} °C^{-1} (Wagner *et al.*, 2017)
- Specific heat of sludge mass: $H_s = 0.000292$ kWh kg^{-1} °C^{-1} (Wagner *et al.*, 2017)
- Energy loss through the walls of the sanitizing tank: $EL_{sanit\text{-}tank} = 15\%$
- Energy loss in the sanitizing drying beds: $EL_{sanit\text{-}drying\text{-}beds} = 24$ kWh m^{-2}
- Thermal efficiency of the boiler: $\eta_{boiler} = 90\%$
- Solids capture in the drying bed: $SC_{drying\text{-}bed} = 95\%$
- Solids application load adopted for the drying beds: $S_a = 15$ kgTS m^{-2}
- Solids application load adopted for the sanitizing beds: $S_b = 22$ kgTS m^{-2}

Solution

(1) Estimate the organic load removed (OL_{rem})

$$OL_{remov} = Q_{s\text{-}avg} \times C_{COD} \times E_{COD} = 3{,}343.7\,m^3\,d^{-1} \times 0.598\,kg\,m^{-3} \times 0.80 = 1{,}600\,kgCOD_{rem}\,d^{-1}$$

(2) Estimate the daily methane production (Q_{CH4})

$$Q_{CH4} = OL_{rem} \times Y_{CH4} = 1{,}600\,kgCOD_{rem}\,d^{-1} \times 0.130\,Nm^3\,CH_4\,kgCOD_{rem}{}^{-1} = 208\,m^3\,CH_4\,d^{-1}$$

(3) Estimate the amount of chemical energy in the biogas (E_{biogas})

$$E_{biogas} = Q_{CH4} \times LCV_{CH4} = 208\,m^3CH_4\,d^{-1} \times 9.9\,kWh\,m^{-3} = 2{,}059\,kWh\,d^{-1}$$

(4) Estimate the daily sludge mass production in the UASB reactors ($P_{sludge\text{-}UASB}$)

$$P_{sludge\text{-}UASB} = Q_{s\text{-}avg} \times C_{COD} \times Y_{sludge}$$
$$= 3{,}343.7\,m^3\,d^{-1} \times 0.598\,kg\,m^{-3} \times 0.15\,kgTS\,kgCOD_{app}^{-1} = 300\,kgTS\,d^{-1}$$

(5) Estimate the daily sludge volumetric production in the UASB reactors ($Q_{sludge\text{-}UASB}$)

$$Q_{sludge\text{-}UASB} = P_{sludge\text{-}UASB}/(C_{sludge\text{-}UASB} \times \gamma) = 300\,kgTS\,d^{-1}/(0.04 \times 1{,}020\,kgTS\,m^{-3})$$
$$= 7.35\,m^3\,d^{-1} \text{ with } 4\%\ TS$$

Scenario 1: Use of a sanitizing tank followed by conventional drying beds

Sanitization will take place in a concrete tank. The sludge being circulated by means of pumps and heated to 70°C by means of a boiler (pasteurization process: 70°C and 30 minutes) in a batch mode (Figure 8.15). The sanitizing tank will be fed whenever the excess sludge is discarded (once a day). The design temperature of the tank is 20°C ($\Delta T = 50°C$). Mass of water present in the sludge of the UASB reactor with 4% TS (M_{water}) = 7,000 kg d^{-1}.

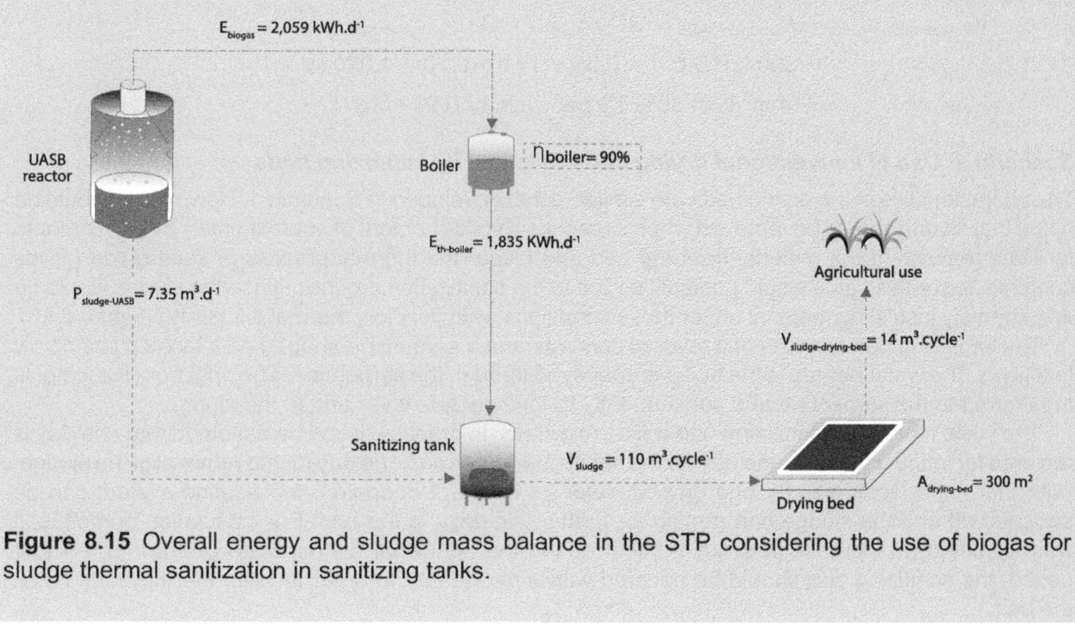

Figure 8.15 Overall energy and sludge mass balance in the STP considering the use of biogas for sludge thermal sanitization in sanitizing tanks.

(6) Energy demand by the sanitizing tank

$E_{th\text{-}needed} = [(P_{sludge\text{-}UASB} \times H_s \times \Delta T) + (M_{water} \times H_w \times \Delta T)] \times EL_{sanit\text{-}tank} = [(300\ kgTS\ d^{-1} \times 0.000292\ kWh\ kg^{-1}{}^\circ C^{-1} \times 50\ deg\,^\circ C) + (7{,}000\ kg\ d^{-1} \times 0.001162\ kWh\ kg^{-1}{}^\circ C^{-1} \times 50^\circ C)] \times 1.15 = (4.38\ kWh\ d^{-1} + 406.7\ kWh\ d^{-1}) \times 1.15 = \mathbf{473\ kWh\ d^{-1}}$

(7) Estimate the amount of heat generated in the boiler ($E_{th\text{-}boiler}$)

$$E_{th\text{-}boiler} = E_{biogas} \times \eta_{th\text{-}boiler} = 2{,}059\ kWh\ d^{-1} \times 0.90 = \mathbf{1{,}853\ kWh\ d^{-1}}$$

(8) Thermal energy balance

It can be concluded that the amount of heat generated in the boiler ($E_{th\text{-}boiler} = 1{,}853\ kWh\ d^{-1}$) is enough to meet 100% of the thermal energy needed to heat (pasteurize) all the sludge produced in the STP ($E_{th\text{-}needed} = 473\ kWh\ d^{-1}$), with a large surplus of energy.

After the sanitization process, the sludge is then directed to the drying beds. As the volume of the sludge has not changed in the sanitizing tank, a drying time (D_t) of 10 days in conventional beds and an operation time (O_t) of 5 days was considered.

(9) Cycle time of the drying beds (C_t)

$$C_t = D_t + O_t = 10\ d + 5\ d = 15\ d$$

(10) Volume of sludge (Vs) directed to the drying beds, per cycle (V_{sludge})

$$V_{sludge} = Q_{sludge\text{-}UASB} \times C_t = 7.35\ m^3\ d^{-1} \times 15\ d = 110\ m^3 \text{ per cycle}$$

(11) Calculate the area necessary for the drying beds ($A_{drying\text{-}bed}$)

$$A_{drying\text{-}bed} = (P_{sludge\text{-}UASB} \times C_t)/S_a = (300\ kgTS\ d^{-1} \times 15\ d)/15\ kgTS\ m^{-2} = \mathbf{300\ m^2}$$

(12) Volume of sludge removed from the drying beds per cycle ($V_{sludge\text{-}drying\text{-}bed}$)

$$V_{sludge\text{-}drying\text{-}bed} = (P_{sludge\text{-}UASB} \times SC_{drying\text{-}bed} \times C_t)/(C_{drying\text{-}bed} \times \gamma)$$
$$= (300\ kgTS.d^{-1} \times 0.95 \times 15\ d)/(0.30 \times 1{,}020\ kg\ m^{-3}$$
$$= 14\ m^3 \text{ with 30\% TS per cycle or } 0.93\ m^3\ d^{-1}$$

Scenario 2: Use of conventional drying beds followed by sanitizing beds

The characteristics of the drying beds are similar to that developed in Scenario 1. However, it should be noted that sludge should be removed when it reaches a solids content of approximately 30%, in order to facilitate removal of the material from the bed and favour the thermal process of sanitization (solids contents above this value would pose difficulties to the sanitization process, since the sludge would be an extremely insulating material under these conditions, with very low thermal diffusivity) (Figure 8.16).

The sanitizing bed consists of a layer of concrete and a serpentine-shaped pipe located just below that layer. The water heated in the boiler is conveyed through the serpentine. The heat from the water is transferred to the serpentine and consequently to the concrete layer and to the sludge.

The cycle time of the sanitizing bed will be four days. In practice, it can be assumed that one day is required for loading the bed, one day for the sanitization itself and one day for the removal of the sludge. With that there would still be one day off. After 24 hours of operation, the heating system can be switched off and the sludge can remain for further two days in the bed. For calculation purposes, it was assumed no increase in solids content during the additional drying days. During this 24-hour period, the sanitizing bed should be covered with a plastic tarp in order to keep the moisture inside the bed.

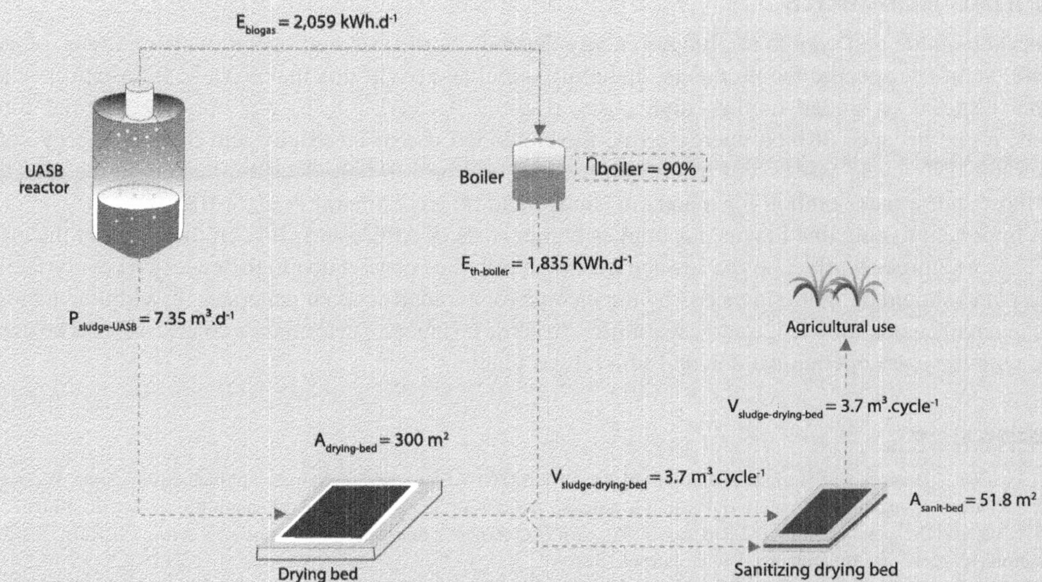

Figure 8.16 Overall energy and sludge mass balance in the STP considering the use of biogas for sludge thermal sanitization in heated drying beds.

(13) Cycle time of the sanitizing beds (C_t)

$$C_t = 4 \text{ d}$$

(14) Volume of sludge directed to sanitizing beds, per cycle ($V_{sludge\text{-}sanit\text{-}bed}$)

$$V_{sludge\text{-}sanit\text{-}bed} = Q_{sludge\text{-}drying\text{-}bed} \times C_t = 0.93 \text{ m}^3 \text{ d}^{-1} \times 4 \text{ d} = 3.7 \text{ m}^3 \text{ per cycle}$$

(15) Calculate the area necessary for the sanitizing beds ($A_{sanit\text{-}bed}$)

$$A_{sanit\text{-}bed} = (P_{sludge\text{-}UASB} \times SC_{drying\text{-}bed} \times C_t)/S_b$$
$$= (300 \text{ kgTS d}^{-1} \times 0.95 \times 4 \text{ d})/22 \text{ kgTS m}^{-2} = 51.8 \text{ m}^2$$

(16) Energy demand by the sanitizing beds

The design temperature of the sanitizing bed is 20°C, the sludge being heated to 70°C ($\Delta \approx 50$°C). For better process efficiency in the sanitizing beds, the serpentine should be copper because of their high thermal conductivity. Additionally, it is recommended that the concrete layer be as thin as possible. Mass of water present in the sludge of the drying beds with 30% TS (M_{water}) = 632 kg d^{-1}.

$$E_{th\text{-}needed} = (P_{sludge\text{-}drying\text{-}bed} \times H_s \times \Delta) + (M_{water} \times H_w \times \Delta) + (EL_{sanit\text{-}bed} \times A_s)$$
$$= (285 \text{ kgTS d}^{-1} \times 0.000292 \text{ kWh.kg}^{-1\circ}\text{C}^{-1} \times 50^\circ\text{C}) + (632 \text{ kg d}^{-1}$$
$$\times 0.001162 \text{ kWh kg}^{-1\circ}\text{C}^{-1} \times 50^\circ\text{C}) + (24 \text{ kWh m}^{-2} \times 51.8 \text{ m}^2)$$
$$= 4.16 \text{ kWh} + 36.7 \text{ kWh} + 1{,}243 \text{ kWh} = \mathbf{1{,}283 \text{ kWh d}^{-1}}$$

(17) Thermal energy balance

It can be concluded that the amount of heat generated in the boiler ($E_{th\text{-}boiler}$ = 1,853 kWh d^{-1}) is enough to meet 100% of the thermal energy needed to heat the sanitizing beds ($E_{th\text{-}needed}$ = 1,283 kWh d^{-1}), with around 30% surplus of energy.

8.9 FINAL REMARKS

The decision on the best way to use biogas energy depends on the size and operational conditions of each STP and on on-site specific requirements, including social and environmental aspects. In instances where the final disposal of sludge implies high costs, it could be advisable to use only the thermal route, whereas, in the instance of high energy costs, the combined use of electricity and thermal energy could be favorable. Life cycle assessment has been proposed as a tool to help decision-makers in the selection of the most appropriate solution for managing sludge and biogas (Amaral *et al.*, 2018).

It is important to note that biogas use implies higher costs (CAPEX and OPEX); therefore, in the effort not to have a significant effect on the sewage treatment bill, the costs related to the energy recovery facility (biogas plant) should be "auto financed" by the income (or avoided costs) it generates. In addition, a biogas plant can require extra knowledge and capabilities. In some instances, performance contracts could be useful to deal with the extra operational demand of a biogas plant.

REFERENCES

ABNT (Associação Brasileira de Normas Técnicas) (2011). NBR 12209 – Elaboração de projetos hidráulico-sanitários de estações de tratamento de esgoto (Elaboration of Hydraulic-Sanitary Designs of Sewage Treatment Plants), Rio de Janeiro, Brazil.

ABNT (Associação Brasileira de Normas Técnicas) (2016). *NBR IEC 60079-0: Equipamentos – requisitos gerais* (NBR IEC 60079-0: Equipment – general requirements), Rio de Janeiro, Brazil.

Amaral K. C., Aisse M. M., Possetti G. R. C. and Prado M. R. (2018). Use of life cycle assessment to evaluate environmental impacts associated with the management of sludge and biogas. *Water Science and Technology*, **77**(9), 2292–2300.

Andreoli C. V., von Sperling M. and Fernandes F. (2001). Lodo de esgoto: tratamento e disposição final (Sewage Sludge: Treatment and Final Disposal). Department of Sanitary and Environmental Engineering of UFMG, Curitiba, Paraná, Brazil, **6**, 484pp.

Andreoli C. V., Ferreira A. C., França M. and Cherubini C. (2002). Desinfecção e secagem térmica de lodo de esgoto anaeróbio pelo uso de biogás (Disinfection and thermal drying of anaerobic sewage sludge by using biogas). XXVII Congresso Interamericano de Ingeniería Sanitária y Ambiental, Cancún, México.

ANEEL (Agência Nacional de Energia Elétrica) (2012). *Resolução Normativa no 482* (Normative Resolution no 482), Brasília, Brazil.

ANEEL (Agência Nacional de Energia Elétrica) (2015) *Resolução Normativa n° 687* (Normative Resolution n° 687), Brasília, Brazil.

Barton L., Kiese R., Gatter D., Butterbach-Bahl K., Buck R., Hinz C. and Murphy D. V. (2008). Nitrous oxide emissions from a cropped soil in a semi-arid climate. *Global Change Biology*, **14**, 177–192.

Baukal C. E. (2013). Applications. In: The John Zink Hamworthy Combustion Handbook, E. Charles and J. R. Baukal (eds.), 2nd edn, CRC Press, Boca Raton, USA, vol. **3**, pp. 252–330.

Borges E. S. M., Godinho V. M., Bejar D. O. and Chernicharo C. A. L. (2009). Tratamento térmico de lodo anaeróbio com utilização do biogás gerado em reatores UASB: avaliação da autossustentabilidade do sistema e do efeito sobre a higienização e a desidratação do lodo (Thermal treatment of anaerobic sludge by using biogas generated in UASB reactors: evaluation of the self-sustainability of the system and of the effect on sanitation and dewatering of sludge). *Engenharia Sanitária e Ambiental*, **14**, 337–346.

Bressani-Ribeiro T., Lobato L. C. S., Souza S. N., Pegorini E. S. and Chernicharo C. A. L. (2018). Contribuição para o aprimoramento de projeto, construção e operação de reatores UASB aplicados ao tratamento de esgoto sanitário - Parte 2: Tratamento preliminar, bombeamento e distribuição de vazão (Contribution to the improvement of design, construction and operation of UASB reactors treating sewage - Part 2: Preliminary treatment, pumping and distribution of flow). *Revista DAE*, **66**(214), 17–29.

Cabral B. G. C. (2016). Avaliação da produção de biogás para fins energéticos em reatores anaeróbios tratando esgoto sanitário (Evaluation of the Biogas Production for Energy Purposes in Anaerobic Reactors Treating Sewage). MSc Thesis, Environmental Engineering Program, Federal University of Santa Catarina, Florianópolis, Brazil.

Cabral B. G. C., Platzer C. J., Rosenfeldt S., Hoffmann H. and Chernicharo C. A. L. (2015). Caracterização do biogás gerado no tratamento de efluentes domésticos em reatores UASB no Brasil (Characterization of biogas generated in UASB reactors treating sewage in Brazil). Congresso Brasileiro de Engenharia Sanitária e Ambiental 28, ABES, Rio de Janeiro, Brazil.

Cabral B. G. C., Chernicharo C. A. L., Hoffmann H., Neves P. N. P., Platzer C., Bressani-Ribeiro T. and Rosenfeldt S. (2016). Resultados do projeto de medições de biogás em reatores anaeróbios/Probiogás (Results of the biogas measurement project in anaerobic reactors/ Probiogás). Deutsche Gesellschaf für Internationale Zusammenarbeit GmbH (ed.), GIZ/Ministério das Cidades, Brasília, Brazil.

Cabral B. G. C., Chernicharo C. A. L., Platzer C. J., Barjenbruch M. and Filho P. B. (2017a). Evaluation of biogas production and energy recovery potential in 5 full-scale WWTPs with UASB reactors. *Chemical Engineering and Chemical Process Technology*, **3**(3), 1043.

Cabral B. G. C., Rosenfeldt S., Platzer C. J., Hoffmann H. and Barjenbruch M. (2017b). Análise da geração de energia elétrica a partir do biogás produzido por reatores UASB: um estudo de caso em escala real (Analysis of the Generation of Electric Energy from the Biogas Produced in UASB Reactors: A Full-Scale Case Study). Congresso ABES FENASAN – 29° Congresso Brasileiro de Engenharia Sanitária e Ambiental, F. Abes (ed.), São Paulo, Brazil.

Carvalho M. E. and Possetti G. R. C. (2017). Potencial de redução das emissões de gases do efeito estufa provenientes de reatores anaeróbios de estações de tratamento de esgoto: proposição de cenários para o Estado do Paraná (Potential of reduction of greenhouse gases emissions from anaerobic-based sewage treatment plants: proposal of scenarios for the State of Paraná). Congresso ABES FENASAN – 29° Congresso Brasileiro de Engenharia Sanitária e Ambiental, F. Abes (ed.), São Paulo, Brazil, pp. 1–7.

CENBIO (Centro Nacional de Referência em Biomassa) (2005). Projeto, instalação e testes de uma unidade de demonstração de geração de energia elétrica a partir de biogás de tratamento de esgoto – ENERG-BIOG (Design, Installation and Testing of a Demo-Unit For Power Generation from Biogas from Sewage Treatment – ENERG-BIOG), CENBIO, Technical Report, São Paulo, Brazil.

Chambers A. K. and Potter I. (2000). Gas utilization from sewage waste. *Carbon and Energy Management*, **12**, 1–13.

Chernicharo C. A. L. (2007). Reatores Anaeróbios (Anaerobic Reactors). 2nd edn, Belo Horizonte: Departamento de Engenharia Sanitária e Ambiental. Department of Sanitary and Environmental Engineering of UFMG, Curitiba, Paraná, Brazil, 380pp.

Chernicharo C. A. L. and Stuetz R. M. (2008). Improving the design and operation of UASB reactors for treating domestic wastewater: management of gaseous emissions. IX Latin American Workshop and Seminar on Anaerobic Digestion, IWA, Easter Island, Chile.

Chernicharo C. A. L., Lobato L. C. S., Melo G. C. B., Pujatti F. J. P. and Stuetz R. M. (2011). Alternativas para aproveitamento energético do biogás gerado em reatores anaeróbios tratando esgoto doméstico (Alternatives for energy recovery from biogas generated in anaerobic reactors treating sewage). *Engenharia Sanitária e Ambiental*.

Chernicharo C. A. L., van Lier J. B., Noyola A. and Bressani-Ribeiro T. (2015). Anaerobic sewage treatment: state of the art, constraints and challenges. *Environmental Science Biotechnology*, **14**(4), 649–679.

Chernicharo C. A. L., Bressani-Ribeiro T., Garcia G. B., Lermontov A., Pereira C. B., Platzer C. J., Possetti G. R. C., Leites M. A. L. and Rosseto R. (2018a). Panorama do tratamento de esgoto sanitário nas regiões Sul, Sudeste e Centro-Oeste do Brasil: tecnologias mais empregadas (Overview of sewage treatment in the south, south-east and center-west regions of Brazil: most employed technologies). *Revista DAE*, **66**(213), 5–19.

Chernicharo C. A. L., Bressani-Ribeiro T., Pegorini E. S., Possetti G. R. C., Miki M. K. and Souza S. N. (2018b). Contribuição para o aprimoramento de projeto, construção e operação de reatores UASB aplicados ao tratamento de esgoto sanitário - Parte 1: Tópicos de Interesse (Contribution to the improvement of design, construction and operation of UASB reactors treating sewage – Part 1: Topics of Interest). *Revista DAE*, **66**(214), 5–16.

Copasa (Companhia de Saneamento de Minas Gerais) (2011). Secador térmico de lodos da ETE de Montes Claros (Sludge thermal dryer of the Montes Claros STP). http://www.copasa.com.br/wps/portal/internet/imprensa/noticias/releases/filter?area=/site-copasa-conteudos/internet/perfil/imprensa/noticias/releases/2011/junho (accessed 30 November, 2018).

Cortez L. A. B., Lora E. E. S., Gómez E. O. (2008). Biomassa para Energia (Biomass for Energy). UNICAMP (ed.), Campinas, São Paulo, Brazil, 732pp.

Costa D. F. (2006). Geração de energia elétrica a partir do biogás do tratamento de esgoto (Generation of Electricity from Biogas from Sewage Treatment). MSc Thesis, Energy Postgraduate Interunits Program, University of São Paulo, São Paulo, Brazil.

Deublein D. and Steinhauser A. (2008). Biogas from Waste and Renewable Resources. Wiley–VCH, Weinheim, Germany.

Drosg B. (2013). Process Monitoring in Biogas Plants. IEA Bioenergy (ed.), United Kingdom. https://nachhaltigwirtschaften.at/resources/iea_pdf/reports/iea_bioenergy_task37_brochure_biogas_process_montoring.pdf?m=1469661329 (accessed 30 November, 2018).

Duarte O. A. H., Pereira A. M., Cantão M., Possetti G. and Aisse M. M. (2018). Assessment of an in situ Real Time UV/VIS Based Spectrometry System for Chemical Oxygen Demand Measurement in a Wastewater Anaerobic Treatment Reactor. *Water Resources Management*, **1**, 167–180.

DWA 212 (Deutsche Vereinigung für Wasserwirtschaft, Abwasser und Abfall e.V.) (2008). Equipamento Técnico de Usinas de Biogás de esgoto em ETEs – DWA 212 (Technical Equipment of Sewage Biogas Plants in STPs – DWA 212). German Association for Water and Waste, Hennef, Germany.

DWA-M363 (Deutsche Vereinigung für Wasserwirtschaft, Abwasser und Abfall e.V.) (2010). Origem, Tratamento e Utilização do Biogás (Origin, Treatment and Use of Biogas). DWA-M 363, German Association for Water and Waste, Hennef, Germany.

DWA-M361PT (Deutsche Vereinigung für Wasserwirtschaft, Abwasser und Abfall e.V.) (2011). Tratamento de Biogás – DWA-M361PT (Biogas Treatment – DWA-M 361PT), German Association for Water and Waste, Hennef, Germany, 40pp.

Fernandez-Anez N. and Medic-Pejic L. (2014). Flammability properties of thermally dried sewage sludge. *Fuel*, **134**, 636–643.

Ferreira R. A. R. and Nishiyama L. (2003). Uma revisão sobre a disposição controlada de lodo de esgoto no solo. 2° Simpósio Regional de Geografia (A review on the controlled disposal of sewage sludge in soil. In: 2nd Regional Geography Symposium). In: *Simpósio Regional de Geografia – Perspectivas para o cerrado no século XXI*, Uberlândia, Minas Gerais, Brazil.

Filippini R. M. K., Possetti G. R. C., Silva T. C. F. and Amaral K. J. (2017). Perspectivas de redução das emissões de gases de efeito estufa no tratamento de esgoto no Estado do Paraná até o ano de 2030 (Perspectives for Reducing Greenhouse Gas Emissions in the Treatment of Sewage in the State of Paraná by the Year 2030). Congresso ABES FENASAN – 29° *Congresso Brasileiro de Engenharia Sanitária e Ambiental*, ABES, FENASAN, São Paulo, Brazil, pp. 1–7.

FNR (2012). Leitfaden Biogas: Von der Gewinnung zur Nutzung (Guide to Biogas: From Production to Use). Fachagentur für nachwachsende Rohstoffe e.V. (ed.), Leipzig, Germany.

GIZ (Deutsche Gesellschaft für Internationale Zusammenarbeit Cooperación) (2017). Guía técnica para el manejo y aprovechamiento de biogás en plantas de tratamiento de águas residuales (Technical Guide for Management and Recovery of Biogás in Wastewater Treatment Plants). Programa Aprovechamiento Energético de Residuos Urbanos en México, Alemana al Desarrollo Sustentable en México (ed.), Mexico, 236pp.

Glória R. M., Motta T. M., Silva P. V. O., Costa P., Brandt E. M. F., Souza C. L. and Chernicharo C. A. L. (2016). Stripping and dissipation techniques for the removal of dissolved gases from anaerobic effluents. *Brazilian Journal of Chemical Engineering*, **33**, 713–721.

Gomes E. C., Dias I. A., Ferreira L. R. A., Otto R. B., Moreira H. C., Possetti G. R. C., Wagner L. G., Junior J. A. S. and Zeni A. L. (2017). *Guia técnico sobre geração distribuída de energia elétrica por biogás em ETEs/Probiogás (Technical Guide on Distributed Generation of Electricity by Biogas in STPs/Probiogás)*. GIZ/Ministério das Cidades, Brasília, Brazil.

Herndon J. (2008a). Selecting the right biosolids dryers: Part I. *Water & Wastes Digest*, **48**(3), pp. 18–23.

Herndon J. (2008b). Selecting the right biosolids dryers: Part II. *Water & Wastes Digest*, **48**(3), pp. 22–45.

International Energy Agency (2000). Biogas Flares: State of the Art and Market Review IEA Bioenergy, 12pp.

IPCC (Intergovernmental Panel on Climate Change) (2014). Future pathways for adaption, mitigation and sustainable development, Synthesis Report. Contribution of Working Groups I, II and III to the Fifth Assessment Report of the Intergovernmental Panel on Climate Change, R.K. Pachauri, L. A. Meyer (eds.), IPCC, Geneva, Switzerland, 151pp.

Kaminski G. F., Wagner L. G., Silva F. O. M. and and Possetti G. R. C. (2018). Análise crítica acerca da aplicação de queimadores enclausurados em ETEs para destruição de biogás (Critical analysis on the application of enclosed flares in STPs for the destruction of biogas). *Simpósio Meio Ambiente Urbano e Industrial Brasil*, Alemanha, Curitiba, Paraná, Brazil.

Lobato L. C. S. (2011). Aproveitamento Energético de Biogás Gerado em Reatores UASB Tratando Esgoto Doméstico (Energy utilization of biogas generated in UASB reactors treating sewage). PhD Thesis, Postgraduate Program in Sanitation, Environment and Water Resources, Federal University of Minas Gerais, Belo Horizonte, Brazil.

Lobato L. C. S., Chernicharo C. A. L. and Souza C. L. (2012). Estimates of methane loss and energy recovery potential in anaerobic reactors treating domestic wastewater. *Water Science and Technology*, **66**(12), 2745–2753.

Manchester S. J. (2001). The Fire and Explosion Hazards of Dried Sewage Sludge. *IChemE Symposium Series*, **148**, 241–254.

Medic-Pejic L., Fernandez-Anez N., Mateos L. M., Torrent J. G. and Ramírez-Gómez A. (2014). Characterization of spontaneous combustion tendency of dried sewage sludge. Proceedings of Tenth International Symposium on Hazards, Prevention, and Mitigation of Industrial Explosions, Norway, pp. 1453–1462.

Moreira H. C. (2017). Biogás em estações de tratamento de esgotos: os principais legados da cooperação técnica Brasil-Alemanha (Biogas in Sewage Treatment Plants: The Main Legacies of Brazil-Germany Technical Cooperation). MSc Thesis, Department of Civil and Environmental Engineering, Pontifical Catholic University of Rio de Janeiro, Rio de Janeiro, Brazil, 174pp.

Moreno V. C. and Cozzan I. V. (2015). Major accident hazard in bioenergy production. *Journal of Loss Prevention in the Process Industries*, **35**(28), 135–144.

Nelting K., Trautmann N., Caicedo C., Weichgreb D., Rosenwinkel K. H., Costa F. J. O. G. and Possetti G. R. C. (2017). Constraints on the dissolved methane in the effluent of full-scale municipal UASB reactors. Presnted at 14th IWA Leading Edge Conference on Water and Wastewater Technologies, Florianópolis, Santa Catarina, Brazil.

Noyola A., Morgan-Sagastume J. M. and López-Hernández J. E. (2006). Treatment of biogas produced in anaerobic reactors for domestic wastewater: odor control and energy/resource recovery. *Reviews in Environmental Sciences and Bio/Technology*, **5**, 93–114.

Noyola A., Padilla-Rivera A., Morgan-Sagastume J. M. L., Guereca L. P. and Hernandez-Padilla F. (2012). Typology of municipal wastewater treatment technologies in Latin America. *Clean – Soil, Air, Water*, **40**, 926–932.

Park K., Lee H., Phelan S., Liyanaarachchi S., Marleni N., Navaratna D., Jegatheesam V. and Shu L. (2014). Mitigation strategies of hydrogen sulphide emission in sewer networks. *International Biodeterioration & Biodegradation* **95**(A), 251–252.

Perry R. H. and Green D. W. (1997). Perry's Chemical Engineers' Handbook (7th edn), McGraw-Hill, New York, USA.

Pilavachi P. A. (2002). Mini and micro-gas turbines for combined heat and power. *Applied Thermal Engineering*, **22**, 2003–2014.

Possetti G. R. C., Jasinski V. P., Andreoli C. V., Bittencourt S. and Carneiro C. (2012). Sistema térmico de higienização de lodo de esgoto movido a biogás para ETEs de médio e pequeno porte (Thermal system fueled on biogas for sanitization of sewage sludge in medium- and small-scale STPs). Simpósio Luso-Brasileiro de Engenharia Sanitária e Ambiental, Belo Horizonte, Minas Gerais, Brazil.

Possetti G. R. C., Jasinski V. P., Mesquita N. C., Kriguel K. and Carneiro C. (2013). Medições em tempo real do biogás produzido em reatores UASB alimentados com esgoto doméstico (Real-time measurements of biogas produced in UASB reactors treating sewage). Congresso Brasileiro de Engenharia Sanitária e Ambiental, 27th edn, Goiânia, Goiás, Brazil.

Possetti G. R. C., Rietow J. C., Gervasoni R., Althoff C. A. and Carneiro C. (2015). Investigação experimental de um sistema piloto de secagem térmica de lodo movido a biogás (Experimental investigation of a pilot biogas sludge drying system). Congresso Brasileiro de Engenharia Sanitária e Ambiental, 28th edn, Rio de Janeiro, Rio de Janeiro, Brazil.

Possetti G. R. C., Rietow J. C., Costa F. J. O. G., Wagner L. G., Lobato L. C. S., Bressani-Ribeiro T., Melo D. F., Reis J. A. and Chernicharo C. A. L. (2018). Contribuição para o aprimoramento de projeto, construção e operação de reatores UASB aplicados ao tratamento de esgoto sanitário – Parte 5: Biogás e emissões fugitivas de metano (Contribution to the improvement of design, construction and operation of UASB reactors treating sewage – Part 5: Biogas and fugitive emissions of methane). *Revista DAE*, **66**(214), 73–89.

Prata L. L. V., Possetti G. R. C. and Limons R. S. (2017). Método para a seleção de secadores térmicos de lodo proveniente de estações de tratamento de esgoto doméstico (Method for the selection of thermal driers of sludge from sewage treatment plants). Congresso ABES FENASAN – 29° Congresso Brasileiro de Engenharia Sanitária e Ambiental, F. Abes, São Paulo, Brazil.

Rietow J. C., Possetti G. R. C., Fernandes I. R., Lisboa A. M. and Wagner L. G. (2017). Classificações de atmosferas explosivas em estações de tratamento de esgotos com produção de biogás (Classification of explosive atmospheres in sewage treatment plants with biogas production). Congresso ABES FENASAN – 29° Congresso Brasileiro de Engenharia Sanitária e Ambiental, F. Abes (ed.), São Paulo, Brazil.

Rietow J. C., Wagner L. G., Carneiro C., Aisse M. M. and Possetti G. R. C. (2018). Estudo comparativo entre alternativas de sistemas de secagem e higienização térmica de lodo para estações de tratamento de esgoto de médio e pequeno porte (Comparative study between alternatives of sludge drying and thermal hygienization systems for medium and small sewage treatment plants). 1° Seminário Nacional sobre Estações Sustentáveis de Tratamento de Esgoto, INCT ETEs Sustentáveis, Curitiba, Paraná, Brazil.

Rosa A. P., Conesa J. A., Fullana A., Melo G. C. B., Borges J. M. and Chernicharo C. A. L. (2016). Energy potential and alternative usages of biogas and sludge from UASB reactors: case study of the Laboreaux wastewater treatment plant. *Water Science and Technology*, **73**(7), 1680–1690.

Rosenfeldt S., Ho-mann H., Platzer C. and Cabral C. B. G. (2017a). Viabilidade técnico-econômica de produção de energia elétrica em ETEs a partir do biogás/Probiogás (Technical-Economical Feasibility of Electric Power Production from Biogas in STPs/Probiogás). GIZ/Ministério das Cidades, Brasília, Brazil.

Rosenfeldt S., Cabral C. B. G., Platzer C., Hoffmann H., Freitas D. and Cruz A. O. (2017b). Análise da viabilidade econômica de ETE para geração de energia elétrica no Brasil (Economic feasibility analysis of STP for electric power generation in Brazil). Congresso de Inovação Tecnológica em Energia Elétrica, João Pessoa, Paraíba, Brazil.

Rotária do Brasil (2018). Florianópolis, Santa Catarina, Brazil.

RIRDC (Rural Industries Research and Development Corporation) (2008). Using Methane in Intensive Livestock Industries. RIRDC Publication 8/50, 64pp.

Salomon K. R. and Lora E. E. S. (2009). Estimate of the electric energy generating potential for different sources of biogas in Brazil. *Biomass and Bioenergy*, **33**, 1101–1107.

Sanepar (Companhia de Saneamento do Paraná) (2017). Estação de Tratamento de Esgoto Ouro Verde Foz-do-Iguaçu-Paraná (Foz do Iguaçu-Paraná sewage treatment plant). http://site.sanepar.com.br/noticias/sanepar-e-lactec-apresentam-em-congresso-praticas-de-ficiencia-energetica (accessed 22 May, 2017).

Silva T. C. F. (2015). Utilização de biogás de estações de tratamento de esgoto para fins energéticos (Use of biogas from sewage treatment plants for energy purposes). MSc Thesis, Institute of Energy and Environment, University of São Paulo, São Paulo, Brazil.

Silva T. C. F., Possetti G. R. C. and Coelho S. T. (2014). Avaliação do potencial de produção de energia a partir do biogás gerado no tratamento de esgotos domésticos (Evaluation of the energy production potential from the biogas generated in the treatment of domestic sewage). Congresso Brasileiro de Planejamento Energético, 9th edn, Florianópolis, Santa Catarina, Brazil.

Silveira B., Chernicharo C. A. L., Cabral C. B. G., Suidedos C., Platzer C., Silva G., Possetti G. R. C., Hoffmann H., Moreira H. C., Adamatti H. B., Miki M., Silva M., Takahashi R., Miki R., Rosenfeldt S., Araújo V. S. F., Valente V. B. and Villani W. (2015). Guia técnico de aproveitamento energético de biogás em estações de tratamento de esgoto/Probiogás (Technical Guide for Energy Recovery from Biogas in Sewage Treatment Plants/Probiogás). GIZ/Ministério das Cidades, Brasília, Brazil.

Souza C. L. (2010). Estudo das rotas de formação, transporte e consumo dos gases metano e sulfeto de hidrogênio resultantes do tratamento de esgoto doméstico em reatores UASB (Study of the Routes of Formation, Transport and Consumption of

Methane Gas and Hydrogen Sulfide Resulting from the Treatment of Domestic Sewage in UASB Reactors). PhD Thesis, Postgraduate Program in Sanitation, Environment and Water Resources, University Federal of Minas Gerais, Belo Horizonte, Minas Gerais, Brazil.

Strumillo C. and Kudra T. (1986). Drying: Principles, Applications and Design. Gordon and Breach, New York, USA.

Tchobanoglous G., Stensel H. D., Tssuchihashi R., Burton F. L., Abu-Orf M., Bowden G. and Pfrang W. (2013). Wastewater engineering, treatment and reuse, 5th edn, McGraw-Hill, New York.

Umweltbundesamt (2018). Informationspapier zur Sicherheit bei Biogasanlagen (Information paper on safety of biogas plants). http://www.umweltbundesamt.de/ (accessed 16 August, 2018).

USEPA (1996). Turning a Liability into an Asset: A Landfill Gas-to-Energy Project Development Handbook. Landfill Methane Outreach Program, United States Environmental Protection Agency, DC, USA. https://nepis.epa.gov/Exe/ZyPURL.cgi?Dockey=600008CZ.TXT (accessed 16 April, 2009).

USEPA (2002). Powering Microturbines with Landfill Gas – EPA430-F-02-012. United States Environmental Protection Agency, Washington, DC, USA. https://nepis.epa.gov/Exe/ZyPURL.cgi?Dockey=P100112N.TXT (accessed 30 November, 2018).

USEPA (2003). A Brief Characterization of Reciprocating Engines in Combined Heat and Power Applications. United States Environmental Protection Agency, Washington, DC, USA. http://www.epa.gov/lmop/res/pdf/chp_gasturbines.pdf (accessed 16 April, 2009).

USEPA (2008). Adapting Boilers to Utilize Landfill Gas: An Environmentally and Economically Beneficial Opportunity. United States Environmental Protection Agency, Washington, DC, USA. https://www.epa.gov/sites/production/files/2016-07/documents/boilers.pdf (accessed 18 February 2009).

USEPA (2011). Opportunities for Combined Heat and Power at Wastewater Treatment Facilities: Market Analysis and Lessons from the Field. United States Environmental Protection Agency, Washington, DC, USA. https://www.epa.gov/sites/production/files/2015-07/documents/opportunities_for_combined_heat_and_power_at_wastewater_treatment_facilities_market_analysis_and_lessons_from_the_field.pdf (accessed 30 November 2018).

Valente V. B. (2015). Análise de viabilidade econômica e escala mínima de uso do biogás de reatores anaeróbios em Estações de Tratamento de Esgoto no Brasil (Economic Feasibility Analysis and Minimum Scale of Use of Biogas of Anaerobic Reactors in Sewage Treatment Plants in Brazil). MSc Thesis, Energy Planning, University Federal of Rio de Janeiro, Rio de Janeiro, Brazil.

van Haandel A. C. and Lettinga G. (1994). Tratamento anaeróbio de esgoto: um manual para regiões de clima quente (Anaerobic Sewage Treatment: A Manual for Hot Climate Regions). University Federal of Paraíba, Campina Grande, Paraíba, Brazil.

Wagner L. G., Possetti G. R. C., Carneiro C. and Urbanetz J. (2015). Sistema térmico de higienização e secagem térmica de lodo de esgoto movido a energia solar e biogás (Thermal System of Hygienization and Thermal Drying of Sewage Sludge Powered by Solar Energy and Biogas). Congresso sobre Geração Distribuída e Energia no Meio Rural, São Paulo, São Paulo, Brazil.

Wagner L. G., Possetti G. R. C., Carneiro C. and Urbanetz J. (2017). Avaliação do desempenho operacional de duas configurações de um sistema térmico para higienização e secagem de lodos de esgoto movido a biogás e energia solar (Evaluation of the operational performance of two configurations of a thermal system powered by biogas and solar energy for sanitization and drying of sewage sludge). Congresso ABES FENASAN – 29° Congresso Brasileiro de Engenharia Sanitária e Ambiental, F. ABES, (ed.) São Paulo, Brazil.

Chapter 9

Control of diffuse emissions in UASB reactors treating sewage

Emanuel M. F. Brandt[1], Adalberto Noyola[2] and Ewan J. McAdam[3]*

[1]Department of Sanitary and Environmental Engineering, Universidade Federal de Juiz de Fora – Campus da Universidade Federal de Juiz de Fora, Juiz de Fora, Brazil 3 036 330
[2]Instituto de Ingeniería, Universidad Nacional Autónoma de México, Circuito Escolar, Ciudad Universitaria, Coyoacán, 04510, México D.F., México
[3]Cranfield University, Cranfield Water Science Institute, Bedfordshire, United Kingdom
Corresponding author: emanuel.brandt@ufjf.edu.br

9.1 INTRODUCTION

Anaerobic processes for sludge digestion and sewage treatment have been associated frequently with odour problems, which, in some instances, are a major drawback, resulting in the decommissioning of the treatment facilities or in costly process reconfiguration. Moreover, there are several examples of sewage treatment plants (STPs) that after being constructed in a rural context, were assimilated into urbanized areas because of unplanned growth, resulting in complaints from the newly established communities about the emitted malodourous compounds.

It should be recognized that the emission of offensive gases is inherent to the anaerobic degradation of complex organic substrates taking place in digesters or reactors. The main group of odourous compounds produced in an anaerobic system is related to sulphur-containing gases or volatile compounds. The malodourous organic compounds that can be produced in STPs are carbon oxysulfide (COS), carbon disulfide (CS_2), mercaptans of low molecular weight (R-SH), thiophenes (C_4H_4S), dimethylsulfide ($(CH_3)_2S$), dimethyldisulfide ($(CH_3)_2S_2$), and dimethyltrisulfide ($(CH_3)_2S_3$) (Tchobanoglous *et al.*, 2013). Other odourous molecules include hydrogen sulphide (H_2S), ammonia (NH_3), inorganic and organic amines, organic acids, aldehydes, and ketones. In STPs, H_2S has such a characteristic odour that it generally masks the presence of organosulphur volatile compounds (Bhatia 1978; Smet & van

Editor(s): Carlos Augusto de Lemos Chernicharo and Thiago Bressani-Ribeiro
doi: 10.2166/9781780409238_0237

Langenhove, 1998). H_2S is produced in anaerobic environments mainly by sulphate reduction. The sulphate present in municipal wastewater generally derives from natural sources, and its concentration is dependent on the local geology, although the collection of industrial wastes can also contribute to the overall sulphur balance (usually added as sulphuric acid in the production process). Hydrogen sulphide is not only the primary reason for complaints about odour but it is also toxic when inhaled, and is highly corrosive, causing considerable damage to concrete and steel structures at ambient conditions.

The other notable gaseous emission from anaerobic-based STPs is methane (CH_4), a non-odourous, flammable gas produced in anaerobic reactors and digesters but also, to a lesser extent, in poorly operated primary settlers and aeration tanks. The quantitation of fugitive CH_4 emissions has gained significant focus because of the associated global warming potential (GWP), which is 28 times higher than that of carbon dioxide (CO_2) in a 100-year horizon (IPCC, 2014). Point source CH_4 emissions emanate through collection devices (biogas) and are controlled generally through ignition to recover heat and power or are controlled by flaring. Diffuse CH_4 emissions can derive from uncovered anaerobic systems, leaks in the biogas collection system, and from the dissolved fraction contained in the anaerobic effluent. This latter fraction has been ignored generally. However, it will desorb to the atmosphere once released into the environment at a rate dependent on mixing (Chernicharo *et al.*, 2015). The advantages of direct anaerobic sewage treatment (Foresti *et al.*, 2006) and its sustainability could be hindered if CH_4 were not managed properly, a situation often encountered in small STPs in developing countries (Noyola *et al.*, 2006).

Figure 9.1 shows the main aspects of the above-mentioned diffuse emissions, particularly in the anaerobic treatment stage (e.g., UASB reactors). It should be mentioned that malodourous compounds and CH_4 are not associated only with the anaerobic steps in STPs, as these gases can be produced also in the sewerage system. In such instances, the exogenous emissions occur mainly in the headworks (e.g., screening) and primary settlers.

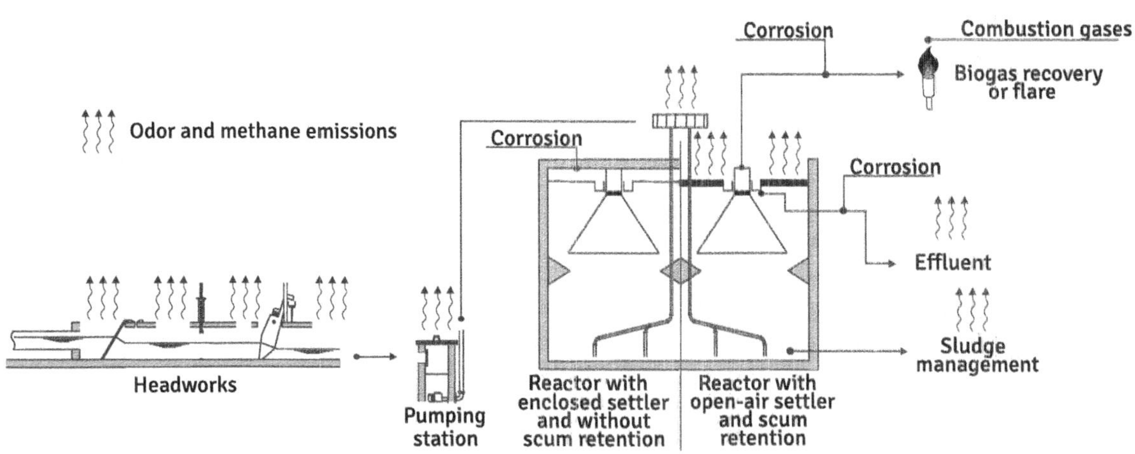

Figure 9.1 Diffuse gaseous emissions in STPs, highlighting the emissions from UASB reactors.

In this chapter, the diffuse (or fugitive) emissions of the two more relevant gaseous compounds produced in anaerobic-based STPs (H_2S and CH_4) are addressed, with emphasis on their capture and the available technological control options.

9.2 FUNDAMENTALS OF THE FORMATION AND EMISSION OF DIFFUSE GASES

In anaerobic-based STPs, sulphide is generated in response to the biological reduction of the sulphate or thiosulphate present in the sewage. The reduction of sulphate can occur through assimilatory or dissimilatory pathways. The assimilatory pathway produces reduced sulphur compounds for the biosynthesis of amino acids and proteins and, therefore, does not lead directly to sulphide, but to organic sulphur. Sulphide is subsequently produced via the desulphurisation pathway, in which anaerobic microorganisms are capable of breaking down proteins into amino acids, with subsequent degradation to sulphides. In the dissimilatory pathway, sulphate or thiosulphate is reduced to sulphide by sulphate-reducing bacteria (SRB).

SRBs are affiliated to more than 40 genera (e.g., *Desulfovibrio*, *Desulfomonas*, *Desulfococcus*, and *Desulfosarcina*) and are a broad group that oxidizes organic matter or hydrogen and uses these as the source of electrons in the reduction of sulphate or thiosulphate to the various dissolved forms of sulphide (Tang *et al.*, 2009). Species of dissolved sulphide diffuse into the liquid medium from the microbial cells and, depending on the pH of the liquid medium, they can be found in the ionized form (the bisulphide, $HS^{-}{}_{(aq)}$, and the sulphide, $S^{2-}{}_{(aq)}$) or associated with hydrogen (hydrogen sulphide, $H_2S_{(aq)}$). For pH below 7 (usually occurring in anaerobic-based STPs), the non-dissociated form (H_2S) is the major dissolved component, which can be desorbed from water as a diffuse emission. The first acid dissociation constant is approximately pH 7, after which HS^- is the predominant sulphur form up to pH 13. Thereafter, the second dissociation is reached to form S^{2-} as the predominant sulphur form. The ionized form is not gaseous and, therefore, is stable in wastewater. At neutral pH, approximately 50% H_2S and 50% HS^- will be dissolved in the wastewater. Some sulphur forms, such as polysulphide anions, can be present in anaerobic reactor effluents because of chemical or biological one-electron sulphide anion oxidation, which is the first step in elemental sulphur formation in oxygen-deficient environments (Steudel, 1996).

In most instances, SRBs use organic compounds both as a source of electrons and of carbon for biosynthesis. In anaerobic-based STPs, these compounds are generated generally from complex organic compounds (carbohydrates, proteins, lipids, and the like) by fermentative and acetogenic bacteria, or become available from the incomplete oxidation of organic matter by the SRB (which produces acetate). However, when hydrogen is used as a source of electrons by the SRB in anaerobic treatment, more restricted carbon sources such as CO_2 and acetate are used for biosynthesis. Because of the thermodynamic preference for $SO_4{}^{2-}$, SRBs compete with the methanogenic archaea (hydrogenotrophic and acetoclastic methanogenesis) for the same substrates (Oude Elferink *et al.*, 1994). In anaerobic environments, CH_4 is an important intermediate in the reactions that lead eventually to the mineralization of the organic matter. When the CH_4 produced by the biodegradation of organic matter is collected in a controlled way to form a point source emission, the gas stream is called biogas. On the other hand, when CH_4 is released into the atmosphere in a diffuse way, the generated gas stream is called waste gas, which contains CH_4 at low concentrations. CH_4 containing diffuse source emissions can occur, for example, in anaerobic reactor effluents, in the surfaces of settlers and thickeners, and in sludge dewatering units.

9.3 CONCERNS ABOUT DIFFUSE GASEOUS EMISSIONS

Depending on the nature of the wastewater feed, pH, and oxidation-reduction potential of the anaerobic process, various odorant compounds are formed biologically. In anaerobic-based STPs, H_2S is the compound most commonly associated with odours, although other sulphur compounds (e.g., mercaptans, dimethyl sulphide), volatile organic acids, aldehydes, alcohols, phenols, ammonia, and amines can contribute to the perception of odour (Table 9.1). In addition to concerns about odour, H_2S is oxidized to sulphuric acid (H_2SO_4) in the presence of moisture and small amounts of oxygen, causing corrosion in concrete and metallic structures. Furthermore, health risks are associated with the exposure of humans to H_2S (Table 9.1). Humans exposed to concentrations of 150–200 ppm_v for 2–15 minutes can lose their sense of smell, which is dangerous, considering that the perception of the odour is a key alarm factor for personal safety assessment. Table 9.2 shows the typical concentrations of H_2S associated with some stages of the sewerage and treatment systems. It should be mentioned that the presence of H_2S and NH_3 in biogas burned in flares or combined heat and power engines will favor the formation of by-products such as sulphur dioxide (SO_2) and nitrogen oxides (NOx) in the exhaust gas, which, in some cases, can be regulated.

Table 9.1 Odour threshold for the most relevant compounds generated in anaerobic-based STPs and the acute symptoms and effects from the exposure of humans to H_2S.

Compound		Threshold (ppm_v)	Characteristic Odour	Acute Symptoms and Effects
Sulphur compounds	Hydrogen sulphide (H_2S)	0.00047–0.1	Rotten eggs	*For H_2S* • Above 20 ppm_v: possible fatigue, loss of appetite, headache, tearing of the eyes, irritability, poor memory, dizziness. • 100–200 ppm_v: coughing, eye irritation, loss of smell, altered breathing, drowsiness (15–30 min); respiratory tract irritation (1 h); pulmonary edema (prolonged exposure). • 500–700 ppm_v: staggering, collapse (5 min); considerable damage to the eyes, and death (30–60 min). • Above 700 ppm_v: rapid unconsciousness or nearly instant death.
	Organic sulphur compounds (methyl and ethyl mercaptans; dimethyl sulphide, and the like)	0.0002–0.01	Rottenness, decayed cabbage	
Nitrogen compounds	Ammonia (NH_3)	0.02–46.8	Pungent, irritating	*For NH_3:* • 300–500 ppm_v: change in respiration rate, and moderate to severe irritation (30–60 min). • 2,500–6,000 ppm_v: life threatening (30 min). • 5,000–10,000 ppm_v: fatal.
	Amines (e.g. methylamine)	0.03–21	Rottenness, decomposed fish	

(Tchobanoglous *et al.*, 2013; United States Department of Labor, Occupational Safety and Health Administration, https://www.osha.gov/SLTC).

Table 9.2 Typical concentrations of H_2S and CH_4 in the gas phase of sewers and treatment systems.

System Unity	Mean or Range of H_2S Concentrations		References	Mean or Range of CH_4 Concentrations		References
	mg m^{-3}	ppm_v		mg m^{-3}	ppm_v	
Sewerage	0–417	0–300	Matos and Aires (1995)	0–16,420	0–23,000	Chaosakul *et al.* (2014)
	70–556	50–400	Jobbagy *et al.* (1994)	0–2,356	0–3,300	Willis *et al.* (2012)
Sewage pumping station	4.8	3.3	Bohn (1993)	1,018–1,437	1,426–2,013	Willis *et al.* (2012)
	0.57	0.4	Silva *et al.* (2007)	2–317	3–444	Eijo-Río *et al.* (2015)
	1–3	0.7–2.0	Silva *et al.* (2007)	10–1,100	14–1,540	Eijo-Río *et al.* (2015)
Preliminary treatment for solids and grit removal	3.5	2.4	Bohn (1993)	700–1,220	980–1,710	Daelman *et al.* (2012)
	2.8–51.5	2–37	Al-Shammiri (2004)	140	196	Czepiel *et al.* (1993)
Sludge dewatering	6.5	4.5	Bohn (1993)	73–156	102–218	Daelman *et al.* (2012)
Waste gas from anaerobic reactors[1]	0–73	0–50	Pagliuso *et al.* (2002); Souza *et al.* (2012)	0–2,430	0–3,400	Souza *et al.* (2012)
Waste gas from anaerobic reactors[2]	146–730	100–500	Pagliuso *et al.* (2002), Souza (2010), Huete *et al.* (2018)	1,071–50,000	1,500–70,000	Souza (2010), Huete *et al.* (2018)

[1]Diffuse gas emissions from the settler compartment of UASB reactors treating sewage.
[2]Waste gas from desorption units receiving effluent from UASB reactors.

Controlling greenhouse emissions in anaerobic-based STPs is needed urgently and has been addressed recently by several authors (Hartley & Lant, 2006; Noyola *et al.*, 2006; Souza *et al.*, 2011). Whereas CH_4 is poorly soluble (22.7 mg L^{-1} at 25°C and 1 atm-CH_4), the high CH_4 partial pressure inside the anaerobic reactor results in anaerobic effluents with dissolved CH_4 concentrations between 6 mg L^{-1} and 25 mg L^{-1} (Nelting *et al.*, 2015; Souza, 2010; Souza *et al.*, 2011). Several studies have indicated CH_4 'supersaturation' of the effluent, implying that the measured value was greater than the thermodynamic expectation, which exacerbated fugitive losses (Heile *et al.*, 2017). Although the concentrations are modest, when considered in the context of the significant hydraulic flows through high-rate anaerobic systems, the losses can amount to 30–41% (Souza *et al.*, 2011) or even more than 50% (Noyola *et al.*, 1988) of the total CH_4 production. As dissolved CH_4 is gaseous, it can partition into the gas phase downstream of the anaerobic reactor. Consequently, if only partially filled pipes or ducts were used for transmission, the dissolved CH_4 could be released and, as the lower explosive limit for CH_4 is only $5\%_{v/v}$, this can present a health and safety risk (Heile *et al.*, 2017). CH_4 concentrations up to $7\%_{v/v}$ have been found in the head space of covered UASB reactors or inside effluent-receiving desorption units (see Table 9.2).

9.4 MANAGEMENT OF DIFFUSE GASEOUS EMISSIONS

Managing diffuse gases involves both the preventive control of the gaseous emissions and the containment and conduction of the waste gases for subsequent abatement (Figure 9.2). Various methods are used for stripping or desorption of gases dissolved in the effluent of anaerobic reactors and these methods also produce waste gases that need subsequent abatement (Figure 9.2). In addition, managing diffuse gaseous emissions should include actions to maximize CH_4 capture in the biogas ductwork and minimize CH_4 diffuse emissions as waste gas.

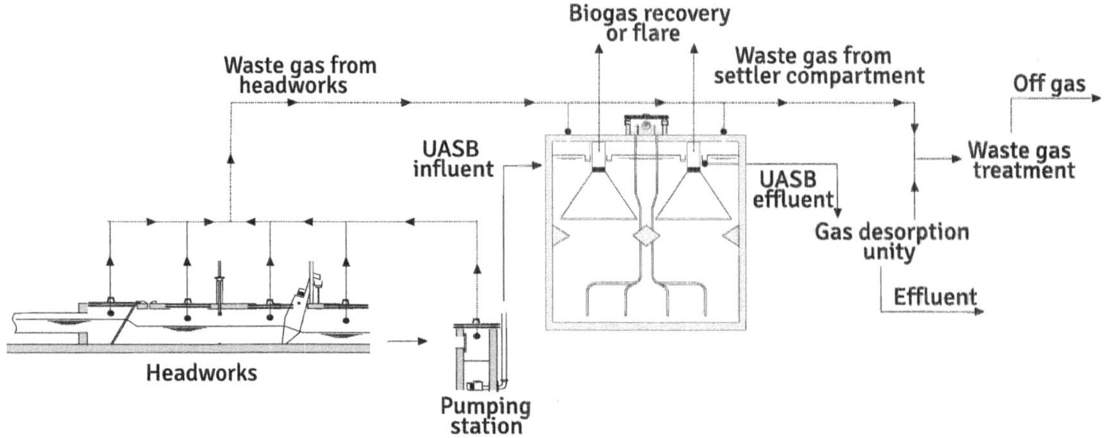

Figure 9.2 Diffuse gaseous emissions in STPs, highlighting the emissions from UASB reactors.

9.4.1 Preventive control of the gaseous emissions in sewer and treatment systems

Minimizing of waste gas emissions should be considered in the engineering design and during the operation of the sewerage and STP, based on good practices (Table 9.3).

Table 9.3 Preventive actions to minimize waste gas emissions from anaerobic reactors.

Step	Rationale	What is Needed
Design	Accumulation of scum in the sewerage and the STP, intensifying odour generation because of organic matter decomposition	• Use of grease traps in homes. • Inclusion of systems to prevent the accumulation of scum or to allow the removal of accumulated scum (e.g., points of water jets in pumping stations, sieves and aerated grit chambers upstream of the anaerobic reactors, and the like).
	Sediment build-up in the sewerage and the STP, intensifying odour generation because of organic matter decomposition	• Accurate quantification of sewage flow and organic loads for appropriate hydraulic design. • Removal of accumulated sediments (e.g., meshes and points of water jets in pumping stations and maintenance holes).
	Odour emissions from sludge streams in the STP	• Using proper units for receiving sludge from septic tanks. • Covering sludge management areas (thickeners, dewatering units, and the like). • Extracting and treating the gases from sludge management sheds.
	Odour dispersing around the STP because of diffuse emissions	• Devices for capturing and conducting emissions in headworks. • Minimizing turbulence during the conduction of anaerobic effluent, preventing hydraulic falls, reducing the volumes of pumping wells. • Submersible incoming pipelines or using turbulent regime and hydraulic falls to force the liberation of dissolved gases in properly covered structures. • Using perimeter vegetation and structures that induce atmospheric turbulence (aromatic trees, such as eucalyptus; 4 m high walls). • Containing, conducting, and treating waste gases.
	Desorption and emission of gases dissolved in the effluent from the anaerobic reactors (odours and CH_4)	• Including systems to prevent the accumulation of scum inside the gas-liquid-solid (GLS) separator of the UASB reactors (e.g., internal devices for scum removal, viewing windows for inspection and manual removal of scum, and the like) see Chapters 4 and 6. • Using downstream devices to allow the controlled desorption and treatment of gases dissolved in the effluent of anaerobic reactors.
Operation		• Implementing specific rules (pretreatment requirements) for receiving non-domestic effluents.

(*Continued*)

Table 9.3 Preventive actions to minimize waste gas emissions from anaerobic reactors (*Continued*).

Step	Rationale	What is Needed
	Odour generation upstream of the STP because of organic matter decomposition (e.g., sediments, scum) and release of precursor compounds in the sewer system	• Restricting the discharge of industrial effluents containing high loads of organic nitrogen, ammonia, organic sulphur, sulphate, thiosulphate, and volatile organic acids, among others. • Chemical dosing for preventative and reactive responses to odours.
	Odour emissions from the headworks	• Frequently removing retained material from the headworks. • Proper guidelines for receiving sludge from septic tanks (preferably at the highest flow of the STP in the morning). • Chemical dosing for preventative and reactive responses to odours.
	Odour emissions from sludge streams in the STP	• Restricting the acceptance of non-stabilized sludge. • Daily removal of dewatered sludge. • Chemical dosing for preventative and reactive responses to odours.
	Odour dispersing around the STP because of diffuse emissions	• Frequently removing scum accumulated in the UASB reactors (GLS separators and settler compartment). • In aeration tanks, secondary settlers, and sludge thickeners, coping with foam and bulking by using chlorinated water jets. • Cleaning and washing units after daily operational interventions.

9.4.2 Control in sewer systems and headworks by chemical dosing

Chemical addition can be applied to raw sewage for both preventative and reactive response to odours. Once sulphides are generated by microorganisms, the reactive addition of chemicals can be used to oxidize or precipitate sulphide, or to ionize $H_2S_{(aq)}$ to HS^- and S^{2-}, thereby preventing its release to the environment as $H_2S_{(g)}$. Preventative chemical dosing includes the introduction of air (aeration)/oxygen (O_2) or nitrate into the liquid stream (e.g., in sewer dosing) to change the oxidation-reduction potential or to provide conditions for inhibiting SRB and for the development of sulphide-oxidizing bacteria. The conventional chemicals/doses are (Brandt *et al.*, 2017; USEPA, 1985, 1991; WERF, 2007; Zhang *et al.*, 2008):

- Reactive addition of chemicals to oxidize the sulphide: 1.3–4.0 $gH_2O_2\ (gS)^{-1}$; 4.0–15.0 $gCl_2\ (gS)^{-1}$; 2.0–15.0 $gNaOCl\ (gS)^{-1}$; 6–7 $gKMnO_4\ (gS)^{-1}$; 7–8 $gNaMnO_4\ (gS)^{-1}$.
- Reactive addition of chemicals to precipitate the sulphide: iron salts, such as $FeCl_3$, $Fe(Cl)_2$, $Fe(NO_3)_3$, and $Fe_2(SO_4)_3$, 1.2–5.3 $gFe\ (gS)^{-1}$.
- Reactive addition of chemicals to ionize the sulphide: approximately 80 $gMg(OH)_2 \cdot m_{sewage}^{-3}$; 90 $gNaOH\ m^{-3}_{sewage}$, or more than 500 $gNaOH \cdot m^{-3}_{sewage}$ for pH shocks.
- Preventative chemical dosing: aeration, 0.3–9.6 $m^3_{air} \cdot m^{-3}_{sewage}$; $NaNO_3$ and $Ca(NO_3)_2$, 0.3–9.6 $gN\text{-}NO_3\ (gS)^{-1}$; permanganates and iron salts preferentially react with the sulphides.

The required dosage of chemicals will depend on the configuration of the sewerage and treatment systems, and the data reported previously should be used as a starting point for more-precise adjustments. Comparing all the techniques, aeration presents lower cost and easier operation. However, when applied in low-pressure pipes, it can be expensive because of the limited oxygen mass transfer. In anaerobic-based STPs, the upstream application can induce undesirable aerobic conditions inside the anaerobic reactors. In addition to sulphide oxidation, H_2O_2 decomposes into H_2O and O_2, resulting in aerobic environments. However, it does not react only with sulphides and is consumed readily by scavengers, requiring excess dosage, which, inevitably, will influence the operating cost. Because permanganates and iron salts react preferentially with sulphides, they are appropriate for sludge streams or streams with a high concentration of organic matter (scavenger for other chemicals). However, they present higher costs compared with other techniques and, as regards the dosages of iron salts, there is an increase in the production of excess sludge in the STP. The reactive addition of chemicals to change pH and ionize the sulphide has a temporary effect and should be applied regularly, resulting in shocks of high pH. Finally, nitrate addition is a promising technique, especially when considering the use of effluents from treatment processes that include nitrification, which are rich in nitrate. In such instance, the dilution factor must be considered.

9.4.3 Capture of gaseous emissions using covers and ventilation systems

Coupled with the preventive actions, several control efforts should be considered, such as the capture of emitting sources to transform diffuse emissions to a point source emission, atmospheric air introduction below the containment coverage (by holes and supply plenums), followed by extraction and treatment of the waste gases for corrosion/odour/greenhouse gas emission control, as well as health and safety management. The closer to the emitting source the containment, the lower would be the flow of waste gas for the treatment, resulting in savings on equipment and energy. However, the collection and conduction of the desorbed gases must be compatible with the need to access the confined area for operation and maintenance.

Non-accessible areas include those the STP operators have limited access to (e.g., confined tanks and closed channels). In such areas, ventilation can be limited and required simply to purge the accumulated gases containing moisture, H_2S, and other sulphur compounds, which can lead to the corrosion of concrete and steel structures. Using covers and ventilation, in turn, prevents fugitive emissions of CH_4 and odourous gases from these areas.

Accessible areas include those the STP operators have unrestricted access to. The ventilation of these areas is essential to ensure appropriate health and safety conditions. Typical odorant gases (H_2S and other sulphur compounds) have specific gravity above that of atmospheric air, remaining in the lower air layers inside confined areas. Consequently, the gases are exhausted from such environments by the bottom preferably being as close as possible to the emitting source (e.g., near the water level).

Completely containing the odours generated in the headworks demands considerable investments, as constructing a shed is required to cover the entire treatment stage. Where strict control of odours is not required in the headworks, an alternative requiring smaller investment should be considered. An alternative to costly sheds is the creation of a zone of negative pressure below the channel slabs near the grids. To promote these negative pressure zones, captor hoods and curtains of PVC strips must be set upstream and downstream of the grids (Figure 9.3). The curtains will reduce the free area of the cross section of the channel, providing higher face velocities of the gas in the openings (WEF, 2004).

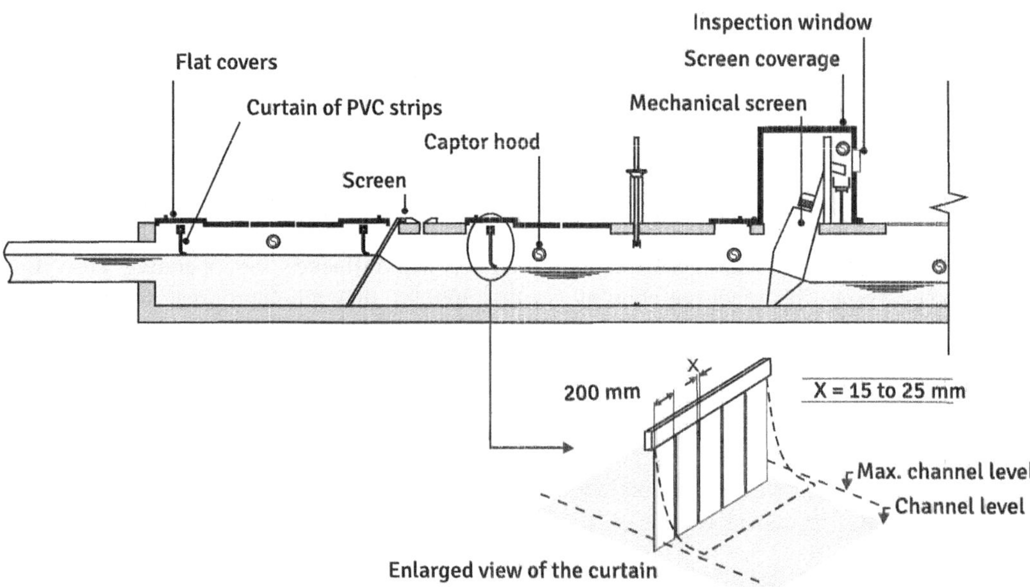

Figure 9.3 Scheme of waste gas containment in the headworks.

Several types of covers are used commonly for waste gas containment. The choice depends on the characteristics listed below (WEF, 1995).

Flat covers (Figure 9.4): These are appropriate for non-accessible areas. They minimize the enclosed volume of waste gas and, consequentially, the extracted volume to be treated. If the thickness necessary to overcome the bending stress is too high, a thinner type of cover (e.g., arch-shaped covers) should be favored. Internal or external structural support can be required, particularly if the covers are used as a walking surface. These covers are applied to channels, distribution boxes, wells, tanks, and the like.

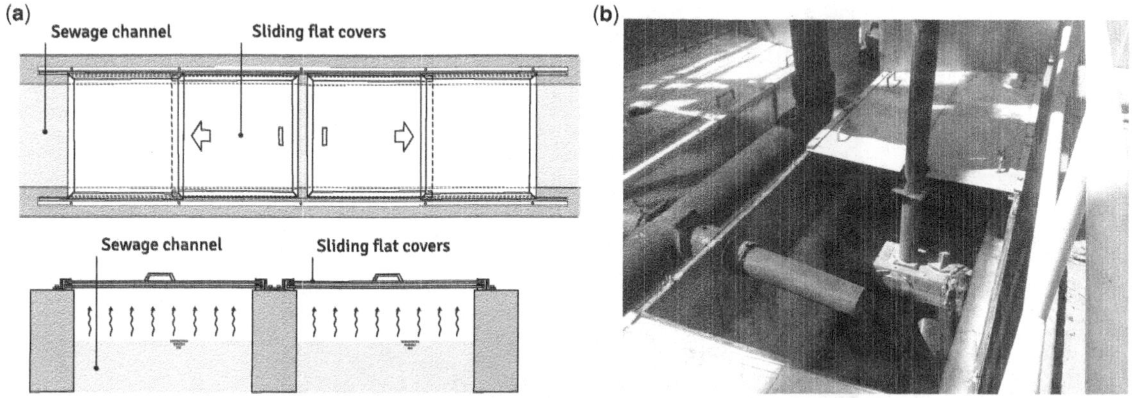

Figure 9.4 Flat covers for containing waste gases: (a) schematic drawing of top and side view in a sewage channel; (b) photograph of a cover placed in a sewage channel at a pumping station.

Arch-shaped covers (Figure 9.5): They are appropriate for non-accessible areas. As these covers are not as thick as flat covers, they could be cheaper. They are applicable to channels and tanks (non-walking surfaces) that are relatively narrow but have a large gap that hampers using flat covers without internal or external support.

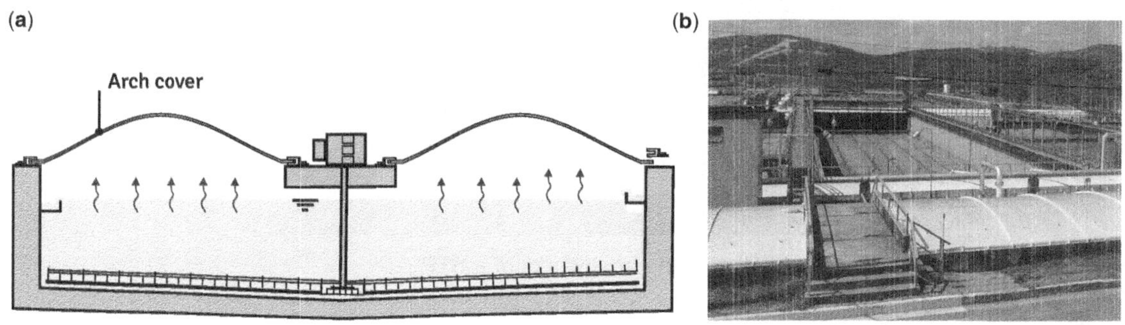

Figure 9.5 Arch-shaped covers to contain waste gases: (a) schematic drawing of a side view in a settler; (b) photograph of a cover placed in a sewage channel in a STP.

Roof-shaped covers (Figure 9.6): They have the same characteristics as arch coverings in terms of applicability, strengths, and weaknesses. However, they have an additional disadvantage, namely, depending on the length of the gap, using additional support and slats in the cover flats could be necessary.

Shed or dome (Figures 9.7 and 9.8): They are appropriate for accessible areas. However, these covers have higher implementation and operational costs. They generate larger flows of waste gas and require more energy for extraction and treatment. Some examples include a building for a certain treatment step, shed to cover a certain area of the STP, dome to cover a settler or a trickling filter, and the like.

In various situations, certain zones of a process unit (e.g., settler outlet channel, effluent discharge from anaerobic reactors) could emit considerably more intense odours, depending on the degree of fluid turbulence. In such areas, using flat or arched covers directly over the emission source can reduce odours considerably without having to cover the treatment unit fully (Figure 9.9).

Figure 9.6 Photograph of a roof-shaped cover placed in a sewage channel in a pumping station.

Figure 9.7 Photograph of a shed covering the headworks of a STP.

Figure 9.8 Dome for waste gas containment: (a) View of a dome over a trickling filter fed by UASB reactor effluent; (b) inside view of the same dome over the trickling filter.

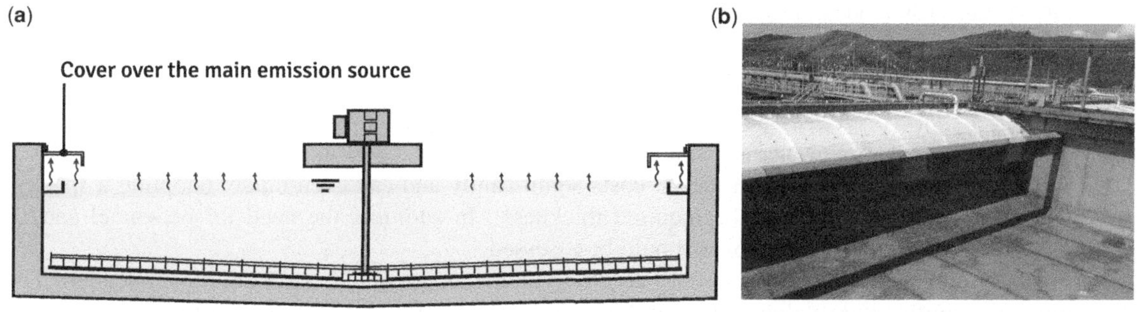

Figure 9.9 Covers for localized waste gas containment: (a) schematic drawing of a side view in a settler outlet channel; (b) photograph of a cover placed in the outlet channel of an empty primary settler in a STP.

Containing the odours and corrosive gases generated in UASB reactors can be done by covering and exhausting the influent distribution box, the effluent collection channel, as well as the settler compartment (Figure 9.10). It is necessary to position makeup air openings (e.g., chimneys) on the reactor slab and captor hoods in the side wall near the water level, aiming at a downflow of gases inside the settler compartment and preventing the build-up of moisture and corrosion under the slab. Alternatively, for small UASB reactors, open joints in precast slab panels are sufficient to allow passive air to enter the settler compartment. Furthermore, when strict control of odours is not required, open-air UASB reactors can be used, combined with scum retention in the settler surface. Recent practical studies have indicated the beneficial effects provided by sulphur-oxidizing microorganisms that grow in scum (Garcia, 2009; Souza *et al.*, 2006).

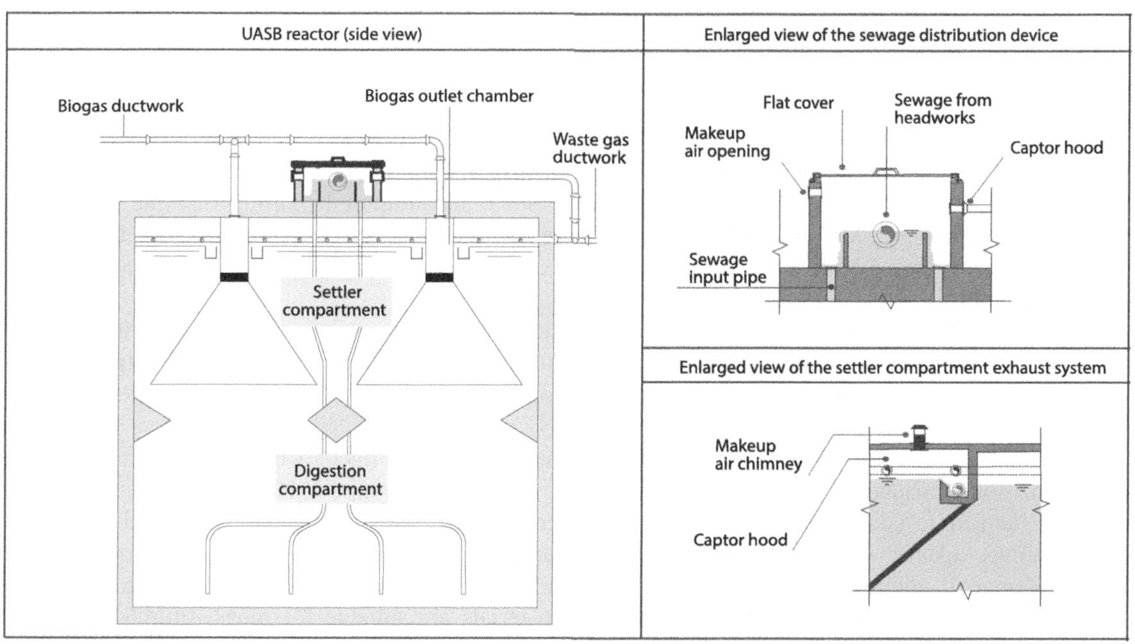

Figure 9.10 Scheme of the containment of corrosive and odorant gases in UASB reactors.

The following factors have to be considered in choosing the different types of coverage (WEF, 2004):

- *Access to the zones to be covered*: Some coverage alternatives do not allow easy access to the covered zones or allow only limited access.
- *Lifetime of the coverage*: The coverage must be corrosion and weather resistant and should be designed to prevent rainwater accumulation on the cover.
- *Structural integrity*: This can influence costs significantly and can even hinder adopting a specific type of coverage (because of the required thickness). In addition, the need for personnel and/or machine transit on top of the cover should be assessed.
- *Costs*: Designing a cover that confines as little volume as possible is strongly recommended. The lower the confined gas volume, the lower would be the investment required for the extraction and treatment systems.
- *Aesthetics*: Maintaining a low-profile arrangement would be aesthetically pleasing and would reduce unwanted attention from the neighborhood.

Selecting appropriate building materials for the covers is crucial in the design of the containment systems, and all the above-mentioned factors are influenced strongly by the selected material. The following materials are commonly used:

- *Concrete*: Features low cost and the best capacity to support weight loads, among other factors. However, concrete modules are heavy and thick, limiting the access of workers. Corrosion is likely to occur and a protective coating is needed (polyurea, epoxy, elastomeric polyurethanes, high-density polyethylene (HDPE) blankets, fibreglass coatings, and the like). Concrete is suitable for the construction of tanks, the structural elements of gas scrubbers, and biological reactors. It should be emphasized, however, that the application of a proper corrosion-proof coating material is as important as the selection of such material, as numerous problems associated with corrosion of concrete structures are related to improper application of the coating material.
- *Aluminium*: Features high cost (not recommended outside the STP because of depredation) but high capacity to resist tensile stresses with thin modules. The modules are lightweight and facilitate worker access. Aluminium has good chemical resistance in headspaces with H_2S, mercaptans, ammonia, and nitrogen compounds. However, it can be affected by mineral acids (HCl, H_2SO_4, and others), as well as caustic soda, in addition to strong alkaline chemicals. Chlorides and hypochlorites (chlorine in general) in the aqueous phase can damage the coating and cause corrosion. It is suitable for covers (doors, windows, and frames) and gas transport elements (pipes and fittings).
- *Stainless steel*: It has roughly the same advantages and disadvantages as aluminium but is more resistant to corrosion and to the degradation caused by most chemicals. Because of its durability and resistance to abrasion and mechanical/thermal/chemical stress, it is suitable for all structures. AISI 316 L stainless steel is recommended highly.
- *Polymeric materials*: These materials feature higher cost but lower weight and extremely high corrosion resistance. They can be produced in different shapes. However, they have low mechanical resistance compared with the other materials, are flammable, and need additional coatings against sun degradation. The most-applicable materials are (in ascending order of preference): high-density polyethylene (HDPE), polypropylene (PP), glass-reinforced plastic (GRP), and polyvinyl chloride (PVC). These materials are suitable for pipes, fittings, channels, tanks, corrosion coatings, casings, and the like.

After confinement/collection, the waste gas should be transported to the treatment units by a system of ducts and exhaust pipes. The design of this system starts with the definition of the exhaust flow, which is a function of three factors:

- *Air changing rate*: A minimum reference range to prevent moisture above the coverage is 3–4 h^{-1} (WEF, 2004). The typical range for non-accessible zones is 4–6 h^{-1} (WEF, 2004). Reference data for accessible zones indicate 10 h^{-1} (WEF, 2004). To suit the exhaust fan models available in the market, in places where the volume to be confined is extremely low (<30 m^3), an air changing rate of 30–60 h^{-1} can be adopted.
- *Enclosure negative differential pressure*: The confined gases should not be allowed to escape to the external atmosphere; therefore, it is necessary to maintain a negative enclosure differential pressure of at least 2.5 mmH_2O. This differential pressure depends on a gas effective face-velocity minimum of 6.4 $m.s^{-1}$ in the cross-sectional area of the cover air inlets and plenums (WEF, 2004).
- *Concentration of H_2S, CH_4, and other gases*: Some treatment systems work better in a specific pollutant concentration range, requiring dilution of the gas to suit the requirements. In addition, the potential flammability because mixing oxygen (from atmospheric air) with CH_4 and H_2S (produced in anaerobic environments) should be evaluated. Therefore, the air change rate should be assessed to maintain the concentration of CH_4 and H_2S outside the rage between the lower and upper explosive limits (LEL/UEL) of 5/15% and 4/44%, respectively. In this assessment, an emission rate of CH_4 and H_2S is considered for the confined zone (some data are shown in Table 9.4) and, subsequently, the exhaust flow benchmarks are estimated.

Table 9.4 Emission factors for H_2S and CH_4 in sewerage and STPs.

Factor	Unity	Data Range	
		Min.	Max.
H_2S released in pumping stations, headworks, and sewage channels in STPs	$mg(H_2S)\ g(S\text{-}SO_4{}^{2-}{}_{inf})^{-1}$	~0	0.5
H_2S released from UASB reactor effluents (from the surface of the settler compartment)	$mg(H_2S)\ g(S\text{-}SO_4{}^{2-}{}_{inf})^{-1}$	2.0	4.2
CH_4 released from UASB reactor effluents (from the surface of the settler compartment)	$mg(CH_4)\ g(COD_{inf})^{-1}$	38	41

(Souza *et al.*, 2012).

After obtaining the exhaust flow, it is necessary to define the capturing hoods, pipe extension, and fittings. Subsequently, the flow velocities and pipe diameters are determined. The ducts must be sloped to allow condensed moisture to be moved to the lower level of the exhaust network. The system comprises a drip-tray, glass trap, air break, and associated drainage pipework. In addition, in-line blast gates, sliding dampers or needle valves have to be set in the pipeline network to allow flow and pressure adjustments. The air-flow rates at each hood/inlet must be checked and, where the system is multi-branched, the air flows must be balanced. Depending on the complexity of the system, the ducting can have ports for in-line flow measurement in the periodic verification of the system. Various instruments, such as several types of anemometers, various Pitot tubes, and various manometers are used to measure air velocity and differential pressure. Several types of materials can be used for the captors, ducts, and fittings, such as aluminium, epoxy coated carbon steel (minimum thickness of coat 400 μm), stainless steel, various polymeric materials (e.g., PP, GRP, and HDPE). The ducts should be sized to maintain the gas face velocity in the range 10–15 m s^{-1} (WEF, 2004).

Finally, it is necessary to calculate frictional head losses along duct walls, dynamic head losses in fittings, and component losses in duct-mounted equipment. In addition, it is necessary to assess head losses in the treatment system (e.g., head losses associated with the packing media of a biofilter, as shown in section 9.5). To determine head losses, ducts and fittings, consulting a reference book on industrial ventilation or air pollution control, e.g., Goodfellow and Tähti (2001), is recommended strongly.

After designing the captor hoods and ducting, it is necessary to select the air mover (fan) that powers the exhaust system. Several types of fan are available:

- *Centrifugal forward inclined blades*: has good extraction capacity, low speeds, and satisfactory performance; however, it is subject to blade abrasion problems (not suitable for high suspended particulate loads).
- *Centrifugal radial blades*: is robust, moves large loads of particles, but has moderate performance and considerable noise level.
- *Centrifugal backward blades*: has high performance, suitable for gases free of particles. It has a low noise level. It is the type of fan most applicable to waste gas containment and extraction systems in STPs.
- *Axial fans*: the impeller is set in-line. They are not used for waste gas exhaustion in STPs but can be used for force-introducing air in the accessible coverings. As regards non-accessible covers, it is not necessary to use forced air in the enclosed zones.

The fan type should be chosen with regard to a set of head losses (static pressure range), waste gas flow, and fan speed. With the fan type selected, the model and the power features can be selected. The model must be chosen using the manufacturers' performance charts. The fan performance chart should be compared with the system characteristic curve, thereby obtaining the operational point represented by the intersection of these curves. To minimize power consumption, the operating point must be located as close as possible to the maximum efficiency point.

The fans should be corrosion resistant (10% sulphuric acid solution) and fitted with a drainage system to prevent condensate from accumulating inside the equipment. However, drains are possible sources that generate odours and they can be controlled by installing a water seal or similar device (e.g., a trap). When the same fan is used for collecting gases at more than one zone, the fan must be designed to provide the conditions necessary for design flow rates to each of the extraction branches. In this way, the head loss in the branch with greater resistance must be calculated, giving in-line blast gates, sliding dampers, or needle valves in the other branches for a controlled balance of the head losses.

9.4.4 Desorption and stripping of gases dissolved in anaerobic reactor effluent

The previously mentioned means for capturing diffuse emissions should be linked with dedicated equipment to effectively remove (oxidize) the desorbed H_2S and CH_4 dissolved in the effluent. Capturing fugitive emissions should be accomplished at certain downstream points where turbulence is likely to occur, favoring their desorption to the atmosphere. However, more-effective measures can be applied to increase the amount of desorbed gases, based on physical separation operations. Therefore, the inclusion of devices that accomplish the controlled stripping or desorption, collection, and treatment of the H_2S and the CH_4 are sound measures to prevent their release to the atmosphere. Desorption or stripping technology facilitates contact between the liquid effluent and gas (usually atmospheric air) to facilitate the separation of dissolved gases from the liquid phase to the gas phase.

Gas desorption from the liquid phase occurs because of a concentration gradient between the two phases. In an anaerobic reactor, CH_4 and H_2S are subject to given partial pressure. Once the effluent is evacuated to atmospheric conditions (zero partial pressure for CH_4 and H_2S), the gases will be transferred to the gaseous phase at a certain rate, depending on the mixing (turbulence) conditions. This is a passive, gradient-driven separation unit operation. In a more effective approach, gas stripping favors desorption by introducing a gaseous (air) stream, generally using a packed column, which increases mass transfer from the liquid to the gas phase.

Both physical operations are favored by a higher temperature because of the increase in the vapor pressure of the dissolved gas and, therefore, its transfer to the gaseous phase. Inversely, a decrease in the system pressure (vacuum) will increase gas transport to the gaseous phase. As regards stripping, higher gas stream flows will improve the removal of the dissolved gases, although at an extremely diluted concentration.

A general description of commonly used equipment for stripping dissolved gases follows. Considering that the physical separation operation, either as a liquid–gas (desorption) or gas–liquid (absorption), is based on the same fundamentals, many equipment arrangements are similar. Figure 9.12 in section 9.4.5 presents three types of gas scrubbers (absorption), mentioned below, which can be regarded as the first three devices for gas stripping. For an extensive overview of technologies for dissolved gas desorption in anaerobic wastewater systems based on various process scales and applications, see Heile *et al.* (2017).

Packed columns: Columns with packing media are used to improve the contact of the liquid and gaseous phases, either in a countercurrent or concurrent pattern in vertical arrangements. The liquid enters at the top of the column and flows down through the packed bed, in such a way that it is exposed to the gaseous stream over a large surface, provided by the packing material. These devices usually have lower pressure drops (around 400 Pa m^{-1}) and lower mass transfer compared with tray columns, and they are suited to low gas to liquid volumetric ratios (G/L).

Tray columns: Tray or plate columns are often considered alternatives to packed towers as a gas to liquid contact device. In a tray or plate column, the liquid is introduced at the top of the vertical device and is brought into contact with a gaseous phase by a series of stages, with each stage being formed by two trays (upper and lower). The gas circulates upward, through orifices in the trays that are flooded to some extent. The overall effect is a multiple countercurrent contact between the gas and the liquid and higher mass transfer, albeit resulting in an increased drop in pressure. This equipment is well suited to handling high G/L ratios.

Spray towers: These devices can be considered a simpler version of a packed or tray tower, as they have a full void volume (no internals other than a water distributor on top and means for air injection and water discharge at the bottom). They have lower mass transfer and gas pressure drops; however, pumping energy is needed to introduce water at a given pressure to produce small droplets and enhanced mass transfer.

Bubble (diffused aeration) tanks: This arrangement consists of a tank (or a series of tanks), in which a stream of air in the form of fine bubbles is introduced at its bottom. The air diffuser should provide bubbles with a high surface to volume ratio to increase the interfacial contact surface. The drop in gaseous pressure is high, related directly to depth of the water column.

Mechanically stirred tanks: In this device, usually associated with a cover at vacuum pressure, desorption is increased because of the high turbulence provided by mechanical mixers. High mixing patterns are required to achieve effective exposure of the bulk liquid to the atmospheric (or head space) conditions.

Membrane contactors: Although commercially available for separation of liquid and high-value gas, membranes have only been applied for anaerobic effluents at pilot scale (Cookney *et al.*, 2016). A compact footprint and concentrated gas phase are the main advantages of these arrangements. In addition, high mass transfer is achieved when using microporous membranes, although fine pre-filtration is needed to prevent fouling.

Data on the desorption and stripping of dissolved CH_4 and H_2S from effluents from anaerobic-based STPs are limited. Matsuura *et al.* (2015) investigated the desorption and biological oxidation of CH_4 dissolved in the effluent of a UASB pilot reactor (155 L volume) treating municipal wastewater. The system consisted of two columns filled with a polyurethane sponge (80 L volume each). In the first one, CH_4 desorption was carried out with recovery efficiencies in the range 57–88%, with CH_4 concentrations greater than $30\%_{v/v}$. This was accomplished with extremely low G/L ratios (0.05 and 0.07). The liquid phase was directed to a second reactor, where the residual dissolved CH_4 and chemical oxygen demand were removed, reaching elimination efficiencies up to 99% and 88%, respectively. Notably, these authors worked with full effluent post-treatment that additionally recovered dissolved CH_4.

Glória *et al.* (2016) evaluated two alternatives for desorption of CH_4 and H_2S dissolved in the effluent of two identical UASB pilot reactors (360 L volume) treating sewage. In the first one, stripping air was injected at four different points in the settler compartment (GLS separator) at 15 cm below the liquid level, with three air flows (213, 160, and 107 L min^{-1}). With this system, approximately 30% of CH_4 and 40–60% of H_2S were desorbed, with concentrations in the gas stream being between $0.5\%_{v/v}$ and $1\%_{v/v}$ CH_4 and 90 ppm_v to 200 ppm_v H_2S. The second reactor had a hydraulic dissipation unit with exhaust air extraction, in which the reactor effluent flowed by gravity (two heights of 0.55 m and 1.10 m). Three air exhaust flows of 1.6 L min^{-1}, 1.2 L min^{-1}, and 0.8 L min^{-1} were evaluated. This system was more efficient, reaching a percentage of recovery greater than 60% for the two gases, with $1\%_{v/v}$ to $4.5\%_{v/v}$ CH_4 and 100 ppm_v to 500 ppm_v H_2S in the exhaust gas. At a height of 1.10 m and an exhaustion rate of 12 h^{-1}, removal up to 73% and 97% for CH_4 and H_2S were achieved, respectively. The author recommends this hydraulic dissipation unit for possible application in real-scale reactors.

In a pilot-scale biofiltration arrangement, Huete *et al.* (2018) employed a packed tower for stripping CH_4 and H_2S from a fraction of the effluent of a full-scale UASB reactor. Two operating conditions were applied, with a G/L ratio of 1 and 0.5 (0.9 m^3 h^{-1} and 1.8 m^3 h^{-1} water flow, respectively). In both conditions, CH_4 recovery was 99%, with lower values for H_2S (30%) because of the high effluent pH (above 8).

As regards membrane contactors, few laboratory-scale studies have been published for degassing anaerobic effluents. Bandara *et al.* (2011) evaluated a hollow-fiber membrane for recovering dissolved CH_4 from the effluent of a laboratory-scale UASB reactor treating synthetic wastewater, with 89% to 97% recovery efficiencies being achieved. The same group (Bandara *et al.*, 2012) applied a similar membrane to a UASB fed with low-strength municipal wastewater, obtaining CH_4 recovery efficiencies of 40% (winter) to 60% (summer). Cookney *et al.* (2012) applied this technique to a pilot-scale expanded anaerobic granular sludge bed (EGSB) fed with municipal sewage. The coupled membrane contactor, in which nitrogen was used as the sweep gas phase, achieved 72% recovery efficiency of dissolved CH_4.

9.4.5 Treatment of collected gaseous emissions

Once the above-mentioned steps have accomplished capture, desorbing or stripping, and extraction, the waste gas produced must be conducted to a treatment area for the abatement of odour and greenhouse gases (CH_4). The selection of the technology will depend on whether the recovered gas will be used to recover heat, electrical power, or to limit GHG emissions and odour complaints. This generally depends on the quality of the gas and the scale of the operation.

A range of technologies can be used for abatement of H_2S in STPs, some of which have the potential to treat CH_4 as well. This range of technologies is based on physical, chemical, and biological methods (Figure 9.11). Current experience in the design and operation of full-scale odour abatement systems indicate physicochemical absorption, the adsorption process, and biofiltration (biofilters and biotrickling filters) as the most widely used technologies. The bioscrubber, a system consisting of a water absorption operation (usually in a packed column), followed by a biological oxidation reactor is seldom applied in wastewater facilities. However, it can be selected for desulphurization of high biogas flows. In a comparative analysis, biological techniques present lower operating costs (up to six times) and lower sensitivity toward H_2S concentration compared with their physical/chemical counterparts (Estrada *et al.*, 2012). Furthermore, in terms of environmental benefits associated with some of these technologies, a life cycle assessment indicated that biological alternatives for odour control attained the best environmental performance (Alfonsín *et al.*, 2015).

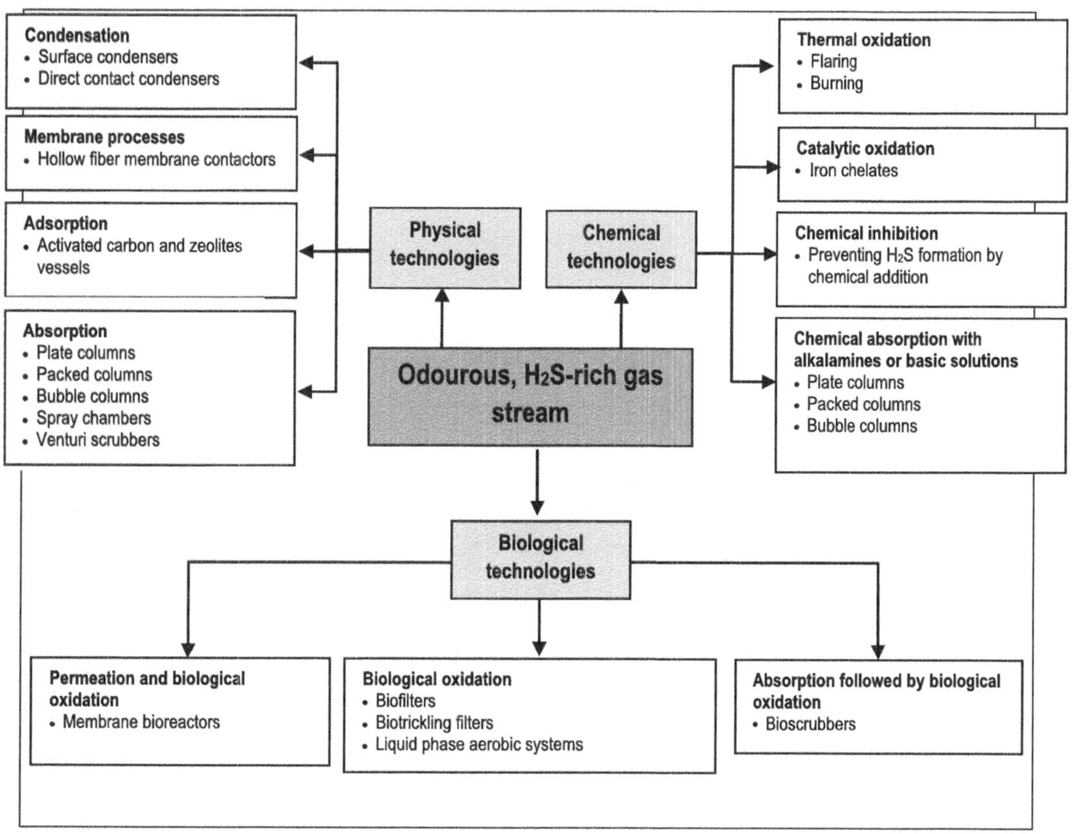

Figure 9.11 Technologies for the abatement of odorous gases in STPs.

Because of the low affinity of CH_4 with the chemical and adsorption approaches used commonly in odour abatement technologies, two-stage systems comprising physical separation followed by biological treatment appears a promising alternative to the treatment of low waste gas flows containing low CH_4 concentrations (Brandt *et al.*, 2016; Huete *et al.*, 2018; Limbri *et al.*, 2013; Melse & van der Werf, 2005; Nikiema *et al.*, 2007; Veillette *et al.*, 2012). For methane-rich streams generated in desorption units (e.g., $>20\%_{v/v}$), direct combustion and recovering the energy can be considered (e.g., by mixing the desorbed stream with the biogas).

Physicochemical absorption systems are designed to transfer (or react to) the pollutants from the gas phase to the liquid (absorbent) phase and are referred to commonly as scrubbers. They are used for removing highly soluble gaseous compounds. Gas scrubbers can be classified as packed columns, tray or plate columns, or spray chambers (Figure 9.12). They are compact systems (extremely low empty-bed residence time) that are stable regardless of load variations, but within prescribed flow limits to avoid flooding and foaming problems. However, the high operational costs because of the chemicals required and the high demand for energy limit their suitability to larger STPs (up to 100,000 inhabitants). The most commonly used absorptive liquid is water. To enhance the ionization and oxidation of the odorant compounds, chemicals must be added to the water (e.g., sodium and potassium hydroxides, sodium hypochlorite, hydrogen peroxide, and the like). Acids are used for the abatement of nitrogen compounds such as ammonia. Bases are used for the abatement of sulphur compounds, and oxidants to react with a wider range of odorant compounds. The demand for chemicals of each absorption system can be estimated by the stoichiometry of the chemical reactions involved or from the speciation diagrams. In packed, tray, or spray towers, the ratio of absorbent flow/gas flow ranges from 0.8 to 2.5 $m^3 H2O \cdot 1{,}000\ m^{-3}$gas. The range of empty-bed residence time is 20–90 s (Kohl & Nielsen, 1997).

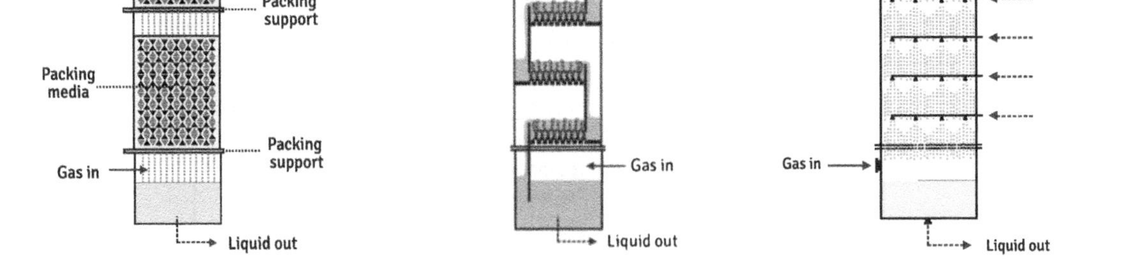

Figure 9.12 Gas scrubbers for waste gas treatment in STPs: (a) packed column; (b) tray or plate column; (c) spray chamber.

Gas adsorption corresponds to the transfer of a pollutant from a gas phase to a solid phase. In adsorption, the pollutant concentrates on the surface of the solid as a result of the chemical forces existing on its surface. The pollutant in the waste gas is concentrated selectively on the adsorbent, whereas the purified gas passes through the bed. Figure 9.13 shows an adsorber vessel for the treatment of waste gas in a STP. The adsorbent quality is related directly to the surface area available for adsorption. Generally, commercial adsorbents are materials with a large surface area per unit weight. The adsorbent used most commonly in waste-gas treatment processes is granular activated carbon impregnated with KI, K_2CO_3, or $KMnO_4$. The adsorbent is used in layers ranging from 0.15 m to 0.90 m in height. The waste gas is fed in the adsorber vessel at face velocities ranging from 0.15 to 0.30 $m \cdot s^{-1}$. Head losses range from 20 to 250 mmH_2O (Kohl & Nielsen, 1997). The duration of use of the activated carbon is a function of the saturation curve and the applied volumetric load (variable, depending on the material, or even between the various sources of the activated carbons). Once saturated, the activated carbon must be replaced or regenerated, which raises the operating costs substantially, although the adsorber vessels present extremely low operational complexity. Because of the need for substitution/regeneration of the adsorbent, the units are usually operated in parallel, thereby preventing interruption of the overall operation.

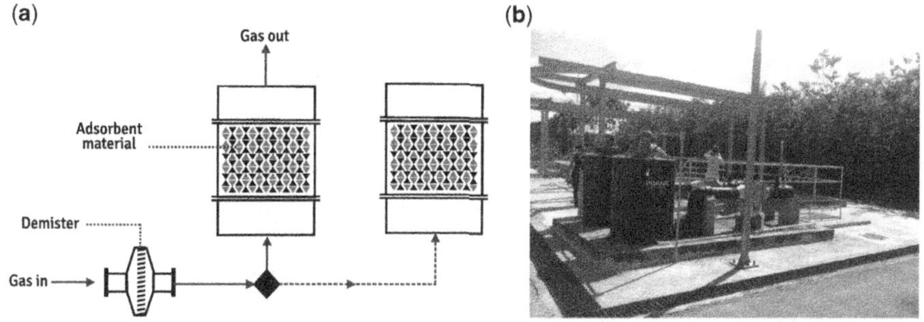

Figure 9.13 Adsorber vessels for waste gas treatment in STPs: (a) schematic drawing; (b) photograph of an adsorber system (impregnated activated carbon) used for odour abatement in a pumping station.

In a broad sense, biofiltration consists of a fixed-bed reactor, in which the waste gas is blown through the packing media (upstream or downstream) containing the adhered biomass that is often humidified. An enriched aerobic group of bacteria develops on the biofilm and converts H_2S to S^0 or SO_4^{2-}. These bacteria are affiliated to diverse genera, such as obligatory chemolithotrophs (*Thiobacillus* and *Thiomicrospira*), facultative chemolithotrophs (*Sulfolobus*, *Thermothrix*, and *Thiobacillus*), and litho-heterotrophs (*Beggiatoa* and *Thiobacillus*). The most reported genera in biofilters for the biological oxidation of H_2S are *Thiobacillus* and *Beggiatoa* (Pérez & Villa, 2005). The oxidation of reduced sulphur compounds can generate acidic conditions to which some *Thiobacillus* species have adapted. As regards biofilters for CH_4 abatement, an enriched aerobic methanotrophs community develops on the biofilm and converts CH_4 via methanol to formaldehyde, which serves as an intermediate in catabolism (resulting in its complete oxidation to carbon dioxide (CO_2) and anabolism) (Jiang *et al.*, 2010). *Methylocystis*, *Methylosinus*, *Methylomonas*, *Methylosarcina*, and *Methylocaldum* are commonly the dominant genera in biofilters for CH_4 abatement (Gebert *et al.*, 2008; Kim *et al.*, 2012, 2013, 2014a; Nikiema *et al.*, 2005; Wang *et al.*, 2008).

Biofiltration of gaseous streams can be accomplished with three main arrangements, as shown in Figure 9.14. In the biofilters (Figure 9.14(a)), the packing media consists of natural/organic materials such as soil (peat) or compost, providing both the microorganisms and the nutrients for effective metabolism. Inorganic and/or synthetic materials (e.g., rocks, clays, plastics, and the like) are added to these natural/organic materials for structuring of the filter bed, thereby increasing its permeability (preventing excessive head losses).

(a)

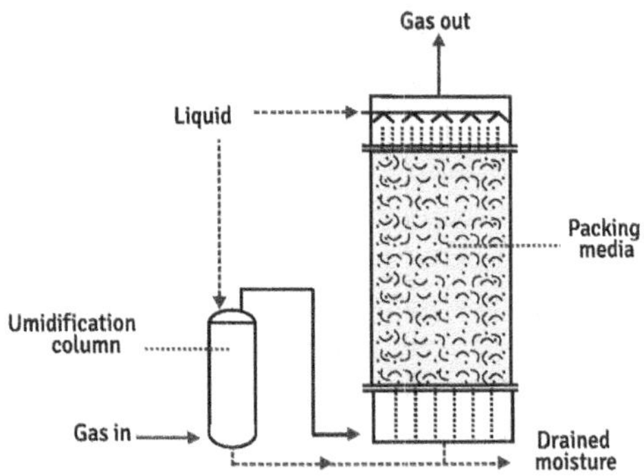

Main characteristics of biofilters:

- Packing media contains organic materials (soil, peat, compost) in size from 2 mm to 20 mm usually mixed with wood chippings, and non-organic or larger materials (up to 35 mm) such as expanded clay, silicates, ceramic material, plastic, fibers etc.
- Filter bed moisture is maintained by pre-humidification of the waste gas stream or by intermittent addition of liquid on the filter bed at a very low rate, generally less than 0.004 $m^3\ m^{-2}\ h^{-1}$.
- Nutrients generally are not added to the liquid phase, depending on the kind and quantity of organic materials (source of nutrients) in the filter bed.
- The filter bed provides high head losses (up to 1,200 $mmH_2O\ m_{height}^{-1}$) so rather small bed packing heights should be used.
- They have high area requirements because the low filter beds depth rarely exceed 1m to avoid high head losses.

(b)

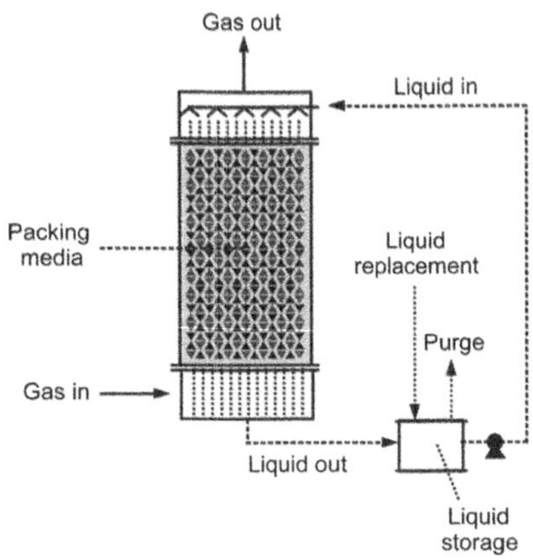

Main characteristics of biotrickling filters:

- Packing media contains mainly non-organic materials (gravel, expanded clay, silicates, ceramic material, plastic, sponges etc.) in variable sizes (usually up to 35mm).
- Packing material should favor biofilm adhesion
- Liquid phase is applied directly to the filter bed at high rates (0.05 to 20 $m^3\ m^{-2}\ h^{-1}$), requiring continuous circulation guaranteed by external pump and tank.
- Usually nutrients are added to the liquid phase.
- The filter bed provides low head losses (up to 30 $mmH_2O\ m_{height}^{-1}$).
- They have low area requirements because the filter beds are designed considering depth greater or equal to 1m.

(c)

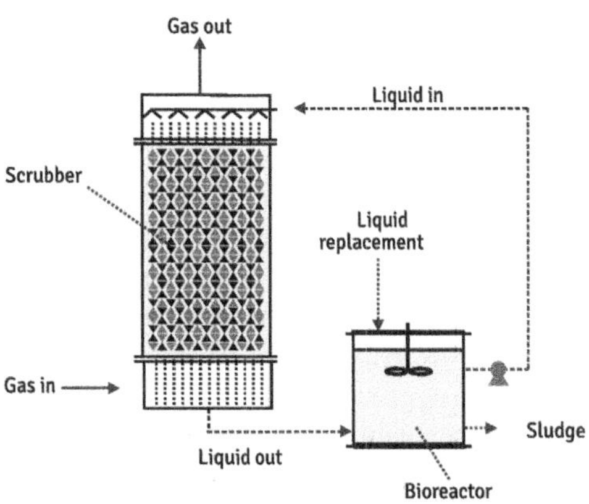

Main characteristics of bioscrubbers:

- The gaseous flow passes through a plate column or a packed column filled with Raschig rings, Pall rings, Intalox saddles (made of plastic, ceramic, or metallic materials).
- Liquid phase (absorbent) is applied in the scrubber in countercurrent flow at high rates (above 10 $m^3\ m^{-2}\ h^{-1}$), requiring continuous circulation guaranteed by external pump and tank. The rate of application of the liquid phase is determined using the absorption capacity of the pollutant by the liquid phase.
- The liquid absorbent is bio-regenerated in a side bioreactor.
- Usually nutrients must be added to the liquid phase.
- The scrubber provides low head losses (up to 30 $mmH_2O\ m_{height}^{-1}$).
- They are indicated for the treatment of biodegradable gases with high solubility in the liquid phase (absorbent).

Figure 9.14 Reactor configurations mostly used for waste gas treatment in STPs: (a) biofilter; (b) biotrickling filter; (c) bioscrubber.

In several studies and practical applications, the natural/organic materials have been replaced by inorganic/synthetic materials, resulting in the biotrickling filter concept (Figure 9.14(b)). In this arrangement, an inoculum is needed because of the inert packing material, i.e., the addition of nutrients by continuous liquid-phase recirculation. This recirculation, in turn, facilitates improved absorption of pollutants because of the continuous mass transfer from the gas stream to the liquid-phase/biofilm (Kennes & Veiga, 2001; Kennes *et al.*, 2009a, b). As regards CH_4, the effect can be hindered, because of its low aqueous solubility. In this regard, in a study using a compost biofilter, diffusional internal mass transport was identified as the rate-limiting step (Gómez-Borraz *et al.*, 2017) for CH_4 oxidation at 4% concentration in air. To deal with this, some researchers have evaluated strategies, such as the addition of mass transfer vectors (e.g., silicone oil, soybean oil, non-ionic surfactants, and the like) in the liquid phase of the biotrickling filters, aiming at increasing the solubility and diffusivity of CH_4 in the biofilm (Kraakman *et al.*, 2011; Lebrero *et al.*, 2015; Muñoz *et al.*, 2007; Ramirez *et al.*, 2012a, b; Rocha-Rios *et al.*, 2009). Data from the literature show that mass transfer vectors can increase CH_4 removal efficiencies in biotrickling filters by up to 20% (Rocha-Rios *et al.*, 2009); however, contradictory studies have reported deleterious effects in using such vectors (Ramirez *et al.*, 2012a, 2012b), which demonstrates the need for scientific in-depth investigation of the subject.

A third biofiltration arrangement suitable for soluble and biodegradable gases is the bioscrubber (Figure 9.14(c)). Bioscrubbers contain two treatment units, one for the absorption of pollutants and the other for their biological degradation. The first unit is usually a tray or packed column. The second is an aerated bioreactor, where the liquid containing the dissolved gases comes into contact with the biomass suspended in or adhered to a submerged medium (Edwards & Nirmalakhandan, 1996; Kennes & Veiga, 2001; Syed *et al.*, 2006).

If the anaerobic-based STP contains an activated sludge system for the post-treatment of the UASB reactor effluent, a simple and economical possibility for odour control in the waste gas is sparging it at the bottom of the activated sludge reactor. The aerobic biomass in the reactor will oxidize the sulphides and other odorant compounds biologically.

9.5 BIOFILTERS AND BIOTRICKLING FILTERS FOR WASTE GAS ABATEMENT

As previously mentioned, the most frequently used techniques for H_2S abatement in STP are absorption, adsorption, and biofiltration. However, for economic considerations (CAPEX/OPEX expenses, and energy and chemical requirements) employing biofilters and biotrickling filters is usually favored, particularly when suitable space is available (Chernicharo *et al.*, 2010; Estrada *et al.*, 2012). As biofilters and biotrickling filters are applicable also to the abatement of CH_4-containing gases (whereas the other techniques, such as bioscrubbers, are limited because of low CH_4 solubility), we present design and constructive guidelines only for these treatment systems in this section.

Biofiltration was used initially in Europe and the United States for the abatement of odorant emissions, using soil filter beds designed employing empirical data (Pomeroy, 1957; Prüss & Blunk, 1941). However, over the last few decades, the scientific deepening of the process has led to the development of new concepts and design parameters, extending the application of the technology to the abatement of other non-odorant compounds. The aim of such studies was to design bioreactors for the removal of one or several components of a gas stream, combining conditions suitable for microbial action at the lowest cost.

Biofiltration setup includes all or some of the following parts: (i) Gas inlet system; (ii) gas distributor in the bottom; (iii) drainage system; (iv) support for the packing media; (v) packing media; (vi) humidification system; (vii) coverage for the collection of the cleaned gas stream. Considering the diverse combinations of the above parts, biofilters can be classified into three main types:

9.5.1 Unstructured (simpler) biofilter

Consists essentially of a basin dug in the soil, which has a network of perforated pipes (waste-gas inlet system) in the bottom, surrounded by gravel or silica-rolled pebbles for the distribution of the gaseous stream (Figure 9.15(a, b)). In addition to the distribution of gases, the gravel/pebbles are necessary to support the packing media (placed on top) and to allow the drainage of moisture. In most instances, the basin is sealed with a polymer geomembrane, concrete, or clay to protect the water table and surrounding ground. Typically, the openings in the distribution network are made in the lower half of the tubing, vertically aligned at 45°. The size and spacing between the openings are determined by the gas flow rate and the expected head losses. In addition, openings in the lower part of the main distribution pipes must be provided to drain the condensed liquid. Normally, this type of biofilter is not covered and the humidification of the packing media is done naturally by rain, but a wetting system (such as spray nozzles) should be considered for improved control. The excess moisture should be collected in the bottom by an air-tight drainage system associated with a trap or other water sealing device. The drained liquid should be returned to some unit of the sewage treatment system (the alkalinity of the sewage is usually sufficient to neutralize it).

(a)

(b)

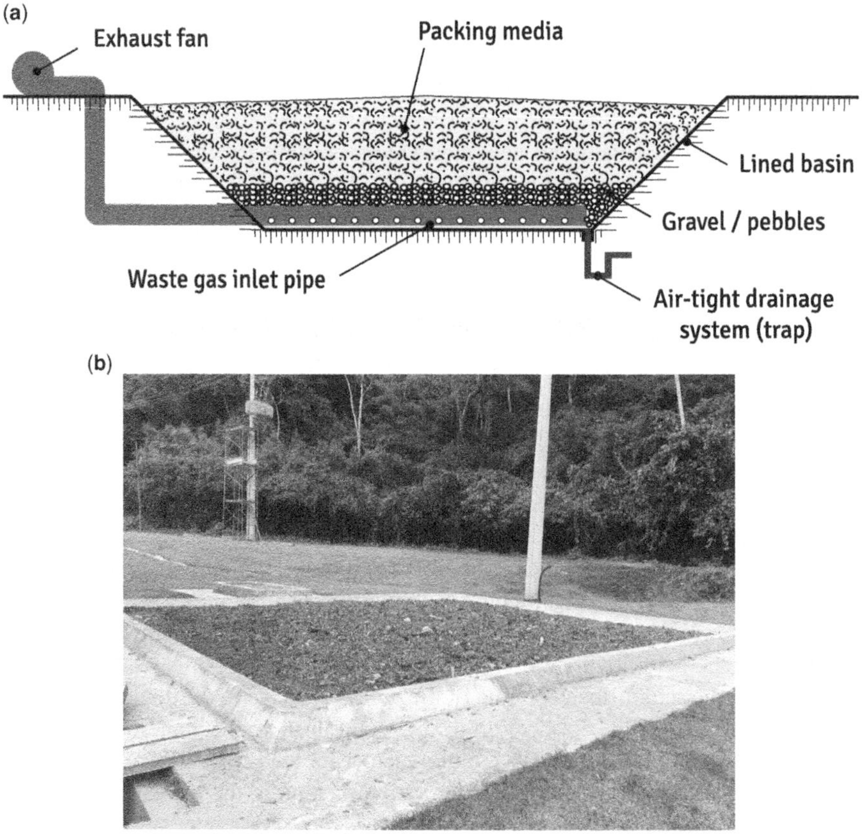

Figure 9.15 Unstructured biofilter: (a) schematic drawing; (b) photograph of a biofilter in a STP.

9.5.2 Structured biofilter

This configuration is quite similar to the unstructured one, but the waste gas is blown through a void plenum (without gravel or pebbles). The packing media is placed over a grid or perforated precast concrete slab (Figure 9.16(a, b)). To support the packing media, the walls and bottom of the biofilter are usually built from concrete coated with a corrosion-resistant material. Likewise, the support for the packing media must be corrosion resistant. This type of biofilter is usually open to the air; however, installing a roof can be considered to improve the control of moisture in the packing media. In such instance, water is added by lawn sprinklers to the top of the packing media or by perforated hoses inside the packing media.

(a)

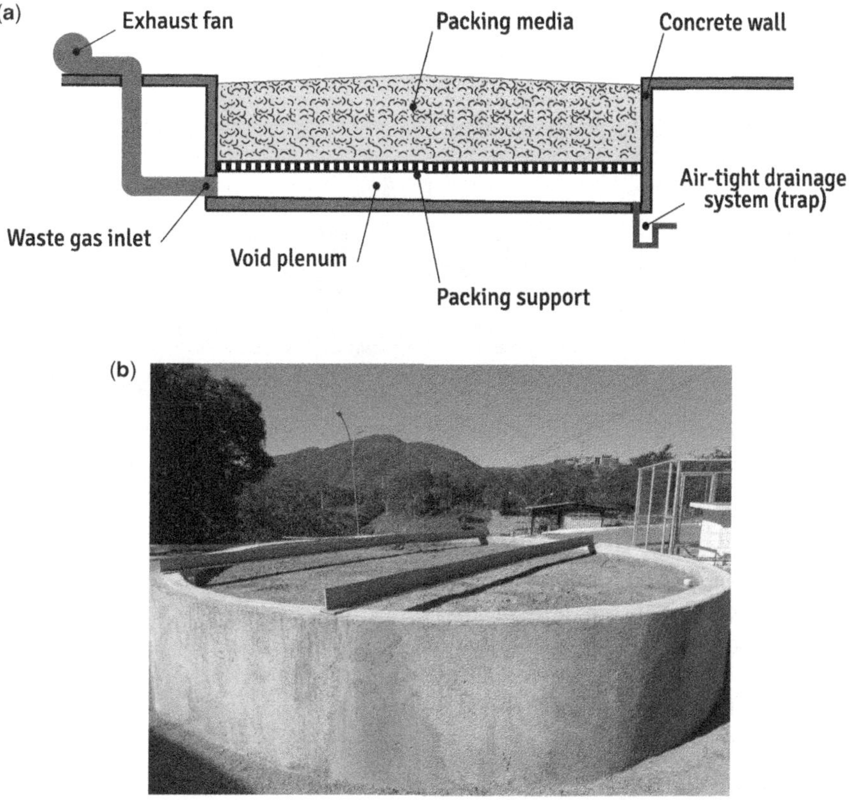

(b)

Figure 9.16 Structured biofilter: (a) schematic drawing; (b) photograph of a biofilter in an STP.

9.5.3 Modular biofilter

Compact structured configuration that also features a void plenum. Consists of a completely closed unit, which facilitates control of the temperature and moisture of the packing media (Figure 9.17(a–c)). Water is added by sprinklers or spray nozzles to the top of the packing media or by perforated hoses inside the packing media. The humidification system is automated generally by a programmable logic controller coupled to a soil hygrometer. Because the system is closed, the exhaust fan can be placed downstream of the biofilter to prevent corrosion.

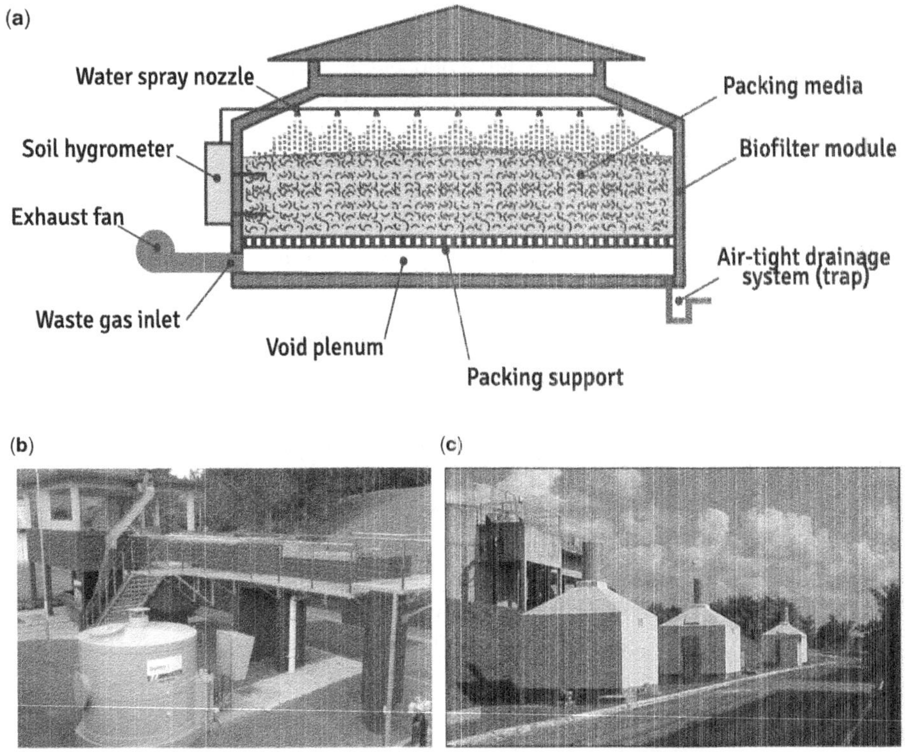

Figure 9.17 Modular biofilter: (a) schematic drawing; (b,c) photographs of biofilters treating waste gases from preliminary treatment unit and UASB reactors.

9.5.4 Features and design of biofilters and biotrickling filters

In bioreactors, the packing media are intended to support biomass growth, retain moisture and, in some instances, provide nutrients for microorganisms. Selecting a filter bed depends on the biological, hydrodynamic, and economic criteria presented in Table 9.5. In addition, various properties of the packing media shown in Table 9.5 are crucial elements in bioreactor operation, of which details are presented in sequence. The depth of the filter bed ranges is:

- From 0.6 to 1.2 m (frequently being 1.0 m) in biofilters
- Up to 3 m in biotrickling filters.

The fibrous/inert/synthetic materials used in biotrickling filters and biofilter beds (to improve porosity and structure) are volcanic rocks, perlite, expanded clays, silicates, ceramic or glass materials, plastic materials, fibrous peat and other synthetic fibrous materials (resins), or natural fibrous materials originating from coconut, sugar cane, and eucalyptus, and wood chips, among others. The mixture of fibrous/inert/synthetic materials and organic/natural substrates provides hydrodynamic conditions to the filter bed, maintaining the source of microorganisms and nutrients. The volumetric ratio ranges (Deshusses *et al.*, 1995; Kennes & Veiga, 2001; Tchobanoglous *et al.*, 2013) are:

- From 40 to 80% of organic/natural substrates
- From 20 to 60% of fibrous/inert/synthetic materials.

Table 9.5 Criteria used in the selection of packing media for biofilters and biotrickling filters for waste gas abatement in STPs.

Rationale	Criteria	Key Factors	Recommendations
Biological performance	Area available for biofilm growth	• Particle size • Specific area	• *Particle size*: Varies widely depending on the materials used. When using compost, peat, or soil, the range is 2–12 mm, and can contain larger composted fragments (10–20 mm) that aid in the hydrodynamics of the filter bed (Caron *et al.*, 2005; Morgan-Sagastume *et al.*, 2003; Plessis *et al.*, 2003; Wang *et al.*, 2011). When inorganic and/or synthetic materials are added to improve the porosity and structure of the packed bed, the range increases up to 35 mm when using Pall Rings, Raschig Rings, Jaeger Tri-Packs, polyurethane sponges, and expanded clays, among others (Cox & Deshusses, 2002; Gabriel & Deshusses, 2003; Ramirez *et al.*, 2012b; Shareefdeen *et al.*, 2003; Zhang *et al.*, 2009). • *Specific area*: 300 to 1,000 $m^2 \cdot m^{-3}$ (Perry & Green, 2007); however, some research groups have already used materials with specific areas of up to 1,360 $m^2 \cdot m^{-3}$ (Nikiema & Heitz, 2010).
	Nutrients for microorganisms	• Chemical composition (C/N/P)	• *Chemical composition*: C/N/P ratio in the range 70-230/5-15/1-3 (Barona *et al.*, 2004; Pawłowska *et al.*, 2011; Rocha, 2007).
	Water holding capacity	• Moisture	• *Moisture*: The materials must maintain 40–60% moisture (Morales *et al.*, 2003), although several studies recommend lower values (10–50%) for the biofiltration of gases with low solubility and diffusivity in water (e.g., CH_4) (Bender & Conrad, 1995; Christophersen & Kjeldsen, 2001; Humer & Lechner, 1999; Visvanathan *et al.*, 1999; Wang *et al.*, 2011).
	Ability to remove by-products from microbial metabolism	• Moisture • Filter bed porosity	• *Filter bed porosity*: 0.5 to 0.9, which are the highest values for inert and/or synthetic materials (Devinny *et al.*, 1998).
	Buffering capacity	• pH	• *pH*: In general, the materials should be capable of buffering the pH in the range 6.0–8.0 (Andres *et al.*, 2014). However, the need for pH buffering depends on

(*Continued*)

Table 9.5 Criteria used in the selection of packing media for biofilters and biotrickling filters for waste gas abatement in STPs (*Continued*).

Rationale	Criteria	Key Factors	Recommendations
			the target microbial consortia. pH values as low as 1.0–4.0 can be used in the biofiltration of sulphur compounds, because acidophilic bacteria (e.g., *Acidithiobacillus thiooxidans* and *Thiobacillus thiooxidans*) can oxidize H_2S in such harsh environments (Chaiprapat *et al.*, 2011; Duan *et al.*, 2006).
Hydrodynamic performance	Packing media compaction resistance	• Mechanical resistance • Biodegradability	• *Mechanical resistance and biodegradability*: There are no indications for typical values. They must be determined empirically by the operation of bench-scale or pilot-scale reactors in long-term studies (Hartikainen *et al.*, 2002).
	Less resistance to gas flow	• Particle size • Filter bed porosity • Head loss	• *Head loss*: Varies widely depending on the particle sizes and the face velocity in the bioreactors. When using smaller organic/natural materials, such as compost, peat or soil, it varies from approximately 15 to 1,200 $mmH_2O \cdot m^{-1}$ (Carlson & Leiser, 1966; Moraes Junior, 1991; Webster *et al.*, 1997; WEF, 2004). When mixing with larger materials such as rocks, expanded clays, wood chips, activated carbon, and the like, it varies from approximately 5 $mmH_2O \cdot m^{-1}$ for particles larger than 12 mm to 600 $mmH_2O \cdot m^{-1}$ for particles smaller than 3 mm (Ramirez *et al.*, 2012b; van Langenhove *et al.*, 1986; Webster *et al.*, 1997; Yang & Allen, 1994). With a mixture of organic/natural materials and larger materials, intermediate values are observed, namely, 10–30 $mmH_2O \cdot m^{-1}$ (Kennes & Veiga, 2001; Wu *et al.*, 1999). Using sponges and fibrous materials can lead to extremely low head losses, ranging from less than 0.5 to 30 $mmH_2O \cdot m^{-1}$ (Estrada *et al.*, 2014; Maia, 2003; Pantoja Filho, 2008; Rocha, 2007; Rodrigues, 2002).
Economics	Ease of purchase and lower costs	• Cost	• *Cost*: Inorganic/synthetic materials generally require investments more than ten times greater than the investments required for the use of organic/natural materials (soil, composted waste, peat, wood chips, and the like) (Kennes & Veiga, 2001).

The moisture of the filter bed is a fundamental parameter in the biofiltration process. It is related directly to microbial activity (which occurs only in the presence of water), transportation of substrates, nutrients, and products of the biological activity in the biofilm, and the hydrodynamic of the filter bed. Lack of moisture can reduce the microbial activity and cause filter bed cracks, leading to preferential flow pathways, particularly in the first layers of the bed. On the other hand, excess moisture can reduce the permeability of the filter bed, increasing head losses and inducing the generation of stagnant zones in the biofilm. For the maintenance of the moisture of the packing media (see data in Table 9.5), the rate of water addition in the biofilters is (WEF, 2004):

- Generally, less than 95 $L \cdot m^{-2} \cdot d^{-1}$
- More frequently in the range 20 to 30 $L \cdot m^{-2} \cdot d^{-1}$.

Both the mass transfer of the pollutant between the gas and the liquid/biofilm and the rate of biological degradation are affected by the composition of the gas stream. The flow rate of the gas stream determines the face velocity and the empty bed residence time (EBRT) and is related directly to the removal efficiency of the pollutant. In general, biofiltration systems in STPs are not technically or economically feasible for treating gaseous stream flows larger than 90,000–100,000 $m^3 \cdot h^{-1}$ (face velocities up to 150 $m^3 \cdot m^{-2} h^{-1}$) or above loads of 5–6 $g \cdot m^{-3}$, obviously depending on the characteristics of the pollutants (Chernicharo *et al.*, 2010; Kennes & Veiga, 2001; Noyola *et al.*, 2006).

In this sense, Table 9.6 shows the main parameters for proper operation of biofilters for H_2S and CH_4 abatement. Table 9.7, in sequence, shows the main parameters for biotrickling filters. However, the data depicted in Table 9.7 for biotrickling filters used in CH_4 abatement should be regarded with caution. Improved performance has been observed in the biotrickling filters of Nikiema and Heitz (2009), Veillette *et al.* (2012), and Kim *et al.* (2013), but all these filters were packed with innovative materials, not disclosed because of industrial/patent property issues. Accordingly, designing biotrickling filters with these parameters appears impractical. Therefore, considering the current stage of technology development, it is not recommended to use biotrickling filters for CH_4 abatement.

Table 9.6 Design/operational parameters and performance of biofilters for the oxidation of H_2S and CH_4 in waste gases generated in STPs.

Parameter	Typical Range	
	H_2S	**CH_4**
Influent concentration	~1–2,650 ppm_v	0.02–4.00$\%_{v/v}$ (200–40,000 ppm_v)
Volumetric load	<0.1–865 $gS\text{-}H_2S \cdot m^{-3} \cdot h^{-1}$	0.1–80.0 $gCH_4 \cdot m^{-3} \cdot h^{-1}$
Empty-bed residence time	15–130 sec	20–80 min
Face velocity	5–115 $m \cdot h^{-1}$	0.6–2.0 $m \cdot h^{-1}$
Abatement performance	90 to ~100%	70 to ~100%
pH	1.0–8.5	Close to neutrality

(Cho *et al.*, 1992; Cho *et al.*, 2000; Elias *et al.*, 2002; Gomez-Cuervo *et al.*, 2016; Hartikainen et al., 2002; Huete *et al.*, 2018; Iranpour *et al.*, 2005; Kim *et al.*, 2002; LeBeau & Milligan, 1994; Lee *et al.*, 2005; Lee *et al.*, 2006; Melse & van der Werf, 2005; Nikiema *et al.*, 2009a; Nikiema & Heitz, 2009; Nikiema *et al.*, 2009b; Oyarzún *et al.*, 2003; Pantoja Filho *et al.*, 2010; Pawłowska *et al.*, 2011; Rocha, 2007; Vaith & Heydorn, 1997; van Langenhove *et al.*, 1986; Wani *et al.*, 1999; Yang & Allen, 1994).

Table 9.7 Design/operational parameters and performance of biotrickling filters for the oxidation of H_2S and CH_4 in waste gases generated in STPs.

Parameter	Typical Range	
	H_2S	CH_4[1]
Influent concentration	~1–3,862 ppm_v	0.13–5.00$\%_{v/v}$ (1,300–50,000 ppm_v)
Volumetric load	<0.1–11,830 $gS\text{-}H_2S \cdot m^{-3} \cdot h^{-1}$	0.1–334.7 $gCH_4 \cdot m^{-3} \cdot h^{-1}$
Empty-bed residence time	1.6–120 sec	6–20 min
Face velocity	20–6,300 $m \cdot h^{-1}$	3.0–10.2 $m \cdot h^{-1}$
Abatement performance	95 to ~100%	60 to ~100%
pH	1.0–9.0	Close to neutrality

[1]The nature of the inorganic packing materials used to achieve the best results is confidential (industrial/patent property); therefore, the data should be employed with caution.
(Chaiprapat *et al.*, 2011; Chung *et al.*, 2000; Cox & Deshusses, 2002; Deshusses *et al.*, 2001; Gabriel & Deshusses, 2003; Girard *et al.*, 2011; Iranpour *et al.*, 2005; Jiang *et al.*, 2009; Kim *et al.*, 2014b; Kim *et al.*, 2013; Nikiema & Heitz, 2009; Ramírez *et al.*, 2009; Sercu *et al.*, 2005; Veillette *et al.*, 2012; Zhang *et al.*, 2009).

Keeping in mind all the characteristics and parameters previously mentioned, it is possible to determine the biofilter or biotrickling filter dimensions. Firstly, it is necessary to estimate the concentrations of odorant compounds (particularly H_2S) and CH_4 (if greenhouse gas abatement is also of interest) in the waste gases. The following list and Example 9.1 shows the sequence of calculations required for the design of a biofilter or a biotrickling filter. If the interest is the abatement of both H_2S and CH_4, it is recommended to set two bioreactors in series, the first one for the treatment of odorant compounds and the second one for the treatment of CH_4, allowing acidic conditions in the first bioreactor and neutral pH in the second one, as methanotrophs are affected by low pH. However, if the concentration of H_2S in the incoming gas is lower than 500 ppm_v, both reactions can take place in the same vessel (Huete *et al.*, 2018).

Sequence of calculations required for the design of a biofilter or a biotrickling filter:

(1) From the emission factors shown in Table 9.4 and the load of SO_4^{2-} and COD in sewage, it is possible to estimate the load of H_2S and CH_4 released to the head spaces of each covered zone (pumping stations, headworks, settler compartments of UASB reactors, channels receiving UASB reactor effluents etc.). As regards stripping and desorption systems set downstream of the UASB reactors, desorption efficiencies have to be considered.

(2) The design of the biofilter/biotrickling filter starts with the definition of the waste-gas exhaust flow, as shown in section 9.4 (for covered zones or confined buildings). Considering the empty-bed residence time (Tables 9.6 and 9.7) and the waste-gas exhaust flow, it is possible to establish the filter bed volume (packing media volume).

(3) Adopting a packing media depth (see previous discussion in this section) and considering the volume estimated in the previous step, it is possible to obtain the required area for the bioreactor.

(4) From the load of H_2S and CH_4 entering the system and the estimated filter-bed working volume, it is possible to verify the volumetric load in accordance with Tables 9.6 and 9.7.

(5) From the waste-gas exhaust flow and the bioreactor area, it is possible to verify the face velocity in accordance with Tables 9.6 and 9.7.

EXAMPLE 9.1

Design a biofiltration system for the abatement of H_2S and CH_4 in a UASB-based sewage treatment plant.

Input data

- Average influent sewage flowrate: 50 $m^3 \cdot h^{-1}$
- Average influent COD concentration: $C_{COD} = 600$ mg L^{-1}
- Average influent SO_4^{2-} concentration in the raw sewage = 50 mg L^{-1}, which is 17 mg(S-SO_4^{2-} $_{inf}$) $\cdot L^{-1}$
- H_2S release from the surface of the UASB settler: 3.1 mg(H_2S) $\cdot$ g(S-SO_4^{2-} $_{inf}$)$^{-1}$ (data from Table 9.4)
- CH_4 release from the surface of the UASB settler: 40 mg(CH_4) $\cdot$ g(COD_{inf})$^{-1}$ (data from Table 9.4)
- Waste gas flow rate: 85 $m^3 \cdot h^{-1}$

Solution

(1) Calculate the loads of H_2S and CH_4:
  - H_2S load = 3.1 mg(H_2S) $\cdot$ g(S-SO_4^{2-} $_{inf}$)$^{-1}$ × 17 g(S-SO_4^{2-} $_{inf}$) $\cdot m^{-3}$ × 50 $m^3 \cdot h^{-1}$
    = 2.6 g(H_2S) $\cdot h^{-1}$
  - CH_4 load: 40 mg(CH_4) $\cdot$ g(COD_{inf})$^{-1}$ × 600 g(COD_{inf}) $\cdot m^{-3}$ × 50 $m^3 \cdot h^{-1}$ = 1,200 g(CH_4) $\cdot h^{-1}$

(2) Empty-bed residence time (EBRT) consideration and calculation of the bioreactor working volumes:
  - EBTR of the biofilter for H_2S abatement: 60 s (from Table 9.6)
  - Packing media working volume for H_2S abatement = 85 $m^3 \cdot h^{-1}$ × 60 s/3,600 s h^{-1} = 1.4 m^3
  - EBRT of the biofilter for CH_4 abatement is 25 min (from Table 9.6)
  - Packing media working volume for CH_4 abatement is 85 $m^3 \cdot h^{-1}$ × 25 min/60 min $\cdot h^{-1}$ = 35.4 m^3

(3) Packing media depth and required area:
  - Packing media depth of the biofilter for H_2S abatement is 0.8 m
  - Required area for H_2S abatement = 1.4 m^3/0.8 m = 1.75 m^2
  - H_2S biofilter dimensions are one column based on a 60-inch reinforced concrete pipe (inner diameter = 1.5 m), packed 0.8 m high (working volume = 1.42 m^3; area = 1.77 m^2)
  - Packing media depth of the biofilter for CH_4 abatement is 0.8 m
  - Required area for CH_4 abatement = 35.4 m^3/0.8 m = 44.2 m^2
  - CH_4 biofilter dimensions are three 20-ft standard shipping containers (2.35 m × 5.9 m width/length) packed 0.8 m high (working volume = 33.3 m^3; area = 41.6 m^2; EBRT = 23.5 min)

(4) Verify the volumetric load:
  - Volumetric load for H_2S abatement is 2.6 g(H_2S) $\cdot h^{-1}$ / 1.42 m^3 = 1.8 g(S-H_2S) $\cdot m^{-3} \cdot h^{-1}$ (acceptable according to Table 9.6)
  - Volumetric load for CH_4 abatement is 1,200 g(CH_4) $\cdot h^{-1}$ / 33.3 m^3 = 36.0 g(CH_4) $\cdot m^{-3} \cdot h^{-1}$ (acceptable according to Table 9.6)

(5) Verify the face velocity:
  - Face velocity for H_2S abatement is 85 $m^3 \cdot h^{-1}$ / 1.77 m^2 = 48.0 m $\cdot h^{-1}$ (acceptable according to Table 9.6)
  - Face velocity for CH_4 abatement is 85 $m^3 \cdot h^{-1}$ / 41.6 m^2 = 2.0 m $\cdot h^{-1}$ (acceptable according to Table 9.6)

Additional information:
  - Biofilter for H_2S abatement: The packing media must be prepared from peat or composted waste, alternatively mixed with large-size materials such as wood chips or perlite at a volumetric ratio of 4:1 (Hartikainen *et al.*, 2002; Iranpour *et al.*, 2005; Wani *et al.*, 1999).

- Biofilter for CH_4 abatement: The packing media must be prepared from a mixture at a volumetric basis of 60% composted leaves and 40% expanded vermiculite (Brandt *et al.*, 2016; Huete *et al.*, 2018).
- The H_2S biofilter must be upstream of the CH_4 biofilter and all the bioreactors must be coated with proper corrosion-proof material.

Based on the previous example, when the abatement of CH_4 is of interest as well, it is recommended highly to set the equipment downstream of the UASB reactor to strip dissolved gases (see section 9.4). In this sense, instead of exhausting the settler compartment, the off-gas flow of the stripping unit is much lower. As in the previous example, the waste gas flow is the key factor limiting the CH_4 biofilter area, and, by using a stripping unit, the CH_4 biofilters will be much smaller.

As shown by Huete *et al.* (2018), it is possible to use a packed column to desorb the CH_4 dissolved in the UASB reactor effluent. Considering the air-to-water ratio of 1:2 used by Huete *et al.* (2018), the CH_4 content in waste gas will be approximately $4.3\%_{v/v}$ resulting from the 99% desorption efficiency. Therefore, in the previous example, the following results are obtained:

(6) Change of waste gas flow and the CH_4 load:
- Waste gas flow = $50\ m^3 \cdot h^{-1}$ (raw sewage flow)/2 (air-to-water ratio) = $25\ m^3 \cdot h^{-1}$
- CH_4 load of $0.043\ m^3(CH_4) \cdot m^{-3} \times 25\ m^3 \cdot h^{-1} \times 0.7\ kg(CH_4) \cdot m^{-3}(CH_4) = 753\ g(CH_4) \cdot h^{-1}$

(7) EBRT consideration, and calculation of the bioreactor working volume:
- EBRT of the biofilter for CH_4 abatement is 25 min (from Table 9.6)
- Packing media working volume for CH_4 abatement is $25\ m^3 \cdot h^{-1} \times 25\ min/60\ min \cdot h^{-1} = 10.4\ m^3$

(8) Packing media depth and required area:
- Packing media depth of the biofilter for CH_4 abatement is 0.8 m
- Required area for CH_4 abatement = $10.4\ m^3/0.8\ m = 13.0\ m^2$
- CH_4 biofilter dimensions are one 20-ft standard shipping container (2.35 m × 5.9 m width/length) packed 0.8 m high (working volume = $11.1\ m^3$; area = $13.9\ m^2$; EBRT = 26.6 min)

(9) Verify the volumetric load:
- $753\ g(CH_4) \cdot h^{-1}/11.1\ m^3 = 67.8\ g(CH_4) \cdot m^{-3} \cdot h^{-1}$ (acceptable according to Table 9.6)

(10) Verify the face velocity:
- Face velocity for CH_4 abatement is $25\ m^3 \cdot h^{-1}/13.9\ m^2 = 1.8\ m \cdot h^{-1}$ (acceptable according to Table 9.6)

9.6 FINAL REMARKS

It should be recognized that odorant gaseous emissions and corrosion are inherent to wastewater generally, but are exacerbated in anaerobic-based STPs. However, as presented in this chapter, effective and low-cost techniques can be used to prevent and control such aspects. Sufficient technical and scientific information is available for the design, construction, and operation of such techniques in existing and new anaerobic-based STPs. For gas phase emissions, the economics and process scale of these ancillary technologies are complementary to anaerobic processes and appear not to detract from the cost and sustainability advantages that anaerobic technology affords. However, a new challenge for full-scale anaerobic treatment (e.g., UASB reactors) is to establish economic and practicable methods to control diffuse CH_4 emissions from the anaerobic effluent (dissolved CH_4) in the context of greenhouse gas mitigation and climate change. Maximizing CH_4 recovery while simultaneously reducing diffuse CH_4 emissions is a challenge, particularly in developing countries, as anaerobic sewage treatment is ideally suited to such regions. As shown in this chapter, various techniques can help to achieve this desired scenario.

REFERENCES

Alfonsín C., Lebrero R., Estrada J. M., Muñoz R., Kraakman N. J. R., Feijoo G. and Moreira M. T. (2015). Selection of odour removal technologies in wastewater treatment plants: a guideline based on Life Cycle Assessment. *Journal of Environmental Management*, **149**, 77–84.

Al-Shammiri M. (2004). Hydrogen sulfide emission from the Ardiyah sewage treatment plant in Kuwait. *Desalination*, **170**, 1–13.

Andres Y., Dumont E. and Gerente C. (2014). Characterization techniques of packing material colonization in gas biofiltration processes. *Journal of Environmental Engineering and Science*, **40**, 29–37.

Bandara W. M. K. R. T. W., Satoh H., Sasakawa M., Nakahara Y., Takahashi M. and Okabe S. (2011). Removal of residual dissolved methane gas in an upflow anaerobic sludge blanket reactor treating low-strength wastewater at low temperature with degassing membrane. *Water Research*, **45**, 3533–3540.

Bandara W. M. K. R. T. W., Kindaichi T., Satoh H., Sasakawa M., Nakahara Y., Takahashi M. and Okabe S. (2012). Anaerobic treatment of municipal wastewater at ambient temperature: analysis of archaeal community structure and recovery of dissolved methane. *Water Research*, **46**, 5756–5764.

Barona A., Elias A., Arias R., Cano I. and Gonzalez R. (2004). Biofilter response to gradual and sudden variations in operating conditions. *Biochemical Engineering Journal*, **22**, 25–31.

Bender M. and Conrad R. (1995). Effect of CH_4 concentrations and soil conditions on the induction of CH_4 oxidation activity. *Soil Biology and Biochemistry*, **27**, 1517–1527.

Bhatia S. P. (1978). Organosulfur emissions from industrial sources. In: Sulfur in the Environment, Part I, J. O. Nriagu (ed.), John Wiley, Canada.

Bohn H. L. (1993). Bioprocessing of organic gases in waste air, Symposium on Bioremediation and Bioprocessing. 205th National Meeting. American Chemical Society, Denver, pp. 287–289.

Brandt E. M. F., Duarte F. V., Vieira J. P. R., Melo V. M., Souza C. L., Araújo J. C. and Chernicharo C. A. L. (2016). The use of novel packing material for improving methane oxidation in biofilters. *Journal of Environmental Management*, **182**, 412–420.

Brandt E. M. F., Souza C. L. and Chernicharo C. A. L. (2017). Alternativas para o controle de odores e corrosão em sistemas de coleta e tratamento de esgoto (Alternatives to odour and corrosion control in sanitary sewerage systems and treatment plants). *Engenharia Sanitária e Ambiental*, **22**, 611–623, in Portuguese.

Carlson D. and Leiser C. (1966). Soil beds for the control of sewage odors. *Journal (Water Pollution Control Federation)*, **38**, 829–840.

Caron J., Rivière L.-M. and Guillemain G. (2005). Gas diffusion and air-filled porosity: effect of some oversize fragments in growing media. *Canadian Journal of Soil Science*, **85**, 57–65.

Chaiprapat S., Mardthing R., Kantachote D. and Karnchanawong S. (2011). Removal of hydrogen sulfide by complete aerobic oxidation in acidic biofiltration. *Process Biochemistry*, **46**, 344–352.

Chaosakul T., Koottatep T. and Polprasert C. (2014). A model for methane production in sewers. *Journal of Environmental Science and Health, Part A*, **49**(11), 1316–1321.

Chernicharo C. A. L., Stuetz R. M., Souza C. L. and Melo G. C. B. (2010). Alternativas para o controle de emissões odorantes em reatores anaeróbios tratando esgoto doméstico (Alternatives for the control of odorous emissions in anaerobic reactors treating domestic wastewater). *Engenharia Sanitária e Ambiental*, **15**, 229–236, in Portuguese.

Chernicharo C. A. L., Van Lier J., Noyola A. and Bressani-Ribeiro T. (2015). Anaerobic sewage treatment: state of the art, constraints and challenges. *Reviews in Environmental Science and Bio/Technology*, **14**, 649–679.

Cho K.-S., Hirai M. and Shoda M. (1992). Enhanced removal efficiency of malodorous gases in a pilot-scale peat biofilter inoculated with Thiobacillus thioparus DW44. *Journal of Fermentation and Bioengineering*, **73**, 46–50.

Cho K.-S., Ryu H. W. and Lee N. Y. (2000). Biological deodorization of hydrogen sulfide using porous lava as a carrier of Thiobacillus thiooxidans. *Journal of Bioscience and Bioengineering*, **90**, 25–31.

Christophersen M. and Kjeldsen P. (2001). Lateral gas transport in soil adjacent to an old landfill: factors governing gas migration. *Waste Management & Research*, **19**, 579–594.

Chung Y.-C., Huang C., Tseng C.-P. and Rushing Pan J. (2000). Biotreatment of H_2S- and NH_3-containing waste gases by co-immobilized cells biofilter. *Chemosphere*, **41**, 329–336.

Cookney J., Cartmell E., Jefferson B. and McAdam E. J. (2012). Recovery of methane from anaerobic process effluent using poly-di-methyl-siloxane membrane contactors. *Water Science and Technology*, **65**(5), 604–610.

Cookney J., Mcleod A., Mathioudakis V., Ncube P., Soares A., Jefferson B. and McAdam E. J. (2016). Dissolved methane recovery from anaerobic effluents using hollow fibre membrane contactors. *Journal of Membrane Science*, **502**, 141–150.

Cox H. H. and Deshusses M. A. (2002). Co-treatment of H_2S and toluene in a biotrickling filter. *Chemical Engineering Journal*, **87**, 101–110.

Czepiel P. M., Crlll P. M. and Harrlss R. C. (1993). Methane emissions from municipal wastewater treatment processes. *Environmental Science & Technology*, **27**, 2472–2477.

Daelman M. R. J., van Voorthuizen E. M., van Dongen U. G. J. M., Volcke E. I. P. and van Loosdrecht M. C. M. (2012). Methane emission during municipal wastewater treatment. *Water Research*, **46**, 3657–3670.

Deshusses M. A., Hamer G. and Dunn I. J. (1995). Behavior of biofilters for waste air biotreatment. 2. Experimental evaluation of a dynamic model. *Environmental Science & Technology*, **29**, 1059–1068.

Deshusses M. A., Cox H. H. J., Schroeder E. D. and Converse B. M. (2001). Controlling Odor and VOCs with Bioreactors, Final Report WERF Project 98-CTS-4, 1 ed. Water Environment Research Foundation, Alexandria, USA.

Devinny J. S., Deshusses M. A. and Webster T. S. (1998). Biofiltration for Air Pollution Control. CRC Press, Boca Raton, USA.

Duan H., Koe L. C., Yan R. and Chen X. (2006). Biological treatment of H_2S using pellet activated carbon as a carrier of microorganisms in a biofilter. *Water Research*, **40**, 2629–2636.

Edwards F. G. and Nirmalakhandan N. (1996). Biological treatment of airstreams contaminated with VOCs: an overview. *Water Science and Technology*, **34**(3–4), 565–571.

Eijo-Río E., Petit-Boix A., Villalba G., Suárez-Ojeda M., Marin D., Amores M., Aldea X. and Rieradevall X. (2015). Municipal sewer networks as sources of nitrous oxide, methane and hydrogen sulphide emissions: a review and case studies. *Journal of Environmental Chemical Engineering*, **3**, 2084–2094.

Elias A., Barona A., Arreguy A., Rios J., Aranguiz I. and Peñas J. (2002). Evaluation of a packing material for the biodegradation of H_2S and product analysis. *Process Biochemistry*, **37**, 813–820.

Estrada J. M., Kraakman N. J. R., Lebrero R. and Muñoz R. (2012). A sensitivity analysis of process design parameters, commodity prices and robustness on the economics of odour abatement technologies. *Biotechnology Advances*, **30**, 1354–1363.

Estrada J. M., Lebrero R., Quijano G., Pérez R., Figueroa-González I., García-Encina P. A. and Muñoz R. (2014). Methane abatement in a gas-recycling biotrickling filter: evaluating innovative operational strategies to overcome mass transfer limitations. *Chemical Engineering Journal*, **253**, 385–393.

Foresti E., Zaiat M. and Vallero M. (2006). Anaerobic processes as the core technology for sustainable domestic wastewater treatment: consolidated applications, new trends, perspectives, and challenges. *Reviews in Environmental Science and Bio/Technology*, **5**(1), 3–19.

Gabriel D. and Deshusses M. A. (2003). Retrofitting existing chemical scrubbers to biotrickling filters for H_2S emission control. *Proceedings of the National Academy of Sciences*, **100**, 6308–6312.

Garcia G. P. P. (2009). Caracterização dos Microrganismos Oxidadores de Sulfeto e Metano em Reator UASB (Characterization of Sulphide and Methane Oxidation Microorganisms in UASB Reactor), Escola de Engenharia. Universidade Federal de Minas Gerais, Belo Horizonte, in Portuguese.

Gebert J., Stralis-Pavese N., Alawi M. and Bodrossy L. (2008). Analysis of methanotrophic communities in landfill biofilters using diagnostic microarray. *Environmental Microbiology*, **10**, 1175–1188.

Girard M., Ramirez A. A., Buelna G. and Heitz M. (2011). Biofiltration of methane at low concentrations representative of the piggery industry – influence of the methane and nitrogen concentrations. *Chemical Engineering Journal*, **168**, 151–158.

Glória R. M., Motta T. M., Silva P. V. O., Costa P., Brandt E. M. F., Souza C. L. and Chernicharo C. A. L. (2016). Stripping and dissipation techniques for the removal of dissolved gases from anaerobic effluents. *Brazilian Journal of Chemical Engineering*, **33**, 713–721.

Gómez-Borraz T. L., González-Sánchez A., Bonilla-Blancas W., Revah S. and Noyola A. (2017). Characterization of the biofiltration of methane emissions from municipal anaerobic effluents. *Process Biochemistry*, **63**, 204–213.

Gomez-Cuervo S., Hernandez J. and Omil F. (2016). Identifying the limitations of conventional biofiltration of diffuse methane emissions at long-term operation. *Environmental Technology*, **37**, 1947–1958.

Goodfellow H. and Tähti E. (2001). Industrial Ventilation Design Guidebook, 1st edn. Academic, San Diego, USA.

Hartikainen T., Martikainen P. J., Olkkonen M. and Ruuskanen J. (2002). Peat biofilters in long-term experiments for removing odorous sulphur compounds. *Water, Air, and Soil Pollution*, **133**(1-4), 335–348.

Hartley K. and Lant P. (2006). Eliminating non-renewable CO_2 emissions from sewage treatment: an anaerobic migrating bed reactor pilot plant study. *Biotechnology and Bioengineering*, **95**, 384–398.

Heile S., Chernicharo C. A. L., Brandt E. M. F. and McAdam E. J. (2017). Dissolved gas separation for engineered anaerobic wastewater systems. *Separation and Purification Technology*, **189**, 405–418.

Huete A., de los Cobos-Vasconcelos D., Gómez-Borraz T., Morgan-Sagastume J. M. and Noyola A. (2018). Control of dissolved CH_4 in a municipal UASB reactor effluent by means of a desorption – biofiltration arrangement. *Journal of Environmental Management*, **216**, 383–391.

Humer M. and Lechner P. (1999). Alternative approach to the elimination of greenhouse gases from old landfills. *Waste Management and Research*, **17**, 443–452.

Intergovernmental Panel on Climate Change (2014). Climate change 2014: synthesis report. In: Contribution of Working Groups I, II and III to the Fifth Assessment Report of the Intergovernmental Panel on Climate Change, Core Writing Team, R.K. Pachauri and L.A. Meyer (ed.), Intergovernmental Panel on Climate Change – IPCC, Geneva, Switzerland, pp. 151.

Iranpour R., Cox H. H., Deshusses M. A. and Schroeder E. D. (2005). Literature review of air pollution control biofilters and biotrickling filters for odor and volatile organic compound removal. *Environmental Progress*, **24**, 254–267.

Jiang H., Chen Y., Jiang P., Zhang C., Smith T. J., Murrell J. C. and Xing X.-H. (2010). Methanotrophs: multifunctional bacteria with promising applications in environmental bioengineering. *Biochemical Engineering Journal*, **49**, 277–288.

Jiang X., Yan R. and Tay J. H. (2009). Simultaneous autotrophic biodegradation of H_2S and NH_3 in a biotrickling filter. *Chemosphere*, **75**, 1350–1355.

Jobbagy A., Szanto I., Varga G. I. and Simon J. (1994). Sewer system odour control in the Lake Balaton area. *Water Science and Technology*, **30**(1), 195–204.

Kennes C. and Veiga M. C. (2001). Bioreactors for Waste Gas Treatment. Kluwer Academic, Dordrecht, Netherlands.

Kennes C., Montes M., López M. E. and Veiga M. C. (2009a). Waste gas treatment in bioreactors: environmental engineering aspects. *Canadian Journal of Civil Engineering*, **36**, 1887–1894.

Kennes C., Rene E. R. and Veiga M. C. (2009b). Bioprocesses for air pollution control. *Journal of Chemical Technology and Biotechnology*, **84**, 1419–1436.

Kim H., Kim Y. J., Chung J. S. and Xie Q. (2002). Long-term operation of a biofilter for simultaneous removal of H_2S and NH_3. *Journal of the Air & Waste Management Association*, **52**, 1389–1398.

Kim T. G., Yi T., Lee E.-H., Ryu H. W. and Cho K.-S. (2012). Characterization of a methane-oxidizing biofilm using microarray, and confocal microscopy with image and geostatic analyses. *Applied Microbiology and Biotechnology*, **95**, 1051–1059.

Kim T. G., Lee E.-H. and Cho K.-S. (2013). Effects of nonmethane volatile organic compounds on microbial community of methanotrophic biofilter. *Applied Microbiology and Biotechnology*, **97**, 6549–6559.

Kim T. G., Jeong S.-Y. and Cho K.-S. (2014a). Functional rigidity of a methane biofilter during the temporal microbial succession. *Applied Microbiology and Biotechnology*, **98**, 3275–3286.

Kim T. G., Jeong S.-Y. and Cho K.-S. (2014b). Characterization of tobermolite as a bed material for selective growth of methanotrophs in biofiltration. *Journal of Biotechnology*, **173**, 90–97.

Kohl A. and Nielsen R. (1997). Gas Purification, 5 edn. Gulf Publishing Company, Houston, Texas, USA.

Kraakman N. J., Rocha-Rios J. and van Loosdrecht M. C. (2011). Review of mass transfer aspects for biological gas treatment. *Applied Microbiology and Biotechnology*, **91**, 873–886.

LeBeau A. and Milligan D. (1994). Control of hydrogen sulfide gas from a wastewater lift station using biofiltration, Odor and Volatile Organic Compound Emission Control for Municipal and Industrial Wastewater Treatment Facilities. Water Environment Federation, Alexandria, USA, pp. 6-49–46-60.

Lebrero R., Hernández L., Pérez R., Estrada J. M. and Muñoz R. (2015). Two-liquid phase partitioning biotrickling filters for methane abatement: exploring the potential of hydrophobic methanotrophs. *Journal of Environmental Management*, **151**, 124–131.

Lee E. Y., Cho K.-S. and Ryu H. W. (2005). Simultaneous removal of H_2S and NH_3 in biofilter inoculated with Acidithiobacillus thiooxidans TAS. *Journal of Bioscience and Bioengineering*, **99**, 611–615.

Lee E. Y., Lee N. Y., Cho K.-S. and Ryu H. W. (2006). Removal of hydrogen sulfide by sulfate-resistant *Acidithiobacillus thiooxidans* AZ11. *Journal of Bioscience and Bioengineering*, **101**, 309–314.

Limbri H., Gunawan C., Rosche B. and Scott J. (2013). Challenges to developing methane biofiltration for coal mine ventilation air: a review. *Water, Air, & Soil Pollution*, **224**, 1–15.

Maia G. D. (2003). Tratamento biológico de gás sulfídrico (H_2S) em biofiltro, Departamento de Engenharia Química (Biological treatment of hydrogen sulphide (H_2S) in biofilter). Universidade Federal de São Carlos, São Carlos, in Portuguese.

Matos J. S. and Aires C. M. (1995). Mathematical modeling of sulphides and hydrogen sulphide gas build-up in the cosas do estoril sewerage system. *Water Science and Technology*, **31**(7), 255–261.

Matsuura N., Hatamoto M., Sumino H., Syutsubo K., Yamaguchi T. and Ohashi A. (2015). Recovery and biological oxidation of dissolved methane in effluent from UASB treatment of municipal sewage using a two-stage closed downflow hanging sponge system. *Journal of Environmental Management*, **151**, 200–209.

Melse R. W. and van der Werf A. W. (2005). Biofiltration for mitigation of methane emission from animal husbandry. *Environmental Science & Technology*, **39**, 5460–5468.

Moraes Jr D. (1991). Metodologia para o estudo de um sistema biológico de tratamento de emissões gasosas e aplicação na depuração do gás amoníaco (Methodology for the study of a biological system for treatment of gaseous emissions and application in the purification of ammonia gas), Escola de Engenharia de São Carlos. Universidade de São Paulo, São Carlos, in Portuguese.

Morales M., Hernández S., Cornabé T., Revah S. and Auria R. (2003). Effect of drying on biofilter performance: modeling and experimental approach. *Environmental Science & Technology*, **37**, 985–992.

Morgan-Sagastume J. M., Noyola A., Revah S. and Ergas S. J. (2003). Changes in physical properties of a compost biofilter treating hydrogen sulfide. *Journal of the Air & Waste Management Association*, **53**(8), 1011–1021.

Muñoz R., Villaverde S., Guieysse B. and Revah S. (2007). Two-phase partitioning bioreactors for treatment of volatile organic compounds. *Biotechnology Advances*, **25**, 410–422.

Myhre G., Shindell D., Bréon F. M., Collins W., Fuglestvedt J., Huang J., Koch D., Lamarque J. F., Lee D., Mendoza B., Nakajima T., Robock A., Stephens G., Takemura T. and Zhang H. (2013). Anthropogenic and Natural Radiative Forcing. In: Climate Change 2013: The Physical Science Basis. Contribution of Working Group I to the Fifth Assessment Report of the Intergovernmental Panel on Climate Change, T. F. Stocker, D. Qin, G. K. Plattner, M. Tignor, S. K. Allen, J. Boschung, A. Nauels, Y. Xia, V. Bex and P. M. Midgley (eds.), Cambridge University Press, Cambridge, United Kingdom and New York, NY, USA.

Nelting K., Caicedo C., Weichgrebe D. and Rosenwinkel K.-H. (2015). Influence of UASB reactor configuration and operation conditions on the dissolved methane quantities in the effluent. 14th World Congress on Anaerobic Digestion. International Water Association, Viña del Mal, Chile, p. 4.

Nikiema J. and Heitz M. (2009). The influence of the gas flow rate during methane biofiltration on an inorganic packing material. *The Canadian Journal of Chemical Engineering*, **87**, 136–142.

Nikiema J. and Heitz M. (2010). The use of inorganic packing materials during methane biofiltration. *International Journal of Chemical Engineering, Article ID*, **573149**, 8pp.

Nikiema J., Bibeau L., Lavoie J., Brzezinski R., Vigneux J. and Heitz M. (2005). Biofiltration of methane: an experimental study. *Chemical Engineering Journal*, **113**, 111–117.

Nikiema J., Brzezinski R. and Heitz M. (2007). Elimination of methane generated from landfills by biofiltration: a review. *Reviews in Environmental Science and Bio/Technology*, **6**, 261–284.

Nikiema J., Girard M., Brzezinski R. and Heitz M. (2009a). Biofiltration of methane using an inorganic filter bed: influence of inlet load and nitrogen concentration. *Canadian Journal of Civil Engineering*, **36**, 1903–1910.

Nikiema J., Payre G. and Heitz M. (2009b). A mathematical steady state model for methane bioelimination in a closed biofilter. *Chemical Engineering Journal*, **150**, 418–425.

Noyola A., Capdeville B. and Roques H. (1988). Anaerobic treatment of domestic sewage with a rotating-stationary fixed-film reactor. *Water Research*, **22**, 1585–1592.

Noyola A., Morgan-Sagastume J. M. and Lopez-Hernandez J. E. (2006). Treatment of biogas produced in anaerobic reactors for domestic wastewater: odor control and energy/resource recovery. *Reviews in Environmental Science and Bio/Technology*, **5**, 93–114.

Oude Elferink S. J. W. H., Visser A., Hulshoff Pol L. W. and Stams S. A. J. (1994). Sulfate reduction in methanogenic bioreactors. *FEMS Microbial Reviews*, **15**, 119–136.

Oyarzún P., Arancibia F., Canales C. and Aroca G. E. (2003). Biofiltration of high concentration of hydrogen sulphide using *Thiobacillus thioparus*. *Process Biochemistry*, **39**, 165–170.

Pagliuso J. D., Passig F. H. and Villela L. C. H. (2002). Odour treatment and energy recovery in anaerobic sewage treatment plants, VII Oficina e Simposio Latino-americano de Digestão Anaeróbia. IWA/FEMISCA, México.

Pantoja Filho J. L. R. (2008). Avaliação da utilização de diferentes materiais suporte na biofiltração de sulfeto de hidrogênio (Evaluation of using different packing materials in the biofiltration of hydrogen sulphide), Escola de Engenharia de São Carlos. Universidade de São Paulo, São Carlos, in Portuguese.

Pantoja Filho J. L. R., Sader L. T., Damianovic M. H. R. Z., Foresti E. and Silva E. L. (2010). Performance evaluation of packing materials in the removal of hydrogen sulphide in gas-phase biofilters: polyurethane foam, sugarcane bagasse, and coconut fibre. *Chemical Engineering Journal*, **158**, 441–450.

Pawłowska M., Rożej A. and Stępniewski W. (2011). The effect of bed properties on methane removal in an aerated biofilter – model studies. *Waste Management*, **31**, 903–913.

Pérez H. and Villa P. (2005). Desulfuración biológica: una alternativa para el tratamiento de emisiones de gases a la atmosfera. *Agua Latinoamericana*, **5**(3), 17–20.

Perry R. H. and Green D. W. (2007). Chemical Engineers' Handbook, 8 edn. McGraw-Hill, New York, USA.

Plessis C. A. d., Strauss J. M., Sebapalo E. M. T. and Riedel K.-H. J. (2003). Empirical model for methane oxidation using a composted pine bark biofilter. *Fuel*, **82**, 1359–1365.

Pomeroy R. D. (1957). Deodorizing gas streams by the use of microbiological growths. *US patent*, **2**, 096.

Prüss M. and Blunk H. (1941). 710.954. German Patent.

Ramirez A. A., García-Aguilar B. P., Jones J. P. and Heitz M. (2012a). Improvement of methane biofiltration by the addition of non-ionic surfactants to biofilters packed with inert materials. *Process Biochemistry*, **47**, 76–82.

Ramirez A. A., Jones J. P. and Heitz M. (2012b). Methane treatment in biotrickling filters packed with inert materials in presence of a non-ionic surfactant. *Journal of Chemical Technology and Biotechnology*, **87**, 848–853.

Ramírez M., Gómez J. M., Aroca G. and Cantero D. (2009). Removal of hydrogen sulfide by immobilized Thiobacillus thioparus in a biotrickling filter packed with polyurethane foam. *Bioresource Technology*, **100**, 4989–4995.

Rocha A. P. (2007). Desempenho de filtro biológico na depuração e desodorização de emissões de sulfeto de hidrogênio (Performance of biofilters in the purification and deodourization of hydrogen sulfide emissions), Escola de Engenharia de São Carlos. Universidade de São Paulo, São Carlos.

Rocha-Rios J., Bordel S., Hernández S. and Revah S. (2009). Methane degradation in two-phase partition bioreactors. *Chemical Engineering Journal*, **152**, 289–292.

Rodrigues K. B. (2002). Desenvolvimento de uma unidade experimental para degradação de gás sulfídrico (H_2S) em filtro biológico (Development of an experimental unit for the degradation of hydrogen sulphide (H_2S) in a biofilter). Departamento de Engenharia Química. Universidade Federal de São Carlos, São Carlos.

Sercu B., Nunez D., van Langenhove H., Aroca G. and Verstraete W. (2005). Operational and microbiological aspects of a bioaugmented two-stage biotrickling filter removing hydrogen sulfide and dimethyl sulfide. *Biotechnology and Bioengineering*, **90**, 259–269.

Shareefdeen Z., Herner B., Webb D. and Wilson S. (2003). Hydrogen sulfide (H_2S) removal in synthetic media biofilters. *Environmental Progress*, **22**, 207–213.

Silva M. B., Lechi R. A. L., Silva H. P. M. and Gonçalves R. F. (2007). Comparação do desempenho de diferentes filtros biológicos na remoção de H_2S do ar em estações de tratamento de esgoto sanitário (Comparison of the performance of different biofilters in the removal of H2S from air in sewage treatment plants). 24° Congresso Brasileiro de Engenharia Sanitária e Ambiental. ABES, Belo Horizonte, in Portuguese.

Smet E. and van Langenhove H. (1998). Abatement of volatile organic sulfur compounds in odorous emissions from the bioindustry. *Biodegradation*, **9**(3-4), 273–284.

Souza C. L. (2010). Estudo das rotas de formação, transporte e consumo dos gases metano e sulfeto de hidrogênio resultantes do tratamento de esgoto doméstico em reatores UASB (Study of the routes of formation, transport and consumption of methane and hydrogen sulfide gases resulting from the treatment of domestic wastewater in UASB reactors). Escola de Engenharia. Universidade Federal de Minas Gerais, Belo Horizonte, in Portuguese.

Souza C. L., Silva S. Q., Aquino S. F. and Chernicharo C. A. L. (2006). Production and characterization of scum and its role in odour control in UASB reactors treating domestic wastewater. *Water Science and Technology*, **54**(9), 201–208.

Souza C. L., Chernicharo C. A. and Aquino S. F. (2011). Quantification of dissolved methane in UASB reactors treating domestic wastewater under different operating conditions. *Water Science and Technology*, **64**(11), 2259–2264.

Souza C., Chernicharo C. and Melo G. (2012). Methane and hydrogen sulfide emissions in UASB reactors treating domestic wastewater. *Water Science and Technology*, **65**(7), 1229–1237.

Steudel R. (1996). Mechanism for the formation of elemental sulfur from aqueous sulfide in chemical and microbiological desulfurization processes. *Industrial & Engineering Chemistry Research*, **35**(4), 1417–1423.

Syed M., Soreanu G., Falletta P. and Béland M. (2006). Removal of hydrogen sulfide from gas streams using biological processes – a review. *Canadian Biosystems Engineering* **48**, 2.1–2.14.

Tang K., Baskaran V. and Nemati M. (2009). Bacteria of the sulphur cycle: an overview of microbiology, biokinetics and their role in petroleum and mining industries. *Biochemical Engineering Journal*, **44**, 73–94.

Tchobanoglous G., Stensel H. D., Tsuchihashi R., Burton F. L., Abu-Orf M., Bowden G. and Pfrang W. (2013). Wastewater engineering, treatment and reuse. 5th ed. McGraw Hill, New York.

USEPA (1985). EPA Design Manual for odor and corrosion control in sanitary sewerage systems and treatment plants. EPA/625/1-85/018, 1st ed. Center for Environmental Research Information, Washington.

USEPA (1991). Hydrogen sulphide corrosion in wastewater collection and treatment system. 430/09-91-010, 1st ed. Office of Water, Washington.

Vaith K. and Heydorn J. (1997). Comparison of biofilter performance for hydrogen sulfide, methyl mercaptan, carbon disulfide, and dimethyl disulfide removal. Control Odors Volatile Organic Chemicals Emissions Conference, Water Environment Federation, Alexandria, USA, pp. 9-1–9-12.

van Langenhove H., Wuyts E. and Schamp N. (1986). Elimination of hydrogen sulphide from odorous air by a wood bark biofilter. *Water Research*, **20**, 1471–1476.

Veillette M., Girard M., Viens P., Brzezinski R. and Heitz M. (2012). Function and limits of biofilters for the removal of methane in exhaust gases from the pig industry. *Applied Microbiology and Biotechnology*, **94**, 601–611.

Visvanathan C., Pokhrel D., Cheimchaisri W., Hettiaratchi J. and Wu J. S. (1999). Methanotrophic activities in tropical landfill cover soils: effects of temperature, moisture content and methane concentration. *Waste Management and Research*, **17**, 313–323.

Wang H., Einola J., Heinonen M., Kulomaa M. and Rintala J. (2008). Group-specific quantification of methanotrophs in landfill gas-purged laboratory biofilters by tyramide signal amplification-fluorescence in situ hybridization. *Bioresource Technology*, **99**, 6426–6433.

Wang J., Xia F.-F., Bai Y., Fang C.-R., Shen D.-S. and He R. (2011). Methane oxidation in landfill waste biocover soil: kinetics and sensitivity to ambient conditions. *Waste Management*, **31**, 864–870.

Wani A. H., Lau A. K. and Branion R. M. (1999). Biofiltration control of pulping odors–hydrogen sulfide: performance, macrokinetics and coexistence effects of organo-sulfur species. *Journal of Chemical Technology and Biotechnology*, **74**, 9–16.

Webster T. S., Devinny J. S., Torres E. M. and Basrai S. S. (1997). Microbial ecosystems in compost and granular activated carbon biofilters. *Biotechnology and Bioengineering*, **53**, 296–303.

WEF (1995). Odor Control in Wastewater Treatment Plants. Manual of Practice 22, 1st edn. Water Environment Federation, Alexandria, USA.

WEF (2004). Control of Odors and Emissions from Wastewater Treatment Plants. Manual of Practice 25, 1st edn. Water Environment Federation, Alexandria.

WERF (2007). Minimization of Odors and Corrosion in Collection Systems. Report 04-CTS-1, 1st edn. Water Environment Research Foundation, Alexandria.

Willis J., Shah A., Yuan Z., Sharma K., Joiner M. and Bocarro R. (2012). Methane Evolution From Wastewater Conveyance. Final Report. Water Environment Research Foundation. IWA Publishing, London, United Kingdom, pp. 4-10–4-11.

Wu G., Conti B., Leroux A., Brzezinski R., Viel G. and Heitz M. (1999). A high performance biofilter for VOC emission control. *Journal of the Air & Waste Management Association*, **49**, 185–192.

Yang Y. and Allen E. R. (1994). Biofiltration control of hydrogen sulfide 1. Design and operational parameters. *Air & Waste*, **44**, 863–868.

Zhang L., Schryver P., Gusseme B., Muynck W., Boon N. and Verstraete W. (2008). Chemical and biological technologies for hydrogen sulfide emission control in sewer systems: a review. *Water Research*, **42**, 1–12.

Zhang L.-H., Meng X.-L., Wang Y. and Liu L.-D. (2009). Performance of biotrickling filters for hydrogen sulfide removal under starvation and shock loads conditions. *Journal of Zhejiang University Science B*, **10**(8), 595–601.

Chapter 10

Post-treatment of anaerobic effluents

Marcos von Sperling, Paulo G. S. Almeida, Thiago Bressani-Ribeiro and Carlos A. L. Chernicharo*

Department of Sanitary and Environmental Engineering (DESA), Universidade Federal de Minas Gerais (UFMG), Av. Antonio Carlos, 6.627, Campus Pampulha, Belo Horizonte, 31270-901, Brazil

**Corresponding author: marcos@desa.ufmg.br*

10.1 INTRODUCTION

In most sewage treatment plants (STPs), particularly in developing countries, the primary and frequently only treatment objectives are related to removing the organic matter, expressed as biochemical oxygen demand (BOD) and chemical oxygen demand (COD). However, as shown in Chapter 3, in UASB reactors, removal of organic matter and suspended solids is limited. Previously, this was considered a limitation of the process but, currently, it is understood as one of its inherent characteristics. The sole implementation of UASB reactors could be sufficient in some specific circumstances, such as (i) in regions of non-stringent discharge standards where the treatment plant is implemented, (ii) high dilution and assimilative capacity of the receiving water body, and (iii) stepwise implementation of the treatment infrastructure, with the UASB reactors representing only the first step.

However, in most instances, superior effluent quality is required in terms of the **organic matter** and **suspended solids**. Accordingly, a successful approach is using UASB reactors as a first stage in the treatment flowsheet, with most of the remaining concentration of these pollutants removed in a second stage, known as **post-treatment**.

In addition, the post-treatment stage can play other important roles as well. In UASB reactors, **ammonia**, **nitrogen**, **phosphorus**, and **pathogens** are removed at small, negligible, or insufficient efficiencies. Therefore, the post-treatment of anaerobic effluents facilitates incorporating the removal of one or more of these constituents.

Editor(s): Carlos Augusto de Lemos Chernicharo and Thiago Bressani-Ribeiro
doi: 10.2166/9781780409238_0275

Ammonia (or, in most instances, ammonium ion NH_4^+) removal is aimed at controlling the following associated problems of water pollution and public health:

- Ammonium ion is a nutrient for algae, leading, under certain conditions, to the phenomenon of eutrophication of lakes and reservoirs;
- Ammonium ion can lead to dissolved oxygen consumption in the receiving water body because of the processes of its conversion to nitrite and nitrite to nitrate (nitrification);
- Free ammonia (NH_3) can be formed under high pH conditions by the shifting of the equilibrium with ammonium ion, and is directly toxic to fish;
- Nitrate, produced in the receiving water following nitrification of ammonia and incorporated further into the public water supply, is associated with illnesses such as methemoglobinemia.

The objective of removing **total nitrogen** (TN) and **total phosphorus** (TP) is associated mainly with the protection of sensitive water bodies (mainly lakes and reservoirs, but also some rivers) from eutrophication.

Pathogen removal is related to public health objectives, being associated with either (i) the use of the receiving water for bathing, animal watering, or other uses with direct contact, or with (ii) the direct use of the treated sewage for irrigation, aquaculture, urban applications, and other forms of water reuse. Pathogenic bacteria and viruses are removed by disinfection or inactivation mechanisms. The efficiency of these processes is usually assessed indirectly, by the removal efficiency of indicator organisms, such as thermotolerant coliforms, or *E. coli*, in the treatment process. Removal of protozoans (in the form of cysts or oocysts) and helminths (in the form of eggs) takes place by physical processes, such as sedimentation and filtration. For these organisms, there is no widely used indicator organism; frequently the removal of helminth eggs is used as an indicator of removal efficiency.

Other constituents, such as organic and inorganic micropollutants, metals, and others, can be incorporated in the treatment objectives, as well. However, despite considerable research, the knowledge is still limited regarding their full-scale applications. They will not be discussed in this Chapter.

In principle, any process used for treating raw sewage or primarily settled effluent can be applied as post-treatment of anaerobic effluents. The following post-treatment processes are mentioned and/or are discussed in more detail in this chapter:

- Extensive (natural) processes:
  - ponds (facultative and maturation)
  - overland flow (land disposal in slopes)
  - constructed wetlands (horizontal and vertical flows)
- Intensive (compact or mechanized) processes:
  - activated sludge
  - aerated biofilter
  - biofilm reactors (trickling filters [TF] and variants)
  - anaerobic filter
  - physical-chemical process (e.g., flotation)

Emphasis is placed on the **biological** treatment processes. In most instances, there is a suitable combination of an **anaerobic** stage (UASB reactor) with an **aerobic** process, taking advantage of various environmental conditions and microbiological routes, allowing the removal of a wider range of pollutants. However, in some instances, this combination is not possible (e.g., post-treatment by horizontal wetlands and anaerobic filters). Disinfection (chlorination, ozonation, and ultraviolet (UV) radiation is not discussed in this chapter but is addressed well in Tchobanoglous *et al.* (2013). Membrane reactors are discussed in Chapter 11.

The approach here, is to present post-treatment options that have been used widely at full scale or, because of their potential, are close to being implemented more widely, particularly in developing countries. The considerable experience accumulated in Brazil, where anaerobic treatment and post-treatment of anaerobic effluents are implemented widely, serves as a guide to the post-treatment options discussed here. Of course, it is understood that this field is developing dynamically and new approaches are most welcome.

In this Chapter, post-treatment has not been classified as secondary, tertiary, or advanced treatment; herein it is simply referred to as **post-treatment**. Although UASB is sometimes considered primary treatment because it is the first unit in the treatment line, our understanding is that it is a secondary level treatment process, as it involves biological mechanisms, which is a common feature of the other treatment processes at secondary level, regardless of their treatment efficiency. Despite post-treatment following after the secondary level treatment in UASB reactors, this does not mean it should be considered tertiary treatment either. As per our understanding, post-treatment remains within the domain of **secondary treatment** because of the utilization of biological processes, which allow higher removal efficiencies of organic matter and the removal of additional constituents while remaining at secondary level. Accordingly, we consider post-treatment an additional stage at a secondary level treatment plant.

As will be seen in the subsequent sections in this Chapter, the main advantages associated with anaerobic treatment followed by post-treatment, in comparison with the adoption of the same processes treating raw sewage, are:

- Lower total volume and area of the treatment units
- Lower construction costs
- Substantially lower operating costs
- Lower energy consumption (in aerated systems)
- Lower production of excess sludge to be treated and disposed of.

The organization of this Chapter is as follows:

- Overview of the main post-treatment options (short description and main flowsheets);
- Comparison of the main post-treatment options (in terms of removal efficiencies of the main constituents, land requirements, sludge production, energy consumption, and qualitative elements);
- Separate description of the main post-treatment options, including design criteria, construction elements, operational aspects, and a short design example (preliminary design for conception studies, leading to the calculation of the main dimensions of reactors and tanks only, without details).

10.2 OVERVIEW OF THE MAIN POST-TREATMENT OPTIONS

A short description of the main post-treatment options is presented in Table 10.1 and the main flowsheets are shown in Figure 10.1. Other processes do exist, including recent and promising technologies, but only the main options are shown here. For the sake of clarity, the processes are separated into extensive (natural) and intensive (compact or mechanized) ones.

10.3 COMPARISON OF THE MAIN POST-TREATMENT OPTIONS

10.3.1 Expected removal efficiencies in the systems comprising UASB reactors followed by post-treatment

Various mechanisms can act separately or simultaneously in the removal of the major pollutants, depending on the process being used. The main mechanisms are listed in Table 10.2, showing that the post-treatment

Table 10.1 Overview of the main options for the post-treatment of anaerobic effluents.

Post-Treatment Process	Short Description
Extensive systems	
Facultative polishing ponds	Anaerobic effluent flows continuously through a pond especially constructed for sewage treatment. The liquid enters in one side and leaves from the other side, with the passage lasting several days. The remaining soluble and fine particulate BOD are stabilized aerobically by bacteria that grow dispersed in the liquid medium, whereas the settleable particulate BOD undergoes sedimentation, being converted anaerobically at the sludge layer at the bottom of the pond. The oxygen required by the aerobic bacteria in the top layer is supplied by algae through photosynthesis. The land requirements are high. The alternative of using maturation polishing ponds is preferred, as they are more efficient in removing pathogenic organisms.
Maturation polishing ponds	The main objective of maturation ponds is the removal of pathogenic organisms. Environmental conditions that are detrimental to these organisms prevail in maturation ponds, such as ultraviolet radiation, high pH, high DO, lower temperature (compared with the human intestinal tract), lack of nutrients, and predation by other organisms. Maturation ponds are a traditional name for these ponds, but they are also called polishing ponds, as they polish (improve) the quality of the anaerobic effluent. They are usually designed as a series of ponds or a single-baffled pond. Coliform removal efficiency is extremely high, and helminth eggs are removed effectively by sedimentation. In addition to pathogens, ammonia is removed. Complementary BOD removal is achieved, of which the efficiency will depend mainly on the concentration of algae (representing particulate BOD) in the effluent. A final stage of clarification can be implemented (e.g., coarse filters).
Overland flow	Anaerobic effluent is distributed in the upper part of vegetated slopes, flows over them, and is collected in ditches at the lower part. Infiltration is negligible, and evapotranspiration reduces the liquid volume. Treatment occurs in the root-soil-stem system. The application is intermittent. Sewage can be distributed by gated or perforated pipes or channels.
Horizontal subsurface flow constructed wetlands	This variant does not resemble natural wetlands, as there is no free water on the surface. The treatment unit comprises a filter bed composed of small stones, gravel, or sand, supporting the growth of emergent plants (macrophytes). The water level remains below the surface of the bed, and the liquid flow is in contact with the roots and rhizomes of the plants, where a bacterial biofilm develops. Biological, chemical, and physical mechanisms act on the root-soil system. In the subsurface horizontal flow mode, the liquid is fed continuously, flowing through the pores of the bed until it reaches the outlet. Aerobic conditions are usually not encountered, because of the saturated hydraulic conditions inside the bed and the negligible input of oxygen. Nutrients (N and P) are removed by the plants, but is generally a small amount compared with the applied nutrient load from the anaerobic reactor. Clogging may take place after some years of operation.
Vertical flow constructed wetlands	This variant is typically a sand or gravel filter, planted with vegetation (emergent macrophytes). In the most common variant of vertical wetlands (downflow), the influent is applied to the top of the bed. Liquid flows through the bed and is collected at the bottom by underdrain pipes. The operation resembles the routine of a

(*Continued*)

Table 10.1 Overview of the main options for the post-treatment of anaerobic effluents (*Continued*).

Post-Treatment Process	Short Description
	filter, with batch cycles of dosage and drainage and, therefore, it differs from the operation of horizontal flow systems. As the dosage is intermittent, the flow is usually in a hydraulically unsaturated medium. After the liquid passes, the pores are filled with air, which allows the maintenance of aerobic conditions. Feeding is alternated between the beds in parallel; therefore, some units are in operation and others at rest.
Intensive systems	
Anaerobic filter	The remaining BOD from the UASB reactor enters this second anaerobic reactor, where it is converted anaerobically by microorganisms that grow attached to a support medium (usually stones) and also grow in the pore spaces. The filter is submerged (saturated) and the flow is upward, with the inlet in the bottom and the outlet in the upper part. As this is an anaerobic reactor, sludge production is low, and the excess sludge is already digested. Although there are treatment plants incorporating anaerobic filters as post-treatment, this is currently not considered a recommendable choice, because of the sequence of two anaerobic stages, without the benefits of aerobic conditions.
High-rate trickling filter	The remaining BOD from the UASB reactor is stabilized aerobically by bacteria that grow attached to a support medium (commonly stones or plastic material). The liquid is applied on the surface of the tank through rotating distributors, percolates (trickles) through the tank, and leaves from the bottom. During this passage, organic matter is retained and subsequently removed further by the bacteria. The medium is hydraulically unsaturated, that is, the water level is at the bottom of the tank. The void spaces allow for the circulation of air. Biofilm develops on the surface of the stones or plastic material. When the biofilm thickness is large, it detaches from the support medium and is removed in the secondary sedimentation tank, comprising the excess sludge. This excess sludge can be sent to the UASB reactors to undergo thickening and digestion, to be removed together with the anaerobic sludge. In some systems, the liquid effluent from the secondary sedimentation tank is recirculated to the filter to dilute the influent and to guarantee a homogeneous hydraulic load. Partial nitrification can be achieved, depending on the applied loads.
Sponge-bed trickling filters	The sponge-bed trickling filter is conceptually designed and operated as a conventional trickling filter unit, considering a sponge-based packing media as a filling material. The system is designed to operate as a post-treatment of effluents of UASB reactors, at low hydraulic loadings. Recirculation is often considered to provide a better wetting efficiency. High sludge retention time, endogenous respiration and a significant presence of microfauna within the system are factors influencing low excess sludge production. Thus, for sponge-bed trickling filters following UASB reactors, the secondary settler tends to be smaller, or even removed from the flowsheet in some cases.
Submerged aerated biofilter	The submerged aerated biofilter comprises a tank filled with a porous material (usually submerged), through which the liquid and air flow permanently. The air flow is always upward, while the liquid flow can be downward or upward. The biofilters with granular material conduct, in the same reactor, the removal of the remaining

(*Continued*)

Table 10.1 Overview of the main options for the post-treatment of anaerobic effluents (*Continued*).

Post-Treatment Process	Short Description
	soluble organic compounds and particulate matter. In addition to being a support medium for biomass growth, the granular material acts as a filter medium. Periodic backwashing is necessary to eliminate the accumulation of excess biomass, reducing the head loss through the medium. Energy consumption, associated with aeration, is high. Nitrification is expected to occur in the tank.
Conventional activated sludge	The biological stage comprises two units, namely, the aeration tank (reactor) and the secondary sedimentation tank. The biomass concentration in the reactor is extremely high because of the recirculation of the settled solids (bacteria) from the bottom of the secondary sedimentation tank. The biomass remains in the system longer than the liquid, which guarantees high BOD removal efficiency. It is necessary to remove a quantity of the sludge (biomass) that is equivalent to what is produced. This removed excess sludge needs to be stabilized, and, in many instances, is sent to the UASB reactors to undergo further thickening and digestion, and to be removed together with the anaerobic sludge. Oxygen supply is provided by mechanical aerators or by diffused air. Energy consumption, associated with the aeration, is high. Nitrification is expected to occur in the aeration tank.
Activated sludge with biological nitrogen removal	The biological reactor incorporates an anoxic zone (absence of oxygen, but presence of nitrates). The anoxic zone can be upstream and/or downstream of the aerated zone. The nitrates formed in the nitrification process that takes place in the aerobic zone are used in the respiration of facultative microorganisms in the anoxic zones, being reduced to gaseous molecular nitrogen, which escapes to the atmosphere.
Activated sludge with biological nitrogen and phosphorus removal	In addition to the aerobic and anoxic zones, the biological reactor incorporates an anaerobic zone, situated at the upper end of the tank. Internal recirculation successively exposes the biomass to anaerobic and aerobic conditions. With the alternation, a certain group of microorganisms absorbs phosphorus from the liquid medium in quantities that are much higher than those normally necessary for their metabolism. The withdrawal of these organisms in the excess sludge results in the removal of phosphorus from the biological reactor.
Dissolved air flotation	Treatment takes place by physical-chemical methods. The tank receives the input of fine air bubbles from the bottom. These bubbles are produced by air that is saturated under high pressure in a tank (saturation tank), and the liquid from this tank, with high concentrations of dissolved air, flows to the treatment reactor where the pressure is decreased suddenly, leading to the formation of small air bubbles. These bubbles, in their upward movement, attach to the particulate matter, dragging and transporting it to the tank surface from where it is removed. Chemical products, such as metallic salts, are added to the anaerobic effluent, which can undergo coagulation and flocculation, rendering the organic matter and phosphorus in the form of suspended solids that will be removed by the bubbles. The consumption of energy and chemical products is high. The excess sludge produced (removed from the upper part of the tank) needs to be treated further and disposed of.

Source: Adapted from von Sperling and Chernicharo (2005).

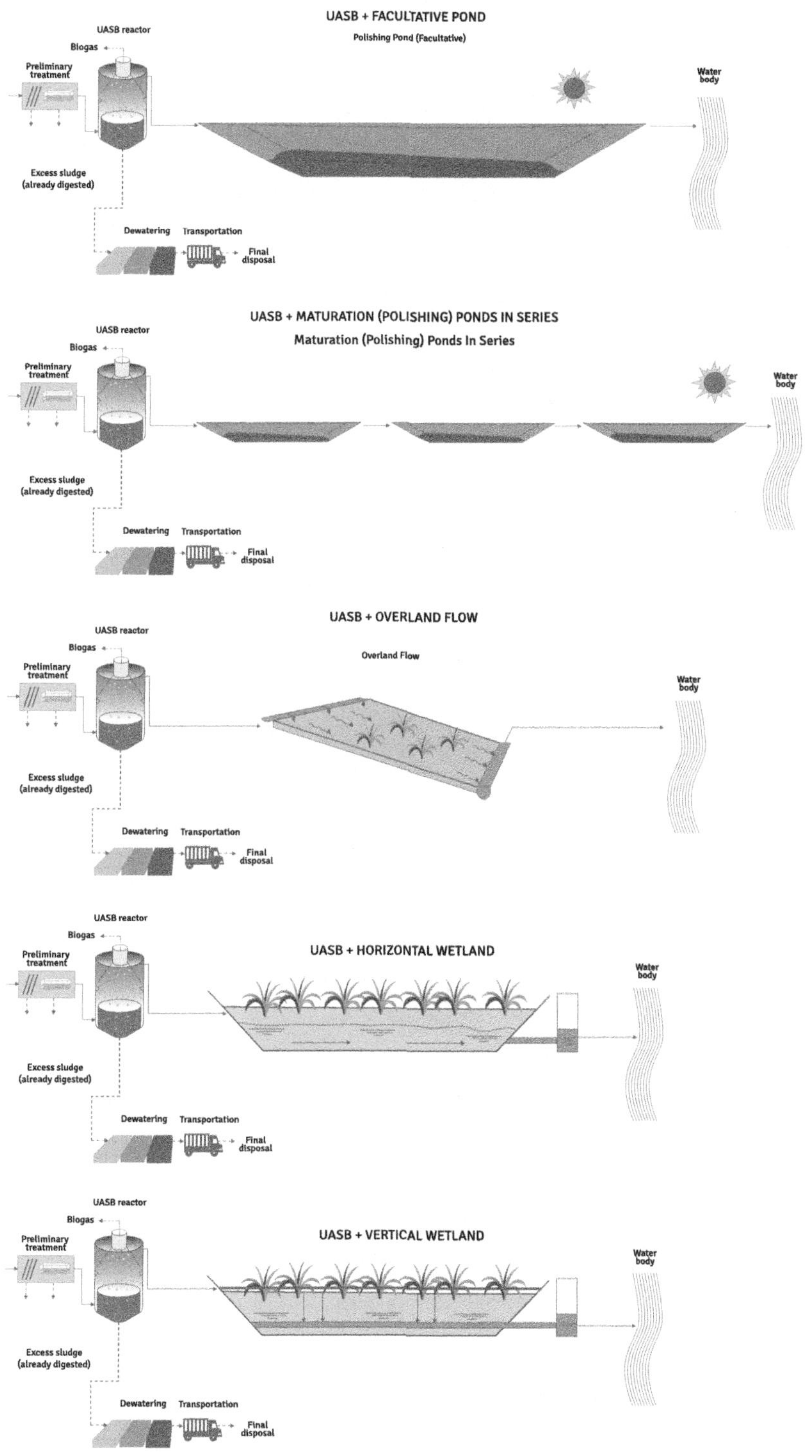

Figure 10.1 (a) Flowsheets of UASB reactors, followed by extensive post-treatment processes.

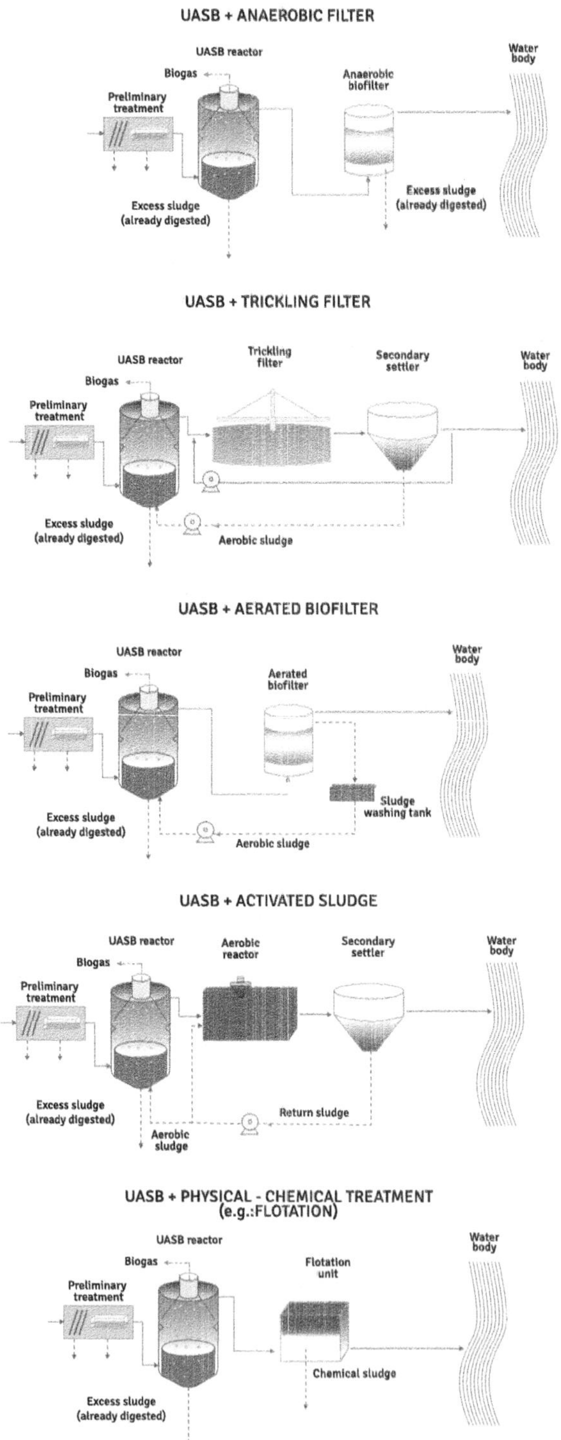

Figure 10.1 (b) Flowsheets of UASB reactors, followed by intensive post-treatment processes.

Table 10.2 Main mechanisms for the removal of pollutants in the post-treatment of effluents from UASB reactors.

Pollutant	Subdivision	Main Removal Mechanisms		Post-treatment Process, in which Mechanism is Important
Solids	Suspended solids (>~1 μm)	*Sedimentation*	Separation of the particles with density greater than the liquid	• Ponds • Wetlands • Overland flow • Activated sludge, aerated biofilter, trickling filter (sedimentation of biomass) • Chemical precipitation
		Filtration	Retention of particles by size, adsorption, and other mechanisms	• Wetlands • Trickling filters • Anaerobic filters • Granular filters • Processes with membranes: microfiltration, ultrafiltration
	Dissolved solids (<~1 μm)	*Adsorption*	Retention on the surface of biomass flocs or biofilms	• Ponds • Wetlands • Overland flow • Activated sludge, aerated biofilter, trickling filter, anaerobic filter • Chemical precipitation (adsorption involving physical-chemical interactions)
		Physical barrier		• Processes with membranes: nanofiltration, reverse osmosis
Organic matter	BOD in suspension (particulate BOD) (>~1 μm)	*Sedimentation*	Separation of the particles with density greater than the sewage	• Ponds • Wetlands • Overland flow • Activated sludge, aerated biofilter, trickling filter (sedimentation of biomass) • Chemical precipitation
		Adsorption	Retention on the surface of biomass flocs or biofilms	• Ponds • Wetlands • Overland flow • Activated sludge, aerated biofilter, trickling filter, anaerobic filter • Chemical precipitation (adsorption involving physical-chemical interactions)

(*Continued*)

Table 10.2 Main mechanisms for the removal of pollutants in the post-treatment of effluents from UASB reactors (*Continued*).

Pollutant	Subdivision	Main Removal Mechanisms		Post-treatment Process, in which Mechanism is Important
		Hydrolysis	Conversion of the BOD in suspension into soluble BOD by means of enzymes, allowing its stabilization	• Most of the hydrolysis of particulate matter from sewage occurs in the UASB reactor itself
		Stabilization	Utilization by biomass as food, with conversion into gases, water, and other inert compounds	• All biological post-treatment processes
		Filtration	Retention of particles by size, adsorption, and other mechanisms	• Wetlands • Trickling filters • Anaerobic filters • *Granular filters* • Processes with membranes: microfiltration, ultrafiltration
	Soluble BOD ($<\sim 1\ \mu m$)	*Adsorption*	Retention on the surface of biomass flocs or biofilms	• Ponds • Wetlands • Overland flow • Activated sludge, aerated biofilter, trickling filter, anaerobic filter • Chemical precipitation (adsorption involving physical-chemical interactions)
		Stabilization	Utilization by biomass as food, with conversion into gases, water, and other inert compounds	• All biological post-treatment processes
		Physical barrier		• Processes with membranes: nanofiltration, reverse osmosis
Pathogens	Larger dimensions and/or organisms with protective layer (protozoan cysts and helminth eggs)	*Sedimentation*	Separation of pathogens with larger dimensions and density greater than the liquid	• Ponds • Wetlands • Overland flow
		Filtration	Retention of pathogens in a filter medium with adequate pore size	• Wetlands • Trickling filters (with limitations) • Anaerobic filters (with limitations) • Processes with membranes: microfiltration, ultrafiltration, nanofiltration, reverse osmosis

(*Continued*)

Table 10.2 Main mechanisms for the removal of pollutants in the post-treatment of effluents from UASB reactors (*Continued*).

Pollutant	Subdivision	Main Removal Mechanisms		Post-treatment Process, in which Mechanism is Important
	Lower dimensions (bacteria and viruses)	*Adverse environmental conditions*	Temperature, pH, lack of food, competition with other species, predation	• Ponds • Other extensive treatment processes (with limitations) • Sponge-bed trickling filters
		Ultraviolet radiation	Radiation from the sun (direct or indirect damage, mediated by oxygen)	• Ponds
		Physical barrier	Retention of organisms by a physical barrier, such as a membrane	• Post-treatment processes that include membranes • Bacteria: microfiltration (variable performance), ultrafiltration, nanofiltration, reverse osmosis • Viruses: nanofiltration, reverse osmosis
Nitrogen	Organic nitrogen	*Ammonification*	Conversion of organic nitrogen into ammonia	• Most of the ammonification occurs in the UASB reactor itself
	Ammonia	*Nitrification*	Conversion of ammonia into nitrite and nitrite into nitrate by nitrifying bacteria	• Ponds (with limitations), assimilation by algae • Overland flow (with limitations) • Wetlands (vertical) • Conventional trickling filters (with limitations) • Sponge-bed trickling filters • Aerated biofilters • Activated sludge
		Bacterial assimilation	Incorporation of ammonia into the composition of bacterial cells	• All biological post-treatment processes (but contribution is small)
		Stripping	Release of free ammonia (NH_3) into the atmosphere, under high pH conditions	• Ponds (with limitations)
		Plant assimilation	Assimilation by vegetation	• Wetlands (with limitations)
	Nitrate	*Denitrification*	Conversion of nitrate into molecular nitrogen (N_2), which escapes into the atmosphere under anoxic conditions	• Activated sludge (with anoxic zones and internal recirculation) • Other treatment processes that achieve nitrification, and where anoxic niches occur in flocs or biofilms

(*Continued*)

Table 10.2 Main mechanisms for the removal of pollutants in the post-treatment of effluents from UASB reactors (*Continued*).

Pollutant	Subdivision	Main Removal Mechanisms		Post-treatment Process, in which Mechanism is Important
		Physical barrier		• Processes with membranes: nanofiltration, reverse osmosis
Phosphorus	Phosphate	*Bacterial assimilation*	Excess assimilation of the phosphate from the liquid by phosphate-accumulating organisms, occurring when aerobic and anaerobic conditions are alternated	• Activated sludge (with anaerobic zones and internal recirculation) • Some assimilation occurs in the other biological treatment processes (but contribution is small)
		Precipitation	Phosphorus precipitation under conditions of high pH by adding metallic salts or by reaction with active media	• Dissolved air flotation and other chemical precipitation methods • Wetlands (reactions with active media composed of steel slag)
		Filtration, flotation, membranes	Retention of phosphorus-rich biomass, after stage of biological excessive P assimilation	• Activated sludge (with anaerobic zones and internal recirculation) associated with efficient solid–liquid separation (granular filtration, flotation, membranes)

Source: Adapted from von Sperling and Chernicharo (2005) and Tchobanoglous *et al.* (2013) to incorporate post-treatment methods.

methods afford multiple combinations of mechanisms that can facilitate the removal of major pollutants. However, to remove nutrients (nitrogen and phosphorus) with high efficiency, the choice of biological technologies is more limited, although phosphorus can be removed with chemical precipitation. Similarly, for pathogens, the choice of processes that facilitate high removal efficiencies is limited (disinfection by chlorination, ozonation, and UV-radiation by lamps is not included here, and membranes are mentioned only briefly).

As regards the expected effluent concentrations, *in most instances, the effluent quality is dictated mainly by the post-treatment process.* The previous anaerobic stage removes a large fraction of organic matter and prepares the post-treatment units to use the other mechanisms, as listed in Table 10.2. The effluent concentration of a system comprising, e.g., a UASB reactor followed by activated sludge process, will not differ significantly from the effluent concentration of an activated sludge process treating raw sewage or primary settled effluent. Obviously, the interaction of two stages in the treatment flowsheet imparts robustness to the treatment plant.

Table 10.3 presents the typical expected ranges of removal efficiencies from UASB reactors alone and with different types of post-treatment. In many instances, the ranges have been established by actual monitoring data. In other instances, when the monitoring data did not represent the plants under normal operating conditions, the values in the ranges have been adjusted. For other processes and combinations, with no consolidated data from full-scale plants, the values have been assumed, considering the potential from each unit operation and process involved, and referring to the literature for treatment processes in general, and not specifically for post-treatment. Therefore, the table is a reflection of the expectation of values and cannot be regarded as a fully consolidated guideline. Note: there is still much room to expand the database on the performance of UASB reactors followed by post-treatment.

Table 10.3 has been interpreted in conjunction with the analysis of Figures 10.2 to 10.7. In the Figures, the removal efficiencies have been classified into three ranges (low, intermediate, and high), for easy viewing the categories into which each treatment combination may fall. These values consider the reality of developing countries, whereas different classifications may be made for countries with more stringent discharge requirements. Each graph (Figures 10.2, 10.3, 10.5, 10.6, and 10.7) depicts only the UASB reactor at the bottom, followed by a combination of the anaerobic reactor and the extensive systems for post-treatment, and lastly in combination with the intensive systems. Specific different post-treatment technologies are presented with a different legend.

10.3.1.1 Removal of organic matter

UASB reactors, without post-treatment, have BOD removal efficiencies in the range 60 to 75%, and COD removal efficiencies of 55 to 70%, as mentioned in Chapter 4. Including post-treatment raises these efficiencies to 75–95% for BOD and 70–90% for COD, depending on the specific post-treatment process adopted. Intensive aerobic processes tend to lead to higher removal efficiencies. Frequently, including a post-treatment step is aimed at increasing the efficiency of organic matter removal, and several options are available, covering a wide range of efficiency values, all in the categories ‘intermediate’ and ‘high’ removal efficiencies.

Although not shown in the (Figure 10.2), removing organic matter is associated also with the removal of suspended solids, as shown in Table 10.3. Usually, systems that efficiently remove organic matter are also efficient in removing suspended solids, as the particulate fraction of BOD and COD is associated with the suspended solids in the effluent. Generally, most of the effluent suspended solids are associated with the biological solids (biomass) produced in the system and escape along with the final effluent. An exception is ponds, in which the effluent also contains algae, of which the presence in the final effluent cannot be controlled (unless there is a specific final step involving filtration or flotation, after the pond treatment).

Table 10.3 Expected ranges of average removal efficiencies for the major pollutants in sewage treatment and for different combinations of UASB reactors and post-treatment methods.

System	Expected Average Removal Efficiency (UASB and Post-treatment)							
	BOD_5 (%)	COD (%)	SS (%)	Ammonia (%)	Total N (%)	Total P (%)	*E. coli* (log units)	Helm. eggs (log units)
UASB reactor alone	60–75	55–70	65–80	0	0–5	0–5	0.5–1	0.5–1
UASB + facultative ponds	77–87	70–83	73–83	20–30	20–40	10–30	2–3	2–3
UASB + maturation ponds	77–87	70–83	73–83	50–65	50–65	20–50	3–5	3–4
UASB + overland flow	77–90	70–85	80–93	30–50	20–40	10–30	2–3	1–2
UASB + horizontal wetlands	80–90	75–85	87–93	20–30	20–40	10–30	2–3	2–3
UASB + vertical wetlands	83–93	75–88	87–93	60–85	20–30	10–30	2–3	2–3
UASB + activated sludge	83–95	75–90	87–93	60–85	20–40	20–40	1–2	0.5–1.5
UASB + aerated biofilter	83–93	75–88	87–93	60–85	20–40	10–35	1–2	0.5–1.5
UASB + trickling filter	80–90	73–88	87–93	20–40	20–40	10–30	1–2	0.5–1.5
UASB + sponge-bed trick. filter	90–95	85–90	90–95	~80	65–70	10–30	2–4	1–2
UASB + anaerobic filter	75–87	70–80	80–90	0–10	0–10	5–20	1–2	0.5–1.5
UASB + flotation	83–93	83–90	90–97	10–20	10–20	75–90	1–2	2–3

Source: Adapted mainly from von Sperling and Chernicharo (2005), including a table based on several references.

- These data do not represent absolute values, but rather the expected values from normally operating treatment plants.
- Values outside the ranges are possible and can, indeed, occur frequently.
- Not all ranges are based on actual monitoring data. Some, with limited full-scale data, are based on expected removal, the characteristics of the unit processes, and the operations involved in the system.
- Log units removed $= -\log_{10}$ (1-Efficiency/100).
- Efficiency (%) $= 100.\ (N_o - N)/N_o = 100.\ (1 - 10^{-\text{log units removed}})$.

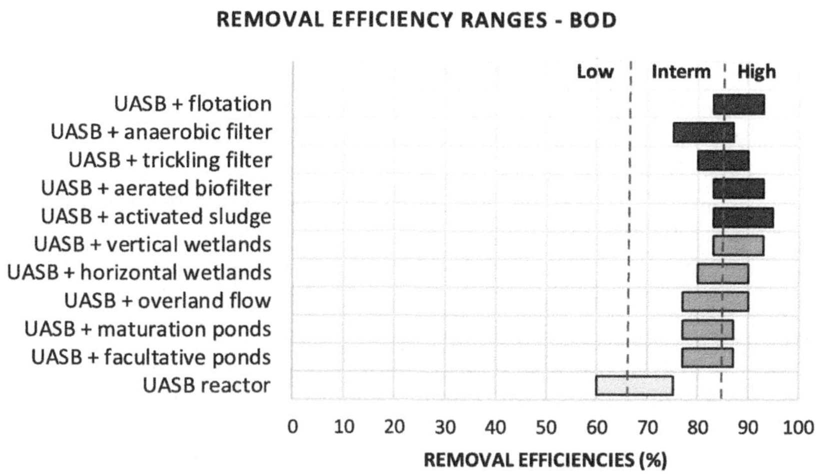

Figure 10.2 Visualization of expected typical ranges of BOD removal efficiency, according to the specific post-treatment method employed.

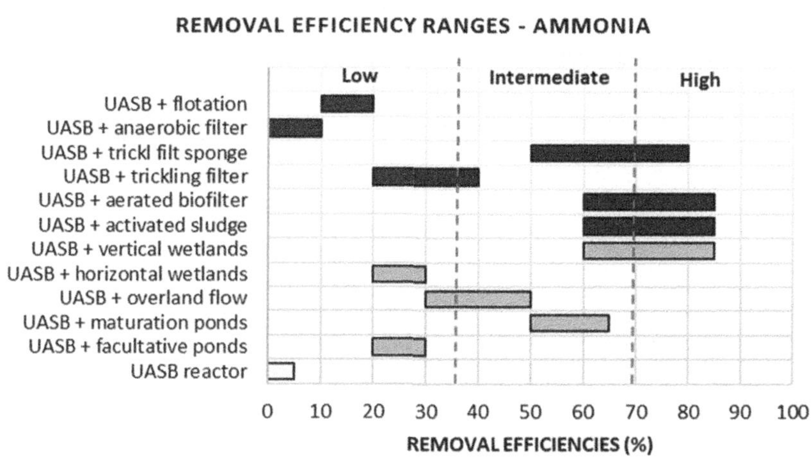

Figure 10.3 Visualization of expected typical ranges of ammonia-N (NH_4^+) removal efficiency, according to the specific post-treatment method employed.

10.3.1.2 Removal of ammonia

Frequently, ammonia removal is the second objective of incorporating a post-treatment stage. Several countries specifically require the removal of ammonia in sewage treatment, even if total nitrogen removal is not demanded.

It should be noted that when ammonia removal is done by nitrification, no actual removal of total nitrogen occurs, as ammonia is simply converted into another form of nitrogen; in this instance, nitrate. For the other removal routes (biomass assimilation, plant assimilation, or stripping), ammonia removal leads to total nitrogen removal.

UASB reactors have negligible ammonia removal efficiencies and, frequently, even negative removals, as ammonification (conversion of organic nitrogen to ammonia) occurs inside the reactor; therefore, higher ammonia concentrations are common in the UASB effluent compared with the influent (raw sewage).

Table 10.3 and Figure 10.3 show that many post-treatment processes have intrinsic low ammonia removal efficiencies. Another group of processes can facilitate only intermediate removal efficiencies, with high removals (above 70%) achieved by only a few post-treatment processes.

Figure 10.4 complements this information, presenting in the same graph the removal ranges of BOD and ammonia. The figure shows that BOD removal remains in an efficiency range that is more confined, whereas ammonia removal ranges from virtually zero to high values, close to 90%. As many different mechanisms are involved in the removal of ammonia (see Table 10.2), it is difficult to generalize a group of processes that can lead to high removals. Activated sludge, aerated biofilters, and vertical flow wetlands are highly aerobic processes that are capable of achieving high levels of nitrification (although the nitrification rates are lower compared with systems treating raw sewage or primary settled effluent, possibly because of toxicity by hydrogen sulphide in the UASB effluent). Trickling filters are also aerobic and capable of sustaining nitrifying activity but, usually, they are designed for BOD removal only and, therefore, nitrification is partial (higher efficiencies would require larger tank volumes). A special case in terms of trickling filters is when plastic and special media are adopted, as with sponge-bed filters, in which instance, nitrification could be intense. Shallow maturation ponds can also bring about intermediate to high efficiencies by the combined action of various mechanisms in the liquid and sludge columns.

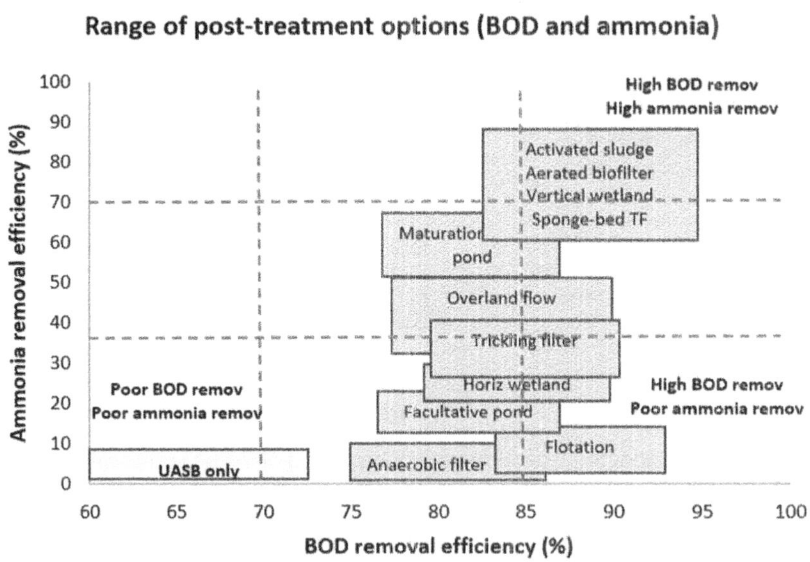

Figure 10.4 Visualization of expected typical ranges of BOD and ammonia-N (NH_4^+) removal efficiencies, according to the specific post-treatment method employed.

10.3.1.3 Removal of total nitrogen

In some countries, total nitrogen (TN) removal is required only when discharging the treated effluent in sensitive water bodies (control of eutrophication). Other countries require more systematic removal, whereas some other countries do not impose any limits on the effluent concentrations or removal efficiencies of total nitrogen. However, when using the treated effluent for agricultural use, the presence of nitrogen can be considered a bonus in the irrigation of the land (see also Chapter 12).

As Table 10.3 and Figure 10.5 show, most of the post-treatment processes usually adopted achieve extremely poor removal efficiencies of total nitrogen. This indicates that removing nitrogen is difficult and, particularly in developing countries, requirements for high removals of TN should be implemented extremely judiciously, considering the real needs of the receiving water body.

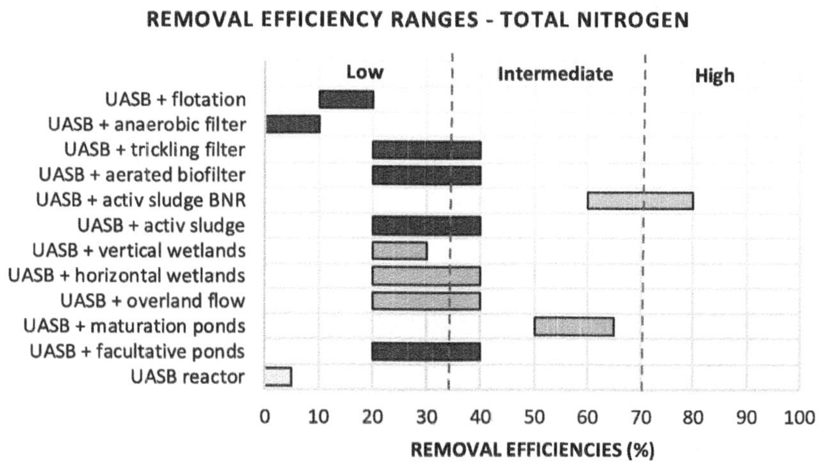

Figure 10.5 Visualization of expected typical ranges of total nitrogen (TN) removal efficiency, according to the specific post-treatment method employed.

Shallow maturation ponds have shown satisfactory removal efficiencies because of the multitude of mechanisms occurring in the water column, liquid-sludge interface, and sludge layer. Activated sludge systems, when designed to perform biological nutrient removal (BNR), can achieve high levels of TN removal; however, for this, anoxic zones and internal recirculation are necessary. One problem of having a previous major fraction of BOD and COD removed at the UASB reactor is that organic carbon can be insufficient for the heterotrophic denitrifying organisms converting nitrate into N_2 in the anoxic zones. A solution frequently adopted is the provision of a by-pass line around the UASB reactor, allowing the feeding of raw sewage (with organic carbon) to the biological reactor in the activated sludge stage. Other treatment systems that involve nitrification will achieve enhanced TN removal if anoxic niches are present inside the flocs or in the inner layer of biofilms.

10.3.1.4 Removal of total phosphorus

The reasoning for total phosphorus (TP) removal is similar to that described for TN (control of eutrophication and differences in regulatory approaches in various countries). As regards the agricultural use of the treated effluent, the presence of phosphorus is beneficial in the irrigation of the land.

The limited availability of post-treatment technologies that can achieve efficient TP removal is evident from Table 10.3 and Figure 10.6. Of the biological processes, only activated sludge with BNR can achieve high P removals via the mechanisms of the luxury uptake of phosphate. These mechanisms are promoted by the alternation between aerobic and anaerobic conditions inside the activated sludge reactor, achieved with specific internal recirculation. For extremely high TP removal, activated sludge with BNR needs an extraordinarily effective stage of solid–liquid separation, as the biological solids that escape from the secondary sedimentation tank have an extremely high phosphorus content that could deteriorate the final effluent quality. Phosphorus removal can be achieved efficiently by physical-chemical methods. One of these is conducting post-treatment with dissolved air flotation, including the addition of metallic salt (e.g., ferric chloride or aluminium sulphate that will cause the precipitation of the dissolved P into ferric phosphate or aluminium phosphate, which is removed by solid–liquid separation in the flotation tank). In a broader sense, any of the biological post-treatment options can be followed by a physical-chemical step that includes the addition of the metallic salt and a suitable solid–liquid separation step (e.g., dissolved air flotation or sand filtration). The phosphorus removal can be high, but the chemical dosages are also high, and there is a substantial increase in the sludge production.

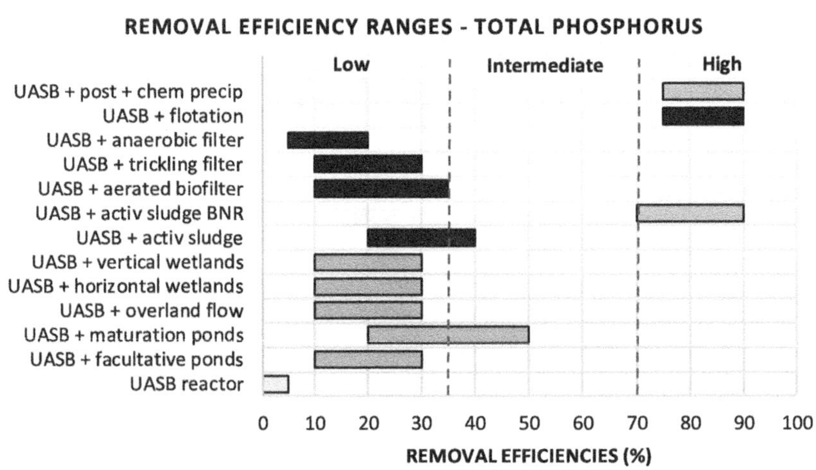

Figure 10.6 Visualization of expected typical ranges of total phosphorus (TP) removal efficiency, according to the specific post-treatment method employed.

*10.3.1.5 Removal of coliforms (*E. coli*)*

Table 10.3 presents the expected removal efficiencies of *E. coli* (as indicators of pathogenic bacterial removal and, with limitations, virus removal) and helminth eggs (as indicators of their own removal and, with limitations, of protozoan cyst removal). Figure 10.7 shows the typical efficiencies of *E. coli* removal.

A UASB reactor alone can achieve low *E. coli* removal efficiencies, as no inherent mechanisms inside the reactor are present that could bring about substantial decay, inactivation, or separation. Extensive (natural) post-treatment systems have coliform removal efficiencies in the range of 2 to 3 log units (99 to 99.9%) because they have long hydraulic retention times and facilitate some inactivation or removal mechanisms. Intensive post-treatment options have lower efficiencies of approximately 1 to 2 log units (90 to 99%), because of the lower retention times and the absence of important mechanisms that bring about inactivation or removal of coliforms and pathogens. Although a removal efficiency of, e.g., 99.9% can appear high, it is not sufficient to comply with the quality requirements for the effluent for discharge or reuse. High removal efficiencies employing natural processes can be achieved with the inclusion of maturation ponds, especially shallow ponds in series, or with baffles. In this instance, in addition to the climatic conditions, the removal efficiency will depend on the total hydraulic retention time, the number of ponds in series, and the shape and depth of the ponds. If high removal efficiencies are required, the inclusion of a specific disinfection stage (chlorination, ozonation, or UV lamp radiation) or a physical barrier (membranes) can be necessary and can be implemented after any of the post-treatment processes. However, in general, higher disinfection efficiencies will be achieved with well-clarified effluents (low suspended solids content).

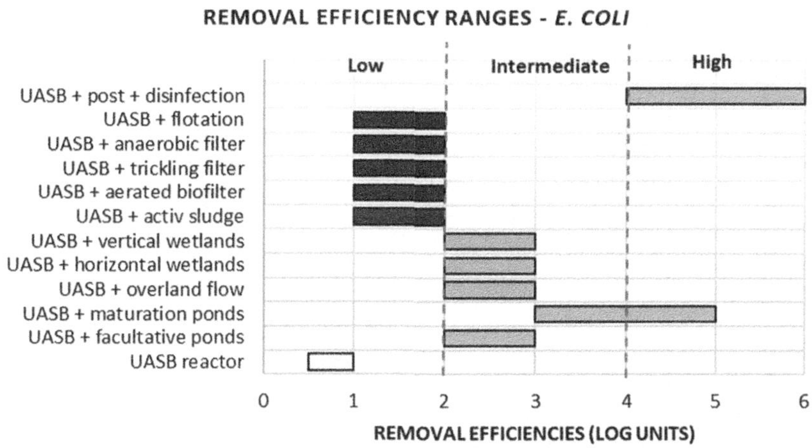

Figure 10.7 Visualization of expected typical ranges of *E. coli* removal efficiency, according to the specific post-treatment method employed. Values are expressed as $\log_{10}$ units removed (1 log unit = 90%, 2 log units = 99%, 3 log units = 99.9%, 4 log units = 99.99%, 5 log units = 99.999%, 6 log units = 99.9999%).

10.3.2 Comparison of other characteristics of the different post-treatment technologies

Table 10.4 presents a comparison of the main post-treatment alternatives related to land requirements, power for aeration, and sludge production, all of which are expressed in simple per capita terms. As expected, natural systems have substantial land requirements, no power consumption for aeration, and no need for continuous or periodic removal and treatment of the sludge generated in the system. In this instance, sludge accumulates in the biological reactor for several years, undergoing thickening and digestion. Intensive systems, on the other hand, require much less area and, if they involve an aerated reactor, they require power for aeration. Because the units in the system are compact, the sludge produced cannot be stored in the units, and needs to be continuously or periodically removed, subsequently being treated and disposed of or used in a productive manner (e.g., soil conditioning).

Von Sperling and Chernicharo (2005) and Andreoli *et al.* (2007) present a summary of the characteristics shown in Table 10.4 but applied to the treatment of raw sewage or primary settled effluent. The following conclusions can be drawn in comparing these values with those shown in Table 10.4.

- Systems composed of UASB reactors followed by post-treatment occupy much smaller land areas. The decrease is more important in the extensive systems that are designed based on organic matter load (e.g., $\text{kgBOD} \cdot \text{ha}^{-1}\ \text{d}^{-1}$ or $\text{gCOD m}^{-2}\ \text{d}^{-1}$), because of the previous organic matter removal that takes place in a compact unit, such as a UASB reactor (of which the efficiency is more than double that of a primary sedimentation tank).
- Systems composed of UASB reactors followed by post-treatment require less power for aeration. However, the decrease is not proportional to the BOD or COD reduction that occurs in the UASB

Table 10.4 General characteristics of different biological post-treatment options (land and power requirements, sludge production), expressed on a per capita basis.

System	Land Requirements ($m^2 \cdot inhab^{-1}$)*	Power for Aeration		Sludge Volume	
		Installed Power ($W \cdot inhab^{-1}$)	Consumed Power ($kWh \cdot inhab^{-1} \cdot y^{-1}$)	Liquid Sludge to be Treated ($L \cdot inhab^{-1} \cdot y^{-1}$)	Dewatered Sludge to be Disposed of ($L \cdot inhab^{-1} \cdot y^{-1}$)
UASB reactor	0.03–0.10	0	0	70–220	10–35
UASB + facultative ponds	0.5–1.5	0	0	**	**
UASB + maturation ponds	1.5–5.0	0	0	**	**
UASB + overland flow	1.5–3.5	0	0	**	**
UASB + horizontal wetlands	1.5–3.0	0	0	**	**
UASB + vertical wetlands	0.5–2.5	0	0	**	**
UASB + activated sludge	0.08–0.2	1.8–3.5	14–20	180–400	15–60
UASB + aerated biofilter	0.05–0.15	1.8–3.5	14–20	180–400	15–55
UASB + trickling filter	0.1–0.2	0	0	180–400	15–55
UASB + anaerobic filter	0.05–0.15	0	0	150–300	10–50

Source: Adapted from von Sperling and Chernicharo (2005).

Notes:

*Land requirements (extensive systems): based on warm climatic conditions.

**Sludge production in extensive (natural) systems is not shown here, because the sludge accumulates very slowly, is digested, is stored in the biological reactor and is removed at extremely long intervals (several years).

reactor, as oxygen (and therefore power) will still be required to oxidize ammonia in the aerated systems (activated sludge and aerated biofilter), as ammonia is not removed in the UASB reactor.

- Systems composed of UASB reactors followed by post-treatment produce much less excess sludge that needs to be treated and disposed of. UASB reactors produce small quantities of sludge, and the remaining organic matter will produce only a small amount of sludge in the post-treatment stage. This is one of the significant advantages of these systems, implying saving associated with the sludge management.

Table 10.5 presents, in a semi-qualitative structure, a comparison of additional elements in the selection criteria of the process to be adopted for post-treatment. A clear distinction in the two major groups (extensive and intensive post-treatment processes) is observed. Interestingly, some characteristics are dictated mainly by the UASB reactor itself, whereas other elements are related to the post-treatment process *per se*.

10.4 DESCRIPTION OF THE MAIN TECHNOLOGIES FOR THE POST-TREATMENT OF ANAEROBIC EFFLUENTS

10.4.1 Introduction

As mentioned previously, the description of the main post-treatment processes is presented from the perspective of sewage treatment. Although many of the concepts are applicable also to industrial wastewater, the design parameters presented are specific to UASB reactors treating sewage.

Although physical-chemical post-treatment can play an important role under some specific circumstances, only biological post-treatment processes are described in this section. Of the latter, only the major alternatives are presented, either because more consolidated knowledge is available to sustain their design and operation, or because they represent widely adopted processes, such as those used extensively in Brazil (see Chapter 1). Because of their substantial potential, some processes are included although they might not be used widely yet in the post-treatment of anaerobic effluents.

The description is oriented to the post-treatment process, and not to the treatment objective. The processes start with the extensive (natural) systems and subsequently move to the intensive (compact or mechanized) systems. The following processes are discussed:

Extensive Systems	Intensive Systems
• Polishing ponds	• Trickling filters
• Overland flow	• Aerated biofilters
• Constructed wetlands	• Activated sludge

For each process, the following elements are presented: short description, design criteria, construction, operation, design example (preliminary design for conception studies, with the calculation of only the main dimensions, and without detailed design).

10.4.2 Polishing ponds (facultative ponds and maturation ponds)

10.4.2.1 Description

Conventional stabilization ponds are discussed in several international textbooks (Arceivala, 1981; Mara *et al.*, 1992; Yanez, 1993; Mara, 2003; von Sperling & Chernicharo, 2005; Shilton, 2006; von Sperling,

Table 10.5 Semi-qualitative comparison between different biological post-treatment options in terms of several inherent characteristics.

Treatment System	Removal Efficiency			Economy					Resistance Capacity to Influent Variations and Shock Loads			Reliability	Simplicity in O&M.	Independence of Other Charact. for Good Perform.		Lower Possibility of Environmental Problems			
	BOD or COD	Nutrients (N and P)	Coliforms	Requirements		Costs		Generation	Flow	Quality	Toxic Comp.			Climate	Soil	Bad Odours	Noise	Aerosols	Insects and Worms
				Land	Energy	Constr.	O & M	Sludge											
UASB reactor	++	+	+	+++++	+++++	++++	+++	++++	++	++	++	+++	++++	++	+++++	+	+++++	+++++	++++
UASB + facultative pond	+++	++	+++	+	+++++	+++	+++++	+++++	+++++	+++++	+++++	++++	+++++	+	++	+++	+++++	+++++	++
UASB + maturation pond	+++	+++	+++++	+	+++++	++	+++++	+++++	+++++	+++++	+++++	+++++	+++++	+	++	++++	+++++	+++++	++
UASB + overland flow	++++	+++	+++	+	+++++	+++	+++++	+++++	+++++	+++++	+++++	++++	+++++	++	++	+++	+++++	+++++	++
UASB + horizontal wetlands	++++	++	+++	+	+++++	+++	+++++	+++++	++++	+++++	+++++	+++++	+++++	++	++	++++	+++++	+++++	++
UASB + vertical wetlands	++++	+++	+++	+	+++++	+++	+++++	+++++	++++	+++++	+++++	++++	+++++	++	++	++++	+++++	+++++	++
UASB + activated sludge	+++++	++/++++	++	++++	+	+	+	+	+++	+++	++	++++	+	++++	+++++	+++++	+	+/+++++	+++++
UASB + aerated biofilter	+++++	++/+++	++	++++	+	+	+	+	+++	+++	++	++++	+	++++	+++++	+++++	+	+++++	+++++
UASB + trickling filter	++++	+/+++	++	+++	++++	++	+++	++	+++	+++	+++	++++	+++	+++	+++++	+++	++++	+++++	+++
UASB + anaerobic filter	+++	+	++	++++	++++	++	+++	+++	+++	+++	++	+++	+++	+++	+++++	+	++++	+++++	+++
Comment	(a)	(a)	(a)	(a)	(a)	(a)	(a)	(a)	(b)	(b)	(b)	(a)	(a)	(a)	(a)	(b)	(a)	(a)	(a)

Source: Adapted from von Sperling and Chernicharo (2005).

Notes:

+++++ : most favourable; +: least favourable; ++++, +++, ++: intermediate grades, in decreasing order.

+/+++++: variable with the type of process, equipment, variant or design.

The grading is relative only in each column and is not generalized for all the items. The grading can vary widely with the local conditions.

Comment: (a) post-treatment characteristics prevail; (b) UASB reactor characteristics prevail.

2007a; USEPA, 2011) and are, therefore, not described here. For further details, these publications should be consulted.

Ponds acting as post-treatment of anaerobic effluents are not "stabilization" ponds *per se*, as most of the stabilization of the organic matter has already occurred in the preceding step, i.e., the anaerobic reactor. However, they do contribute to the additional removal of the remaining fraction of the organic matter originally present in the raw sewage, as well as facilitate an additional general improvement in the effluent quality. Although not representing an internationally adopted nomenclature, such ponds have also been called 'polishing' ponds. Depending on how they are conceived and designed, they can accomplish complementary objectives:

- Design as **facultative ponds**. Under this conception, the aim is mainly to remove the additional organic matter and suspended solids. The design criteria are similar to those adopted for conventional facultative ponds (either treating raw sewage or, more specifically, effluents from anaerobic ponds). The surface organic loading rate is the main element in the design. These ponds correspond to the first generation of polishing ponds.
- Design as **maturation ponds**. In this instance, the ponds are designed with the explicit objective of pathogen removal and they use design criteria applicable to this configuration, i.e., design based on hydraulic retention time, incorporation of the concepts of ponds in series, high length-to-width ratio, and shallow depth. The additional removal of nitrogen also occurs. These ponds correspond to the second generation of polishing ponds.

It should be understood that high-level improvement in the removal of organic matter and suspended solids is not possible with polishing ponds. This is because although they do remove the fraction of organic matter remaining in the anaerobic effluent, they simultaneously produce another type of organic matter that is associated with the algae naturally produced in the ponds. Algae occur in suspended solids and organic matter (particulate BOD and COD) and, unless there is a **final stage of clarification** (coarse filters, dissolved air flotation, or other process; refer to Barros *et al.*, 2013; Dias *et al.*, 2018; von Sperling *et al.*, 2007), the effluent SS, BOD, and COD concentrations will be relatively high.

Therefore, additional treatment objectives need to be included when incorporating polishing ponds in the treatment scheme. Usually, these incorporate pathogens removal, and it is important to emphasize that well designed ponds can achieve extremely high removal efficiencies of the four major categories of pathogenic organisms, namely, bacteria, viruses, protozoan cysts, and helminth eggs.

The first generation of polishing ponds was based on the concept of facultative ponds, with some being even more simple, aiming only at the additional settlement of solids. The adoption of facultative ponds, with their usually larger area, is ascribed to concerns about the possible generation of bad odours emanating from an excessive applied surface organic load, which could be higher than the oxygen produced by the photosynthetic activity of algae. Photosynthesis needs sunlight penetration in the pond, and, for this, large surface areas were thought to be compulsory. However, later experience showed that the organic matter present in the effluent from anaerobic reactors had slow biodegradability, with oxygen consumption proceeding at a low rate. Therefore, sufficient compensation is achieved by photosynthetic oxygen production, even in a pond with a smaller surface area. This finding led to the application of smaller ponds in series, as the first pond in the series would probably not be organically overloaded and would not release hydrogen sulphide under normal operation (Catunda *et al.*, 1994; Cavalcanti *et al.*, 2001; von Sperling, 2007a). The significant advantage of having ponds in series is that pathogen removal (bacteria and viruses), which follows a first-order reaction, can benefit substantially from treatment units arranged in series. Therefore, maturation ponds, and not necessarily facultative ponds, could be applied after the anaerobic reactors. In this second arrangement, involving maturation ponds, the main design parameter would be hydraulic retention time (for pathogen removal), and not surface organic loading rate (used for facultative ponds).

One alternative to ponds in series is ponds with baffles, which have a large length/width (L/W) ratio and, therefore, a flow pattern that seeks to approach plug-flow, which leads to a more efficient reactor than completely mixed reactors for removing constituents that decay according to a first-order reaction.

Another distinguishing element in both configurations is pond depth. As regards facultative ponds, the depth is generally between 1.5 and 2.0 m, whereas maturation ponds are much shallower (lower than 1.0 m). The lower depth is intended to maximize the factors that bring about pathogen inactivation (high pH, high DO, and high UV penetration), which are all related to small water columns.

Figure 10.8 presents two typical flowsheets involving UASB reactors and polishing ponds.

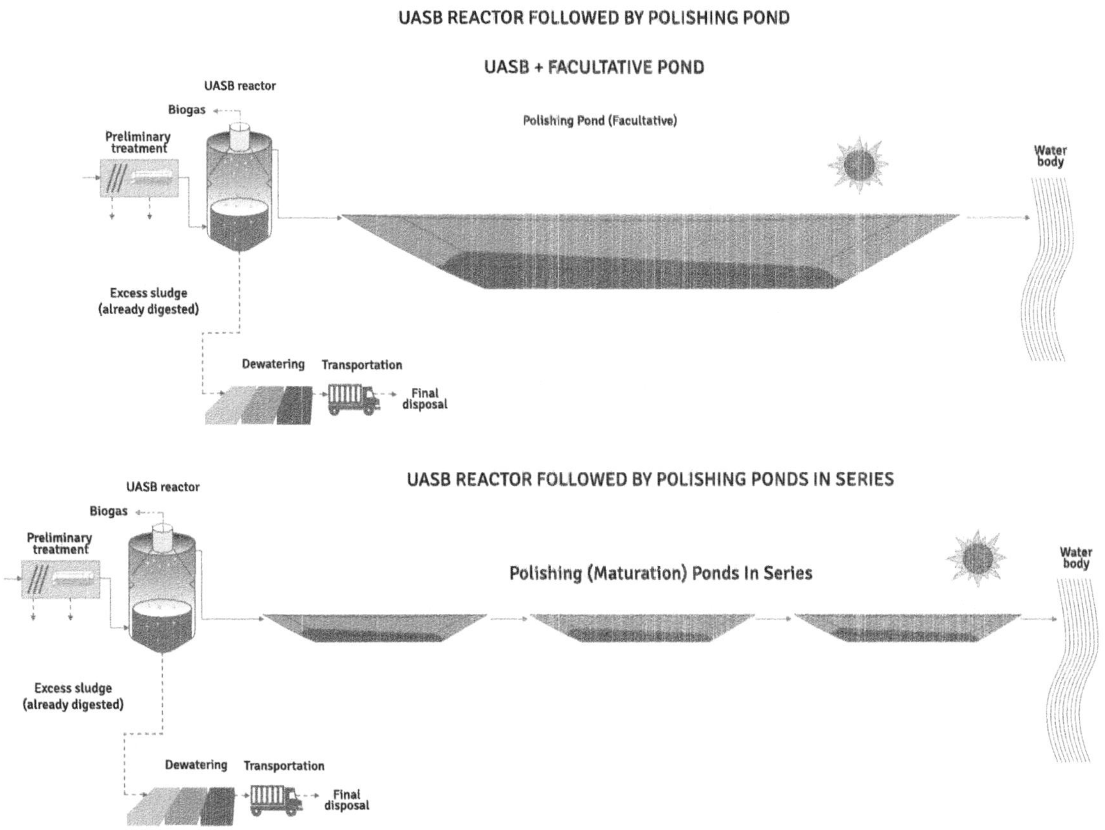

Figure 10.8 Flowsheets of two different arrangements involving UASB reactors and polishing ponds.

10.4.2.2 Design criteria

Table 10.6 presents the main design criteria for ponds acting as post-treatment of anaerobic effluents. Most of these are guidelines consolidated for conventional ponds but they have been adapted for this particular application.

Table 10.6 Typical design criteria for ponds treating effluent from UASB reactors (warm-climate regions).

Design Parameter	Facultative	Maturation
Total hydraulic retention time HRT (d)	5–15*	**
Surface loading rate L_s ($kgBOD_5 \cdot ha^{-1}\ d^{-1}$)	120–350	–
Depth H (m)	1.5–2.0	0.6–1.0
L/W ratio (length/width)	2–4	***
Number of ponds in the series	1	1–5****

Source: Adapted from von Sperling (2007a).
For details of the parameters and relevant equations, see text and the cited publications. L = length (m); W = width (m).
*The retention time in a facultative pond is a value resulting from the design based on surface organic loading rates, and not a direct design parameter. As such, the resulting values of HRT are much lower than are the conventional ones in facultative ponds treating raw sewage (primary facultative ponds), because of the BOD reduction of approximately 70% that occurs in the preceding UASB reactors.
**The retention time in maturation ponds is a combined function of the number of ponds, pond shape, and the required coliform removal efficiency.
***Single pond with baffles: $L/W > 10$; ponds in series without baffles: L/W between 3 and 5.
****For maturation ponds without baffles, the number of ponds in series is usually between 3 and 5. Where the ponds are baffled (large L/W ratio), 1–2 ponds in series can be used.

It is difficult to estimate the effluent quality in terms of organic matter (BOD and COD concentrations). The soluble fraction can be estimated using existing models but, probably, their concentrations will be low compared with the particulate BOD and COD, particularly considering that substantial removal has already taken place in the previous anaerobic reactor. On the other hand, the effluent particulate BOD and COD concentrations are associated directly with the algae present in the final effluent, which is difficult to estimate. Therefore, for design purposes, it is probably acceptable to estimate the effluent organic matter concentrations based on simple ranges of removal efficiencies, such as those presented in Table 10.3.

If coarse rock filters are adopted after the ponds to remove algae (particulate BOD and COD), they could be designed using volumetric hydraulic loading rates between 0.5 and 1.5 $m^3\ d^{-1}$ per m^3 of filter bed. Typical stone dimensions are between 50 and 200 mm (Mara, 2003; von Sperling, 2007a; von Sperling *et al.*, 2007).

When polishing ponds are designed as maturation ponds, aiming primarily at pathogen removal, the estimation of effluent concentrations of indicator organisms, such as thermotolerant coliforms or *E. coli*, is important because these will dictate the required retention times, number of ponds in series, and pond shape. Several complex mechanisms act simultaneously toward the inactivation and removal of the pathogenic organisms (Verbyla *et al.*, 2017) associated with complex kinetic and hydraulic models; however, this is outside the scope of this Chapter. Table 10.7 presents a simple estimation procedure of the coliform removal efficiency, proposed by von Sperling (2005), based on the actual removal efficiency of 186 facultative and maturation ponds worldwide, modelled using the dispersed-flow hydraulic model. These estimates have been shown to apply well also to polishing ponds treating anaerobic effluents. The log removal values in Table 10.7 refer to each individual pond in the series, and the values from each pond can be added to calculate an estimation of the overall removal efficiency. The table presents log removal values pertaining to a liquid temperature of 20°C. For a temperature of 25°C, a similar table presented in von Sperling (2005) can be consulted but, usually, the removal efficiency values are between 20 and 35% higher for the same retention time (HRT), depth (H), and L/W ratio shown in Table 10.7.

Table 10.7 Expected average coliform removal efficiencies from facultative and maturation ponds, expressed in terms of logarithmic units removed, for different values of the hydraulic retention time (HRT), depth (*H*), and *L*/*W* ratio. Liquid temperature = 20°C.

HRT (d)	*H* (m)	Log Units Removed								
		L/*W* Ratio								
		1	2	3	4	6	8	10	12	16
3	0.6	0.80	0.90	0.97	1.02	1.10	1.16	1.20	1.23	1.28
	1.0	0.48	0.52	0.55	0.57	0.60	0.62	0.63	0.64	0.66
	1.5	0.31	0.32	0.33	0.34	0.35	0.36	0.37	0.37	0.38
	2.0	0.22	0.23	0.23	0.24	0.24	0.24	0.25	0.25	0.25
5	0.6	1.09	1.26	1.39	1.48	1.63	1.73	1.81	1.88	1.97
	1.0	0.69	0.76	0.82	0.86	0.92	0.96	0.99	1.01	1.04
	1.5	0.45	0.49	0.51	0.53	0.56	0.57	0.59	0.60	0.61
	2.0	0.33	0.35	0.36	0.37	0.39	0.39	0.40	0.40	0.41
10	0.6	1.62	1.94	2.18	2.36	2.65	2.87	3.05	3.19	3.41
	1.0	1.06	1.22	1.34	1.43	1.57	1.67	1.74	1.80	1.89
	1.5	0.73	0.82	0.88	0.93	0.99	1.04	1.08	1.10	1.14
	2.0	0.55	0.60	0.64	0.67	0.70	0.73	0.75	0.76	0.78
20	0.6	2.34	2.89	3.30	3.64	4.16	4.57	4.90	5.19	5.65
	1.0	1.58	1.88	2.11	2.29	2.56	2.77	2.94	3.07	3.29
	1.5	1.13	1.31	1.44	1.54	1.69	1.80	1.89	1.96	2.06
	2.0	0.88	0.99	1.07	1.14	1.23	1.30	1.35	1.39	1.45
30	0.6	2.88	3.62	4.17	4.61	5.33	5.89	6.36	6.77	7.44
	1.0	1.96	2.39	2.71	2.96	3.35	3.66	3.91	4.11	4.44
	1.5	1.43	1.69	1.88	2.03	2.26	2.43	2.57	2.68	2.85
	2.0	1.12	1.30	1.42	1.52	1.67	1.78	1.87	1.93	2.04

Source: von Sperling (2005).

- HRT = hydraulic retention time in each pond (d); *H* = pond depth (m); *L*/*W* ratio = length-to-width ratio; *L* = pond length (m); *W* = pond breadth or width (m).
- Log units removed = $-\log_{10}$ (1 – Efficiency/100).
- Efficiency (%) = 100 $(N_o\text{-}N)/N_o$ = 100 $(1 - 10^{-\text{log units removed}})$.
- Total log units removed in a system with ponds in series = sum of the log units removed in each individual pond in the series.
- Values with *H* = 0.6 m are based on a smaller database, but most of the ponds in this category are polishing ponds.

The removal of helminth eggs can be estimated by using the model proposed by Ayres *et al.* (1992), also explained in von Sperling (2007a), which has been shown to apply satisfactorily to polishing ponds treating anaerobic effluents (von Sperling *et al.*, 2005). For the same retention times as used in Table 10.7, Table 10.8

Table 10.8 Expected average helminth eggs removal efficiencies from facultative and maturation ponds, expressed in terms of logarithmic units removed for different values of the hydraulic retention time (HRT), using the Ayres *et al.* (1992) model.

HRT (d)	3	5	10	20	30
Log_{10} units removed	1.35	1.68	2.50	4.15	5.80

presents the expected values of the average removal efficiencies, also expressed as log unit values, calculated using the equation in Ayres *et al.* (1992).

It is clear that significant efficiencies can be achieved in the removal of helminth eggs, as the ponds act as large sedimentation tanks. In fact, if a pond system is designed for high coliform removal, it is expected indirectly that extremely high removal of helminth eggs will be achieved.

10.4.2.3 Construction details

The construction of polishing ponds for the post-treatment of anaerobic effluents is similar to that of conventional facultative and maturation ponds, as described in the general literature on ponds (Arceivala, 1981; Mara *et al.*, 1992; Yanez, 1993; Mara, 2003; Shilton, 2006; von Sperling, 2007a; USEPA, 2011). Sufficient attention must be given to the hydraulic details pertaining to the inlet and outlet structures to guarantee the uniform distribution and collection of the liquid and to avoid short circuits, particularly when high-level coliform removal efficiency is desired. Employing bottom and internal embankment lining, using compacted clay or geomembranes is important to prevent groundwater pollution and pond emptying by severe exfiltration.

10.4.2.4 Operational aspects

Because of their inherent simplicity, there are no control variables in ponds, and their operation is simple; however, maintenance of the ponds is essential and their simplicity is, therefore, no excuse for lax or absent operational staff. Preliminary treatment units (screens and grit chambers), together with pond inlet and outlet structures need to be inspected and cleaned regularly. In addition, vegetation growth must be prevented on the internal embankments.

An important element in pond operation is the management of the bottom sludge. The accumulation of sediment is extremely slow, and ponds can operate for several years without the need to remove the sludge. This is particularly relevant to polishing ponds acting as post-treatment of anaerobic reactors. Most of the original suspended solids from the raw sewage are removed in the anaerobic reactor, and the growth of heterotrophic biomass is low because of the lower availability of organic matter, again, because of the previous removal in the anaerobic reactor. However, algae grow in ponds, and part of their mass can settle and be incorporated in the sediments. The bottom sludge undergoes thickening and digestion, and is spread over a large area, justifying the low rates of volume increase.

In conventional facultative ponds, sludge accumulates at rates between 0.03 and 0.08 m^3 $year^{-1}$ per population equivalent (PE). Information is scarce on full-scale ponds acting as post-treatment of anaerobic reactors, but values of approximately 0.02 m^3 $year^{-1}$ per PE have been found for the first maturation pond in the series or the first polishing pond after the UASB reactor, whereas values between 0.004 and 0.009 m^3 $year^{-1}$ per PE have been found for the second maturation or polishing pond (Possmoser-Nascimento *et al.*, 2014). Therefore, sludge accumulation proceeds at a low rate in polishing ponds, and the need for its removal is considered only when the volume occupies approximately 1/3 of the pond volume. Sludge removal is a major operational issue, with different possibilities, either with the pond in operation, or after draining of the pond and rest for several months, which allows the sludge to dry.

EXAMPLE 10.1

Design of a system of polishing (maturation) ponds for the post-treatment of the effluent from anaerobic reactors, with the main objective being removing coliforms. The expected *E. coli* concentration in the final effluent is 1,000 MPN 100 mL^{-1}.

Input data

- Population equivalent: PE = 2,000 inhabitants
- Average flow: Q = 250 $m^3\ d^{-1}$
- Influent *E. coli* concentration (raw sewage): 1.00×10^8 MPN 100 mL^{-1}
- Influent BOD concentration (raw sewage): 400 mg L^{-1}
- Liquid temperature: 20°C

Solution

(1) Estimate the *E. coli* concentration in the effluent from the UASB reactors
From Table 10.3, assume a removal efficiency of 1 log unit (90%) in the UASB reactor. Therefore, the effluent from the UASB reactor will have a concentration of:

$$N_{effluent} = N_{influent} \times (1 - \text{Efficiency}) = 1.00 \times 10^8 \text{ MPN 100 mL}^{-1} \times (1 - 0.9)$$
$$= 1.00 \times 10^7 \text{ MPN 100 mL}^{-1}$$

This value is the influent to the polishing ponds, which will be designed as maturation ponds.

(2) Calculate the required removal efficiency at the polishing ponds
As the influent concentration to the ponds is 1.00×10^7 MPN 100 mL^{-1} and the desired effluent concentration is 1.00×10^3 MPN 100 mL^{-1}, the required removal efficiency at the pond stage is:

$$\text{Required efficiency (log units)} = \log10(N_{influent}) - \log10(N_{effluent})$$
$$= \log_{10}(1.00 \times 10^7) - \log_{10}(1.00 \times 10^3)$$
$$= 7.0 - 3.0 = 4.0 \text{ log units (or 99.99\%)}$$

(3) Combination of hydraulic retention time, pond depth, length/width ratio, and number of ponds in series
To obtain the desired efficiency of 4.0 log units removed, different combinations of hydraulic retention time, pond depth, length/width ratio, and number of ponds in series can be adopted (there is no single solution, as several variables act simultaneously). If four ponds in series ($n = 4$) are adopted, the required efficiency at each pond will be:

$$\text{Efficiency one pond} = \text{Efficiency all ponds}/n = 4.0/4 = 1.0 \text{ log unit per pond}$$

From Table 10.7, if the following are adopted, namely, depth $H = 1.0$ m, L/W ratio = 6, then, for HRT = 5, the expected removal efficiency is 0.92 log units. If, for the same depth H and L/W ratio, HRT = 10 d is adopted, the expected removal efficiency is 1.57 log units. With simple linear interpolation between these values, the removal efficiency increases (1.57–0.92)/(10–5) = 0.13 log units for each day under these conditions. Therefore, a retention time of 6 days in each pond will be sufficient for achieving 1.0 log units in each pond or 4.0 log units in all four ponds in the series.
Different combinations of n, HRT, H, and L/W can be tried, using Table 10.7, and assuming that simple linear interpolations are possible. Obviously, this is a simplification, and an improved approach is to use models (see von Sperling, 2005, in which the model is presented, and the derivation of the table is explained). Although derived from a smaller database, information shows that ponds with shallow depths (e.g., $H = 0.6$ m) have the advantage of a higher coliform decay coefficient, which compensates for the smaller depth and HRT for the same surface area. Therefore, an option in this design could be to adopt a shallower depth, with the resulting design being less conservative.

(4) Determine the physical dimensions of each pond
Given the average flow of 250 $m^3\ d^{-1}$ and the hydraulic retention time in each pond equal to 6 days, the required volume for each pond will be:

$$V = Q \times HRT = 250\ m^3\ d^{-1} \times 6\ d = 1{,}500\ m^3$$

Given the proposed depth of 1.0 m, the required surface area of each of the four ponds will be:

$$A = V/H = (1{,}500\ m^3)/(1.0\ m) = 1{,}500\ m^2$$

With the adoption of a L/W ratio of 6, the length and width will be given by:

$$A = L \times W = [(L/W) \times W] \times W = (L/W) \times W^2$$

$$W = \sqrt{\frac{A}{(L/W)}} = \sqrt{\frac{1{,}500\ m^2}{6}} = 15.8\ m$$

$$L = A/W = (1{,}500\ m^2)/(15.8\ m) = 94.9\ m$$

In summary, the pond dimensions will be:

- Number of ponds in the series: $n = 4$
- Depth of each pond: $H = 1.0$ m
- Length of each pond: $L = 94.9$ m
- Width (breadth) of each pond: $W = 15.8$ m

The dimensions of length and width are at mid-depth. The dimensions at the bottom and top of the pond will depend on the internal slope of the embankment, of which the definition is a function of the soil type.
The area of each pond will be 1,500 m^2 and, therefore, the total area of the four ponds will be 4 × 1,500 m^2 = 6,000 m^2. For a population of 2,000 inhabitants, this represents a net per capita area of 6,000 m^2/2,000 inhabitants = 3.0 m^2 inhabitant^{-1}. A large total area will be required to account for embankments, internal roads, urbanization, and the like.

(5) Verify the applied surface organic loading rate to the first pond in the series
From the input data, the influent flow is 250 $m^3\ d^{-1}$ and the influent BOD (raw sewage) is 400 mg L^{-1}. For design purposes, assuming a BOD removal efficiency of 65% in the UASB reactor, the effluent BOD concentration from the UASB reactor (influent to the first pond) will be:

$$BOD_{effluent\ UASB} = BOD_{influent\ UASB} \times (1 - Efficiency_{UASB}) = 400 \times (1 - 0.65)$$
$$= 140\ mg\ L^{-1} = 140\ g\ m^{-3}$$

The influent load to the first pond will be:

$$Load = Flow \times Concentration = (250\ m^3\ d^{-1}) \times (140\ g\ m^{-3})/(1000\ g\ kg^{-1})$$
$$= 35\ kg\ BOD\ d^{-1}$$

The resulting applied loading rate over the surface of the first pond (1,500 m^2 = 0.15 ha) will be:

$$Surface\ organic\ loading\ rate = Load/Area = (35\ kg\ BOD\ d^{-1})/(0.15\ ha)$$
$$= 233\ kgBOD\ ha^{-1}d^{-1}$$

This value is within the typical accepted range for facultative ponds (see Table 10.6) and, therefore, it is expected that no organic overloading will occur in the first pond. This conclusion is for the specific climatic conditions considered in this example (warm climate, temperature in the liquid = 20°C), but it should be remembered that the decomposition of the organic matter will proceed at lower rates, given the lower biodegradability of anaerobic effluents (compared with raw sewage). Therefore, a lack of oxygen is not expected in the first pond, which is the most critical in the series, as it receives a higher BOD load.
Accordingly, the first pond has the mixed characteristics of a facultative pond (because of the surface organic loading rate) and a maturation pond (because of the hydraulic retention time). This highlights the difficulty in the nomenclature on these ponds and points to the reason they are called polishing ponds.

(6) Dimensioning of the coarse rock filter for the removal of algae from the pond effluent
In accordance with Section 10.4.2.2, volumetric hydraulic loading rates (HLR_v) applied for coarse filters range between 0.5 and 1.5 $m^3\ m^{-3}\ d^{-1}$. Adopting an intermediate value of 1.0 $m^3\ m^{-3}\ d^{-1}$, the resulting required filter volume will be:

$$V = Q/HLR_v = (250\ m^3\ d^{-1})/(1.0\ m^3\ m^{-3}\ d^{-1}) = 250\ m^3$$

Adopting the same liquid depth of the ponds (H = 1.0 m), the required surface area will be:

$$A = V/H = 250\ m^3/1.0\ m = 250\ m^2$$

Adopting the same width for the ponds (W = 15.8 m), the resulting filter length will be:

$$L = A/B = (250\ m^2)/(15.8\ m) = 15.8\ m$$

Therefore, a square filter with side dimensions of 15.8 m can be adopted. As the filter has a small area compared with the ponds, it can be inserted inside the last pond. The useful volume of the last pond will decrease, but the overall influence will be small.
The height of 1.0 m is the liquid height. An additional height of the filter media of, e.g., 0.1 m can be adopted, leading to a total filter height of 1.1 m. The size of the stones should be between 50 and 200 mm.

10.4.3 Overland flow

10.4.3.1 Description

The following section is based, with adaptations, on von Sperling and Chernicharo (2005) and Chernicharo (2007). Additional details can be found in these publications.

In the overland flow method, the vegetation associated with the top soil layer acts as a filter and support medium for the growth of microorganisms, facilitating nutrient removal and providing conditions for the retention and transformation of the organic matter contained in the liquid to be treated. Furthermore, it protects the soil against erosion and creates a support layer on which the microorganisms settle. The main mechanisms through which organic matter and solids are removed are biological oxidation, sedimentation, and filtration. The main characteristic that differentiates this method from other methods of land disposal of sewage is that the effluent flows downward on a slightly inclined vegetated ramp and the remaining water (effluent), which is neither absorbed nor evapotranspirated, is collected downstream and directed for disposal. For more permeable soils, the process is similar to that of irrigation, but with the generation of effluent.

The method comprises applying the liquid in the highest part of the ramp. The effluent drains by gravity all over the slope, where part of it is lost by evapotranspiration and the remaining part is collected at the base of the ramp. Percolation can be insignificant because this system was conceived initially for low-permeability soils. However, its use has been reported also in soils with medium permeability and impermeable underground (USEPA, 1981).

In comparison with other land disposal methods, overland flow presents the following characteristics as its main advantages (Coraucci Filho *et al.*, 2001):

- Appropriate for the treatment of sewage from rural communities and from seasonal industries that generate organic wastewater;
- Provides advanced secondary treatment, with relatively simple and cheap operation;
- The covering vegetation can be reused or commercially used after cutting;
- Presents the smallest restriction regarding the characteristics of the land, requiring only relatively impermeable soil for installation and an adequate slope.

The disadvantages are:

- The method is limited by the climate, culture tolerance in relation to water, and slope of the land;
- The application can be limited by wet weather;
- The loading rates can be restricted by the growth pattern of the culture;
- Flat or very steep land is not suitable for this type of treatment.

The typical configuration of a STP consisting of a UASB reactor and post-treatment by overland flow has an extremely simple flowsheet (Figure 10.9). In addition to the preliminary treatment units (screens and grit chambers), the flowsheet comprises the anaerobic treatment unit, the land treatment system, and the dewatering unit for the sludge produced in the UASB reactor. The same considerations for the other systems discussed in this book are valid here in respect of the characteristics of the anaerobic sludge that is already thickened and stabilized. Dewatering units using drying beds can be used in small-size plants.

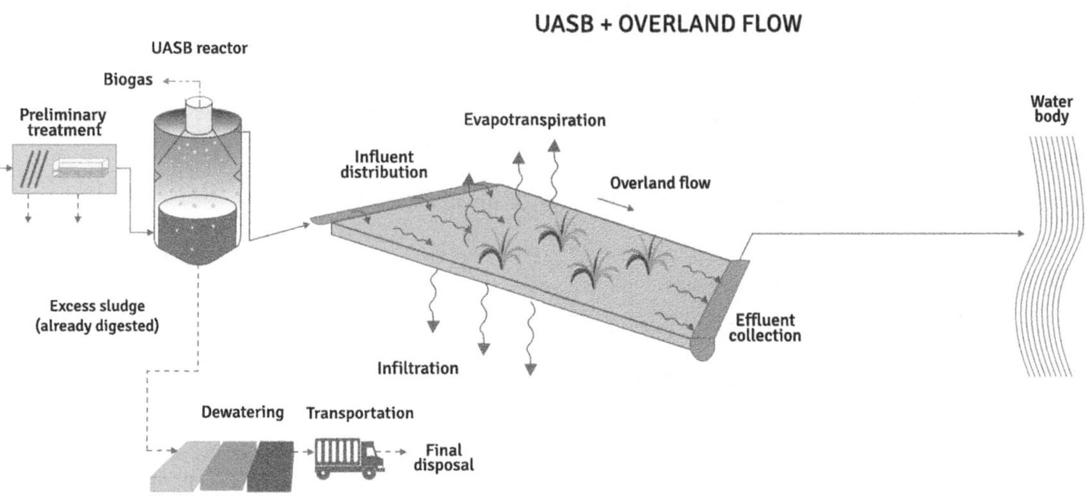

Figure 10.9 Typical configuration of a treatment plant with a UASB reactor and overland flow system.

10.4.3.2 Design criteria

According to Coraucci Filho *et al.* (2001), the main criteria for the design of overland flow systems applied to the post-treatment of effluents from anaerobic reactors are as follows:

Length of the slope. The length is the longitudinal dimension of the physical surface of the soil, defined by the flow direction of the effluent. The length of the slope usually ranges from 30 to 45 m.

Hydraulic loading rate. The loading rate is considered the main parameter in the design of the system, and is defined as the volume applied to the treatment module, divided by the loading period in hours. There is a tendency to standardize this parameter, expressing it in terms of the unit width of the module, in $m^3\ h^{-1}\ m^{-1}$ (Paganini, 1997; Coraucci Filho *et al.*, 2001). This parameter is dependent on the effluent discharge regime, the sewage pre-treatment level, depth and slope of the ground, and the climate. Usually, for the post-treatment of anaerobic effluents, loading rates between 0.2 and 0.4 $m^3\ h^{-1}$ per metre of width of the slope are employed.

10.4.3.3 Construction details

The following main aspects in relation to the construction of overland flow systems should be taken into consideration (Coraucci Filho *et al.*, 2001):

Ground slope. Ground slope between 1 and 12% (vertical/horizontal) is recommended, with an optimal interval between 2 and 8%. A slope lower than 1% is not recommended because of the possible formation of pools with sewage and the consequent proliferation of flies. Extremely high slopes cause a decrease in the flow time and the treatment efficiency, in addition to favouring the development of erosive processes.

Classification of the soil. The overland flow system was initially developed for soils with low permeability, lower than 15 mm h^{-1}. However, the system can be used in locations with moderate permeability (15 to 50 mm h^{-1}). This is because the void spaces of the soil can become filled with influent solids (clogging) and vegetation growth over time. The permeability can also be changed by soil compaction during the construction of the system.

Distribution of the sewage. The uniform distribution of the sewage on the entire width of the ramp is a critical point in the performance of the system. The influent can be distributed using, e.g., pipes with openings or channels with spaced triangular weirs.

Selection of the vegetation. The covering vegetation is essential to the satisfactory performance of the system. Perennial and water-resistant grasses are those that adapt better to overland flow systems. Their main functions are protection against erosion, redistribution of the flow (which prevents short circuits), support for microorganisms, and removal of nutrients.

10.4.3.4 Operational aspects

Operation is intermittent, with a feeding period between 8 and 12 h d^{-1}, followed by a dry period ranging from 16 to 24 h d^{-1}. Operational cycles with four days of feeding and two days of resting (dry) prevent the propagation of insects. However, it should be noted that other feeding and resting cycles have also been used successfully.

The flow, applied rates, period and frequency of sewage loading, and the quality of the influent and effluent should be monitored constantly. If there is significant infiltration of the soil, the quality of the groundwater should be monitored as well.

EXAMPLE 10.2

Design of an overland flow system acting as post-treatment of the effluent from an UASB reactor.

Input data (see Section 10.4.3.2)

- Population equivalent: PE = 2,000 inhabitants
- Average flow: $Q = 250\ m^3\ d^{-1}$
- Influent BOD concentration (raw sewage): 400 mg L^{-1}
- Loading rate: $q_L = 0.35\ m^3\ h^{-1} m^{-1}$
- Length of the slope: $Z = 35$ m
- Ground slope: 4%
- Feeding periods (feeding hours in each slope, per day): $L_p = 8\ h\ d^{-1}$
- Feeding frequency (loading days per week): $f = 5$ d week^{-1}

Solution

(1) Calculate the required area
Net area required:

$$A = \frac{Q \times Z}{q_L \times L_p} \times \left(\frac{7}{f}\right) = \frac{(250\,m^3\,d^{-1}) \times (35\,m)}{(0.35\,m^3\,m^{-1}\,h^{-1}) \times (8\,h\,d^{-1})} \times \left(\frac{7\,d\,week^{-1}}{5\,d\,week^{-1}}\right) = 4{,}375\,m^2$$

This is the net area required for the overland flow system. A large total area will be required, considering internal roads, urbanization, and the like.
Per capita net area requirement = $(4{,}375\ m^2)/(2{,}000\ inhab) = 2.2\ m^2\ inhabitant^{-1}$

(2) Dimensions of each slope
Number of slopes (initial trial value, which can be revised to allow more favourable dimensions and improved adjustment among units in terms of load, daily rest, and weekly rest): $n = 6$
Area of each slope: $A_u = A/n = (4{,}375\ m^2)/6 = 729\ m^2$
Length of each slope: $Z = 35.00$ m (previously defined design parameter)
Width of each slope: $W = A_u/Z = (729\ m^2)/(35.00\ m) = 20.80$ m
Gradient of the slopes: $s = 4\%$ (design parameter)

Level difference between the upper and lower part of each ramp: $H = (Z \cdot s/100) = 35.00\ m \times 4/100 = 1.40$ m

(3) Operational regime of the slopes
Weekly cycle:
- Number of slopes in rest: $n_r = n(1 - f/7) = 6 \times (1-5/7) = 1.71$ (adopted 2 units in rest)
- Daily cycle:
- Number of slopes in operation: $n_{op} = n - n_r = 6-2 = 4$
- Number of slopes in loading (in each instant): $n_{load} = n_{op}\ L_p/24 = 4 \times 8/24 = 1.33$ (adopted 1 unit in loading)
- Number of slopes in resting (in each instant): $n_r = n_{op} - n_{load} = 4-1 = 3$

(4) Concentration of effluent BOD
For design purposes, assuming a BOD removal efficiency of 65% in the UASB reactor, the effluent BOD concentration from the UASB reactor (influent to the overland flow system) will be:

$$BOD_{effluent\ UASB} = BOD_{influent\ UASB} \times (1 - Efficiency_{UASB}) = 400 \times (1 - 0.65) = 140\ mg\ L^{-1}$$

Effluent concentration of the overland flow (assuming E = 50%):

$$BOD_{effluent} = 140\ mg\,L^{-1} \times (1 - 50/100) = 70\ mg\,L^{-1}$$

Overall efficiency of the system:

$$E = (400 - 70)/400 = 0.83 = 83\%$$

This is within the typical removal range, as shown in Table 10.3.

10.4.4 Constructed wetlands (horizontal and vertical)

10.4.4.1 Description

Thousands of units of constructed wetlands (also called treatment wetlands) are implemented worldwide, mainly treating sewage from single-family homes or small communities. In most instances, the wetlands are used as secondary treatment, receiving the effluent from septic tanks. In other applications, wetlands are used as tertiary treatment, polishing the effluent from the secondary treatment systems.

Anaerobic baffled reactors, coupled with anaerobic processes, have been used in the first stage of the treatment and, to a more limited extent, also as UASB reactors. Although the consolidated reported experience of utilizing wetlands for the post-treatment of anaerobic effluents is limited, the basic working principles are the same, taking into account that the wetlands, in this instance, function in between the two well-known applications of secondary and tertiary treatment. Therefore, the kinetic coefficients and applied loading rates can be intermediate to these two situations.

The following books and manuals address constructed wetlands and should be consulted for further details: Kadlec and Knight (1996); Campos (1999); OPS/OMS (1999); USEPA (2000); IWA Specialist Group on Use of Macrophytes in Water Pollution Control (2000); Philippi and Sezerino (2004); Crites *et al.* (2006); Kadlec and Wallace (2008); Hoffman *et al.* (2011); Matos and Matos (2017); Lombard Latune and Molle (2017); Dotro *et al.* (2017); Alarcón-Herrera *et al.* (2018). The removal of pathogenic organisms is detailed in Maiga *et al.* (2017). The international literature is based on experience mainly in temperate countries, and, consequently, consolidated design and operation criteria pertaining to warm weather conditions based on numerous full-scale treatment plants operating for many years remain scarce.

This section covers only the following two wetland variants, which have potential for treating anaerobic effluents:

- Horizontal subsurface flow constructed wetlands, also known simply as horizontal wetlands;
- Vertical flow constructed wetlands.

10.4.4.1.1 Horizontal wetlands

These do not resemble natural flooded systems, as there is no free water on the surface. Horizontal wetlands consist of a bed of small stones, gravel, steel slag, or other similar material, supporting the growth of aquatic plants. The water level remains below the surface of the bed, and the sewage flow is in contact with the roots and rhizomes of the plants, where the bacterial biofilm develops. Examples of plants that have been used include *Typha*, reeds, *Scirpus*, *Carex*, *Phragmites*, *Cyperus* (papyrus), *Helicornia*, *Zizanopsis*, *Pennisetum* (elephant grass), and *Canna*, among others.

The total height of the bed is often between 0.5 and 0.9 m and that of the water inside the bed between 0.4 and 0.8 m, i.e., the water level is approximately 0.10 m below the upper level of the filter bed. The granulometric composition of the support medium should allow for the continuous flow of the liquid without clogging. A large part of the subsurface area can be anaerobic or anoxic, with aerobic sites immediately adjacent to the rhizomes.

As there is no free water flow, there is less potential for generating unpleasant odours, attracting rodents, and the breeding of mosquitoes. These are vital factors, particularly in considering the post-treatment of anaerobic effluents. Wetlands present an attractive alternative for treating sewage, given the landscape composition with plants of aesthetic value, combined with concealed treatment and, therefore, no direct perception of sewage being treated. However, over time, the beds tend to become clogged, increasing the head loss and probably leading to superficial runoff or overland flow (sewage flowing above the upper level of the bed). This can cause deterioration in the aesthetic element of the process. Therefore, it is important that operational measures for cleaning or replacing the bed are implemented before such undesirable consequences occur. In addition, if the sewage treatment plant has been designed to have a visually pleasant appearance, it is important that there is adequate maintenance, performing the cuttings of the plants and the control of invasive species.

Horizontal flow wetlands are suitable for receiving sewage from septic tanks, anaerobic reactors, and other sewage treatment processes. The overall BOD removal efficiency is high (85 to 90%, or more) and the final effluent is well clarified. Because no aerobic conditions prevail in most of the bed, the expectation of nitrification is low. Nutrient removal is done by plants, but this route is usually not particularly important compared with the N and P values in the influent load. Phosphorus can also be removed by using a reactive medium, although the efficiency is expected to decrease with time after the saturation of the medium has occurred.

10.4.4.1.2 Vertical wetlands

This variant is typically a filter composed of sand (most frequently), gravel, or crushed stone and planted with vegetation. In the most common variant of a downflow vertical wetland, sewage is applied at the top of the bed. Pipes are installed at the bottom of the filter medium to collect the treated liquid. The operation resembles the routine of an intermittent filter, with batch cycles of dosing and drainage; therefore, it differs from the operation of horizontal flow wetlands that have continuous feeding. Because the dosage is intermittent, the flow is usually in a hydraulically unsaturated medium. After the passage of the liquid, the pore spaces are occupied by air, facilitating the maintenance of aerobic conditions. The feed is switched between the beds in parallel, with some units in operation and others at rest.

Vertical wetlands are efficient in the removal of BOD (efficiencies between 85 and 90%, or more) and, because they have an aerobic environment after the passage of the liquid, when the bed is not saturated, nitrification is possible. Certain denitrification occurs in anoxic niches of the biofilm that grows attached to the support medium. The removal of nutrients by incorporation into the biomass of the plants is low.

10.4.4.2 Design criteria

The design guidelines described here derive from the synthesis of standards and guidelines from several countries, presented in Dotro *et al.* (2017), and adapted to the conditions of post-treatment of anaerobic effluents, operating mainly under warm climate conditions. However, it should be recognized that these design criteria are influenced significantly by the international literature and were developed based mainly on the application of wetlands in temperate or cold climate regions.

The support medium of wetlands can be gravel, crushed stone, steel slag, sand, or other materials that have physical resistance and that do not suffer breakdown because of the chemical and biological activities that occur in the bed. Each wetland variant has a particle-size specification. In the present text, grain sizes are presented in millimetre (mm), associated with typical standard sieving: 0 to 4.8 mm; 4.8 to 9.5 mm; 9.5 to 19 mm; and 19 to 25 mm.

10.4.4.2.1 Horizontal wetlands

Table 10.9 presents the main design and operational criteria for horizontal wetlands treating anaerobic effluents. The main physical dimensions are shown in Figure 10.10.

Table 10.9 Main design and operational criteria for horizontal wetlands treating anaerobic effluents.

Parameter	Value
Maximum surface organic loading rate OLR_S (g BOD m^{-2} d^{-1})	6–15
Maximum surface hydraulic loading rate HLR_S (m^3 m^{-2} d^{-1})	0.04–0.12
Total height of filter medium (m)	0.50–0.90
Water level (m)	0.40–0.80
Distance between the water level and the top of the medium (m)	~0.10
Length/width (*L*/*W*) ratio	2:1–4:1
Maximum width of each unit (m)	25–30
Longitudinal bottom slope (%)	0–1
Medium grain size (mm)	Gravel 9.5–19 or 19–25
Feeding mode	Continuous

Note: Calculations are based on average influent flow (m^3 d^{-1}) and average influent BOD load (g d^{-1}).

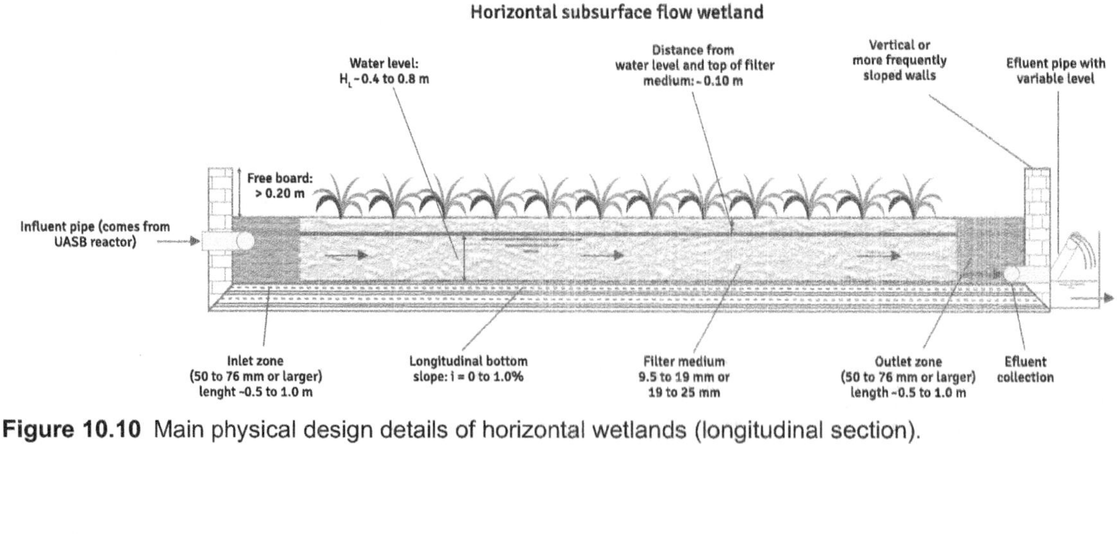

Figure 10.10 Main physical design details of horizontal wetlands (longitudinal section).

10.4.4.2.2 Vertical wetlands

Table 10.10 presents the main design and operational criteria for vertical wetlands treating anaerobic effluents. The main physical dimensions are shown in Figure 10.11.

Table 10.10 Main design and operational criteria for vertical wetlands treating anaerobic effluents.

Parameter	Value
Maximum surface organic loading rate OLR_s (g BOD m^{-2} d^{-1})	10–40
Maximum surface hydraulic loading rate HLR_s (m^3 m^{-2} d^{-1})	0.05 to 0.20
Height of upper protection layer (optional) (m)	0.05–0.10
Height of the main upper layer of the filter medium (m)	0.40–0.60
Height of middle transition layer (m)	0.10
Height of bottom drainage layer (m)	0.10–0.30
Free board (m)	0.15–0.30
Length/width (*L*/*W*) ratio	No restrictions
Number of units	Multiples of 2 or 4
Units in rest and operation	1 in rest for every 1 in operation, or 1 at rest for every 3 in operation
Maximum area of each unit (m^2)	400
Grain size of upper protection layer (optional) (mm)	Gravel 4.8–9.5 or 9.5–19
Grain size of the main upper layer of the filter medium (mm)	Sand 0.06 to 4.8
Grain size of middle transition layer (mm)	Gravel 4.8–9.5
Grain size of bottom drainage layer (mm)	Gravel 4.8–9.5 or 9.5–19
Feeding mode	Intermittent
Number of batches per day in the unit (batch/day)	3–12

Note:

- Calculations are based on average influent flow (m^3 d^{-1}) and average influent BOD load (g d^{-1}).
- Using organic and hydraulic application rates facilitates the calculation of the area of the beds in operation. The total area of the system should include the area of the resting beds.

10.4.4.3 Construction details

The main construction details are presented in Tables 10.9 and 10.10 and in Figures 10.10 and 10.11. The construction details for horizontal wetlands are presented in the design example.

10.4.4.4 Operational aspects

Horizontal wetlands are fed continuously. As this is a simplified natural system, there are no additional operational control variables, such as recirculation and aeration. The only variation is a change in the output pipe level.

The operation of vertical wetlands follows an intermittent feeding scheme, in batches, with feeding in some units and resting in others. As shown in Table 10.10, one unit is resting for each one or three units in operation. The number of batches per day is between 3 and 12, i.e., on average, one batch per 2 to 8 hours.

The maintenance of constructed wetlands is fairly straightforward. In addition to the activities related to the previous anaerobic treatment, maintenance is usually associated with the control (cutting, removal, and disposal) of the vegetation.

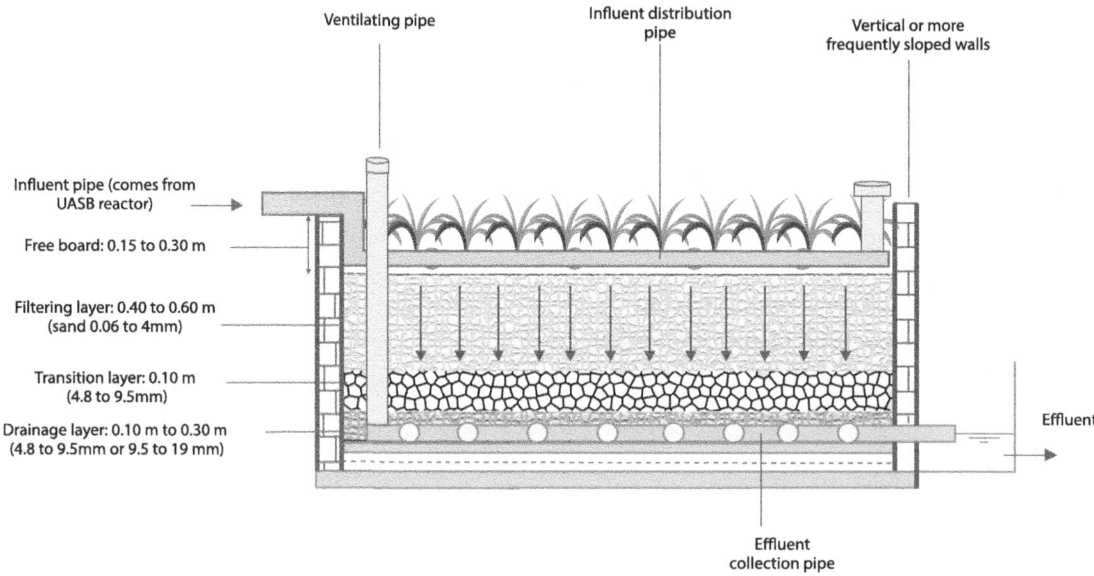

Figure 10.11 Main physical design details of vertical wetlands (longitudinal section).

The main concern in horizontal wetlands is the clogging of the bed, which can occur after a few years of operation, depending on the characteristics of the influent, the applied loads, and the granulometry of the support medium. However, with careful design, the expectation is that problems arising from clogging (such as surface flow above the filter medium) will occur only after several years of operation. The surface flow of the liquid can cause aesthetic problems for the treatment plant, but this does not necessarily imply the deterioration of the quality of the treated effluent.

In vertical flow wetlands, the intermittent operation and the alternation between the beds allow digestion and drying of the sludge layer on the surface of the filter medium, which minimizes the problems of surface clogging. In addition, the plants play a fundamental role, as the stems move in the wind, exerting a lever effect, which opens spaces in the top sludge layer through which the influent penetrates and moves to the lower layers.

In horizontal wetlands, the sludge accumulates inside the units, which can cause clogging eventually, but this does not imply that the sludge must be removed periodically. In vertical wetlands, most of the sludge accumulates on the surface and needs to be removed only after several years of operation.

EXAMPLE 10.3

Design of a horizontal subsurface flow constructed wetland system acting as post-treatment of the effluent from an UASB reactor.

Input data

- Population equivalent: PE = 2,000 inhabitants
- Average flow: $Q = 250\ m^3\ d^{-1}$
- Influent BOD concentration (raw sewage): 400 mg L^{-1}
- Design parameters listed in Table 10.9

Solution

(1) UASB reactor effluent data (influent to wetlands)

For design purposes, assuming a BOD removal efficiency of 65% in the UASB reactor, the effluent BOD concentration from the UASB reactor (influent to the wetland system) will be:

$$BOD_{effluent\ UASB} = BOD_{influent\ UASB} \times (1 - Efficency_{UASB}) = 400 \times (1 - 0.65) = 140\ mg\ L^{-1}$$

The influent BOD load to the wetlands will be:

Influent BOD load to wetlands = Influent flow × Influent BOD concentration = 250 $m^3\ d^{-1}$ × 140 g m^{-3} = 35,000 g BOD d^{-1}

(2) Determine the surface area of the horizontal wetland

Adopting a surface organic loading rate (OLR_s) of 8 g BOD $m^{-2}\ d^{-1}$ (value within the range shown in Table 10.9, 6–15 g BOD $m^{-2}\ d^{-1}$) and with a calculated influent organic load of 35,000 g BOD d^{-1}, the surface area of the horizontal wetland system is obtained by:

$$\text{Area } A = \text{BOD Load}/OLR_s = (35{,}000\ \text{g BOD d}^{-1})/(8\ \text{g BOD m}^{-2}\text{d}^{-1}) = 4{,}375\ \text{m}^2$$

The net surface area of 4,375 m^2 of the filter bed corresponds to a per capita area of 4,375/2,000 = 2.2 m^2/inhabitant^{-1}. The total area will be larger, as it would include the entire area of the treatment plant, and not only the net area of the wetland.

(3) Number of units in parallel

For flexibility, adopt 4 units in parallel ($n = 4$).

Therefore, the surface area of each unit will be 4,375/4 = 1,094 m^2.

(4) Determine the dimensions of each unit

Adopting a length:width ratio (L:W) equal to 2:1, giving:

$$A = L \times W = \left[\left(\frac{L}{W}\right) \cdot W\right] \times W = \left(\frac{L}{W}\right) \cdot W^2 \rightarrow W = \sqrt{A/\left(\frac{L}{W}\right)}$$

With $L/W = 2$, giving:

$$\text{Width}\,W = \sqrt{\frac{1094}{2}} = 23.40\,\text{m} \quad \text{(lower than maximum recommended width)}$$

$$\text{Length}: \quad L = \left(\frac{L}{W}\right) \cdot W = 2 \times 23.40 = 46.80\,\text{m}$$

Water level: $H_{liquid} = 0.60$ m (see Table 10.9; value between 0.4 and 0.8 m).

Total height of filter medium: $H_{Total} = 0.70$ m (0.60 m of liquid height and 0.10 m above the water level)

Total volume of the filter medium: $V_{Total} = n \times (L \times W \times H_{Total}) = 4 \times (46.80 \times 23.40 \times 0.70) =$ 3,066 m^3 (1.5 m^3/inhabitant^{-1})

(5) Verify the resulting surface hydraulic loading rate

Considering the flow rate of 250 $m^3\ d^{-1}$ and the total surface area of 4,375 m^2, the resulting surface hydraulic loading rate (HLR_s) is:

$$HLR_s = \frac{Q}{A} = \frac{250\ \ m^3\,d^{-1}}{4{,}375\,m^2} = 0.057 m^3\ m^{-2}\ d^{-1}$$

This value falls in the range of HLR_s proposed in Table 10.9 (0.04 to 0.12 $m^3\ m^{-2}\ d^{-1}$). In this instance, it is advisable to remain close to the lower limit of the range (more conservative), as the secondary treatment used (UASB reactor) is less efficient than other secondary processes.

10.4.5 Trickling filters

10.4.5.1 Description

Post-treatment of anaerobic effluents in sewage treatment plants typically comprises an aerobic stage, for which trickling filter (TF) technology is an interesting compact process option. The technological choice of trickling filters as a post-treatment unit guarantees low sludge production, relative operational and maintenance simplicity, as well as low operating costs compared with other aerobic post-treatment processes (Chernicharo, 2007). For the removal of residual carbon from UASB effluents, high-rate trickling filters usually provide similar effluent quality when compared with mechanized compact systems as activated sludge (von Sperling & Chernicharo, 2005).

Trickling filters typically consist of a tank filled with a coarse-material bed (e.g., large rocks, blast furnace slag) or plastic-based packing material. Noteworthy, is that sponge-based packing media has been considered recently as a potential filling material, with successful full-scale applications being reported (Okubo *et al.*, 2015; Onodera *et al.*, 2016). Depending on the organic and hydraulic loadings applied to the system, the operation without secondary settlers has been contextualized for TF technology. The types of packing media usually applied to trickling filters post-UASB reactors are discussed further. Figure 10.12 presents the typical flowsheet of the UASB-TF system.

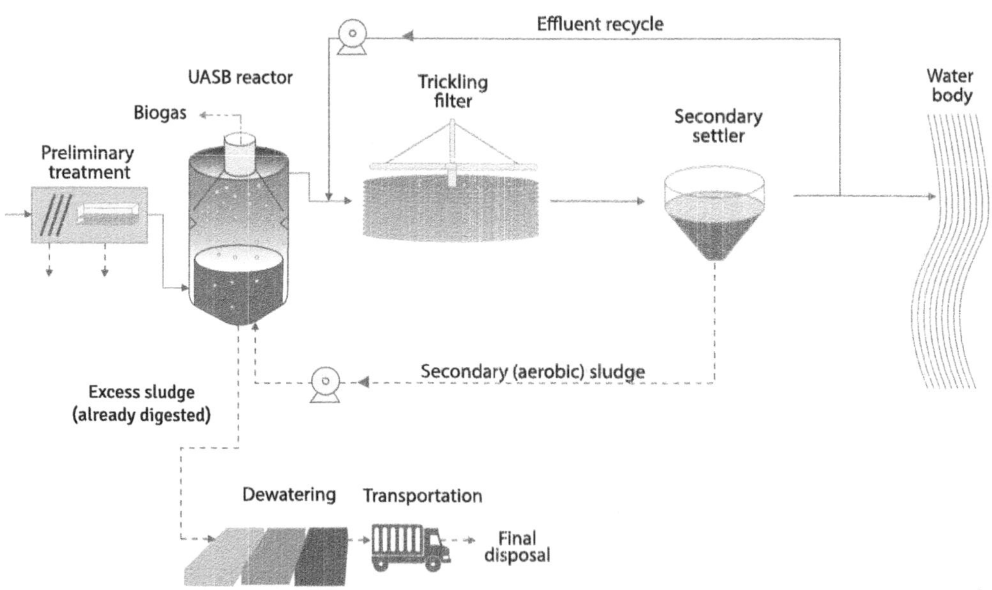

Figure 10.12 Typical configuration of a treatment plant with a UASB reactor followed by trickling filter.

For the operation of the trickling filters, the effluent of the UASB reactor is applied over the surface of the packing media using a stationary or rotary distributor. The liquid moves downward along the reactor and over the surface of unsaturated packing material, which provides a suitable environment for biomass growth and biofilm formation. In sequence, the treated effluent is collected at the bottom of the reactor and routed generally to a secondary settler, aiming to reduce the quantity of solids eluted in the effluent. Recirculation of the final effluent is considered an operational strategy in an UASB/TF flowsheet, as is discussed later (see Section 10.4.5.4).

As in any biofilm system, diffusion is the dominant transport process of soluble substrates into the biological matrix. Outside the biofilm, advection and turbulent dispersion are processes of relevance. The related transport processes allow the contact between microorganisms and substrates at the outer and inner parts of the biofilm, resulting in subsequent biological conversions. As regards sponge-based packing media, the biofilm is formed at the outer side of the sponges and over the inner fibres, which provide noticeable protection against macrofauna overgrazing (Onodera *et al.*, 2016). In general, trickling filter technology is applied to carbon and ammonia oxidation. However, depending on the design and operational conditions, denitrification is observed also within the system (Parker & Richards, 1986).

Packing media. Trickling filters post-UASB reactors are filled generally with crushed stone (diameter: 75–100 mm) or plastic-based packing media (random pack or vertical/angular flow). However, obtaining size homogeneity for rock media is quite difficult in several situations. Other types of filling material are being manufactured under industrial control (e.g., plastic-based packing media). Fundamentally, the surface area is a crucial factor in performance improvement; however, using high surface-area packing media needs to be followed by wetting efficiency. To achieve high levels of wetting efficiency, final effluent recirculation is needed frequently. Table 10.11 presents the main packing media types being adopted as filling material for trickling filters post-UASB reactor.

10.4.5.2 Design criteria

Table 10.12 presents the parameters usually applied in designing rock-bed trickling filters following UASB reactors. In general, the full-scale units have been designed as high-rate trickling filters. The reason for adopting higher organic loading rates is based on the less stringent discharge standards considered in developing countries to progressively accomplish the goal of expanding sewage treatment. Consequently, in several instances, ammonia oxidation might not be a priority. The listed criteria and parameters for the design of trickling filters as a post-treatment of UASB reactor effluents are based on the outputs of the Brazilian Research Program on Basic Sanitation (PROSAB) (Chernicharo *et al.*, 2001).

An aspect for consideration when designing trickling filters following UASB reactors, is the biodegradability of UASB reactor effluents. The readily biodegradable organic matter present in such effluents appears to be lower compared with that of raw sewage (as also addressed in Chapter 3). Therefore, the adjustments (or even the revision) of the predictive equations used to estimate the performance of post-treatment units should be considered in principle.

Comments on the use of plastic- and sponge-based packing media. Despite the advantages of using plastic- and sponge-based packing media, their application is an emerging perspective in developing countries. Therefore, as full-scale applications are being consolidated, the design of trickling filters considering such filling materials needs to be assessed carefully.

As observed in trickling filters as post-primary settlers, recent experiments with random plastic rings have indicated that when the applied organic loading was lower than 1.0 kgBOD m^{-3} d^{-1} (hydraulic loading rates approximately 18–20 m^3 m^{-2} d^{-1}), the performance of rock- and plastic-based trickling filters for organic matter removal appeared similar. Regardless of recirculation being an operational strategy, performance improvement of plastic-based trickling filters following UASB reactors remains a topic for further research. In addition, developing predictive equations for these systems is an important aspect. Interestingly, the application of plastic packing media following UASB reactors has been reported only sparsely.

Superior performance is expected when using sponge-based packing media in comparison with other types of filling material. At an organic loading rate of 0.25 kgBOD m^{-3} d^{-1}, performance of approximately 80% has been reported for ammonia removal at full-scale operation (Onodera *et al.*,

Table 10.11 Packing media types applied as filling material for trickling filters following UASB reactors.

Filling Material	Dimensional Characteristics	Weight (kg m^{-3})	Specific Surface area (m^2 m^{-3})	Approx. void Space (%)	Image	Brief Description and General Comments
Crushed stone	Diameter: 7.5–10 cm	1350	50–70	50–60	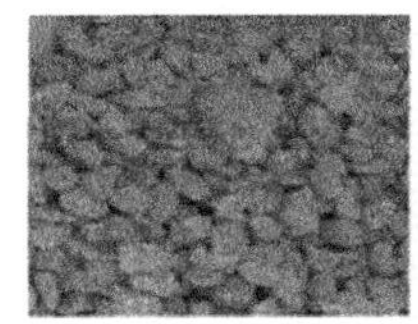	Presents irregular surface and high specific weight. The low cost of this filling material is a convincing advantage for its use in several systems in Brazil, where UASB/TF systems are applied widely. The main application disadvantages are (i) difficulty to maintain uniform sizing, and (ii) weight, which tends to limit the design of deep trickling filters (>2 m). Clogging, plugging, and ponds can occur mainly in high-rate trickling filters and when unqualified operation of the UASB reactor is a factor.
Blast furnace slag	Diameter: 5–10 cm	1350	50–70	50–60		Constituted of residuals from the steel making industry. Presents irregular surface and high specific weight as crushed stones. Using this filling material presents the same advantages and disadvantages as those of crushed stones in terms of cost, quality control, and weight.

(*Continued*)

Table 10.11 Packing media types applied as filling material for trickling filters following UASB reactors *(Continued)*.

Filling Material	Dimensional Characteristics	Weight (kg m^{-3})	Specific Surface area (m^2 m^{-3})	Approx. void Space (%)	Image	Brief Description and General Comments
Random plastic rings	Diameter: 84 mm Height: 9 cm/unit	50	80	96	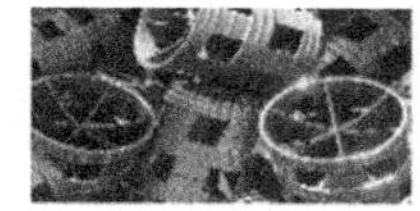	Generally constituted of polyvinyl chloride. Higher specific surface area (allowing the application of higher loading rates), higher void space (better air circulation and lower chances of clogging), and also the possibility to design taller reactors are remarkable advantages related to the packing media. For trickling filters following UASB reactors with no recirculation, the performance of the system tends to be similar to those filled with crushed stones or blast furnace slag. Effluent recirculation is recommended to guarantee wetting efficiency, depending on the hydraulic loading rate.
Vertical flow Cross flow (45°) Cross flow (60°)	Height: 30–60 cm/unit Width: 30–60 cm/unit Length: 130–320 cm	30	80 Vertical flow 98 Cross flow (45°) 98 Cross flow (60°)	95–97		Generally constituted of polyvinyl chloride. The advantages related to the packing material are generally similar to those observed for random plastic rings. This packing material can be used also to enhance water distribution at the top reactor, as well as for retrofitting of rock-bed trickling filters. Because of the high void space, the possibility of local assembly can reduce transportation costs.

(Continued)

Table 10.11 Packing media types applied as filling material for trickling filters following UASB reactors (*Continued*).

Filling Material	Dimensional Characteristics	Weight ($kg\ m^{-3}$)	Specific Surface area ($m^2\ m^{-3}$)	Approx. void Space (%)	Image	Brief Description and General Comments
Sponge-based packing media	Several packing media types (from random filter packing to hanging sponge-based curtains)	The technology has an ability to retain 25 gSS L^{-1}, on average	≈5500 (based on geometric estimations, taking into account the diameter of sponge fibres)	96–98	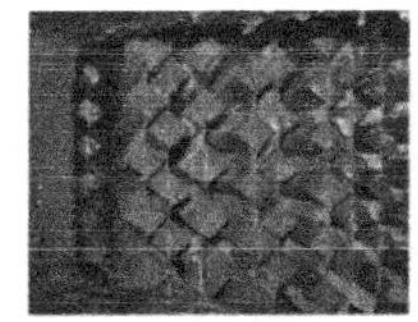 DHS curtain-type 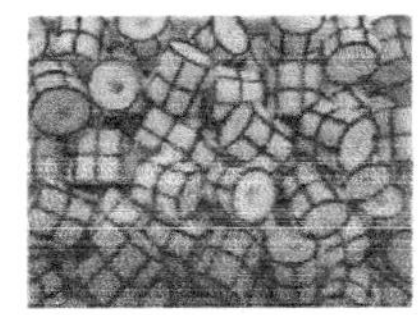 DHS random-type	Sponge-based packing media (currently named ‘downflow hanging sponge’) is an extremely promising technology used in trickling filters following UASB reactors. The main advantages related to the packing material are the extremely high hydraulic percolation time (≈1.5–2.5 hours, in general) and sludge retention time (approximately 100 days). The ability to retain biomass in the system and the presence of oxygen even at the inner parts of the sponge tend to provide a suitable environment for slow growers (e.g., nitrifiers and anaerobic ammonium oxidation [anammox] bacteria). In addition, a high sludge concentration, sufficient oxygen supply, adequate endogenous respiration rate, and the high density and diversity of the microfauna result in low excess sludge production (approximately 0.10 gss $gBOD_{removed}^{-1}$). Curtain type and random packing have been applied in full-scale applications.

Table 10.12 Criteria and parameters for designing high-rate rock-bed trickling filters following UASB reactors.

Parameter	Operational Range		
	Average Flow	**Maximum Daily Flow**	**Maximum Hourly Flow**
Organic loading rate (kgBOD m^{-3} d^{-1})	0.5–1.0	0.5–1.0	0.5–1.0
Hydraulic loading rate (m^3 m^{-2} d^{-1})	15–30	18–22	25–30
Filter depth (m)	2.0–3.0	2.0–3.0	2.0–3.0

Note: For rock-bed trickling filters following UASB reactors, the adopted sludge production rate is 0.8–1.0 kg SS $kgBOD_{removed}^{-1}$.

2016), whereas 50% is expected when using rock media in similar operational conditions (Tchobanoglous *et al.*, 2013). In principle, larger hydraulic and sludge retention times within the sponges are factors of relevance for improving the performance of such systems after UASB reactors, as indicated in Table 10.11. For organic matter removal, recirculation might not be mandatory depending on the local discharge standards. In addition, operation without secondary settlers is being mentioned currently as a promising perspective on sponge-bed trickling filters. This system is also called the downflow hanging sponge (DHS) system.

The DHS system is designed generally for carbon and ammonium removal in a single reaction volume as a low-rate bioreactor. In a full-scale application (Okubo *et al.*, 2015), the organic loading applied to the system was approximately 2.80 kgCOD m_{sponge}^{-3} d^{-1} (0.23 kgBOD $m_{reactor}^{-3}$ d^{-1}), considering a theoretical hydraulic percolation time of 1.5 h. The sponge occupancy was approximately 25% of the reactor. In another instance (Onodera *et al.* 2014), a higher organic loading was applied to the system (approximately 0.40 kgBOD m^{-3} d^{-1}), with comparable results in terms of BOD removal (approximately 85% BOD removal in the post-treatment step). In this instance, sponge occupancy and hydraulic percolation time were 40% and 2.0 hours, respectively. Most of the research related to the use of the DHS system indicates a hydraulic percolation time ranging from 1.5 to 3.0 h.

Empirical equations for the estimation of BOD removal. The equation developed by the National Research Council (NRC) has been used to estimate the BOD removal in rock trickling filters post-UASB reactors (Eq. 10.1). Because the NRC formula was defined from statistical analysis of operating data related to rock-bed trickling filters treating sewage (thus, considering higher concentrations of readily biodegradable organic matter), it should be used with caution for TFs post-UASB reactors. NRC formula is represented considering the effect of volumetric loading and recycle, as follows:

$$E = \frac{1}{1 + 0.4432\sqrt{\frac{OLR}{F}}} \tag{10.1}$$

in which

E = BOD removal efficiency ratio at 20°C
OLR = organic loading rate (kgBOD m^{-3} d^{-1})
F = recirculation factor $(1 + Q_{recycle}/Q_{influent})$, where $F = 1$, if $Q_{recycle} = 0$.

The effect of the temperature can be included as traditionally represented:

$$E_T = E_{20^\circ\text{C}}(1.035)^{T-20} \tag{10.2}$$

in which

E_T = BOD removal efficiency ratio at temperature T in °C
$E_{20^\circ\text{C}}$ = BOD removal efficiency at 20°C (%).

The application of the Eckenfelder formula (Eq. 10.3) is also an option for BOD removal estimations. From a preliminary analysis this first-order and semi-empirical model seems to provide excellent data fitting for rock-bed trickling filters post-UASB reactors. However, more intense calibration efforts are needed aiming to determine a proper range for k and n values, considering high- and low-rate trickling filters. The equation considers wastewater treatability (k) and filter medium (n). Thus, if properly calibrated the equation might be potentially applied to design trickling filters preceded by UASB reactors, even when innovative packing media are considered (e.g.: sponge-based medium). Eckenfelder formula is given as follows:

$$S_{efl} = S_{in} e^{\frac{-k \cdot h^{(1-m)}}{(Q/A)^n}} \tag{10.3}$$

in which

S_{in} = influent BOD concentration (g m^{-3})
S_{efl} = effluent BOD concentration (g m^{-3})
k = factor for water treatability (typically 0.088 for TFs treating effluents from primary settlers)
n = filter medium exponent (typically 0.33 to 0.67)
m = slime distribution factor ($m = 0$, if evenly distributed)
h = packing depth (ft)
Q = flow rate (gal min^{-1}).
A = plan-view area (ft^2).

Comments on the performance of trickling filters following UASB reactors. Rock-bed trickling filters following UASB reactors tend to produce effluents with BOD concentrations lower than 40 mgBOD L^{-1} (65% BOD removal efficiency, approximately), considering organic and hydraulic loading rates of approximately 0.5–1.0 kgBOD m^{-3} d^{-1} and 20–30 m^3 m^{-2} d^{-1}, respectively. For trickling filters filled with sponge-based packing media, 6–10 mgBOD L^{-1} is expected in the final effluent, considering the typical design as a low-rate trickling filter (applied organic loading rate approximately 0.20–0.40 kgBOD m^{-3}). In this case, ammonium nitrogen removal (80%, approximately) is expected.

Excess sludge production. For high-rate trickling filters (applied organic loadings approximately 0.5–1.0 kgBOD m^{-3} d^{-1}), the excess sludge production generally varies within the range 0.8–1.0 kgSS kgBOD$^{-1}_{\text{removed}}$, with the VS/TS ratio between 0.75 and 0.85. For low-rate trickling filters filled with sponge-based packing media, extremely low excess sludge production has been reported (approximately 0.10 kgSS kgBOD$^{-1}_{\text{removed}}$) when the TSS effluent from the UASB reactor is approximately 80 mgTSS L^{-1}. The VS/TS ratio is typically 0.60, indicating that the sludge could be routed to a dewatering unit with the anaerobic sludge. Such an operational strategy tends to reduce the operational demands related to sludge management in UASB/trickling filter technology considerably.

10.4.5.3 Construction details

The construction details are similar to those applied for trickling filters post-primary settlers, as discussed in detail by Tchobanoglous *et al.* (2013). Therefore, a summary of only some relevant aspects related to trickling filters following UASB reactors are reported in this section. Table 10.13 presents constructive and related operative details that can be considered to design the components of trickling filters.

10.4.5.4 Operational aspects

Effluent recirculation. To maintain a proper hydraulic loading, a fraction of the post-treated effluent can be recycled to the top of the reactor and distributed with the effluent from the UASB reactor. For high-rate trickling filters following UASB reactors, the main reason to adopt recirculation is the control of the hydraulic loading, and to ensure the rotation of the flow-driven distribution system during periods of low influent flowrate. Because the BOD concentration of effluents from UASB reactors is usually approximately 70–100 mgBOD L^{-1}, recirculation controlling organic loadings (0.5–1.0 kgBOD m^{-3} d^{-1}) is not needed. For using plastic-based packing media, such an operational strategy should be a design recommendation increasing the wetting efficiency and, consequently, the surface area colonized by active biofilm.

Periodic flooding. As discussed before, most of the trickling filters following UASB reactors have been designed as a high-rate system with crushed stones as filling material. In this instance, periodic flooding is recommended to avoid excessive accumulation of particulates eluted in the UASB effluent and from biofilm growth. To eliminate flies from the filter, effective periodic flooding is used, although higher-level flushing by lowering the distribution speed and alternating the operation of the trickling filters (in the case of using two systems in parallel) can be a potential operational strategy. However, full-scale experience has indicated that sludge management is crucial to reducing clogging within the reaction volume and ponds on the trickling filter surface. It must be pointed out that periodic flooding could be ineffective in overcoming operational issues related to high solid concentrations in the UASB effluent.

Secondary sludge return. To simplify the flowsheet of UASB/trickling filter systems, the sludge removed from the secondary clarifiers is forwarded to the UASB reactors for thickening and digestion. Although sludge return has been investigated successfully, operating issues have been reported recurrently in full-scale plants, usually attributed to the low capacity of the secondary sludge to be retained in the UASB reactors. In such instance, excessive solids washing out from the UASB reactors can bring about mass transfer limitations, clogging, and short-circuiting within the trickling filter.

From the outcomes of a full-scale experience on the practice of returning secondary sludge into UASB reactors, a noticeable reduction in the settleable solid concentration of the effluent was observed when the TS concentrations in the UASB reactor were below 1.5% at a distance of 2.30 m from the bottom of the digestion compartment. In this instance, more frequent sludge discharge from the UASB reactors was practiced. In addition, the excessive concentration of the return sludge from the secondary clarifiers (2.5 to 4.0 times higher than predicted in the design ~1.0%), clearly tended to overload the operation of the UASB reactors. Therefore, to prevent operational issues in the post-treatment step, rigorous sludge management control in the anaerobic and settling units needs to be considered. The control of the solid phase is related closely to the careful estimation of the effective sludge yield coefficient, allowing for adequate design of the dewatering unit. Secondary (aerobic) sludge return is discussed in detail in Chapters 6 and 7.

Table 10.13 Constructive details and brief orientation related to trickling filters following UASB reactors.

Construction Detail	Comments and Brief Orientation
Packing media	• Despite the use of crushed stones being common, the possibility of using plastic packing media should be considered for sizing optimization (application of higher organic and hydraulic loadings). In addition, the higher porosity of the plastic medium could increase the robustness of the system in terms of clogging, when the operation of the UASB reactors is a factor (e.g., process operation with low-skilled staff). • For improved effluent quality, sponge-based packing media can be an option. The significant lower excess sludge production provided by the use of sponges remains an important aspect in reducing the operational demands on sludge management.
Distribution systems and distribution speed of rotary distributors	• Fixed distribution tends to be employed less often in current practice. • Depending on the UASB reactor operation (anaerobic sludge management), the obstruction of nozzles occurs often. In such instance, ease of cleaning must be considered. • The currently adopted rotational speed is approximately 0.1 and 2.0 rpm, not exceeding 1.2 $m\,s^{-1}$ for the maximum flow rate. Lowering the rotational speed can prevent events of sloughing by controlling the biofilm thickness, which results also in biofilm that is more aerobic. • For low-rate trickling filters, reducing the distributor speed tends to be even more relevant than improving the wetting efficiency, as well as washing out the macrofauna and fly eggs.
Underdrain system and airflow	• To maintain the predominance of aerobic conditions in the system, the trickling filters following UASB reactors have been designed with natural ventilation. • The open area related to the underdrain blocks should be not smaller than 15% of the surface area of the filter. • The open area allowing the airflow in the bottom of the filter should be approximately 5% of the surface area of the filter. Ventilation must be designed to enable periodic flooding, mainly when the trickling filter is filled with crushed stones. • The underdrain system needs to be designed in consideration of the natural ventilation of the filter, i.e., with no more than 50% of the sectional area filled with the collected effluent.
Settling tanks	• The system is designed in consideration of hydraulic surface loading criteria. For high-rate trickling filters, the surface loading should be 16–32 $m^3\,m^{-2}\,d^{-1}$ for average flow, and not exceed 40–48 $m^3\,m^{-2}\,d^{-1}$ for the maximum flowrate. • The lower settleability of the detached biomass (the solids detached from the biofilm are flocculated poorly) should be considered in designing the settling tanks. For settlers receiving the effluents from UASB/trickling filters, the adopted useful depth is 3.5 m.

Note: The constructive aspects presented are based on the information extracted from Tchobanoglous *et al.* (2013); Rittmann and McCarty (2001); Chernicharo *et al.* (2001).

EXAMPLE 10.4

Design of trickling filters filled with crushed stones and sponge-based packing media, respectively, treating effluents from a UASB reactor.

Input data

- Population equivalent: PE = 50,000 inhabitants
- Average flow: $Q_{average} = 8{,}000\ m^3\ d^{-1}$
- Maximum daily flowrate: $Q_{max\text{-}d} = 9{,}600\ m^3\ d^{-1}$
- Maximum hourly flowrate: $Q_{max\text{-}h} = 12{,}000\ m^3\ d^{-1}$
- Influent BOD concentration (raw sewage): 333 mg L^{-1}
- BOD removal efficiency in the UASB reactor: 65%
- Operational conditions applied to the post-treatment systems:

Post-treatment System	Organic Loading Rate (kgBOD m^{-3} reactor d^{-1})	Hydraulic Loading Rate ($m^3\ m^{-2}\ d^{-1}$)	Packing Media
High-rate trickling filter	0.70 (adopted)	15-30 (see Table 10.12)	Crushed stones
Downflow hanging sponge (DHS)	0.23 (observed)	10–12	Sponge-based medium (curtain type)
Downflow hanging sponge (DHS)	0.38 (observed)	10–12	Sponge-based medium (random type)

Note: Sponge occupancy ratio for DHS curtain type and random type is 0.25 (Uemura *et al.*, 2016) and 0.34 (Onodera *et al.*, 2014), respectively.

Solution 1

Using a high-rate trickling filter as a post-treatment step

(1) BOD concentration in the effluent of the UASB reactor (influent to the trickling filter system)

$$BOD_{effluent\ UASB} = BOD_{influent\ UASB} \times (1 - Efficiency_{UASB}) = 333 \times (1 - 0.65) = 117\ mgBOD\ L^{-1}$$

(2) Determine the reactor volume using *loading criteria*
Volumetric organic loading rate (OLR) is given as follows:

$$\frac{OLR(kgBOD\ m^{-3}d^{-1}) = Q_{average}(m^3\ d^{-1}) \cdot BOD_{concentration}(g\ m^{-3})}{V_{reactor}(m^3)}$$

$$V_{reactor}(m^3) = \frac{8000\ m^{-1}\ d^{-1} \cdot [(117\ g\ m^{-3})/1000]}{0.70\ kgBOD\ m^{-3}\ d^{-1}} = 1337\ m^3$$

Note: For this example, an organic loading rate of 0.70 kgBOD $m^{-3}\ d^{-1}$ was adopted (according to Table 10.12, the admissible range for high-rate trickling filters is 0.5–1.0 kgBOD $m^{-3}\ d^{-1}$). At maximum daily and hourly flowrates, an OLR of 0.85 and 1.0 kgBOD $m^{-3}\ d^{-1}$ were obtained, respectively

(3) Determine the trickling filter cross-section area ($A_{cross\text{-}section}$), verification of hydraulic loading rates (HLR) and filter diameter
A common range for the reactor height for high-rate trickling filters is 2.0–3.0 m. In this instance, *2.8 m height was adopted.*

$$A_{cross-section}(m^2) = \frac{V_{reactor}(m^3)}{H_{reactor}(m)} = \frac{1337\,m^3}{2.8\,m} = 478\,m^2$$

Hydraulic loading rate is the amount of flow distributed over the packing media area at the top of the reactor. Then,

- HLR (for $Q_{average}$): $(8000\ m^3\ d^{-1})/(478\ m^2) \approx 17\ m^3\ m^{-2}\ d^{-1}$ (admissible range: $15\text{–}30\ m^3\ m^{-2}\ d^{-1}$)
- HLR (for $Q_{max\text{-}d}$): $(9600\ m^3\ d^{-1})/(478\ m^2) \approx 20\ m^3\ m^{-2}\ d^{-1}$ (admissible range: $18\text{–}22\ m^3\ m^{-2}\ d^{-1}$)
- HLR (for $Q_{max\text{-}h}$): $(12000\ m^3\ d^{-1})/(478\ m^2) \approx 25\ m^3\ m^{-2}\ d^{-1}$ (admissible range: $25\text{–}30\ m^3\ m^{-2}\ d^{-1}$)

Considering the admissible HLR range for high-rate trickling filters (Table 10.12), the conceptual design of the system is acceptable under the related perspective.
For constructive and operational reasons (e.g., rotary distributor sizing and maintenance of the post-treatment unit) two trickling filters will be adopted. Then, the diameter (D_{filter}) of each trickling filter is:

$$D_{filter} = [4.(478\ m^2/2)/\pi]^{1/2} = 17.4\ m$$

(4) Expected performance and final effluent BOD concentration of the post-treatment step with rock-bed trickling filters
As discussed in Section 10.4.5.2, the expected performance of rock-bed trickling filters following UASB reactors for BOD removal is approximately 65%. Therefore, the expected final effluent concentration is:

$$BOD_{effluent} = BOD_{influent}[1 - BOD_{removal}(\%)/100] = 117\ g\ m^{-3}[1 - 0.65] = 41\ gBOD\ m^{-3}$$

- When using an empirical formula (*NRC equation*), the expected performance and effluent BOD concentration are:

$$E = \frac{1}{1 + 0.4432\sqrt{\frac{OLR}{F}}} = \frac{1}{1 + 0.4432\sqrt{\frac{0.70}{1}}} = 0.73$$

Expected performance: 73%

$$BOD_{effluent} = BOD_{influent} \cdot [1 - BOD_{removal,NRC}(\%)/100] = 117\ g\ m^{-3} \cdot [1 - 0.73]$$
$$= 32\ gBOD\ m^{-3}(\text{expected final effluent concentration})$$

- When using a semi-empirical formula (*Eckenfelder formula*), the expected performance and effluent BOD concentration are (conversion factors: 1 m = 3.2808 ft and $1\ m^3\ d^{-1} = 0.1834$ gal min^{-1}):

$$S_{efl} = S_{in} e^{\frac{-k \cdot h^{(1-m)}}{(Q/A)^n}}$$

$$= 117 \cdot e^{\frac{-0.088 \cdot (2.5\,m \cdot 3.2808\,ft)^{1-0}}{[(8.000\,m^3\,d^{-1} \cdot 0.1834\,gal\,min^{-1})/(434\,m^2 \cdot 10.7639\,ft^2)]^{0.67}}}$$

$BOD_{effluent} = 24$ g BOD m^{-3}(expeced final effluent concentration), ($k = 0.088$, $n = 0.67$ *and* $m = 0$)

Expected performance: $[(117\ g\ m^{-3} - 24\ g\ m^{-3})/(117\ g\ m^{-3})] \cdot 100 = 79\%$

For trickling filters following UASB reactors, NRC and Eckenfelder equations appear to *overestimate* the performance of the system. As discussed before, the NRC equation was developed for rock-bed trickling filters treating sewage and, therefore, receiving higher concentrations of readily biodegradable organic matter. As regards the Eckenfelder equation, despite the treatability factor (k) and filter medium exponent (n) not being calibrated to the design and operational conditions imposed on trickling filters following UASB reactor treatment, it could be a good option to consider its calibration when the conversion of BOD along the system follows a first-order kinetics.

Solution 2

Using a low-rate system as post-treatment (filling material: sponge-based packing media)

Based on full-scale and experimental applications, the DHS system is also designed by applying *loading criteria*. In this instance, the OLR applied to curtain- and random-type filling material were 0.23 and 0.38 kgBOD m^{-3} d^{-1}, respectively, as reported in Onodera *et al.* (2014) and Okubo *et al.* (2015). Therefore, the theoretical hydraulic percolation time is verified. A minimum of 1.5–2.0 hours was adopted, considering the sponge occupancy ratio ($\%_{sponge}$) based on the sponge volume ($DHS_{curtain\text{-}type}$: 25%; $DHS_{random\text{-}type}$: 34%). Therefore, the design of the trickling filters considering sponge-based packing media as filling material is indicated below.

The expected performance of the DHS systems is based on the applications reported in Onodera *et al.* (2014) and Okubo *et al.* (2015).

DHS System	Reactor Volume (m^3)	Reactor Height (m)	HLR (m^3 m^{-2} d^1)	Expected BOD Removal (%)	Effluent BOD Concentration (g m^{-3})
DHS (curtain type)	4070	5	10	90	13
DHS (random type)	2463	3	10	90	15

Verify the theoretical hydraulic percolation time

Curtain-type DHS filter

$$HRT_{curtain\text{-}type} = V_{curtain\text{-}type}\ (m^3_{reactor}) \cdot (\%_{sponge}/100)/Q_{average}\ (m^3\ d^{-1})$$
$$= [4070\ m^3 \cdot (25\%/100)]/(8000\ m^3\ d^{-1})$$

$HRT_{curtain\text{-}type} = 0.13$ d (≈3.0 hours). *Typical range: 1.5–2.5 hours*

Random-type DHS filter

$$HRT_{curtain\text{-}type} = V_{curtain\text{-}type}\ (m^3_{reactor}) \cdot (\%_{sponge}/100)/Q_{average}\ (m^3\ d^{-1})$$
$$= [2463\ m^3 \cdot (34\%/100)]/(8000\ m^3\ d^{-1})$$

$HRT_{curtain\text{-}type} = 0.10$ d (≈2.5 hours). *Typical range: 1.5–2.5 hours*

10.4.6 Aerated biofilters

10.4.6.1 Description

The aerated biofilter is a submerged attached growth process, conventionally treating pre-settled sewage, consisting in a tank filled with a high specific surface packing material constituted by granular medium, such as vitrified clay particles (3–5 mm in size/particle) called *Biolite*. The oxygen is supplied by diffused aeration, considering co-current or counter-current in-bed air sparging. Because solids are efficiently retained within the biological reactor (TSS effluent concentrations around 10 mg L^{-1}), secondary clarifier is not considered in the flowsheet. Backwashing to avoid excessive head loss by particulate components (including biofilm) accumulation is needed. Treated effluent is generally used for this related operational step. One of the remarkable characteristics of the technology is the admissible organic loadings applied to the system. Because of the high specific surface of the medium, the applied volumetric loading is 5–10 kgCOD m^{-3} d^{-1} (Rittmann and McCarty, 2001). Depending on the process design, the effluent is applied in downward or upward direction. Co-current effluent and airflow tends to avoid clogging, also providing higher oxygen-transfer (oxygen supply at the location of maximum COD concentration). Typical submerged aerated biofilters are named Biocarbone®, Biofor® and Biostyr® process, and are described in detail in Tchobanoglous *et al.* (2013), and Rittmann and McCarty (2001).

Submerged aerated biofilters following UASB reactors. Submerged aerated biofilters have been used in Brazil for post-treatment of UASB reactors effluents. The technology is also considered for the retrofitting of trickling filters to meet more stringent discharge standards in terms of ammonium removal. The typical mode of operation is the upflow submerged attached growth process (anaerobically treated effluent and airflow are co-current). As previously observed for the Biofor® process, the advantage of co-current configuration is that the oxygen is supplied at the location of maximum concentration of the organic matter. A primary sedimentation tank and separate unit for sludge stabilization are not included in the flowsheet, and the excess aerobic sludge from the biofilter is returned to the UASB reactor for thickening and digestion. In this instance, using a secondary clarifier is considered in the flowsheet. A crucial difference in comparison with conventional submerged aerated biofilters is related to the packing media used as a filling material. Submerged aerated biofilters preceded by UASB reactors are filled with crushed stones or plastic-based packing media. Because of the low ability to retain solids with the use of high void-ratio packing media and lower concentration of solids eluted in UASB effluents, head loss and hydraulic shortcuts occur less frequently in the system. Consequently, the operational demand for backwashing procedures is reduced considerably for submerged aerated biofilters following UASB reactors. For submerged aerated biofilters preceded by primary settlers, backwashing needs to be performed every 24–48 hours, whereas the interval for biofilters following UASB reactors is approximately 7 days.

Based on the outputs of the Brazilian Research Program on Basic Sanitation (PROSAB), submerged aerated biofilters (specific surface of the plastic-based packing media ≈140 m^2 m^{-3}), preceded by UASB reactors tends to provide BOD removals of approximately 65–85% (final effluent: 15–35 mgBOD L^{-1}). For the reported applications (Aisse *et al.*, 2001; Motta, 1995), the organic loadings were 8–15 gBOD m^{-2} d^{-1} (Figure 10.13). When nitrification is not required, hydraulic and organic loadings of approximately 30 m^3 m^{-2} d^{-1} and 1–2 kgBOD m^{-3} d^{-1} can be applied, respectively. In this instance, effluent BOD concentrations below 30 mg L^{-1} are expected (Chernicharo *et al.*, 2001). In terms of energy demand, 0.40 kWh $kgCOD^{-1}_{removed}$ is observed for air supply ranging from 25 to 40 Nm^3 $kgBOD^{-1}_{applied}$. Therefore, as post-treatment of UASB effluents, the energy consumption is 3–4 times lower compared with that of submerged aerated biofilters treating pre-settled sewage. It is important to point out that nitrification is not expected in this instance. The excess sludge production in the biofilters following UASB reactors is approximately 0.75 $kgTS\ kgBOD_{removed}^{-1}$.

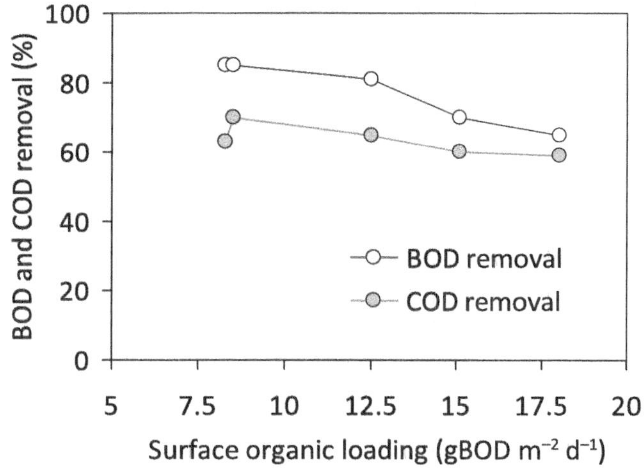

Figure 10.13 Expected performances for BOD and COD removal of submerged aerated biofilters post-UASB reactors filled with plastic-based packing media (138 $m^2\ m^{-3}$) (Motta, 1995).

10.4.6.2 Design criteria

The design of submerged biofilters is based on the organic and hydraulic *loading criteria*. Because tertiary treatment is seldom applied in a flowsheet for the post-treatment of UASB effluents, the required operational conditions for tertiary nitrification and N-removal will not be discussed in context.

Table 10.14 presents typical design loadings for submerged aerated biofilters preceded by primary settler and UASB reactors, considering only BOD removal and BOD removal and nitrification. When the reactor is filled with a single layer of crushed stones or plastic-based packing media, the surface loading is used to estimate the resulting volumetric loading and the effective volume of the reactor. Volumetric loading below 2.0 $kgBOD\ m^{-3}$ is recommended for submerged aerated biofilters following UASB reactors. The design of the system should consider at least two units operating in parallel to properly perform operational steps (see operational details). For the design of the secondary clarifiers, the same procedure as that used for trickling filters is adopted.

10.4.6.3 Construction details

Sewage treatment plants with UASB reactors followed by submerged aerated biofilters can be designed in extremely compact arrangements, generally built from concrete, fibreglass, or even with steel coated with anti-corrosive material. The depth of the effective volume (volume filled with the packing material) is typically 2 to 3 m. The UASB reactor effluent is collected using *Venturi tubes* connected to a pump, and subsequently distributed at the bottom of the biofilter. The effluent collection system enables the passage of air, and oxygen is supplied simultaneously with the UASB effluent. Such a solution tends to control the dispersion of odorant gases, as the air is captured at the location of the odour emissions. Because of the direct contact with dissolved sulphide from the UASB effluent, the perforated plates sustaining the packing material need to be manufactured from or protected with corrosion-resistant materials.

Table 10.14 Design parameters for the design of submerged aerated biofilters.

Pre-treatment	System	Specific Surface Area	Loading Range (kgBOD m^{-3} d^{-1})	
			BOD Removal	BOD Removal and Nitrification
Primary settler	Biocarbone®	Biolite: 1000 m^2 m^{-3}	3.5–4.5	2.0–2.75
Primary settler	Biofor®	Biolite: 1000 m^2 m^{-3}	5.0–6.0	–
Primary settler	Biostyr®	Polyestyrene beads: 400 m^2 m^{-3}	4.0–5.0	2.0–2.5
UASB reactor	Aerated biofilter	Crushed stones: 50–70 m^2 m^{-3} Plastic medium: 100–200 m^2 m^{-3}	1.0 (crushed stones) 2.0 (plastic medium)	–

Loading range considered for the design of ***sewage*** treatment plants.
Note: Information adapted from Tchobanoglous *et al.* (2013) and Chernicharo *et al.* (2001).

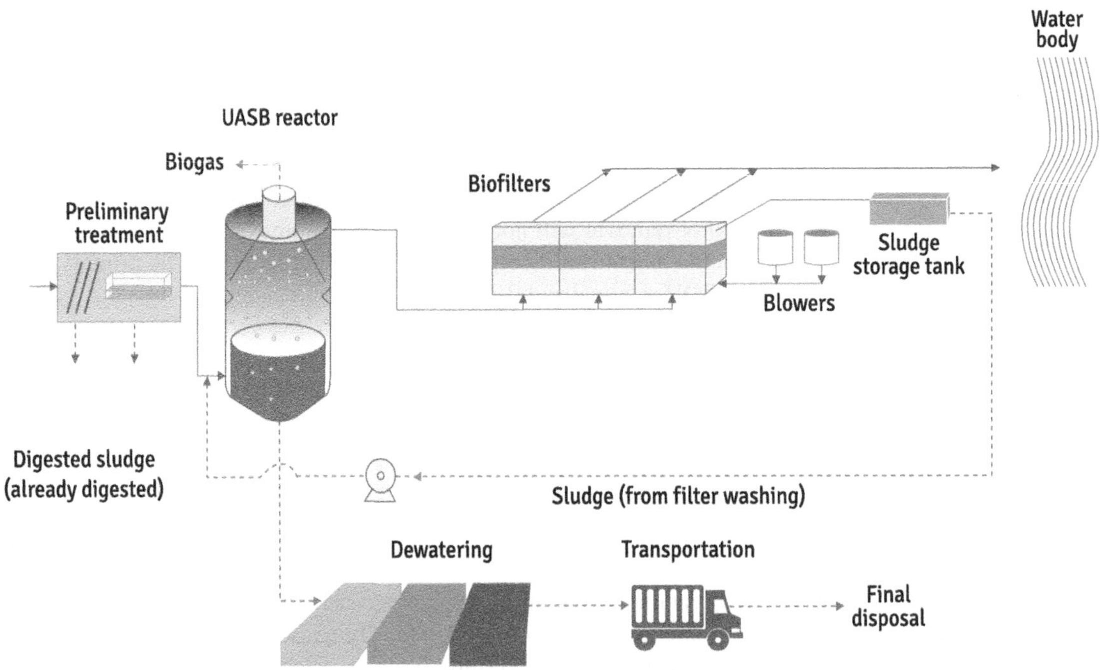

Figure 10.14 Typical flowsheet for UASB-submerged aerated biofilters.

Packing media. As discussed before, conventional submerged aerated biofilters are filled with *biolite*. When preceded by UASB reactors, the reactor bed is usually filled with crushed stones or a plastic medium. Single or multiple layers of crushed stones can be used to fill the reactor. With a single layer, an average granulometry of 50–75 mm is recommended. For multiple layers, a distinct granulometry is used at each layer. Considering up-flow systems, the recommended dimensions (upwardly ordered) are 25–50 mm (layer 1); 20–25 mm (layer 2); 10–20 mm (layer 3); 5–10 mm (layer 4) at bed depths of 30 cm (layers 1 and 2), 40 cm (layer 3), and 100 cm (layer 4), respectively. When a plastic medium is used, cross-flow modules and random plastic media can be adopted for systems following UASB reactors. As mentioned previously, when a single layer of crushed stones or plastic-based medium is used, the system is followed by a secondary clarifier. Figure 10.14 presents a typical flowsheet for UASB-submerged aerated biofilters.

10.4.6.4 Operational aspects

The operational aspects reported in this section are related to submerged aerated biofilters preceded by UASB reactors. The implementation of the backwashing procedure is the most important step for adequate operation of submerged aerated biofilters. The backwashing intervals intended to decrease the excess biomass in the reactor are determined mainly by the applied organic loading and the packing media used. Such a process contributes also to maintaining the system performance in terms of mass transfer, metabolic activity, and flux of substrates to the biofilm. As previously mentioned, for submerged aerated biofilters following UASB reactors, backwashing is performed generally every seven days or at even longer intervals, depending on the operation of the UASB reactor in terms of sludge management. The backwashing procedure should be performed during periods of lower influent flow-rate. Table 10.15 presents in detail the steps for the backwashing procedure.

Table 10.15 Description of backwashing procedures for submerged aerated biofilters following UASB reactors.

Step	Time (min)	Description	Procedure
1. Interrupt unit operation for the treatment of UASB effluent	2	Post-treatment shutdown	Interrupt UASB effluent feeding and air supply
2. Backwashing	2	Intense down-flow treated water discharge (hydraulic loading: $20\ m^3\ m^{-2}\ h^{-1}$)	Open the water logs located at the bottom of the biofilter
3. Interrupt backwashing	0.5	Interruption of treated water discharge	Close the water logs located at the bottom of the biofilter
4. Aeration	2	Intense aeration: $20\ m^3\ m^{-2}\ h^{-1}$	Increase air supply
5. Interrupt aeration	0.5	Air supply interruption	Interrupt air supply
6. Repeat previous procedures	15–20	Repeat the previous steps at least 2–3 times	Restart procedure from step 1
7. Restart operation for the treatment of UASB effluent	5	Restart the operation of the biofilter	UASB effluent feeding and air supply

Note: Information adapted from Chernicharo *et al.* (2001).

Main operational troubleshooting and possible solutions. Table 10.16 shows the operational issues frequently reported during the operation of submerged aerated biofilters and the related possible solutions.

Table 10.16 Main operational issues and possible solutions for submerged aerated biofilters following UASB reactors.

Observation	Problem Cause	Possible Solution
High effluent solids concentration and/or excessive head loss	• Hydraulic and/or organic overloading	• Control organic loadings in accordance with the parameters used to design the system
	• Inadequate backwashing	• Increase backwashing procedure time and frequency, and air and hydraulic loadings
	• Excessive aeration	• Decrease aeration rate, avoiding excessive head loss and biofilm detachment
	• Deficiency in air supply	• Check possible clogging in the air sparging system
	• UASB reactor operation	• Improve sludge management in the anaerobic step and secondary settlers, and the operation of sludge return
Low performance for organic matter removal	• Organic and hydraulic overloading	• Control organic and hydraulic loadings in accordance with the parameters used to design the system
	• Oxygen transfer	• Oxygen transfer to the liquid could be improved by modifying the aeration system to produce coarse bubbles instead of fine bubbles
	• Low temperature	• Evaluate the feasibility to cover the reactor

Note: Information adapted from Chernicharo *et al.* (2001).

EXAMPLE 10.5

Design of submerged aerated biofilters filled with crushed stones and plastic medium, respectively, treating effluents from a UASB reactor.

Input data

- Population equivalent: PE = 50,000 inhabitants
- Average flow: $Q_{average} = 8{,}000\ m^3\ d^{-1}$
- Maximum daily flowrate: $Q_{max\text{-}d} = 9{,}600\ m^3\ d^{-1}$
- Maximum hourly flowrate: $Q_{max\text{-}h} = 12{,}000\ m^3\ d^{-1}$
- Influent BOD concentration (raw sewage): 333 mg L^{-1}
- BOD removal efficiency in the UASB reactor: 65%
- Requirement for final effluent BOD concentration: <30 mgBOD L^{-1}
- Operational conditions applied to the submerged aerated biofilters for organic matter removal

Post-treatment System	Surface Organic Loading Rate ($gBOD\ m^{-2}\ d^{-1}$)	Hydraulic Loading Rate ($m^3\ m^{-2}\ d^{-1}$)	Packing Media (Specific Surface Area)
Option 1	13.5	<30	Crushed stones ($60\ m^2\ m^{-3}$)
Option 2	13.5	<30	Cross-flow medium ($98\ m^2\ m^{-3}$)

Solution 1

Using one layer of crushed stone as filling material

(1) BOD concentration in the effluent of the UASB reactor (influent to the submerged aerated biofilters)
$BOD_{effluent\ UASB} = BOD_{influent\ UASB} \times (1 - Efficiency_{UASB}) = 333 \times (1–0.65) = 117\ mgBOD\ L^{-1}$

(2) Determine the volumetric organic loading rate (OLR)
Considering the application of 13.5 $gBOD\ m^{-2}\ d^{-1}$, the performance of the biofilter for BOD removal is approximately 75% (Figure 10.13). Therefore, the final effluent BOD concentration is:
$BOD_{effluent} = BOD_{influent} \cdot [1 - BOD_{removal}\ (\%)/100] = 117\ g\ m^{-3} \cdot [1–0.75] \approx 29\ gBOD\ m^{-3} <$ $30\ gBOD\ m^{-3}$ (required final effluent BOD concentration)

The determination of volumetric organic loading rate is based on the adopted surface loading and the specific surface area of the packing media.
$OLR\ (kgBOD\ m^{-3}\ d^{-1})$ = surface organic loading rate $(gBOD\ m^{-3}\ d^{-1})$. specific surface area $(m^2\ m^{-3}) = 13.5\ (gBOD\ m^{-2}\ d^{-1})/1000] \times 60\ m^2\ m^{-3} \approx 0.80\ kgBOD\ m^{-3}\ d^{-1}$

(3) Determine the reactor volume

$$OLR(kgBOD\,m^{-3}\,d^{-1}) = \frac{Q_{average}(m^{-3}\,d^{-1}) \cdot BOD_{concentration}(g\,m^{-3})}{V_{reactor}(m^3)}$$

$$V_{reactor}(m^3) = \frac{8000\,m^{-3}\,d^{-1} \cdot [(117\,g\,m^{-3})/1000]}{0.80\,kgBOD\,m^{-3}\,d^{-1}} = 1156\,m^3$$

(4) Determine the biofilter cross-section area ($A_{cross\text{-}section}$) and verification of the hydraulic loading rate (HLR), adopting a reactor height of 3 m

$$A_{cross-section}(m^2) = \frac{V_{reactor}(m^3)}{H_{reactor}(m)} = \frac{1156\,m^3}{3\,m} = 385\,m^2$$

Hydraulic loading rate is the amount of flow distributed by the biofilter cross-section area. Then,

HLR (for $Q_{average}$): $(8000\ m^3\ d^{-1})/(385\ m^2) \approx 21\ m^3\ m^{-2}\ d^{-1}$ (recommended HLR $< 30\ m^3\ m^{-2}\ d^{-1}$)

Considering the admissible HLR $< 30\ m^3\ m^{-2}\ d^{-1}$, the design of the biofilter is acceptable.
Note: For operational reasons (e.g., implementation of backwashing procedures), two aerated submerged biofilters will be adopted. Then, the diameter (D_{filter}) of each biofilter is:

$$D_{filter} = [4.(385\,m^2/2)/\pi]^{1/2} = 16\,m$$

(5) Determine the air flow
Considering organic removal only (nitrification not expected), an aeration rate of 40 $Nm^3\ kgBOD_{applied}^{-1}$ will be adopted. Air flow will be estimated based on the organic loading from the UASB reactor (Loading $_{UASB}$).

Q_{air} = Aeration rate ($Nm^3\ kgBOD_{applied}^{-1}$). Loading $_{UASB}$ ($kgBOD\ d^{-1}$) = 40 ($Nm^3\ kgBOD_{applied}^{-1}$) × [8000 $m^3\ d^{-1}$. 117 ($g\ m^{-3}$)/1000] = 37,440 $Nm_{air}^3\ d^{-1}$ (18,720 $Nm_{air}^3 \cdot reactor^{-1}\ d^{-1}$)

Solution 2

Using plastic medium (cross-flow) as filling material
Considering the same pre-design procedure, the volumetric organic loading applied to the biofilter would be approximately 1.3 $kgBOD\ m^{-3}\ d^{-1}$ with a reactor volume of 707 m^3. In this instance, the volume of the biofilter filled with plastic medium represents 60% of the biofilter volume filled with crushed stones. The reactor height could be reduced also to 2 m for a hydraulic loading rate of approximately 30 $m^3\ m^{-2}\ d^{-1}$ (total cross-sectional area: 354 m^2)

10.4.7 Activated sludge

10.4.7.1 Description

Activated sludge is employed worldwide for treating raw or primary settled sewage and, as such, is described in several international textbooks. Therefore, a general description of the system will not be presented here, and this section will concentrate only on highlighting the specifics of the activated sludge process when acting as post-treatment of anaerobic effluents. The following section is based, with adaptations, on von Sperling and Chernicharo (2005) and von Sperling (2007b). For further detail, the cited publications should be consulted.

The essence of the *continuous flow* activated sludge process is the integration of the aeration tank (aerobic biological reactor), secondary sedimentation tank, and sludge recirculation line. These components are maintained when activated sludge systems are used as post-treatment of effluents from anaerobic reactors. The *intermittent flow* activated sludge system (sequencing batch reactors) can be adopted also as post-treatment, requiring in such instance that the tanks are used alternately for the functions of reaction and sedimentation.

When the activated sludge system acts as post-treatment of anaerobic effluents, *the anaerobic reactor is used instead of the primary sedimentation tank* (which is an integral part of the conventional activated sludge system). The aerobic sludge is recirculated in the usual manner, i.e., from the bottom of the secondary sedimentation tank to the entrance of the aerobic reactor (aeration tank).

The excess aerobic sludge generated in the activated sludge stage, not yet stabilized, is sent to the UASB reactor, where it undergoes thickening and digestion, together with the anaerobic sludge. As the return flow of the excess aerobic sludge is extremely low compared with the influent flow, in principle, no operational disturbances in the UASB reactor are expected when a proper strategy is in place. This discussion is similar to the return of the excess sludge from the trickling filters (see Section 10.4.5.4) and requires adequate planning. With this configuration, the sludge treatment can be simplified significantly, and there is no need for separate thickeners and digesters, with only the dewatering stage being necessary. The mixed sludge removed from the anaerobic reactor is digested, has solid concentrations similar to those from sludge thickeners, and presents adequate dewaterability. Figure 10.15 presents the flowsheet of this configuration.

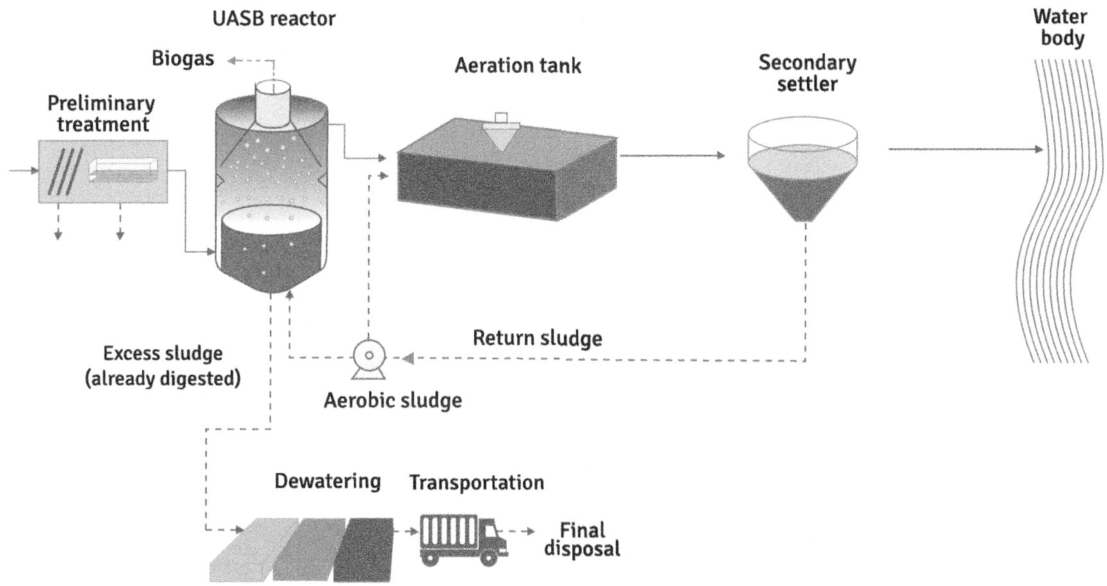

Figure 10.15 Typical configuration of a treatment plant with UASB reactor and activated sludge system.

10.4.7.2 Design criteria

The main design parameters that determine the behaviour of the system and the required volumes and areas are: (a) *reactor*: sludge age (θ_c) and mixed liquor suspended solids (MLSS) concentration; (b) *secondary sedimentation tank*: hydraulic loading rate (HLR) and solids loading rate (SLR).

Table 10.17, adapted from von Sperling and Chernicharo (2005) and von Sperling (2007b), lists the main design parameters used for the activated sludge process as post-treatment of effluents from anaerobic reactors. Further details can be found in these publications.

Comments on design criteria for activated sludge systems acting as post-treatment of anaerobic effluents are as follows:

Kinetic and stoichiometric coefficients. As regards the values of the kinetic and stoichiometric coefficients (mainly Y and K_d), the same common values of the classical configurations of the activated sludge system can be used in principle. However, the values of these coefficients, applied specifically to activated sludge as post-treatment of effluents from anaerobic reactors, must be investigated further, because of the possible influence of the previous anaerobic treatment on the process kinetics.

Design parameters for the reactor. The design parameters for the activated sludge reactor as post-treatment are similar to those for the conventional activated sludge systems. The main difference is in the lower MLSS concentration usually assumed for the post-treatment activated sludge. When higher values are adopted, the volume of the aeration tank is reduced significantly (hydraulic retention time shorter than 2.0 hours; scarce full-scale operational experience so far to demonstrate the process stability of such small tanks). The sludge age can be compatible with the conventional activated sludge, without the need to enter the range of extended aeration activated sludge, as there is no need for sludge digestion in the aeration tank, as is the case in the extended aeration mode.

Design parameters for secondary sedimentation tanks. The loading rates in the secondary sedimentation tanks of post-treatment activated sludge systems are presumably different from those in conventional

Table 10.17 Design parameters of activated sludge systems for the post-treatment of effluents from UASB reactors.

Item	Parameter	Value
Aeration tank	Sludge age (d)	6–12
	F/M ratio (kg BOD · kgMLVSS^{-1} d^{-1})	0.20–0.40
	Hydraulic retention time (h)	3–5
	MLVSS concentration (mg L^{-1})	1100–1500
	MLSS concentration (mg L^{-1})	1500–2000
	VSS/SS ratio in the reactor (–)	0.73–0.77
Oxygen requirements	Average O_2 requirements: carbonaceous demand (kgO_2 kgBOD$^{-1}_{remov}$)	0.80–0.94
	Average O_2 requirements: nitrogenous demand (kgO_2 kgTKN$^{-1}_{applied}$)	3.8–4.3
Sludge production	Product: excess AS sludge (returned to UASB) (kgSS kgBOD$^{-1}_{remov}$ from AS)	0.78–0.90
	Per capita product: of excess AS sludge (returned to UASB) (gSS inhab^{-1} d^{-1})	8–14
	Concentration of SS in the AS sludge returned to the UASB (mg L^{-1})	2500–5000
	Removal efficiency of VSS from the AS sludge in the UASB reactor	0.25–0.45
	Production of anaerobic sludge (kgSS kgBOD^{-1} applied to the UASB)	0.28–0.36
	Per capita production of anaerobic sludge (gSS inhab^{-1} d^{-1})	14–18
	Production of total mixed sludge (to be treated) (kgSS kgBOD$^{-1}_{applied}$)	0.40–0.60
	Per capita production of total mixed sludge (to be treated) (gSS inhab^{-1} d^{-1})	20–30
	Per capita volumetric product: total mixed sludge (to be treated) (L inhab^{-1} d^{-1})	0.5–1.0
	Concentration of mixed sludge (AS+anaerobic) removed from the UASB (%)	3.0–4.0
Secondary sedimentation tank	Hydraulic loading rate (Q/A) (m^3 m^{-2} d^{-1})	24–36
	Solids loading rate [$(Q+Q_r) \cdot X/A$] (kgSS m^{-2} d^{-1})	100–140
	Side water depth (SWD) (m)	3.0–4.0
	Recirculation ratio (Q_r/Q)	0.6–1.0
	Concentration of SS in the sludge recirculated to the aeration tank (mg L^{-1})	2500–5000
Sludge treatment	Per capita production of SS in the sludge to be disposed of (gSS inhab^{-1} d^{-1})	20–30
	Per capita volum. production of sludge to be disposed of (L sludge inhab^{-1} d^{-1})	0.05–0.15
	Solids content (centrifuge, belt press) (%)	20–30
	Solids content (filter press) (%)	25–40
	Solids content (sludge drying bed) (%)	30–45

AS: activated sludge.

activated sludge systems, as the former operate with lower MLSS concentrations and with sludge with slightly different characteristics. In addition, the UASB reactors provide a certain smoothing in the flow to be treated, reducing the $Q_{maximum}/Q_{average}$ ratio in the influent to the sedimentation tanks. These are factors that deserve continued investigation with full-scale STPs to obtain specific design parameters for this configuration.

Nitrification. In respect of the removal of ammonia in the UASB-activated sludge system, it should be mentioned that operational difficulties have occurred in maintaining full nitrification in the aerobic reactor. This is associated apparently with toxicity problems with the nitrifying bacteria, possibly caused by the presence of sulphides. Therefore, if nitrification is desired, even in warm-climate regions, aerobic sludge ages equal to or greater than 8 days should be adopted.

Biological nutrient removal. Post-treatment activated sludge systems are not particularly efficient in the removal of nitrogen, as there is little availability of organic carbon for the denitrifying bacteria, because a large fraction of the organic matter has been removed previously in the UASB reactor. A means of supplying organic carbon to the activated sludge reactor is by a partial by-pass to the UASB reactor, supplying raw sewage to the anoxic zone in the aeration tank. A similar comment can be made about the biological removal of phosphorus, namely, the previous removal of a large fraction of the organic carbon in the UASB reactor hinders the biological P removal process. Similarly, a partial by-pass of the raw sewage can be helpful.

Sequencing batch activated sludge reactor. The design of a sequencing batch activated sludge reactor following an anaerobic reactor should consider an operational cycle that is suitable for low organic loads in the influent to the aerobic stage. In this regard, designs that do not consider optimization could bring about large fill volumes compared with the reaction volumes, which could result in an uneconomical large volume of the aerobic reactor.

In contrast with the other sections in this Chapter, no design example is presented here for the activated sludge process acting as post-treatment because of the inherent complexity of presenting the activated sludge-regulating equations in a relatively short space. The reader can follow the typical design sequences provided in several publications available on activated sludge and consider the utilization of the basic design parameters presented here.

10.4.7.3 Construction details

Activated sludge is a complex system composed not only of the two major process units (aeration tank and secondary sedimentation tank), but also of sludge lines, aeration equipment, pumps, and several accessories. The construction details, as described in the major textbooks on activated sludge and wastewater treatment, also apply here, without the need for specific modifications for treating anaerobic effluents.

As with the other post-treatment options, proper precautions against corrosion must be implemented, as high concentrations of hydrogen sulphide are present in the effluent from the anaerobic reactor. Further details are discussed in Chapters 5 and 9.

10.4.7.4 Operational aspects

The operation is similar to that typically adopted for conventional activated sludge. Obviously, the absence of sludge thickeners and digesters will simplify the operation of the sludge treatment line.

The return of the excess sludge from the activated sludge stage (aerobic sludge) needs to be controlled. Even though, in terms of flow, this represents an addition of only approximately 1 to 2% of the inflow to the UASB reactor, in terms of load this could amount to 10% of the influent BOD load and 15% of the influent SS load. Although this can be considered a small increment, instances of solids flotation and washout have been reported in the effluent from UASB reactors receiving excess aerobic sludge. In-depth investigation is needed into the relevant causes; however, it is advisable that the return of the excess aerobic sludge is distributed over 24 hours, or to times of low influent flow (night hours). Obviously, the return of high volumes of excess sludge distributed over a short period of time during the peak hours of inflow should be avoided. Further details are discussed in Chapter 7.

10.5 FINAL REMARKS

A successful approach for sewage treatment, particularly in warm climate regions, is using UASB reactors as a first stage in the STP flowsheet, with most of the remaining concentration of organic matter and suspended solids removed in a post-treatment step. Such post-treatment facilities typically comprise an aerobic stage, which allows the achievement of the necessary efficiencies to comply with discharge standards, mainly in developing countries.

The selection of a post-treatment process should take into account several variables that affect the treatment levels required and the overall cost. There is no ideal solution applicable to all cases neither generalized formulas to define the best solution. The decision should be based on a comprehensive balance between economic, technical and social criteria, as addressed in Chapter 2. Nevertheless, such aspects may be intangible and, therefore, the final selection process may still be based on a subjective balance of the relative degree of importance of all elements involved in the decision-making process.

Although the post-treatment of anaerobic effluents is developing dynamically, the technological choices presented in this chapter have been used widely at full scale or are close to being implemented more widely, particularly in developing countries.

REFERENCES

Aisse M. M., Jürgensen D., Reali M. A. P., Penetra R., Florencio L. and Alem Sobrinho P. (2001). *Capítulo 6: Pós-tratamento de efluentes de reatores anaeróbios por sistemas de flotação* (Chapter 6: Post-treatment of effluents from anaerobic reactors by flotation systems). In: C. A. L. Chernicharo (coordenador). Pós-tratamento de efluentes de reatores anaeróbios. PROSAB/ABES, Rio de Janeiro, Brazil.

Alarcón-Herrera M. T., Martínez F. Z., Lara-Borrero J. A. and Vidal G. (eds.), (2018). Humedales de tratamiento. Alternativa de saneamiento de aguas residuales aplicable en América Latina (Constructed wetlands. Wastewater treatment alternative for Latin America). Pontificia Universidad Javeriana, Bogotá. 2018. Available at https://repository.javeriana.edu.co/handle/10554/34519 (in Spanish).

Andreoli C. V., von Sperling M. and Fernandes F. (eds.), (2007). Sludge Treatment and Disposal. IWA Publishing, London, UK. Available at http://www.iwapublishing.com/open-access-ebooks/3567.

Arceivala S. J. (1981). Wastewater Treatment and Disposal, Marcel Dekker, New York, USA.

Ayres R. M., Alabaster G. P., Mara D. D. and Lee D. L. (1992). A design equation for human intestinal nematode egg removal in waste stabilization ponds. *Water Research*, **26**, 863–865.

Barros I. P. A. F., Queiroga L. L., von Sperling M. and Pereira C. E. B. (2013). Assessment of the implementation of final polishing by flotation at a large pond system in Brazil. In: 10th IWA Specialist Group Conference on Waste Stabilisation Ponds, Cartagena de las Indias, Colombia.

Campos J. R. (coord.). (1999). Tratamento de esgotos sanitários por processo anaeróbio e disposição controlada no solo (Sewage treatment via anaerobic process and overland flow). PROSAB/FINEP, Rio de Janeiro (in Portuguese).

Catunda P. F. C., van Haandel A. C. and Lettinga G. (1994). Post treatment of anaerobically treated sewage in waste stabilization ponds. In: Anaerobic Digestion, 7th Symposium, Cape Town, South Africa, pp. 405–415.

Cavalcanti P. F. F., van Haandel A. C., Kato M. T., von Sperling M., Luduvice M. L. and Monteggia L. O. (2001). Capítulo 3: Pós-tratamento de efluentes de reatores anaeróbios por lagoas de polimento (Chapter 3: Post-treatment of effluents from anaerobic reactors by polishing ponds). In: C. A. L. Chernicharo (coordenador). Pós-tratamento de efluentes de reatores anaeróbios. PROSAB/ABES, Rio de Janeiro, Brazil (in Portuguese).

Chernicharo, C. A. L. (coordenador) (2001). Pós-tratamento de efluentes de reatores anaeróbios (Post-treatment of effluents from anaerobic reactors). Belo Horizonte: FINEP/PROSAB - Programa de Pesquisa em Saneamento Básico, 544 p.

Chernicharo C. A. L. (2007). Anaerobic Reactors. (Biological Wastewater Treatment Series, Volume 4). IWA Publishing, London, UK. Available at http://www.iwapublishing.com/open-access-ebooks/3567.

Coraucci Filho B., Andrade Neto C. O., Melo H. N. S., Souza J. T., Nour E. A. A. and Figueiredo R. F. (2001). Capítulo 2: Pós-tratamento de efluentes de reatores anaeróbios por sistemas de aplicação no solo (Chapter 2: Post-treatment of effluents from anaerobic reactors by overland flow). In: C. A. L. Chernicharo (coordenador). Pós-tratamento de efluentes de reatores anaeróbios. FINEP/PROSAB, Rio de Janeiro, Brazil (in Portuguese).

Crites R. W., Middlebrooks E. J. and Reed S. C. (2006). Natural Wastewater Treatment Systems, CRC Press, Boca Raton, Florida, USA.

Dias D. F. C., Passos R. G., Rodrigues V. A. J., Matos M. P., Santos C. R. S. and von Sperling M. (2018). Performance evaluation of a natural treatment system for small communities, composed of a UASB reactor, maturation ponds (baffled and unbaffled) and a granular rock filter in series. *Environmental Technology*, **39**, 490–502.

Dotro G., Langergraber G., Molle P., Nivala J., Puigagut J., Stein O. and von Sperling M. (2017). Treatment Wetlands. (Biological Wastewater Treatment Series, Volume 7). IWA Publishing, London, UK. Available at: http://www.iwapublishing.com/open-access-ebooks/3567

Hoffmann H., Platzer C., Winker M. and von Muench E. (2011). Technology review of constructed wetlands: subsurface flow constructed *wetlands* for greywater and domestic wastewater treatment. Deutsche Gesellschaft für Internationale Zusammenarbeit (GIZ) GmbH, Eschborn. Sustainable sanitation – Ecosan Program.

IWA Specialist Group on Use of Macrophytes in Water Pollution Control (2000). Constructed Wetlands for Pollution Control. Processes, Performance, Design and Operation, IWA Publishing, London. Available at: http://www.iwapublishing.com/open-access-ebooks/3567.

Kadlec R. H. and Knight R. L. (1996). Treatments Wetlands, vol. 1, Lewis Publishers, Boca Raton, Florida, USA.

Kadlec R. H. and Wallace S. D. (2008). Treatment Wetlands, vol. 2, CRC Press, Boca Raton, Florida, USA.

Lombard Latune R. and Molle P. (2017). Les filtres plantés de végétaux pour le traitement des eaux usées domestiques en milieu tropical. Guide de dimensionnement de la filière tropicalisée (Constructed wetlands for the treatment of domestic wastewater in tropical environments. Design guide for tropical conditions). Agence Française pour la Biodiversité, Collection Guides et Protocoles. (in French).

Maiga Y., von Sperling M. and Mihelcic J. (2017). Constructed Wetlands. In: Global Water Pathogens Project, J.B. Rose and B. Jiménez-Cisneros (eds.) Part 4. Management of Risk from Excreta and Wastewater C. Haas J. Mihelcic and M. Verbyla (eds.). Michigan State University E. Lansing, MI/ UNESCO, Michigan, USA.

Mara D. D. (2003). Domestic Wastewater Treatment in Developing Countries, Earthscan/James & James, London, UK.

Mara D. D., Alabaster G. P., Pearson H. W. and Mills S. W. (1992). Waste Stabilisation Ponds: A Design Manual for Eastern Africa. Lagoon Technology International, Leeds, UK.

Matos A. T. and Matos M. P. (2017). Disposição de águas residuárias no solo e em wetlands construídos (Wastewater treatment in overland flow and constructed wetlands). Editora UFV, Viçosa (in Portuguese).

Motta S. L. (1995) Tratamento de esgoto doméstico em bio-reator aeróbio de leito fixo submerso (Sewage treatment in submerged aerobic fixed bed reactor). PhD thesis. COPPE/UFRJ, Brazil (in Portuguese).

Okubo T., Onodera T., Uemura S., Yamaguchi T., Ohashi A. and Harada H. (2015). On-site evaluation of the performance of a full-scale down-flow hanging sponge reactor as a post-treatment process of an up-flow anaerobic sludge blanket reactor for treating sewage in India. *Bioresource Technology*, **194**, 156–164.

Onodera T., Tandukar M., Sugiyana D., Uemura S., Ohashi A. and Harada H. (2014). Development of a sixth-generation down-flow hanging sponge (DHS) reactor using rigid sponge media for post-treatment of UASB treating municipal sewage. *Bioresource Technology*, **152**, 93–100.

Onodera T., Okubo T., Uemura S., Yamaguchi T., Ohashi A. and Harada H. (2016). Long-term performance evaluation of down-flow hanging sponge reactor regarding nitrification in a full-scale experiment in India. *Bioresource Technology*, **204**, 177–184.

OPS (Organización Panamerica de la Salud)/OMS (Organización Mundial de la Salud) (1999). *Sistemas de tratamiento de aguas servidas por medio de humedales artificiales (Wastewater treatment systems by constructed wetlands)*, OPS/OMS, Santafé de Bogotá (in Spanish).

Paganini W. S. (1997). Disposição de esgoto no solo (escoamento à superfície). (Overland flow for sewage treatment). Fundo editorial da AESABESP, São Paulo (in Portuguese).

Parker D.S., Richards T. (1986). Nitrification in trickling filters. *Journal WPCF*, **58**(9), 896–901.

Philippi L. S. and Sezerino P. L. (2004). Aplicação de sistemas tipo wetlands no tratamento de águas residuárias: utilização de filtros plantados com macrófitas (Constructed wetland for wastewater treatment: planted systems with macrophytes), Florianópolis, Brazil (in Portuguese).

Possmose-Nascimento T. E., Rodrigues V. A. J., von Sperling M. and Vasel J. L. (2014). Sludge accumulation in shallow maturation ponds treating UASB reactor effluent: results after 11 years of operation. *Water Science and Technology*, **70**(2), 321–328.

Rittmann B. and McCarty P. (2001). Environmental Biotechnology: Principles and Applications, McGraw-Hill International, New York, USA.

Shilton A. (ed.) (2006). Pond Treatment Technology. IWA Publishing, London, UK.

Tchobanoglous G., Stensel H. D., Tssuchihashi R., Burton F. L., Abu-Orf M., Bowden G., Pfrang W. (2013). Wastewater engineering, treatment and reuse, 5th edn, Mc Graw Hill, New York..

Uemura S., Okubo T., Maeno K., Takahashi M., Kubota K., Harada H. (2016). Evaluation of water distribution and oxygen mass transfer in sponge support media for a down-flow hanging sponge reactor. *International Journal Environmental Research*, **10** (2), 265–272.

USEPA (1981). Process design manual: Land treatment of municipal wastewater. United States Environmental Protection Agency, Cincinnati, Ohio, USA.

USEPA (2000). Manual. Constructed Wetlands Treatment of Municipal Wastewaters. United States Environmental Protection Agency. EPA/625/R-99/010. Cincinnati, Ohio, USA.

USEPA (2011). Principles of Design and Operations of Wastewater Treatment Pond Systems for Plant Operators, Engineers, and Managers. United States Environmental Protection Agency. EPA/600/R-10/088, Cincinnati, Ohio, USA.

Verbyla M., von Sperling M. and Maiga Y. (2017). Waste stabilization ponds. In: Global Water Pathogens Project, J. B. Rose and B. Jiménez-Cisneros (eds.), Part 4. Management of Risk from Excreta and Wastewater C. Haas J. Mihelcic and M. Verbyla (eds.). Michigan State University E. Lansing, MI/ UNESCO.

von Sperling M. (2005). Modelling of coliform removal in 186 facultative and maturation ponds around the world. *Water Research*, **39**, 5261–5273.

von Sperling M. (2007a). Waste Stabilization Ponds (Biological Wastewater Treatment Series, Volume 3), IWA Publishing, London, UK. Available at http://www.iwapublishing.com/open-access-ebooks/3567.

von Sperling M. (2007b). Activated Sludge (Biological Wastewater Treatment Series, Volume 5). IWA Publishing, London, UK. Available at http://www.iwapublishing.com/open-access-ebooks/3567

von Sperling M. and Chernicharo C. A. L. (2005). Biological Wastewater Treatment in Warm Climate Regions, IWA Publishing, London, UK. Available at http://www.iwapublishing.com/open-access-ebooks/3567.

von Sperling M., Bastos R. K. X. and Kato M. T. (2005). Removal of *E. coli* and helminth eggs in UASB – polishing pond systems. *Water Science and Technology*, **51**(12), 91–98.

von Sperling M., Andrada J. G. B. and Melo Júnior W. R. (2007). Coarse filters for pond effluent polishing: comparison of loading rates and grain sizes. *Water Science and Technology*, **55**(11), 121–126.

Yanez F. (1993). Lagunas de estabilizacion. Teoria, diseño y mantenimiento (Stabilization ponds: theory, design and maintenance). ETAPA, Cuenca, Equador (in Spanish).

Chapter 11

Upgrading anaerobic sewage treatment applying membranes: AnMBR and UF post filtration

Jules B. van Lier[1], Aurora Seco[2], Bruce Jefferson[3], Mustafa E. Ersahin[4] and Angel Robles[2]

[1]Section Sanitary Engineering, Department of Water Management, Delft University of Technology, Stevinweg 1, 2628CN, Delft, The Netherlands
[2]Departament d'Enginyeria Química, Universitat de València, Avinguda de la Universitat s/n, 46100 Burjassot, València, Spain
[3]Cranfield Water Science Institute, Vincent Building, Cranfield University, Bedfordshire, MK43 0AL, United Kingdom
[4]Istanbul Technical University, Civil Engineering Faculty, Environmental Engineering Department, Ayazaga Campus, Maslak, 34469 Istanbul, Turkey

11.1 INTRODUCTION

Anaerobic high-rate technology has improved significantly in the last few decades, mainly due to the applications of differently configured high-rate reactors. Indeed, high-rate anaerobic treatment of sewage can be considered a consolidated technology in various countries in the (sub)tropical climate zone (Chernicharo *et al.*, 2015; Skouteris *et al.*, 2012; van Lier *et al.*, 2010). To date, there are many full-scale UASB reactors treating sewage in Latin America and India (Chernicharo *et al.*, 2015). Reflection of the experiences gained across this large array of sites enables the technology to be critically evaluated in order to assess the locally attuned merits that are realized.

Nonetheless, the application of anaerobic systems for sewage treatment worldwide is still very limited outside India, Brazil, and some other Latin American countries and certainly in countries with cold or mild climates, since required solids retention times (SRTs) for stabilizing the sludge are not met in the current process designs. Primary clarification preceding the anaerobic reactor solves this problem offering opportunity for retrofitting or upgrading existing sites, especially those with access to existing

Editor(s): Carlos Augusto de Lemos Chernicharo and Thiago Bressani-Ribeiro
doi: 10.2166/9781780409238_0339

sludge processing centres (McAdam *et al.*, 2011). However, if such infrastructure does not pre-exist, its inclusion increases process complexity, e.g. requirement of a separate primary sludge digester. Therefore, alternative anaerobic technologies are sought for and different studies aim at increasing process efficiencies, particularly when operating at low ambient temperature in order to enhance effluent quality, process economics and environmental impact.

Different types of enhanced high-rate anaerobic reactor configurations have been developed, especially in the last two decades (Ersahin *et al.*, 2011; van Haandel *et al.*, 2006; van Lier, 2008; van Lier *et al.*, 2015). However, despite the improvements achieved in anaerobic treatment technology and reactor configurations, a polishing post-treatment step is generally required to meet stringent discharge standards.

11.2 UPGRADING ANAEROBIC SEWAGE TREATMENT

Different anaerobic membrane-based alternatives are being regarded as a core technology for moving towards new, highly efficient, and cost-effective, green alternatives for sewage treatment. These alternatives are increasingly being researched, addressing specific constraints associated to the commonly used high-rate anaerobic process configurations.

11.2.1 Removal of organics and effluent quality

A constraint of major interest in anaerobic digestion (AD) processes is the relatively low removal efficiencies, which are obtained for both COD and suspended solids. Generally, sewage is characterized by a low organic strength, where a high fraction of organics are present as suspended solids and require a hydrolysis step before conversion and full mineralization (see also Chapter 3). In fact, the hydrolysis step can be considered the rate limiting step for anaerobic conversion when treating sewage containing high amounts of particulate organics (Ince *et al.* 2018), particularly at moderate to low temperatures. Next to hydrolysis, the conversion of dissolved organic matter into biogas is rate limiting the process.

Other key concern of applying a single stage anaerobic treatment process for sewage treatment is related to the effluent quality (Chernicharo *et al.*, 2015; van Lier *et al.*, 2008). Conversion efficiencies and effluent quality levels are lower than those generally obtained with aerobic treatment using conventional activated sludge (CAS) systems. Particularly, high variations in organic and hydraulic loading, e.g. in touristic regions or in regions generating high-strength sewage, constrains the anaerobic process performance, particularly at moderate to low temperatures (Gomec, 2010; Mahmoud, 2008). Therefore, the low effluent quality in AD processes requires a post-treatment step since discharge or reuse criteria are generally not met with the single stage anaerobic, as also discussed in Chapter 10.

On average, a UASB reactor roughly removes about 2/3 of the influent load, whereas 1/3 remains in the effluent; actual values may differ a little (Chernicharo *et al.*, 2015; van Lier *et al.*, 2010). Enhanced COD removal and complete removal of suspended solids can be achieved combing anaerobic treatment with membrane separation. Membranes can be applied for just post-treating anaerobic effluents, or they can be integrated with the anaerobic reactor, creating a so-called anaerobic membrane bioreactor (AnMBR). AnMBRs have been extensively researched in the last decade to produce solids-free effluents with a high degree of pathogen removal, while occupying a small footprint (Ozgun *et al.*, 2013a). Especially for the situations where the commonly applied high-rate sludge bed systems are less successful, AnMBRs present extensive opportunity for effective wastewater treatment (van Lier *et al.*, 2015). AnMBRs provide a complete decoupling of the hydraulic retention time (HRT) and the SRT. The latter is of high importance when slow growing organisms need to be retained in the system. The SRT is then set by the allowable sludge concentration in the reactor and the net sludge yield under the given conditions.

11.2.2 Low-temperature treatment

As commented in previous chapters, a low influent temperature is considered a barrier for anaerobic sewage treatment since the energy requirements for heating the wastewater largely outweighs the potential for energy recovery (Lettinga *et al.*, 2001). Considering the energy requirement of 4.2 MJ m^{-3} $°C^{-1}$ for heating water and the biochemical energy content of about 12.4 MJ kg COD^{-1} (based on the lower heating value of CH_4), one can easily calculate the minimally required COD conversion per m^3 of wastewater, if an increase in the operational temperature becomes indispensable. Based on an average sewage concentration of 0.5 kg COD m^{-3} and an average energy recovery as CH_4 of 4.1–7.0 MJ kg $COD_{removed}{}^{-1}$ (Chernicharo *et al.*, 2015), the maximally attainable temperature increase of the influent using the recovered CH_4 is only 0.6°C per m^3 of wastewater, considering a 100% energy conversion from methane gas to transferred heat. Note that a modern combined heat power (CHP) gas engine is characterized by an energy conversion efficiency of about 40%. Therefore, for sustainably implementing anaerobic treatment of low-strength wastewaters, such as sewage, the applicable process temperature should be close or equal to the ambient sewage temperature (Martin-Garcia *et al.*, 2011a).

On the other hand, the operating temperature is an important issue in the anaerobic treatment of sewage since microbes' growth rates decrease following Arrhenius equation when decreasing the reactor temperatures, worsening the anaerobic conversion capacity of the sludge. In addition, at low temperatures, the solubility of fats and grease drops, resulting in reduced accessibility of these compounds for microbial conversion. This leads to an accumulation of non-stabilized sludge in the reactor, diluting the active methanogenic biomass to low levels. This can only be overcome by drastically increasing the sludge age or solids retention times (SRTs) to safeguard a sufficiently high methanogenic conversion capacity.

For low-temperature operation, Mahmoud *et al.* (2004) proposed to integrate UASB technology with a common sludge digester to maintain the specific methanogenic activity of the sludge bed and to stabilize the incoming solids. In the UASB-Digester concept, the retained solids are conveyed to the digester for enhanced stabilization at 30°C and the digested sludge was recirculated to the UASB reactor for improving the methanogenic capacity (Mahmoud *et al.*, 2004). Later, Ozgun (2015) developed a UASB-Digester system combined with a capillary UF tubular membrane module for producing nutrient-rich and 'pathogen-free' irrigation water, applicable under moderate climate conditions. In the proposed system, only the effluent of the sludge bed reactor is subjected to membrane filtration, reducing the membrane fouling rate.

Additionally, AnMBR systems have been shown to perform successfully at low temperatures (see e.g. Martinez-Sosa *et al.*, 2011). Indeed, different authors (Seghezzo *et al.*, 2003, 2006; van der Last and Lettinga, 1992; van Haandel *et al.*, 2006) already concluded that anaerobic treatment of sewage is possible at low temperature when the system is operated at long SRTs. The latter easily can be attained applying primary settling prior to the anaerobic reactor as suggested above. The application of AnMBR technology for sewage treatment at psychrophilic environments (<20°C) is essential for wide-scale application. More recently, Jefferson (2012) demonstrated the successful operation of an AnMBR at 7°C with pre-settled sewage. In the following sections, the use of membranes for upgrading anaerobic sewage treatment is further elaborated.

11.3 ANAEROBIC MEMBRANE BIOREACTORS (AnMBRs) FOR SEWAGE TREATMENT

AnMBR technology for sewage treatment implies a shift of paradigm within sewage management. Conventional aerobic processes are effective in removing organic matter and nutrients, but they are also constrained by a high consumption of (fossil-based) energy and chemicals and a high sludge production.

Over the past 25 years, the application of membranes in aerobic biological wastewater treatment processes has been well researched and established. Aerobic membrane bioreactors (MBRs) have been installed in more than 200 countries for the treatment of wastewaters (Judd & Judd, 2011) and the aerobic MBR market is still growing. For instance, the estimated market value for MBRs in 2019 is $3 billion compared to around $1.2 billion in 2010 (Judd, 2016). Based on the success of aerobic MBR applications worldwide, one of the most recent developments in high-rate anaerobic treatment is incorporating membranes to separate anaerobic biomass from the effluent. AnMBRs physically ensure complete biomass retention by the application of a membrane filtration process enabling independent control of HRT and SRT (Jeison *et al.*, 2008; Zhang *et al.*, 2010).

In the past decades, AnMBR technology has received much attention and research results indicate its attractiveness and potential feasibility for full-scale sewage treatment (Dereli *et al.*, 2012; Ersahin *et al.*, 2014; Fawehinmi *et al.*, 2007; Liao *et al.*, 2006; Ozgun *et al.*, 2013a). As commented above, AnMBRs produce a high-quality permeate that is free of solids and reduces the pathogen content to a large extent due to the cutoff of the membrane's pore sizes and the subsequent complete retention of the biomass. Apart from the retention of solids and pathogens, AnMBRs can obtain stable and high treatment performances in terms of COD, meeting stringent discharge standards (Ozgun *et al.*, 2013a). The applied long SRTs result in very low amounts of excess sludge production. For instance, Pretel *et al.* (2014) reported biosolids productions of only 0.16 kg TSS kg^{-1} $COD_{removed}$ when operating at 70 days of SRT under mesophilic conditions. This extends to systems treating settled sewage in UASB configured AnMBRs where sludge withdrawal is limited to every 70–100 days to remove undigested feed solids that accumulate in a sludge blanket above the granules (Martin-Garcia *et al.*, 2013). Reported yields in such systems range between 0.003 and 0.051 kg VSS kg^{-1} $COD_{removed}$ (Fawehinmi *et al.* 2007; Shin and Bae, 2018). In fact, AnMBR application ensures high-rate anaerobic treatment with very high treatment efficiencies in small-footprint reactor systems. Overall advantages provided by AnMBRs for sewage treatment compared to conventional aerobic processes can be found elsewhere (see e.g. Lin *et al.*, 2013; Ozgun *et al.*, 2013a; Skouteris *et al.*, 2012).

Despite the advantages offered by the AnMBR process for sewage treatment, there are different issues that should be further evaluated for full-scale application, such as:

- *Dissolved methane*: One key issue in anaerobic digestion is the methane dissolved in the effluent, which increases as the temperature of the mixed liquor decreases. The release of this methane to the atmosphere not only represents an energy potential loss, but also a high environmental impact since methane is a powerful greenhouse gas (GHG) with a global warming potential 28-fold greater than CO_2 for a time horizon of 100 years (IPCC, 2014). Methane recovery via biogas production can decrease with temperature to values dropping from 0.27 to 0.06 m^3 CH_4 kg $COD_{removed}{}^{-1}$ when treating low-loaded wastewaters (Ozgun *et al.*, 2013a), mainly due to the limited conversion of suspended solids and the loss of methane dissolved in the effluent. Indeed, Giménez *et al.* (2014) reported that up to 85% of the methane produced in an AnMBR system can be extracted with the effluent when operating at temperatures of about 15°C. The methane dissolved in the effluent could be stripped from the effluent. However, when this is applied in downstream open-air systems the recovered CH_4 will be emitted to the atmosphere, decreasing the economic and environmental feasibility of this technology. Thus, an appropriate methodology/process should be applied to avoid the release of dissolved methane to the atmosphere (Bandara *et al.*, 2012; Giménez *et al.*, 2012); see Chapter 9.
- *Organic matter to sulfate ratio*: One key issue in the anaerobic treatment of sewage is the competition between sulfate-reducing organisms (SRO) and methanogens since sewage can be characterized by a

low organic matter to sulfate ratio (COD:SO_4). Since SRO have a competitive advantage over methanogens in using COD for sulfate reduction, the SO_4^{2-} present in the influent will always be reduced to sulfide at the expense of methane production. Stoichiometrically, 1 g of SO_4^{2-} requires 0.67 g of COD for its reduction to HS^-. In other words, the presence of 100 mg SO_4^{2-} L^{-1} in the influent will lower the methane production potential by 23.4 L m^{-3} influent. Therefore, considering the low COD content of sewage, the presence of SO_4^{2-} will significantly impact on the methane production potential. In addition, the produced sulfide has a pKa value of 7.2, meaning that part of the sulfide will end up as free H_2S in the biogas, whereas the rest will be solubilized in the effluent, lowering the COD removal efficiency. The produced sulfide negatively impacts the treatment process as it may cause: (i) potential toxicity to anaerobic microorganisms, which is however not very likely considering the prevailing average concentrations in sewage (see also Chapter 3); (ii) reduction of the quality of the produced biogas; (iii) corrosion in pipes, engines and boilers, entailing higher maintenance and replacement costs; and (iv) downstream oxygen demand for oxidizing HS^- (Ince *et al.*, 2018).

- *Membrane fouling and cleaning*: Membrane fouling and cleaning represent a critical issue in AnMBR technology (Lin *et al.* 2013). Membrane fouling is the consequence of filtration of solids-rich effluent over the membrane. The level of fouling depends on the interaction between the membrane surface, the pores, and the bulk suspension. The fouling phenomenon leads to increased transmembrane pressures and/or reduced membrane fluxes and requires periodic cleaning, and thus system downtime for allowing physical and/or chemical cleaning. In case frequent chemical cleaning is required the membrane's lifespan will be reduced, increasing the operational costs. As such, the fouling phenomenon decreases the productivity of the system and needs to be critically assessed.
- *Fate of nutrients*: Anaerobic processes for sewage treatment provide no removal of macronutrients such as nitrogen and phosphorus. In contrast, the anaerobic mineralization process will result in even higher concentrations of nitrogen and phosphorus in the effluent compared to the influent. This effect is commonly potentiated by the use of membranes since enhanced hydrolysis is commonly achieved combing anaerobic treatment with membrane separation in AnMBRs. Depending on the receiving water body, a post-treatment step is required for nutrient recovery or removal, as previously commented (see also Chapter 10).

11.3.1 Membrane post-treatment versus AnMBR

In an AnMBR system, the membrane can be located either inside or outside the bioreactor, which is called submerged or side-stream configuration, respectively (Figure 11.1) (Ersahin, 2015). A side-stream AnMBR, also referred to as cross-flow configuration, consists of a membrane unit externally mounted in a separate membrane skid. Such cross-flow configuration is the general set-up for tubular membranes that make use of high cross-flow velocities to scour the membrane to manage the fouling. In the cross-flow configuration, a feed pump pushes the effluent of the bioreactor into the external membrane skid (Figure 11.1(a)). External membrane modules have some advantages such as the ease of membrane replacement and cleaning. In the submerged AnMBR configuration, the membrane is located inside the bioreactor (Figure 11.1(b)) or in an external membrane tank (Figure 11.1(c)). A submerged configuration is frequently chosen for flat-sheet or hollow fiber membranes, whereby an external membrane tank facilitates membrane maintenance, without the need for altering the bioreactor. The flow to the separate membrane tank is accomplished with solely a permeate pump (as indicated in Figure 11.1(c)), or in combination with a feed pump, creating a loop to the membrane tank and back to the bioreactor.

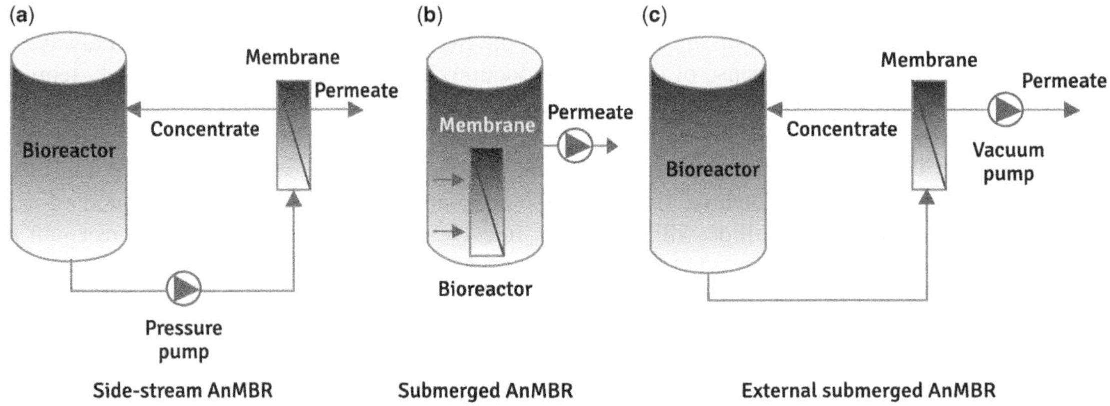

Figure 11.1 Different AnMBR configurations (Source: Ersahin, 2015).

In addition to a hydraulic shear, also biogas recirculation can be used to remove the cake layer from the membrane surface. In both the internal and external submerged AnMBRs, permeate is obtained by vacuum or is driven by gravity, instead of applying pump pressure at the feed side of the membrane. Energy costs of the submerged configuration may be close to one third of the side-stream configuration when operating the reactor below or at the critical flux (Jeison & van Lier, 2008b). However, at elevated fluxes, much less membrane area is required for a side-stream set-up at the expense of increased energy consumption for reaching high cross-flow velocities. In addition, the applied high cross-flow velocity might have a negative impact on biomass activities (Brockmann & Seyfried, 1996, 1997; Lin *et al.*, 2013), although high specific methanogenic sludge activities were found during the treatment of complex wastewaters using the cross-flow filtration set-up (Jeison *et al.*, 2009).

Thus far, full-scale applications are limited to the treatment of highly concentrated wastewaters, applying side-stream reactor configurations. The submerged configurations are commonly applied for research on the treatment of low-concentrated wastewaters like sewage (Shin & Bae, 2018). Efficient fouling control in submerged AnMBRs was achieved applying biogas-assisted membrane scouring combined with vacuum extraction, with or without periodic backwash (Robles *et al.*, 2012a, b, 2013a). In UASB configured AnMBRs, energy demands for fouling control are very low, as the solids flux rate is minimized. Decreased membrane fouling rates permit periodic sparging of the membrane, which has been shown to be effective; gas sparging was required for only 10% of the operating cycle (Martin-Garcia *et al.*, 2013). Optimal performance is observed when the periodic sparging is performed during periods of non-flow (relaxation), i.e. when there is no drag force towards the membrane and thus less binding of the accumulated solids to the membrane (Wang *et al.*, 2018a). This also permits the operating flux to be elevated for short periods of time so that the flux can mirror flow variations (Wang *et al.* 2018b). The importance of this is that the membrane-based treatment plant can be sized at the average rather than peak flow, significantly reducing capital costs, operating costs and energy demand due to the reduced total membrane area.

Different types of bioreactors are used in combination with membrane filtration, including completely stirred tank reactor (CSTR), UASB, expanded granular sludge bed (EGSB), fluidized bed (FB), and hybrid reactors, in different configurations (Ozgun *et al.*, 2013a). It must be realized that membrane integration eliminates the hydraulic selection pressure required for granulation, whereas flocculent biomass with poor immobilization characteristics is retained instead of being washed-out. Moreover, by applying cross-flow membranes, the prevailing shear forces will minimize the diameter of particles. Therefore, no granulation is expected in AnMBRs, which would decrease the settleability of the biomass on the long-term operation (van Lier *et al.*, 2015). However, sustained operation of a granular UASB configuration, treating settled sewage, has shown that the granules remain stable with more than three years of operation (Jefferson 2012). CSTR is the most commonly applied bio-reactor in an AnMBR system (Giménez *et al.*, 2011; Ozgun *et al.*, 2013a; Smith *et al.*, 2012). Nevertheless, coupling of filtering blankets with membranes offers new perspectives for sewage treatment. For instance, in a UASB reactor coupled membrane system, the UASB reactor provides a pre-elimination of suspended solids (SS) by entrapment and biodegradation in the sludge bed and thus, reduces SS load to the membrane. In this way, fouling related to cake layer formation could be minimized and increased fluxes, reaching 10.5–65 $L\,m^{-2}\,h^{-1}$ could be achieved in comparison to CSTR coupled membrane systems (Ozgun *et al.* 2015a; Zhang *et al.*, 2010). Long term operation of such systems revealed sustainable flux operation up to about 15–17 $L\,m^{-2}\,h^{-1}$ (Jefferson, 2012; Wang *et al.*, 2018a). Selection of an appropriate upflow velocity plays a major role in controlling the fouling potential, which is one of the most important challenges for stable operational process performance of membrane coupled UASB reactors (Ozgun *et al.* 2013b).

In contrast to an integrated AnMBR set-up, some researchers used membrane filtration as a polishing step after a UASB reactor without returning the concentrate stream to the bioreactor (Herrera-Robledo *et al.*, 2010; Salazar-Pelaez *et al.*, 2011). Application of membranes for a polishing step can be regarded as tertiary filtration. This approach has the advantage of easier control of hydraulics in the bioreactor compared to a bioreactor that is coupled with a membrane unit and the preservation of dilution rate as bacterial selection criterion. However, the application of a separate membrane feed tank may accumulate SS from the anaerobic bioreactor, resulting in an increased SS load to the membranes and thus more rapid flux decline.

11.3.2 Process stability and treatment efficiencies

AnMBR technology has an extensive applicability potential for the treatment of both high-strength (Dereli *et al.*, 2012; He *et al.*, 2005; Padmasiri *et al.*, 2007) and low-loaded wastewaters (Baek & Pagilla, 2006; Hu & Stuckey, 2006; Kim *et al.*, 2011; Wang *et al.*, 2018a). Tables 11.1 and 11.2 present the major findings obtained from different studies on AnMBRs treating sewage.

Considering Table 11.1, AnMBR processes could typically eliminate >75% COD at defined operational conditions regardless of the configurations. In most of the studies, almost complete suspended solids retention was achieved. This high retention of solids was related to the pore size of the membrane, being this pore size generally between 0.01–0.045 μm (Table 11.2) and the formation of a secondary membrane (cake layer) on top of the real membrane, narrowing the pore size for filtration.

The reported COD and TSS removal efficiencies achieved when using AnMBRs were much higher than those obtained by conventional UASB reactors (see e.g. Chernicharo *et al.*, 2015; Gomec, 2010). When the AnMBR process is not combined with any pre- or post-treatment step for sewage treatment, total nitrogen (TN) and total phosphorus (TP) removal efficiencies are negligible or even negative. However, this situation can be used as an advantage if the treated wastewater would be used for fertirrigation afterwards. For removal of TN and TP, different treatment technologies are being researched, such as forward osmosis (FO) membrane process (Gu *et al.*, 2015; Lin *et al.*, 2013). Besides the removal of conventional pollutants, research on the applicability of AnMBRs for removal of micropollutants is also becoming popular. An AnMBR process achieved an 81% removal efficiency for cephalosporin derivatives through

Table 11.1 Treatment performance of AnMBRs for sewage treatment.

Reactor Type/Module Configuration	Volume (L)	Temperature (°C)	Wastewater Source	Influent COD (mg L^{-1})	HRT (h)	OLR (kg COD m^{-3} d^{-1})	SRT (day)	COD Removal (%)	MLSS (g L^{-1})	Biogas Production (L CH_4 · g $COD_{removed}^{-1}$)	References
CSTR/Submerged	6	10–25	Synthetic	491 ± 112	12–16	0.74–1	–	71–96	7.1–10.5[a]	–	Watanabe *et al.* (2017)
Granular Sludge Bioreactor/Submerged	3	21	Synthetic	330–370	12	0.53–0.59	–	>90	21.30–23.82	0.156	Chen *et al.* (2017)
CSTR (Osmotic AnMBR)/Submerged	4.98	25 ± 0.5	Synthetic	373 ± 7	35–60	–	80	>90[b]	3.2	0.25–0.28	Wang *et al.* (2017)
FB Reactor/Side-stream	2.3	10; 25	Synthetic-Raw	480 ± 50	4.2; 5.6; 9.8	–	–	>90	–	–	Seib *et al.* (2016)
CSTR/Side-stream	550	23 ± 1	Raw	–	8.5	–	40; 70; 100	–	11.3–21.3	–	Dong *et al.* (2016)
UASB Reactor/Side-stream	7	25	Synthetic	530 ± 30	6	2	–	92	–	–	Ozgun *et al.* (2015b)
UASB Reactor/Submerged	284	18	Raw	978 ± 210	12.8–14.2	1.6–2	–	90	–	–	Gouveia *et al.* (2015a, b)
CSTR/Side-stream	2	35 ± 1	Synthetic	400 ± 10	0.96–12	0.8–10	1000	98	4.9–13.3[a]	0.129–0.396	Wei *et al.* (2014)
FB Reactor/Submerged	7.6	35 ± 2	Synthetic	320 ± 44	8	–	–	75.8	–	0.14	Gao *et al.* (2014)
FB Reactor/Submerged	770	Ambient	Pre-settled	207	4.5–6.8	–	6.2–36	81	587–1230	–	Shin *et al.* (2014)
CSTR/Submerged	5	15 ± 0.1	Synthetic-Raw	440	16–24	0.44–0.66	300	92 ± 5	–	–	Smith *et al.* (2013)
CSTR/Submerged	5	25–30	Raw	427 ± 60	10	1.02 ± 0.14	30; 60; 90	84–86	8.03–13.579	0.19–0.50[c]	Huang *et al.* (2013)
CSTR/Submerged	1300	33; 21 ± 1	Raw	410; 720	6–21	0.71; 0.64	76 ± 10; 74 ± 16	–	–	–	Giménez *et al.* (2011)
UASB Reactor/Submerged	4.33	21–24	Synthetic	452 ± 35	8	1.2	–	96.3–98.7	–	–	Ceron-Vivas *et al.* (2012)

(*Continued*)

Table 11.1 Treatment performance of AnMBRs for sewage treatment (*Continued*).

Reactor Type/Module Configuration	Volume (L)	Temperature (°C)	Wastewater Source	Influent COD (mg L^{-1})	HRT (h)	OLR (kg COD m^{-3} d^{-1})	SRT (day)	COD Removal (%)	MLSS (g L^{-1})	Biogas Production (L CH_4 · g $COD_{removed}{}^{-1}$)	References
UASB Reactor/Submerged	10	27–30	Synthetic	500 ± 10	2; 12	6; 1	–	85; 96[b]	1.6; 10.1	–	Liu *et al.* (2012)
CSTR/Submerged	6	–	Synthetic	425	12	–	∞[d]	83	4.6	–	Achilli *et al.* (2011)
CSTR/Submerged	60	30 ± 3	Raw	425 ± 47	10	1	–	88	6.4–9.3	0.26	Lin *et al.* (2011)
CSTR/Side-stream	4	15; 25	Synthetic	500	12	1	–	>85; >95	–	–	Ho and Sung (2010)
UASB Reactor/Side-stream	10	30	Synthetic	500	24	5.1	50	96	–	–	Gao *et al.* (2010)
Upflow Anaerobic Reactor/Side-stream	180	25	Pre-settled	540	12; 6; 4.5	1.08; 2.16; 4.32	∞	88	14–80	–	Lew *et al.* (2009)
CSTR/Submerged	3	35	Synthetic	465	8–20	–	250	99[e]	2–3[f]	–	Vyrides and Stuckey (2009)
UASB Reactor/Side-stream	34	Ambient	Raw	186	5.5–10	0.3–0.9	∞	77–81	12–32	0.062–0.121	An *et al.* (2009)
CSTR/Submerged	3	35	Synthetic	460	3; 6; 12; 24	–	100	90–95	4.3–4.8	–	Hu and Stuckey (2007)
UASB Reactor/submerged	42	16 ± 4	Pre settled	211 ± 78	8	0.6	–	83 ± 7	–	–	Wang *et al.*, 2018a
CSTR/Side stream	850	22	Raw	637	14.4	0.94	∞	92[b]	–	0.12	Grundestam and Hellstrom (2007)

[a]g MLVSS L^{-1};
[b]Total organic carbon (TOC) removal;
[c]L CH_4 d^{-1};
[d]No discharge except sampling;
[e]DOC removal;
[f]g VSS L^{-1}.

Table 11.2 Filtration performance of AnMBRs for sewage treatment

Reactor Type/Module Configuration/Membrane Configuration	Membrane Type	Material	Pore Size (μm)	Filtration Area (m^2)	Flux ($L\ m^{-2}\ h^{-1}$)	TMP (kPa)	Cross-Flow Velocity ($m\ s^{-1}$)	Gas Sparging ($L\ min^{-1}$)	References
CSTR/Submerged/Flat Sheet	MF[a]	Chlorinated Polyethylene	0.2	0.116	–	–	–	5	Watanabe *et al.* (2017)
Granular Sludge Bioreactor/Submerged/Hollow Fiber	MF	PVDF[b]	0.22	0.06	5.3	up to 30	–	–	Chen *et al.* (2017)
CSTR (Osmotic AnMBR)/Submerged/Flat Sheet	FO[c]/MF	Cellulose Triacetate/PVDF	0.2	0.025	<6	–	–	2	Wang *et al.* (2017)
FB Reactor/Side-stream/Tubular	–	PVDF, Seramic	0.018; 0.05	0.059; 0.05	5.9; 6.3; 7.4	9 ± 4; 22 ± 8	0.018–0.024	–	Seib *et al.* (2016)
CSTR/Side-stream/Hollow Fiber	–	PVDF	0.04	5.4	17	1.5–30	–	13.1	Dong *et al.* (2016)
UASB Reactor/Side-stream/-	UF[d]	PES[e]	0.03	0.11	12.3	8.5	1	–	Ozgun *et al.* (2015b)
UASB Reactor/Submerged/Hollow Fiber	UF	–	0.045	0.93	10–14	40–55	–	9–16[f]	Gouveia *et al.* (2015a, b)
CSTR/Side-stream/Hollow Fiber	UF	PVDF	0.03	0.031	6–12	<10	0.1–0.3	–	Wei *et al.* (2014)
FB Reactor/Submerged/Hollow Fiber	–	–	0.4	0.19	11.25	22.5	–	–	Gao *et al.* (2014)
FB Reactor/Submerged/Hollow Fiber	UF	PVDF	0.03	39.5	6.1–7.5	10–50	–	–	Shin *et al.* (2014)
CSTR/Submerged/Flat Sheet	MF	PES	0.2	0.0387	>7	10–80	–	4.67	Smith *et al.* (2013)
CSTR/Submerged/Flat Sheet	–	PES	0.45	0.118	–	0–30	–	–	Huang *et al.* (2013)
CSTR/Submerged/Hollow Fiber	UF	–	0.05	–	–	–	–	–	Giménez *et al.* (2011)
UASB Reactor/Submerged/Tubular	UF	PVDF	100[g]	0.0085	4.2 ± 0.2; 5.2 ± 0.4	<40	–	–	Ceron-Vivas *et al.* (2012)

(*Continued*)

Table 11.2 Filtration performance of AnMBRs for sewage treatment (*Continued*).

Reactor Type/Module Configuration/Membrane Configuration	Membrane Type	Material	Pore Size (μm)	Filtration Area (m^2)	Flux ($L\ m^{-2}\ h^{-1}$)	TMP (kPa)	Cross-Flow Velocity ($m\ s^{-1}$)	Gas Sparging ($L\ min^{-1}$)	References
UASB Reactor/Submerged/Hollow Fiber	MF	PVDF	0.1	1	5	<35	–	–	Liu *et al.* (2012)
CSTR/Submerged/Flat Sheet	MF	Polyolefine	0.4	0.1	10.5	–	–	8	Achilli *et al.* (2011)
CSTR/Submerged/Flat Sheet	MF	PVDF	140[g]	0.6	11	–	–	–	Lin *et al.* (2011)
CSTR/Side-stream/Tubular	MF	PTFE[h]	1	0.09	5	6.9–55.2	–	–	Ho and Sung (2010)
UASB Reactor/Side-stream/Flat Sheet	UF	PVDF coated with a polyether block amide	100 g; 30[g]	0.052	8–12	–	–	–	Gao *et al.* (2010)
Upflow Anaerobic Reactor/Side-stream/Hollow Fiber	MF	–	0.2	4	3.75; 7.50; 11.25	<19.6[i]	–	–	Lew *et al.* (2009)
CSTR/Submerged/Flat Sheet	–	Polyethylene	0.4	0.1	5–10	30–45	–	5	Vyrides and Stuckey (2009)
UASB/Side-stream/Tubular	–	Polyacrylonitrile	–	0.2	10.5	<50	0.4–1.45	–	An *et al.* (2009)
CSTR/Submerged/-	MF	–	0.4	–	10–20	up to 30	–	–	Hu and Stuckey (2007)
UASB Reactor/submerged	UF	PVDF	0.04	0.9	15	–	–	3	Wang *et al.*, 2018a
CSTR/Side-stream/Vibratory Shear Enhanced Processing (VSEP)	–	PTFE Teflon	0.45	1.59	–	–	–	–	Grundestam and Hellstrom (2007)

[a]MF: Microfiltration;
[b]PVDF: Polyvinylidene fluoride;
[c]FO: Forward Osmosis;
[d]UF: Ultrafiltration;
[e]Polyether sulfone;
[f]$m\ h^{-1}$;
[g]kDa, molecular weight cut off.;
[h]PTFE: Polytetrafluoroethylene;
[i]Driven by water level difference.

bioaugmentation (Saravanane & Sundararaman, 2009). Removal efficiencies over 90% could be obtained for some pharmaceutical compounds such as trimethoprim in an AnMBR process. Also addition of powdered activated carbon (PAC) into AnMBR improved the removal of different pharmaceutical compounds (Xiao *et al.*, 2017).

Since the SRT and HRT can be controlled independently in AnMBRs, high SRTs could be applicable, reaching values >100 days; thus, adsorbed particulate organics/refractory compounds would eventually become degraded. HRT in AnMBRs was generally applied longer than 4 h, comparable with aerobic MBRs. Operation of AnMBRs at long HRTs and SRTs enhances methane recovery, pollutant removal performance and reduces sludge production. Applied organic loading rates (OLR) vary between 0.3–10 kg COD m^{-3} d^{-1} (Table 11.1).

Regarding membrane filtration, an operational flux reaching 20 L m^{-2} h^{-1} seems to be feasible. However, fluxes below 15 L m^{-2} h^{-1} were reported in most of the studies which is around the critical flux of anaerobic biomass solids (Martin-Garcia *et al.*, 2011a) (Table 11.2). Apparently, lower membrane fluxes are experienced applying AnMBRs compared to aerobic MBRs, which could represent one of the bottlenecks for large-scale engineering applications and economic feasibility of AnMBRs, especially when applied for sewage treatment. Accordingly, the prevailing low operating fluxes observed in AnMBRs restricts its application for the treatment of sewage collected in combined sewers, which are characterized by large flow fluctuations (e.g. factor 5–6 increase in flow during rain events). Thus, either the system is limited to dry weather or average flow conditions, or the AnMBR should be able to periodically accommodate operating fluxes that may reach values 5–6 times the dry weather flows. Possibly, this might be feasible in systems that experience low solids flux rates, e.g. when only the UASB supernatant with a low solids content is subjected to membrane filtration. However, such huge variation in fluxes is not possible when the membrane unit interacts with the full biomass load (Martin-Garcia *et al.*, 2013; Wang *et al.*, 2018b). In former case, flux rates of up to 25 L m^{-2} h^{-1} have been sustained for 100 hours of operation (Jefferson, 2012).

A modified version of the AnMBR was denominated 'staged anaerobic fluidized membrane bioreactor' (SAF-MBR). Instead of gas sparging as applied in AnMBRs, fluidized granular activated carbon (GAC) can be used to clean the membrane surface through physical contact and movement in the fluidized reactors (Kim *et al.*, 2011). SAF-MBR was tested at pilot scale for the treatment of sewage with promising results at very short HRTs (e.g., 4.5 h) and ambient temperatures ranging between 8–30°C (Shin *et al.*, 2014).

11.3.3 Pilot-scale and prototype examples

The performance of different AnMBR systems at pilot scale for sewage treatment has been evaluated by different authors during the last years (Giménez *et al.*, 2011; Gouveia *et al.*, 2015a, b; Martin-Garcia *et al.*, 2013; Martinez-Sosa *et al.*, 2011; Wang *et al.*, 2018a, b).

As commented before, Kim *et al.* (2011) and Shin *et al.* (2014) operated a SAF-MBR system, where membrane scouring was conducted by fluidized GAC instead of gas sparging. COD removal efficiencies of 81% were achieved even operating at considerable low temperatures in wintertime (8°C), whereas COD removal increased up to 98% during summertime. Due to the fluidization of GAC, membrane fouling was low, which allowed reaching low energy requirements (around 0.23 kWh per m^3 of treated water).

Gouveia *et al.* (2015a) evaluated the long-term performance of an AnMBR consisting of a UASB coupled to an external membrane. The system was operated at 18°C, reaching COD removal efficiencies of 90%. This study proved the reliability of AnMBR technology for sewage treatment under long-term low temperature conditions. Martinez-Sosa *et al.* (2011) evaluated the performance of a CSTR-based AnMBR at 35°C, 28°C and 20°C, obtaining COD removals around 90% and relatively high methane yields of 0.23–0.27 L_{CH4} g $COD_{removed}^{-1}$.

Another example of a CSTR-based AnMBR is the one presented by Giménez *et al.* (2011). This system was fitted with industrial-scale hollow-fiber membranes and was fed with sulfate-rich effluent from the pre-treatment (screening, de-gritting, and grease removal) of a full-scale WWTP, in Valencia, Spain. COD removal efficiencies (based on organic compounds) of about 90% were achieved, although the methane yield was low due to the competition of sulfate reducers and methanogens for the available substrate. The proposed full-scale system should provide (flash) aeration to oxidize the soluble HS^-/S^{2-}. Giménez *et al.* (2014) evaluated the performance at ambient temperature of the same AnMBR system within summer and winter seasons. Sludge production increased and energy recovery potential decreased as the temperature dropped from 20 to 14°C when operating at SRT of 29.6 days in average. Nonetheless, COD removal efficiency and membrane permeability remained nearly stable. In this system, Robles *et al.* (2013b) also evaluated the effect of temperature (20°C, 25°C and 33°C) on membrane filtration, showing that the solids concentration in the mixed liquor is a determining factor affecting membrane permeability. The 20°C standardized membrane permeability was higher at 20°C and 25°C compared to 33°C. It was hypothesized that this was due to a decrease in biomass activity during low temperature operation, which leads to lower amounts of extracellular polymeric substances (EPSs) and soluble microbial products (SMPs) in the mixed liquor. Nevertheless, severe irreversible fouling problems were not detected at any operating temperature.

The LIFE MEMORY project (http://www.life-memory.eu/en/) is taking a further step in the development of AnMBR technology, scaling-up the installation size in order to reach industrial scale. The demonstration project operates an AnMBR plant that consists of an anaerobic reactor of 40 m^3 (34.4 m^3 working volume) connected to three membrane tanks of 0.8 m^3 each (0.7 m^3 working volume). The AnMBR system uses industrial scale hollow-fiber membranes, giving a total filtration area of 123 m^2 and is operated with pre-treated sewage coming from a full-scale WWTP, which is characterized by a very strong load variability. The results obtained in this system showed that it is possible to operate the membranes at 20°C standardized transmembrane fluxes above $20\,L\,m^{-2}\,h^{-1}$, maintaining low membrane fouling propensities by applying specific gas sparging demands per volume of permeate (SGD_P) from around 10 to 14. The applied strategy resulted in low overall energy demands for operating the AnMBR (Robles *et al.*, 2017). Hence, low energy requirements related to filtration were achieved. Regarding the biological process performance, COD removal efficiencies of 92% were achieved while meeting discharge limits and the sludge production was 0.136 kg VSS kg^{-1} $COD_{removed}$ (Durán *et al.*, 2018).

Other examples of projects aimed to demonstrate the viability of a new technological concept for sewage treatment and nutrient recovery based on AnMBR can be found elsewhere, such as Run4Life (http://run4life-project.eu/), watintech (http://watintech.eu/), LIFE Celsius (http://www.lifecelsius.com/en/), LIFE siamec (http://www.life-siamec.eu/siamec-home/), LIFE+WOGAnMBR (http://www.life-woganmbr.eu/), or LIFE Methamorphosis (http://www.life-methamorphosis.eu/en/home).

11.3.4 Fouling mitigation and control

As commented before, the most common and well known bottleneck of any (An)MBR-based technology is membrane fouling, which can result in reduced energy efficiencies for avoiding flux decrease through membrane cleaning (Robles *et al.*, 2018). AnMBRs frequently operate at lower membrane fluxes than aerobic MBRs, which has been attributed to reduced sludge flocculation and increased concentrations of fine particulates and colloids under anaerobic conditions (Liao *et al.*, 2006; Martin-Garcia *et al.*, 2011b). Nevertheless, some studies comparing both aerobic MBRs and AnMBRs for sewage treatment have shown similar fouling propensities (Achilli *et al.*, 2011) or even less fouling propensity in the latter than in the former (Baek & Pagilla, 2006).

Membrane fouling is caused by the accumulation of different organic and inorganic particles present in the wastewater in and/or on the membrane pores and surfaces. Organic compounds rather than inorganic ones have been reported as the main reason of membrane fouling, whereby cake layer formation generally determines the overall membrane resistance (An *et al.*, 2009; Jeison & van Lier, 2007; LeClech *et al.*, 2006; Meng *et al.*, 2009). Influent characterization, reactor operation, membrane specifications, pH, soluble and bound EPS, SMP, and biomass properties and concentration affect the fouling phenomena (Dereli *et al.*, 2012; Gao *et al.*, 2010; Lin *et al.*, 2009; Robles *et al.*, 2013a, b).

Different studies have shown that cake layer formation, which is defined as accumulation of particles on the membrane surface, is the key factor limiting the flux in AnMBRs, irrespective of the applied substrate, configuration (submerged or side-stream) or temperature (Jeison & van Lier, 2008a; Lin *et al.*, 2009; Waeger *et al.*, 2010). In fact, cake layer formation was identified as the most important fouling mechanism in AnMBRs (Choo & Lee, 1998; Jeison & van Lier, 2006, 2007; Xie *et al.*, 2010). Contribution of the cake layer to total filtration resistance could be over 80% in MBRs (Meng *et al.*, 2007). Fouling reduces operational flux, increases transmembrane pressure (TMP), and requires physical (backwash) or chemical cleaning, and possibly membrane replacement (Smith *et al.*, 2012). The intrinsic presence of membrane fouling results in a specific downtime of the installation, which needs to be included calculating the capital and operational costs of AnMBRs (Example 11.1; Pretel *et al.*, 2016).

EXAMPLE 11.1

Design of a submerged anaerobic membrane bioreactor equipped with an external hollow fiber membrane tank for separately collected domestic sewage treatment (no rainwater).

Input data

- Population equivalent: PE = 50,000 inhabitants
- Average flow: $Q = 8{,}000\ m^3 \cdot d^{-1}$
- Average influent COD concentration: $C_{COD} = 598\ mg\ L^{-1}$ ($0.598\ kg\ m^{-3}$)

Solution

(1) Calculate the reactor volume ($V_{reactor}$):
Select a hydraulic retention time (HRT) for the bioreactor of 10 h:

$$HRT(d) = \frac{V_{reactor}(m^3)}{Q(m^3 \cdot d^{-1})}$$

$$V_{reactor}(m^3) = 8{,}000\ m^3 \cdot d^{-1} \cdot 10/24 = 3{,}333\ m^3$$

(2) Estimate the waste sludge production (WSP):
By selecting a solids retention time (SRT) of 70 d for ambient temperature operation (at about 20°C) and a mixed liquor suspended solids (MLSS) concentration of $15\ g\ L^{-1}$, the waste sludge

production (*WSP*) of the system would be:

$$SRT(d) = \frac{V_{reactor}(m^3) \cdot X_{SS}\ (kgSS \cdot m^{-3})}{Q\Delta X_{SS}\ (kgSS \cdot d^{-1})}$$

$$Q\Delta X_{SS}\ (kgSS \cdot d^{-1}) = \frac{3{,}333\ m^3 \cdot 15\ kg \cdot m^{-3}}{70\ d} = 714\ kgSS \cdot d^{-1}$$

WSP = 0.089 kgSS per m^3 of treated sewage (0.15 kgSS $kgCOD_{in}{}^{-1}$).

(3) Calculate the minimum required membrane filtration area (A_{min}):
For a membrane-operating mode entailing 300 s filtration (*F*), 30 s relaxation (*R*), 60 s backwash (*BF*) and a frequency for backwash initiation (BF_{freq}) of 5 F–R basic cycles, the theoretical downtime for physical cleaning is:

$$Downtime(\%) = \frac{BF(s) + BF_{freq} \cdot R(s)}{BF(s) + BF_{freq} \cdot (F(s) + R(s))} \cdot 100 = \frac{60 + 5 \cdot 30}{60 + 5 \cdot (300 + 30)} \cdot 100 = 12.3\%$$

Selecting a conservative design 20°C standardized gross flux (J_{20}) of 15 litres per square metre per hour (LMH):

$$J_{20net}(LMH) = J_{20}(LMH) \cdot (1 - Downtime) = 15 \cdot (1 - 0.123) = 13.2\ LMH$$

$$A_{min}(m^2) = \frac{Q\ (L \cdot h^{-1})}{J_{20net}(LMH)} = \frac{8{,}000 \cdot \frac{1{,}000}{24}}{13.2} = 25{,}252\ m^2$$

Note: Due to operational reasons (e.g., membrane cleaning, peak flowrate treatment), additional membrane area would be required.

(4) Estimate the gas flow for membrane scouring (Q_{gas})
Considering an operating specific gas sparging demand per membrane area (SGD_m) of 0.25 $Nm^3{}_{gas} m^{-2}\ h^{-1}$ the blowing capacity for membrane scouring would be:

$$Q_{gas}(Nm^3{}_{gas}\ h^{-1}) = SGD_m(Nm^3{}_{gas}\ m^{-2}\ h^{-1}) \times A_{min}(m^2) = 0.25 \times 25{,}252 = 6{,}313\ Nm^3{}_{gas}\ h^{-1}$$

11.3.4.1 *Fouling mitigation*

The attainable operational flux in AnMBRs is directly related to the required membrane surface and, therefore, is one of the most important parameters determining the economic feasibility of the technology. In the last two decades, an increasing amount of research focused on flux enhancement. Various additives such as activated carbon, polyelectrolytes, coagulants and flocculants can be used to improve the flux and to reduce fouling in AnMBRs (Hu & Stuckey, 2007; Wu *et al.*, 2009). The flux enhancers can act through a number of different phenomena such as adsorption of SMP and coagulation (Ozgun *et al.*, 2013a). One of the most commonly used additives for membrane fouling control in AnMBRs is powdered activated carbon (PAC). Studies on the effect of PAC on fouling mitigation in AnMBRs have been started in late 1990s. It was reported that both the fouling and cake layer resistances decreased continuously with increasing the PAC dose up to 5 g L^{-1} (Park *et al.*, 1999). The effects of PAC and GAC addition on TMP and flux development were investigated in an AnMBR by Hu and

Stuckey (2007). In terms of TMP, better performance was achieved with activated carbon that continuously scoured deposited particles from the membrane surface in comparison to operation without activated carbon. Besides, activated carbon addition provided a shift to a relatively higher range in particle size distribution. Due to the higher surface area of PAC, a more stable TMP and reduced fouling propensities are achieved with PAC compared to GAC (Ozgun *et al.*, 2013a). Results from a comparison study for different flux enhancer additives showed that poly-aluminium chloride decreased fouling significantly compared to other tested adsorbents/coagulants including PAC, zeolite, and polyamide (Wu *et al.*, 2009). Akram and Stuckey (2008) observed a significant flux improvement from $2\ L\ m^{-2}\ h^{-1}$ to $9\ L\ m^{-2}\ h^{-1}$ with PAC addition of $1.67\ g\ L^{-1}$ due to the adsorption of fine colloids and dissolved organics, and the formation of a thin cake layer on the membrane surface. However, the operating flux decreased with the PAC addition to $3.4\ g\ L^{-1}$ due to the viscosity increase. Additives have significant effects on mass transfer properties, biochemical environment and operational costs. Therefore, the optimum dosage of additives should be determined (Ozgun *et al.*, 2013a). Nanoparticles such as magnesium oxide, titanium oxide and copper-based particles also present an expected opportunity for biofouling control in MBRs. For instance, it was shown that fullerene C60 nanoparticle addition could decrease microbial attachment and thus it may be useful as an anti-fouling agent to prevent biofouling of membranes (Chae *et al.*, 2009).

11.3.4.2 On-line fouling control

On-line control is a desirable tool for the optimization of membrane performance in AnMBR technology (Smith *et al.*, 2012). Indeed, one key operating challenge of any AnMBR process is how membrane performance can be optimized whilst minimizing any kind of membrane fouling.

Different on-line control strategies have been applied for fouling control in AnMBRs. Jeison and van Lier (2006) developed an on-line cake-layer management protocol using an automated routine that continuously monitored the TMP which was combined with periodic critical flux assessment, preventing the build-up and compaction of an excessive cake layer on the membrane surface. Among others, Robles *et al.* (2013c) and Kalboussi *et al.* (2017) proposed different on-line control strategies for backwashing initiation, while Robles *et al.* (2014) optimized the frequency of backwashing through a model-based optimization algorithm. Regarding gas-assisted membrane scouring, Robles *et al.* (2013c, 2014) optimized gas sparging intensity in an AnMBR pilot plant with industrial-scale membranes, resulting in energy savings of up to 25%. Development of an online direct assessment of the sludge filterability combined with online TMP measurements will provide in-situ data for periodic corrective fouling control, such as the addition of flux enhancers (Lousada-Ferreira *et al.*, 2017a, b). Hence, on-line fouling control strategies are valuable tools for reducing membrane fouling, decreasing energy demand and increasing membrane lifetime and deserve more research attention.

11.3.5 Economic and environmental sustainability

Although membrane material costs have significantly decreased during the last years with the improvements in membrane manufacturing technology, the cost of the membranes is still considered a critical issue when treating low-strength wastewater, such as sewage, at full scale. Nevertheless, recent studies have illustrated AnMBR as a cost-effective technology for sewage treatment. For instance, Smith *et al.* (2014) reported total costs for AnMBR treatment being €0.12 per m^3, while values between €0.07 and €0.10 per m^3 were reported by Ferrer *et al.* (2015). Other studies evaluated the life cycle cost (LCC) of different platforms including AnMBR as core technology for sewage treatment. For instance, Pretel *et al.* (2016) evaluated the LCC of an AnMBR system coupled to a post-treatment based on conventional activated sludge (CAS) or aerobic MBR for nutrient removal, where part of the influent was bypassed to the post-treatment for denitrification purposes. LCC studies revealed overall costs of about €0.13–0.17 per m^3. Pretel *et al.* (2015) combined AnMBR as core technology for sulfate-rich sewage treatment with primary settling and side-stream AD, observing a minimum cost of €0.05 per m^3.

Aerobic MBRs have become cost competitive up to quite large scales, where the circumstances are favourable with the largest MBR currently treating 864,000 $m^3\,d^{-1}$ (Krzeminski *et al.*, 2017). Equivalent assessment for AnMBRs is currently not possible but costs assessments for proposed new flowsheets utilizing AnMBRs to treat settled sewage coupled with sorption processes for nutrient removal reveal overall cost equity with traditional biological nutrient removal activated sludge systems providing a number of limiting conditions are met. This includes minimizing the HRT of the UASB reactor to a maximum of 7 hours at average flow and a minimum design flux of 12 $L\,m^{-2}\,h^{-1}$ at average flow (Jefferson *et al.*, 2018). However, the membrane remains the key capital and operating cost component which will inhibit uptake. Replacing the UF membranes with coarse membrane filters such as cloth filters has the potential for significant savings with recent estimates of a reduction of between 30–60% on the capital cost of the membrane systems. However, coarse filtration systems may alter the downstream management requirements, especially with regards to dissolved methane and their technical viability in an integrated flowsheet stills needs to be validated.

Pretel *et al.* (2013) performed a life cycle assessment (LCA) study of an AnMBR system for sewage treatment at different temperatures. The results revealed the improved environmental performance of the system when: (i) operating at ambient temperature (20–33°C), (ii) using the effluent for fertirrigation purposes (environmental impact in eutrophication can be reduced significantly), (iii) recovering the methane dissolved in the effluent as energy carrier; the latter even leading to a positive environmental impact. Smith *et al.* (2014) showed that AnMBRs could recover up to 49% more energy as biogas than a combined high-rate activated sludge and sidestream AD for excess sludge treatment. However, if dissolved methane is not recovered, the global warming impact of AnMBR is higher due to the dissolved methane emissions in the effluent. Pretel *et al.* (2016) observed that significant reductions in several environmental impact categories (global warming potential, marine aquatic ecotoxicity, abiotic depletion and acidification) could be achieved when combining AnMBR with CAS for nutrient removal compared to aerobic MBR and CAS.

11.3.6 Potentials for water reuse

Water scarcity represents a serious problem worldwide. For instance, at least 11% of the EU population and 17% of the EU territory is currently affected by water scarcity, and it is projected that the extension of areas facing water scarcity problems will grow up to 30% in 2030, particularly in summer conditions. The use of treated wastewater is highlighted in EU policy dealing with resource efficiency. Especially, water reuse is most notably highlighted in the 2015 Communication on circular economy which states that "*the use of treated wastewater in safe and cost-effective conditions is a valuable but under-used means of increasing water supply and alleviating pressure on over-exploited water resources in the EU25*" (COM/2015/0614).

AnMBR equipped with ultrafiltration membranes produces a high quality permeate that can reach disinfection levels depending on the cut-off of the membrane (Bair *et al.*, 2015), while reducing the costs associated to sewage treatment due to the energy output in the form of biogas and the absence of energy input related to aeration for organic matter removal. Thus, AnMBR can be considered a cost-effective technology with advanced purification capacity, which could help to alleviate water scarcity through water reuse and recycling. Moreover, this effluent contains levels of nitrogen and phosphorus of interest for agricultural applications, thus representing a valuable water source for fertirrigation purposes depending on the corresponding policy. By way of example, existing European laws affecting this issue are Water Framework Directive 2000/60/EC, Waste Framework Directive 2008/98/EC, Regulation on Food Safety (EC) No 178/2002, Fertilizer Regulation (EC) No 2003/2003, and Soil Framework Directive COM/2006/0232.

The application of circular economy principles within the water management sector by using reclaimed sewage water are prominent issues that can provide important environmental, economic and societal benefits (see also Chapter 12).

11.3.7 Potentials for nutrient recovery

The nutrient content in sewage often remains unexploited. The EU WssTP report highlights that "*The estimated value of some of these nutrients that are available in common wastewaters is between $300–400/mT (N products) and $500–600/mT (P products) considering the huge demand for these products in agriculture*" (WssTP, 2016). At present, Western Europe is highly dependent on fertilizer import (e.g. 80% of its phosphate requirements), which illustrates the potential market opportunities for nutrient recovery (Kok *et al.*, 2018). Hence, the recycle of nutrients from sewage to farmland, particularly phosphorus, is of great interest in the foreseen circular economy, since this component is an essential element irreplaceable in the production of crops and food (Ashley *et al.*, 2011). In addition, nitrogen recovery from sewage for agricultural purposes can help to close cycles at the local scale, reducing (fossil) energy consumption for the production of ammonia-based fertilizers (Do Nascimento *et al.*, 2015) as well as for transport purposes.

The combination of AnMBR and water reclamation in agriculture results in a more sustainable concepts for sewage treatment, where sewage turns into a source of not only energy and reclaimed water but also nutrients (nitrogen and phosphorus). The use of the generated solids-free and (partially) disinfected effluent for agricultural purposes not only can help to alleviate water scarcity problems but also to reduce environmental impacts related to the use of inorganic fertilizers. However, the exact required nutrient composition will vary depending on soil characteristics, crops, season and applied regulations, likely requiring an additional step after the AnMBR.

When fertirrigation is a non-possible practice, the nutrient content of the AnMBR permeate will hamper the direct emission of this effluent to receiving water bodies. Alternative to fertirrigation, the mentioned nutrients can be recovered by different techniques, such as by the cultivation of phototrophic organisms (Viruela *et al.*, 2016, 2018) or through sorption processes (Jefferson, 2012). Nutrient recovery from wastewater by means of microalgae-based processes is attracting interest from the scientific community (see e.g. Gao *et al.*, 2015; Tsolcha *et al.*, 2017). The interest of this hybrid system is focused not only on moving towards water resource recovery facilities (WRRFs), but also on the possibility of offsetting the costs of fertilizers and the water needed to cultivate microalgae biomass. In addition, the carbon footprint is reduced by carbon dioxide biofixation (Mata *et al.*, 2010), while harvesting solar energy in the form of algal biomass. This biomass is considered a high-value product since it can be converted, for instance, into fertilizers or into biogas through anaerobic digestion (see e.g. Jankowska *et al.*, 2017).

Other upcoming cost-effective but resource-destructive techniques for treating AnMBR permeate include denitrification by methanotrophs. Sánchez-Ramírez *et al.* (2015) studied the post-treatment of an AnMBR permeate by an activated sludge system, where denitrification was conducted by methanotrophs. Pelaz *et al.* (2017) also evaluated the efficiency of using alternative electron donors for denitrification, observing that dissolved methane was the main electron donor used to remove nitrites and nitrates, with a removal rate of over 70%. Same authors (Pelaz *et al.*, 2018) also evaluated the performance of denitrification over nitrite in a pilot plant treating AnMBR permeate. Methane and sulfide were used as alternative electron donors and nitrite removal efficiencies of 92.9 and 98.3% were achieved depending on nitrite concentrations.

The alternative to biological processes for nutrient removal and recovery is to use sorption processes with example media for both ammonia and phosphorus currently being explored (Jefferson *et al.*, 2018). The sorption processes are low energy and low footprint with empty bed contact times of 5–10 minutes for phosphorus and 10–20 minutes for ammonia utilizing energy levels similar to those of depth filters. Trials on a pilot AnMBR based flowsheets in the UK enabled stable delivery of effluent total phosphorous of 1 mg L^{-1} and ammonia of 3 mg L^{-1} (Jefferson, 2012). Lower effluent levels have been demonstrated in aerobic flowsheets by adjusting the empty bed contact time such that any required effluent nutrient level is possible (Martin *et al.*, 2013). The media requires regular regeneration with either caustic and/or brine solutions which must be reused to ensure economic suitability. Accordingly, the liquid must be locally cleaned up resulting in the recovery of the captured nutrients. In the case of phosphorus, a hybrid ion exchange resin embedded with iron oxide nanoparticles is preferred (Acelas *et al.*, 2015) and results in the formation of calcium phosphate at a purity level greater than found in equivalent produced calcium phosphate generated from mined resources (Jefferson, 2012). In the case of ammonia, zeolites are used and selectivity is generated through degassing the captured ammonia from the brine liquid generating ammonium sulphate (Mcleod *et al.*, 2015). Alternative systems exist, including trials with polonite in a rapid filter (contact time of 2 hours) post a UASB (Rodriguez-Gomez and Renman, 2017) or reactive media in constructed wetlands. The sorption processes provide additional value in that they remove a further 30–50% of the residual organic matter which provides an effective source water for (in)direct potable reuse post reverse osmosis. Overall the use of AnMBRs appears particular relevant for delivery of resource factory concepts where water, energy and nutrients can be recovered. Recent research of van Linden et al. (2018) focused on the recovery of the energy content of ammonia by extracting the ammonium from the UF permeate to be used in a solid oxide fuel cell (SOFC).

11.3.8 Potentials of dynamic membranes in AnMBR technology

Although the price of membranes has been distinctly reduced in the past decades, membrane costs are still considered one of the major capital costs factors of MBR technology, especially for full-scale applications. The application of dynamic membrane (DM) technology may resolve this problem and lower the capital exploitation costs (Ersahin *et al.*, 2012; Hu *et al.*, 2018). A DM, which is also called a secondary membrane, is formed on an underlying support material, e.g. a mesh, or a filter cloth, when the filtered solution contains suspended solids, such as microbial cells and flocs. Since a DM layer can easily be removed from the surface of the support material and can be re-established again in a short time, this layer is termed “dynamic membrane”. One of the most important potential benefits of DMs is that the actual filter (the membrane) itself may be no longer necessary, because solids rejection is accomplished by the secondary membrane layer that can be formed and re-formed as a self-forming DM in situ (Ersahin *et al.*, 2012). The DM filtration concept can turn one of the most important disadvantages of MBRs, membrane fouling, into an advantage (Ersahin *et al.*, 2014).

The application of DM technology in AnMBR, referred to as anaerobic dynamic membrane bioreactor (AnDMBR), can be used as a cost-effective filtration technology because it employs low-cost filter materials in place of high-cost membranes. Different kinds of materials such as mesh, woven or nonwoven fabric instead of microfiltration (MF) and ultrafiltration (UF) membranes can be used as support layer for creating the DM. In this way, replacement of the membrane by a low-cost filter material is possible. Support material properties are critical for the formation of an effective DM (cake) layer over the filter surface. In a comparison study, the critical fluxes obtained with staple and monofilament filter cloths were higher than those obtained with a multifilament material. The reported results indicated that staple filter cloth resembled depth filtration, whereas a mono-monofilament filter was more suitable for surface (cake) filtration. Thus, the mono-monofilament filter was much easier to clean and thus was considered more appropriate for DM technology (Ersahin *et al.*, 2013).

By decreasing membrane cost and generating energy, AnDMBRs would become an attractive alternative for sewage treatment. AnDMBRs could achieve over 99% organic matter and suspended solids removal and particulate matter retention (Ersahin *et al.*, 2016a). In fact, the DM layer had an important role in organic matter removal in AnDMBRs. Both organic and inorganic materials, such as sludge particles, SMP, EPS, and Ca, N, P, Mg precipitations could contribute to the DM layer formation (Ersahin *et al.*, 2016b). Similar to AnMBRs, different configurations can be applied in the AnDMBR process. Comparison of different membrane configurations showed that an external AnDMBR set-up requires more time for establishing an effective DM layer, enabling a stable removal efficiency and low soluble COD concentration in the permeate. Therefore, submerged AnDMBR configuration appears more suitable when a short start-up period is necessary (Ersahin *et al.*, 2017). Higher methane production rates and methane yields were obtained in the submerged configuration compared to the external configuration reflecting the negative effect of sludge recirculation in the external AnDMBR configuration. Conversely, sludge recirculation in the external configuration was more effective in decreasing DM thickness, and thus TMP, than the bottom biogas sparging in the submerged configuration.

Sewage treatment was satisfactory even at ambient temperature (around 15–20°C) applying AnDMBRs, but with limited biogas production (Hu *et al.*, 2018). Ho *et al.* (2007) reported that AnDMBRs could be operated at low TMP and cross-flow velocity in order to maintain a DM layer for efficient particle removal from sewage. Zhang *et al.* (2010) investigated the formation process of the DM layer at the upper part of a UASB reactor and reported a high flux of 65 $L\ m^{-2}\ h^{-1}$ in an AnDMBR treating sewage at low temperatures (10–15°C). They achieved to have a DM layer over a Dacron mesh with a pore size of 61 μm. Zhi-Guo *et al.* (2005) compared the filtration performances of two support layers, non-woven material and a conventional hollow fiber membrane in a submerged dynamic MBR treating sewage. They obtained similar effluent qualities with both filters. They determined that the formation of a dynamic layer on the non-woven filter surface was effective on the organic carbon removal. Coupling of DM technology to a UASB reactor for the treatment of sewage has been tested and high removal performances were obtained (Quek *et al.*, 2017). The DM-coupled UASB reactor system achieved removal efficiencies of over 64% and 86% for COD and TSS when operating at ambient temperature of 25–30°C, respectively. About 25% increase in fouling rate was observed with an increase in operating flux by 67%. By using a mesh support layer, an efficient treatment was achieved with a flux of 100 $L\ m^{-2}\ h^{-1}$ and HRT of 6 h (Quek *et al.*, 2017). In submerged AnDMBRs coupled with Dacron mesh having a pore size of 61 μm, a flux of over 60 $L\ m^{-2}\ h^{-1}$ and COD removal efficiencies between 63% and 79% were obtained for the treatment of sewage at ambient temperature ranging from 10 to 30°C (Ma *et al.*, 2013; Zhang *et al.*, 2011). DM technology has the capability to achieve a stable and high-quality permeate. Thus, AnDMBRs are regarded as a promising alternative to be used as a reliable and satisfactory treatment technology for sewage treatment. Particularly the low capital costs of support material, low operational costs, and energy generation, while acquiring a high treatment performance on COD and suspended solids, are certainly of major interest. Treated effluents could be beneficially used for fertirrigation of agricultural fields.

11.4 AnMBR AS AN ALTERNATIVE FOR CONVENTIONAL MBR SYSTEMS

The implementation of AnMBR technology in general and the switch from aerobic MBRs to AnMBRs in particular, broadens the recovery potential that can be achieved from sewage. AnMBR eliminates the energy input related to aeration while generating methane as energy carrier. It is important to highlight that aerobic MBRs increase the energy consumption of sewage treatment by 50–70% compared to CAS. Specifically, aerobic MBRs typically require energy within the range of 0.4–1.6 kWh per m^3 of treated water, which is 20–50% higher than the energy consumed in CAS (Krzeminski *et al.*, 2012). Energy demand in AnMBRs currently ranges from a net production of about 0.10 to a consumption of about 0.20 kWh per m^3 depending on operating conditions and dissolved methane recovery. The latter is a key factor determining the potential of AnMBRs to contribute to an energy-neutral solution or even to net energy production when treating sewage.

Smith *et al.* (2014) showed that the environmental impact related to GHG emissions in AnMBR can be up to 5 times higher than the combination of aerobic MBR + side stream AD when dissolved methane is not recovered. However, AnMBR could recover up to 49% more energy as biogas than a combined high-rate activated sludge and a sidestream AD. On the other hand, Pretel *et al.* (2016) observed that a significant reduction in global warming potential could be achieved in AnMBR when recovering the methane dissolved in the effluent as energy carrier; which could lead even to a positive environmental impact (Pretel *et al.*, 2013).

Input energy savings would also result in a significant decrease in the carbon footprint of AnMBR. In this respect, one key factor determining the energy feasibility of AnMBR is the SGD_P, which represents the ratio of required gas flow rate for membrane scouring per volume of treated effluent. Although lower membrane fluxes are still frequently applied in AnMBRs compared to aerobic MBRs, gas sparging intensity for membrane scouring has been optimized during the last years thanks to advances on membrane materials and membrane fouling control strategies, resulting in enhanced energy efficiency. Moreover, it is important to highlight that compared to lab-scale, the energy efficiency of the filtration process increases when scaling-up the membrane size due to a better use of gas sparging through the membrane length, especially for hollow-fiber membranes. However, there is a significant amount of energy data available in literature regarding aerobic MBRs that was obtained using industrial-scale membranes, while most of the data available for AnMBRs was obtained at laboratory scale, complicating the comparison of the results. Recent results obtained using industrial-size hollow-fiber membranes in AnMBRs revealed the potential for maintaining low membrane fouling propensities by applying SGD_P from around 10 to 14 $Nm^3\ m^{-3}$ (Durán *et al.*, 2018; Robles *et al.*, 2017). These values are close to those obtained in full-scale aerobic MBR plants (Judd & Judd, 2011). Hence, it is expected that the energy feasibility of AnMBR for sewage treatment will increase in the next years.

The combination of anaerobic treatment and ultrafiltration membranes allows producing a high quality permeate that can reach disinfection levels depending on the pore size of the membrane, which could help to alleviate water scarcity problems through water reuse and recycling. In addition, ammonium and orthophosphate are not removed in AnMBRs. Therefore, although both aerobic MBR and AnMBR produce a solids-free permeate that can be used for irrigation purposes, the latter permits to recover the nutrient content of the effluent in a post-treatment step for nutrient caption or even through direct fertirrigation, depending on the corresponding policy. Fertirrigation is of great interest in the foreseen circular economy due to the huge nutrient demand in agriculture. As commented before, phosphorus is an essential element irreplaceable in the production of crops and food, whilst the production of ammonia-based fertilizers is an energy-intensive activity. Thus, the use of AnMBR effluent in farmland can help not only to alleviate water scarcity problems but also to reduce environmental impacts related to the use of inorganic fertilizers.

REFERENCES

Acelas N. Y., Martin B. D., Lopez D. and Jefferson B. (2015). Selective removal of phosphate from wastewater using hydrated metal oxides dispersed within anionic exchange media. *Chemosphere*, **119**, 1353–1360.

Achilli A., Marchand E. A. and Childress A. E. (2011). A performance evaluation of three membrane bioreactor systems: aerobic, anaerobic, and attached-growth. *Water Science and Technology*, **63**(12), 2999–3005.

Akram A. and Stuckey D. C. (2008). Flux and performance improvement in a submerged anaerobic membrane bioreactor (SAMBR) using powdered activated carbon (PAC). *Process Biochemistry*, **43**(1), 93–102.

An Y., Wang Z., Wu Z., Yang D. and Zhou Q. (2009). Characterization of membrane foulants in an anaerobic non-woven fabric membrane bioreactor for municipal wastewater treatment. *Chemical Engineering Journal*, **155**(3), 709–715.

Ashley K., Cordell D. and Mavinic D. (2011). A brief history of phosphorus: from the philosopher's stone to nutrient recovery and reuse. *Chemosphere*, **84**(6), 737–746.

Baek S. H. and Pagilla K. R. (2006). Aerobic and anaerobic membrane bioreactors for municipal wastewater treatment. *Water Environment Research*, **78**(2), 133–140.

Bair R. A., Ozcan O., Calabria J., Dick G. H. and Yeh D. (2015). Feasibility of anaerobic membrane bioreactors (AnMBR) for onsite sanitation and resource recovery (nutrients, energy and water) in urban slums. *Water Science and Technology*, **72**(9), 1543–1551.

Bandara W. M. K. R. T. W., Kindaichi T., Satoh H., Sasakawa M., Nakahara Y., Takahashi M. and Okabe S. (2012). Anaerobic treatment of municipal wastewater at ambient temperature: analysis of archaeal community structure and recovery of dissolved methane. *Water Research*, **46**(17), 5756–5764.

Brockmann M. and Seyfried C. F. (1996). Sludge activity and cross-flow microfiltration – a non-beneficial relationship. *Water Science & Technology*, **34**(9), 205–213.

Brockmann M. and Seyfried C. F. (1997). Sludge activity under the conditions of cross-flow microfiltration. *Water Science & Technology*, **35**(10), 173–181.

Ceron-Vivas A., Morgan-Sagastume J. M. and Noyola A. (2012). Intermittent filtration and gas bubbling for fouling reduction in anaerobic membrane bioreactors. *Journal of Membrane Science*, **423–424**, 136–142.

Chen C., Guo W. S., Ngo H. H., Liu Y., Du B., Wei Q., Wei D., Nguyen D. D. and Chang S. W. (2017). Evaluation of a sponge assisted-granular anaerobic membrane bioreactor (SG-AnMBR) for municipal wastewater treatment. *Renewable Energy*, **111** (C), 620–627.

Chae S., Wnag S., Hendren Z. D., Wiesner M. R., Watanabe Y. and Gunsch C. K. (2009). Effects of fullerene nanoparticles on Escherichia coli K12 respiratory activity in aqueous suspension and potential use for membrane biofouling control. *Journal of Membrane Science*, **329**(1–2), 68–74.

Chernicharo C. A. L., van Lier J. B., Noyola A. and Ribeiro T. B. (2015). Anaerobic sewage treatment: state of the art, constraints and challenges. *Reviews in Environmental Science and Bio/technology*, **14**(4), 649–679.

Choo K. H. and Lee C. H. (1998). Hydrodynamic behavior of anaerobic biosolids during crossflow filtration in the membrane anaerobic bioreactor. *Water Research*, **32**(11), 3387–3397.

COM 0614 – European Commission (2015). Closing the loop – An EU action plan for the Circular Economy. Communication from the commission to the European Parliament, the Council, the European Economic and Social Committee and the Committee of the Regions. https://www.eea.europa.eu/policy-documents/com-2015-0614-final (accessed: 18 November 2018)

Dereli R. K., Ersahin M. E., Ozgun H., Ozturk I., Jeison D., van der Zee F. and van Lier J. B. (2012). Potentials of anaerobic membrane bioreactors to overcome treatment limitations induced by industrial wastewaters. *Bioresource Technology*, **122**, 160–170.

Do Nascimento M., Sanchez Rizza L., Arruebarrena Di Palma A., Dublan M. A., Salerno G., Rubio L. M. and Curatti L. (2015). Cyanobacterial biological nitrogen fixation as a sustainable nitrogen fertilizer for the production of microalgal oil. *Algal Research*, **12**, 142–148.

Dong Q., Parker W. and Dagnew M. (2016). Long term performance of membranes in an anaerobic membrane bioreactor treating municipal wastewater. *Chemosphere*, **144**, 249–256.

Durán F., Robles A., Giménez J. B., Vazquez-Padín J., Serralta J., Ribes J., Seco A., Ferrer J. and Rogalla F. (2018). AnMBR technology: boosting circular economy in sewage treatment. IWA World Water Congress & Exhibition, Tokio, Japan.

Ersahin M. E. (2015). Application of Dynamic Membranes in Anaerobic Membrane Bioreactor Systems. PhD thesis, Delft University of Technology/Istanbul Technical University, Delft, Netherlands/Istanbul, Turkey.

Ersahin M. E., Ozgun H., Dereli R. K. and Ozturk I. (2011). Anaerobic treatment of industrial effluents: an overview of applications. In: Waste Water-Treatment and Reutilization, F. S. G. Einschlag (ed.), InTech, India, pp. 415–456.

Ersahin M. E., Ozgun H., Dereli R. K., Ozturk I., Roest K. and van Lier J. B. (2012). A review on dynamic membrane filtration: materials, applications and future perspectives. *Bioresource Technology*, **122**, 196–206.

Ersahin M. E., Ozgun H. and van Lier J. B. (2013). Effect of support material properties on dynamic membrane filtration performance. *Separation Science and Technology*, **48**(15), 2263–2269.

Ersahin M. E., Ozgun H., Tao Y. and van Lier J. B. (2014). Applicability of dynamic membrane technology in anaerobic membrane bioreactors. *Water Research*, **48**, 420–429.

Ersahin M. E., Giménez J. B., Ozgun H., Tao Y., Spanjers H. and van Lier J. B. (2016a). Gas-lift anaerobic dynamic membrane bioreactors for high strength synthetic wastewater treatment: effect of biogas sparging velocity and HRT on treatment performance. *Chemical Engineering Journal*, **305**, 46–53.

Ersahin M. E., Tao Y., Ozgun H., Spanjers H. and van Lier J. B. (2016b). Characteristics and role of dynamic membrane layer in anaerobic membrane bioreactors. *Biotechnology and Bioengineering*, **113**(4), 761–771.

Ersahin M. E., Tao Y., Ozgun H., Giménez J. B., Spanjers H. and van Lier J. B. (2017). Impact of anaerobic dynamic membrane bioreactor configuration on treatment and filterability performance. *Journal of Membrane Science*, **526**, 387–394.

Fawehinmi F., Jefferson B., Chan T. and Rogalla F. (2007). Submerged anaerobic membrane bioreactors (SAnMBR): ready for the big ball? *Proc Water Environment Federation – WEFTEC 2007*, **9**, 6393–6401.

Ferrer J., Pretel R., Durán F., Giménez J. B., Robles A., Ruano M. V., Serralta J., Ribes J. and Seco A. (2015). Design methodology for submerged anaerobic membrane bioreactors (AnMBR): a case study. *Separation and Purification Technology*, **141**, 378–386.

Gao D. W., Hu Q., Yao C., Ren N. Q. and Wu W. M. (2014). Integrated anaerobic fluidized-bed membrane bioreactor for domestic wastewater treatment. *Chemical Engineering Journal*, **240**, 362–368.

Gao D., Zhang T., Tang C. Y., Wu W., Wong C., Lee Y. H., Yeh D. and Criddle C. S. (2010). Membrane fouling in an anaerobic membrane bioreactor: differences in relative abundance of bacterial species in the membrane foulant layer and in suspension. *Journal of Membrane Science*, **364**(1–2), 331–338.

Gao F., Yang Z. H., Li C., Zeng G. M., Ma D. H. and Zhou L. (2015). A novel algal biofilm membrane photobioreactor for attached microalgae growth and nutrients removal from secondary effluent. *Bioresource Technology*, **179**, 8–12.

Giménez J. B., Robles A., Carretero L., Duran F., Ruano M. V., Gatti M. N., Ribes J., Ferrer J. and Seco A. (2011). Experimental study of the anaerobic urban wastewater treatment in a submerged hollow-fibre membrane bioreactor at pilot scale. *Bioresource Technology*, **102**, 8799–8806.

Giménez J. B., Marti N., Ferrer J. and Seco A. (2012). Methane recovery efficiency in a submerged anaerobic membrane bioreactor (SAnMBR) treating sulphate-rich urban wastewater: evaluation of methane losses with the effluent. *Bioresource Technology*, **118**, 67–72.

Giménez J. B., Martí N., Robles A., Ferrer J. and Seco A. (2014). Anaerobic treatment of urban wastewater in membrane bioreactors (MBRs): evaluation of seasonal temperature variations. *Water Science & Technology*, **69**, 1581–1588.

Gomec C. Y. (2010). High-rate anaerobic treatment of domestic wastewater at ambient operating temperatures: a review on benefits and drawbacks. *Journal of Environmental Science and Health Part A*, **45**, 1169–1184.

Gouveia J., Plaza F., Garralon G., Fdz-Polanco F. and Peña M. (2015a). Long-term operation of a pilot scale anaerobic membrane bioreactor (AnMBR) for the treatment of municipal wastewater under psychrophilic conditions. *Bioresource Technology*, **185**, 225–233.

Gouveia J., Plaza F., Garralon G., Fdz-Polanco F. and Peña M. (2015b). A novel configuration for an anaerobic submerged membrane bioreactor (AnSMBR). Long-term treatment of municipal wastewater under psychrophilic conditions. *Bioresource Technology*, **198**, 510–519.

Grundestam J. and Hellstrom D. (2007). Wastewater treatment with anaerobic membrane bioreactor and reverse osmosis. *Water Science and Technology*, **56**(5), 211–217.

Gu Y., Chen L., Ng J. W., Lee C., Chang V. W. C. and Tang C. Y. (2015). Development of anaerobic osmotic membrane bioreactor for low-strength wastewater treatment at mesophilic condition. *Journal of Membrane Science*, **490**, 197–208.

He Y. L., Pei X., Li C. J. and Bo Z. (2005). High-concentration food wastewater treatment by an anaerobic membrane bioreactor. *Water Research*, **39**, 4110–4118.

Herrera-Robledo M., Morgan-Sagastume J. M. and Noyola A. (2010). Biofouling and pollutant removal during long-term operation of an anaerobic membrane bioreactor treating municipal wastewater. *Biofouling*, **26**, 23–30.

Ho J. and Sung S. (2010). Methanogenic activities in anaerobic membrane bioreactors (AnMBR) treating synthetic municipal wastewater. *Bioresource Technology*, **101**(7), 2191–2196.

Ho J. H., Khanal S. K. and Sung S. (2007). Anaerobic membrane bioreactor for treatment of synthetic municipal wastewater at ambient temperature. *Water Science and Technology*, **55**(7), 79–86.

Hu A. Y. and Stuckey D. C. (2006). Treatment of dilute wastewaters using a novel submerged anaerobic membrane bioreactor. *Journal of Environmental Engineering*, **132**, 190–198.

Hu A. Y. and Stuckey D. C. (2007). Activated carbon addition to a submerged anaerobic membrane bioreactor: effect on performance, transmembrane pressure, and flux. *Journal of Environmental Engineering*, **133**, 73–80.

Hu Y., Wang X. C., Ngo H. H., Sun Q. and Yang Y. (2018). Anaerobic dynamic membrane bioreactor (AnDMBR) for wastewater treatment: a review. *Bioresource Technology*, **247**, 1107–1118.

Huang Z., Ong S. L. and Ng H. Y. (2013). Performance of submerged anaerobic membrane bioreactor at different SRTs for domestic wastewater treatment. *Journal of Biotechnology*, **164**, 82–90.

Ince O., Cetecioglu Z., Ozbayram E. G., Iglesias M. M., Ince B., Massalha N., Robles A., Sabbah I. and Seco A. (2018). Anaerobic treatment of municipal wastewater. Innovative Wastewater Treatment & Resource Recovery Technologies: Impacts on Energy, Economy and Environment, J. M. Lema and S. Suarez (eds.), IWA Publishing, London, UK, pp 40–60.

Intergovernmental Panel on Climate Change (2014). *Climate Change 2014: Synthesis Report.* Contribution of Working Groups I, II and III to the Fifth Assessment Report of the Intergovernmental Panel on Climate Change, Core Writing Team, R.K. Pachauri and L. A. Meyer (ed.), Intergovernmental Panel on Climate Change – IPCC, Geneva, Switzerland, pp. 151.

Jankowska E., Sahu A. K. and Oleskowicz-Popiel P. (2017). Biogas from microalgae: review on microalgae's cultivation, harvesting and pretreatment for anaerobic digestion. *Renew and Sustainable Energy Reviews*, **75**, 692–709.

Jefferson B. (2012). Self Sufficient wastewater treatment: an achievable aspiration within 5 years? In: Leading-Edge Conference on Water and Wastewater Technologies, International Water Association (ed.), 9th edn, Brisbane, Australia.

Jefferson B., McAdam E. and Pidou M. (2018). Membrane based processes. In: Advances in Wastewater Treatment, G. Mannina, G. Ekama, H. Odegaard and G. Olsson (eds.), IWA Publishing, London, England.

Jeison D. and van Lier J. B. (2006). On-line cake-layer management by trans-membrane pressure steady state assessment in anaerobic membrane bioreactors for wastewater treatment. *Biochemical Engineering Journal*, **29**, 204–209.

Jeison D. and van Lier J. B. (2007). Cake formation and consolidation: main factors governing the applicable flux in anaerobic submerged membrane bioreactors (AnSMBR) treating acidified wastewaters. *Separation and Purification Technology*, **56**, 71–78.

Jeison D. and van Lier J. B. (2008a). Anaerobic wastewater treatment and membrane filtration: a one night stand or a sustainable relationship? *Water Science and Technology*, **57**(4), 527–532.

Jeison D. and van Lier J. B. (2008b). Feasibility of thermophilic anaerobic submerged membrane bioreactors (AnSMBR) for wastewater treatment. *Desalination*, **231**, 227–235.

Jeison D., van Betuw W. and van Lier J. B. (2008). Feasibility of anaerobic membrane bioreactors for the treatment of wastewaters with particulate organic matter. *Separation Science and Technology*, **43**, 3417–3431.

Jeison D., Telkamp P. and van Lier J. B. (2009). Thermophilic side stream anaerobic membrane bioreactors: the shear rate dilemma. *Water Environment Research*, **81**, 2372–2380.

Judd S. J. (2016). The status of industrial and municipal effluent treatment with membrane bioreactor technology. *Chemical Engineering Journal*, **305**, 37–45.

S. Judd and C. Judd (eds.) (2010). The MBR Book: Principles and Applications of Membrane Bioreactors for Water and Wastewater Treatment, 2nd edn, Butterworth–Heinemann, Oxford, UK.

Kalboussi N., Rapaport A., Bayen T., Ben Amar N., Ellouze F. and Harmand J. (2017). Optimal control of a membrane filtration system. *IFAC Papers Online*, **50**, 8704–8709.

Kim J., Kim K., Ye H., Lee E., Shin C., McCarty P. L. and Bae J. (2011). Anaerobic fluidized bed membrane bioreactor for wastewater treatment. *Environmental Science and Technology*, **45**, 576–581.

Kok D.-J. D., Pande S., van Lier J. B., Ortigara A. R. C., Savenije H. and Uhlenbrook S. (2018). Global Phosphorus Recovery for Agricultural Reuse. *Hydrology Earth System Sciences*, **22**, 5781–5799.

Krzeminski P., van der Graaf J. H. J. M. and van Lier J. B. (2012). Specific energy consumption of membrane bioreactor (MBR) for sewage treatment. *Water Science and Technology*, **65**(2), 380–392.

Krzeminski P., Leverette L., Malamis S. and Katsou E. (2017). Membrane bioreactors – a review on recent developments in energy reduction, fouling control, novel configurations, LCA and market penetration. *Journal of Membrane Science*, **527**, 207–227.

LeClech P., Chen V. and Fane T. A. G. (2006). Fouling in membrane bioreactors used in wastewater treatment. *Journal of Membrane Science*, **284**, 17–53.

Lettinga G., Rebac S. and Zeeman G. (2001). Challenge of psychrophilic anaerobic wastewater treatment. *Trends in Biotechnology*, **19**, 363–370.

Liao B. Q., Kraemer J. T. and Bagley D. M. (2006). Anaerobic membrane bioreactors: applications and research directions. *Critical Reviews in Environmental Science and Technology*, **36**, 489–530.

Lew B., Tarre S., Beliavski M., Dosoretz C. and Green M. (2009). Anaerobic membrane bioreactor (AnMBR) for domestic wastewater treatment. *Desalination*, **243**, 251–257.

Lin H. J., Xie K., Mahendran B., Bagley D. M., Leung K. T., Liss S. N. and Liao B. Q. (2009). Sludge properties and their effects on membrane fouling in submerged anaerobic membrane bioreactors (SAnMBRs). *Water Research*, **43**, 3827–3837.

Lin H., Chen J., Wang F., Ding L. and Hong H. (2011). Feasibility evaluation of submerged anaerobic membrane bioreactor for municipal secondary wastewater treatment. *Desalination*, **280**, 120–126.

Lin H., Peng W., Zhang M., Chen J., Hong H. and Zhang Y. (2013). A review on anaerobic membrane bioreactors: applications, membrane fouling and future perspectives. *Desalination*, **314**, 169–188.

Liu Y., Liu H., Cui L. and Zhang K. (2012). The ratio of food-to-microorganism (F/M) on membrane fouling of anaerobic membrane bioreactors treating low-strength wastewater. *Desalination*, **297**(3), 97–103.

Lousada-Ferreira M., Odriozola M., Spanjers H., van Lier J. B. (2017a). Accurately quantifying fouling in anaerobic membrane bioreactors through filterability measurements. In: IWA Anaerobic Digestion World Conference, International Water Association (ed.), 15th edn, Beijing, China.

Lousada-Ferreira M., Odriozola M., Spanjers H., van Lier J. B. (2017b). On-line fouling control tool for Anaerobic Membrane Bioreactors. In: IWA Specialized Conference on Instrumentation, Control and Automation, International Water Association (ed.), 12th edn, Québec, Canada.

Ma J., Wang Z., Zou X., Feng J. and Wu Z. (2013). Microbial communities in an anaerobic dynamic membrane bioreactor (AnDMBR) for municipal wastewater treatment: comparison of bulk sludge and cake layer. *Process Biochemistry*, **48**, 510–516.

Mahmoud N. (2008). High strength sewage treatment in a UASB reactor and an integrated UASB-digester system. *Bioresource Technology*, **99**, 7531–7538.

Mahmoud N., Zeeman G., Gijzen H. and Lettinga G. (2004). Anaerobic sewage treatment in a one-stage UASB reactor and a combined UASB-Digester system. *Water Research*, **38**, 2348–2358.

Martin B. D., De Kock L., Stephenson T., Parsons S. A. and Jefferson B. (2013). The impact of contactor scale on a ferric nanoparticle adsorbent process for the removal of phosphorous from municipal wastewater. *Chemical Engineering Journal*, **215–216**, 209–215.

Martinez-Sosa D., Helmreich B., Netter T., Paris S., Bischof F. and Horn H. (2011). Anaerobic submerged membrane bioreactor (AnSMBR) for municipal wastewater treatment under mesophilic and psychrophilic temperature conditions. *Bioresource Technology*, **102**, 10377–10385.

Martin-Garcia I., Mokosch M., Soares A., Pidou M. and Jefferson B. (2013). Impact on reactor configuration on the performance of anaerobic MBRs: treatment of settled sewage in temperate climates. *Water Research*, **47**, 4853–4860.

Martin-Garcia I., Pidou M., Soares A., Judd S. and Jefferson B. (2011a). Modelling the energy demands of aerobic and anaerobic membrane bioreactors for wastewater treatment. *Environmental Technology*, **32**, 921–932.

Martin-Garcia I., Monsalvo V., Pidou M., Le-Clech P., Judd S. J., McAdam E. J. and Jefferson B. (2011b). Impact of membrane configuration on fouling in anaerobic membrane bioreactors. *Journal of Membrane Science*, **382**, 41–49.

Mata T., Martins A. A. and Caetano N. S. (2010). Microalgae for biodiesel production and other applications: a review. *Renewable and Sustainable Energy Reviews*, **14**, 217–232.

McAdam E. J., Lu D., Eusebi A. L., Lester J. N., Jefferson B. and Cartmell E. (2011). Integrating anaerobic processes into wastewater treatment. *Water Science and Technology*, **63**(7), 1459–1466.

McLeod A., Jefferson B. and McAdam E. J. (2015). Biogas upgrading by chemical absorption using ammonia rich absorbents derived from wastewater. *Journal of Membrane Science*, **67**, 175–186.

Meng F., Chae S. R., Drews A., Kraume M., Shin H. S. and Yang F. (2009). Recent advances in membrane bioreactors (MBRs): membrane fouling and membrane material. *Water Research*, **43**, 1489–1512.

Meng G., Zhang H., Yang F. and Liu L. (2007). Characterization of cake layer in submerged membrane bioreactor. *Environmental Science and Technology*, **41**, 4065–4070.

Ozgun H. (2015). Anaerobic Membrane Bioreactors for Cost-Effective Municipal Water Reuse. PhD thesis, Delft University of Technology/Istanbul Technical University, Delft, Netherlands/Istanbul, Turkey.

Ozgun H., Dereli R. K., Ersahin M. E., Kinaci C., Spanjers H. and van Lier J. B. (2013a). A review of anaerobic membrane bioreactors for municipal wastewater treatment: integration options, limitations and expectations. *Separation and Purification Technology*, **118**, 89–104.

Ozgun H., Ersahin M. E., Tao Y., Spanjers H. and van Lier J. B. (2013b). Effect of upflow velocity on the effluent membrane fouling potential in membrane coupled upflow anaerobic sludge blanket reactors. *Bioresource Technology*, **147**, 285–292.

Ozgun H., Tao Y., Ersahin M. E., Zhou Z., Giménez J. B., Spanjers H. and van Lier J. B. (2015a). Impact of temperature on feed-flow characteristics and filtration performance of an upflow anaerobic sludge blanket coupled ultrafiltration membrane treating municipal wastewater. *Water Research*, **83**, 71–83.

Ozgun H., Giménez J. B., Ersahin M. E., Tao Y., Spanjers H. and van Lier J. B. (2015b). Impact of membrane addition for effluent extraction on the performance and sludge characteristics of upflow anaerobic sludge blanket reactors treating municipal wastewater. *Journal of Membrane Science*, **479**, 95–104.

Padmasiri S. I., Zhang J., Fitch M., Norddahl B., Morgenroth E. and Raskin L. (2007). Methanogenic population dynamics and performance of an anaerobic membrane bioreactor (AnMBR) treating swine manure under high shear conditions. *Water Research*, **41**, 134–144.

Park H., Choo K. H. and Lee C. H. (1999). Flux enhancement with powdered activated carbon addition in the membrane anaerobic bioreactor. *Separation Science and Technology*, **34**, 2781–2792.

Pelaz L., Gómez A., Garralón G., Letona A. and Fdz-Polanco M. (2017). Denitrification of the anaerobic membrane bioreactor (AnMBR) effluent with alternative electron donors in domestic wastewater treatment. *Bioresource Technology*, **243**, 1173–1179.

Pelaz L., Gómez A., Garralón G., Letona A. and Fdz-Polanco M. (2018). Recirculation of gas emissions to achieve advanced denitrification of the effluent from the anaerobic treatment of domestic wastewater. *Bioresource Technology*, **250**, 758–763.

Pretel R., Robles A., Ruano M. V., Seco A. and Ferrer J. (2013). Environmental impact of submerged anaerobic MBR (SAnMBR) technology used to treat urban wastewater at different temperatures. *Bioresource Technology*, **149**, 532–540.

Pretel R., Robles A., Ruano M. V., Seco A. and Ferrer J. (2014). The operating cost of an anaerobic membrane bioreactor (AnMBR) treating sulphate-rich urban wastewater. *Separation and Purification Technology*, **126**, 30–38.

Pretel R., Durán F., Robles A., Ruano M. V., Ribes J., Serralta J. and Ferrer J. (2015). Designing an AnMBR-based WWTP for energy recovery from urban wastewater: the role of primary settling and anaerobic digestion. *Separation and Purification Technology*, **156**, 132–139.

Pretel R., Robles A., Ruano M. V., Seco A. and Ferrer J. (2016). Economic and environmental sustainability of submerged anaerobic MBR-based (AnMBR-based) technology as compared to aerobic-based technologies for moderate-/high-loaded urban wastewater treatment. *Journal of Environmental Management*, **166**, 45–54.

Quek P. J., Yeap T. S. and Ng H. Y. (2017). Applicability of upflow anaerobic sludge blanket and dynamic membrane-coupled process for the treatment of municipal wastewater. *Environmental Biotechnology*, **101**, 6531–6540.

Robles A., Ruano M. V., García-Usach F. and Ferrer J. (2012a). Sub-critical filtration conditions of commercial hollow-fibre membranes in a submerged anaerobic MBR (HF-SAnMBR) system: the effect of gas sparging intensity. *Bioresource Technology*, **114**, 247–254.

Robles A., Ruano M. V., Ribes J. and Ferrer J. (2012b). Sub-critical long-term operation of industrial scale hollow-fibre membranes in a submerged anaerobic MBR (HF-SAnMBR) system. *Separation and Purification Technology*, **100**, 88–96.

Robles A., Ruano M. V., Ribes J. and Ferrer J. (2013a). Factors that affect the permeability of commercial hollow-fibre membranes in a submerged anaerobic MBR (HF-SAnMBR) system. *Water Research*, **47**, 1277–1288.

Robles A., Ruano M. V., Ribes J. and Ferrer J. (2013b). Performance of industrial scale hollow-fibre membranes in a submerged anaerobic MBR (HF-SAnMBR) system at mesophilic and psychrophilic conditions. *Separation and Purification Technology*, **104**, 290–296.

Robles A., Ruano M. V., Ribes J. and Ferrer J. (2013c). Advanced control system for optimal filtration in submerged anaerobic MBRs (SAnMBRs). *Journal of Membrane Science*, **430**, 330–341.

Robles A., Ruano M. V., Ribes J., Seco A. and Ferrer J. (2014). Model-based automatic tuning of a filtration control system for submerged anaerobic membrane bioreactors (AnMBR). *Journal of Membrane Science*, **465**, 14–26.

Robles A., Giménez J. B., Durán F., Jiménez E., Vazquez-Padín J., Rogalla F., Seco A., Ferrer J., Ribes J., Serralta J. (2017). Scaling-up of AnMBR technology: moving from wastewater to source of resources. In: IWA World Conference on Anaerobic Digestion, International Water Association (ed.), 15th edn, Beijing, China.

Robles A., Ruano M. V., Charfi A., Lesage G., Heran M., Harmand J., Seco A., Steyer J. P., Batstone D. J., Kim J. and Ferrer J. (2018). A review on anaerobic membrane bioreactors (AnMBRs) focused on modelling and control aspects. *Bioresource Technology*, **270**, 612–626.

Rodriguez-Gomez R. and Renman G. (2017). Phosphorous removal from UASB reactor effluent by reactive media filtration. *Environmental Technology*, **38**, 2024–2031.

Salazar-Pelaez M. L., Morgan-Sagastume J. M. and Noyola A. (2011). Influence of hydraulic retention time on UASB post-treatment with UF membranes. *Water Science and Technology*, **64**(11), 2299–2305.

Sánchez-Ramírez J. E., Seco A., Ferrer J., Bouzas A. and García-Usach F. (2015). Treatment of a submerged anaerobic membrane bioreactor (SAnMBR) effluent by an activated sludge system: the role of sulphide and thiosulphate in the process. *Journal of Environmental Management*, **147**, 213–218.

Saravanane R. and Sundararaman S. (2009). Effect of loading rate and HRT on the removal of cephalosporin and their intermediates during the operation of a membrane bioreactor treating pharmaceutical wastewater. *Environmental Technology*, **30**, 1017–1022.

Seghezzo L., Trupiano A. P., Liberal V., Todd P. G., Figueroa M. E., Gutierrez M. A., Wilches A. S., Iribarnegaray M., Guerra R. G., Arena A., Cuevas C. M., Zeeman G. and Lettinga G. (2003). Two-step upflow anaerobic sludge bed system for sewage treatment under subtropical conditions with posttreatment in waste stabilization ponds. *Applied Biochemistry and Biotechnology*, **109**, 167–180.

Seghezzo L., Cuevas C. M., Trupiano A. P., Guerra R. G., González S. M., Zeeman G. and Lettinga G. (2006). Stability and activity of anaerobic sludge from UASB reactors treating sewage in subtropical regions. *Water Science and Technology*, **54**(2), 223–229.

Seib M. D., Berg K. J. and Zitomer D. H. (2016). Low energy anaerobic membrane bioreactor for municipal wastewater treatment. *Journal of Membrane Science*, **514**, 450–457.

Shin C. and Bae J. (2018). Current status of the pilot-scale anaerobic membrane bioreactor treatments of domestic wastewaters: a critical review. *Bioresource Technology*, **247**, 1038–1046.

Shin C., McCarty P. L., Kim J. and Bae J. (2014). Pilot-scale temperate-climate treatment of domestic wastewater with a staged anaerobic fluidized membrane bioreactor (SAF-MBR). *Bioresource Technology*, **159**, 95–103.

Skouteris G., Hermosilla D., Lopez P., Negro C. and Blanco A. (2012). Anaerobic membrane bioreactors for wastewater treatment: a review. *Chemical Engineering Journal*, **198–199**, 138–148.

Smith A. L., Stadler L. B., Love N. G., Skerlos S. J. and Raskin L. (2012). Perspectives on anaerobic membrane bioreactor treatment of domestic wastewater: a critical review. *Bioresource Technology*, **122**, 149–159.

Smith A. L., Skerlos S. J. and Raskin L. (2013). Psychrophilic anaerobic membrane bioreactor treatment of domestic wastewater. *Water Research*, **47**, 1655–1665.

Smith A. L., Stadler L. B., Cao L., Love N. G., Raskin L. and Skerlos S. J. (2014). Navigating wastewater energy recovery 1 strategies: a life cycle comparison of anaerobic membrane bioreactor and conventional treatment systems with anaerobic digestion. *Environment Science Technology*, **48**, 5972–5981.

Tsolcha O. N., Tekerlekopoulou A. G., Akratos C. S., Aggelis G., Genitsaris S., Moustaka-Gouni M. and Vayenas D. V. (2017). Biotreatment of raisin and winery wastewaters and simultaneous biodiesel production using a Leptolyngbya-based microbial consortium. *Journal Cleaner Production*, **148**, 185–193.

van der Last, Lettinga G. (1992). Anaerobic treatment of domestic sewage under moderate climatic (Dutch). Conditions using upflow reactors at increased superficial velocities. *Water Science and Technology*, **25**(7), 167–178.

van Haandel A., Kato M. T., Cavalcanti P. F. F. and Florencio L. (2006). Anaerobic reactor design concepts for the treatment of domestic wastewater. *Reviews in Environmental Science and Bio/Technology*, **5**, 21–38.

van Lier J. B. (2008). High-rate anaerobic wastewater treatment: diversifying from end-of-the-pipe treatment to resource oriented techniques. *Water Science and Technology*, **57**(3–4), 1137–1148.

van Lier J. B., Mahmoud N. and Zeeman G. (2008). Anaerobic biological wastewater treatment. In: Biological Wastewater Treatment: Principles, Modeling and Design, M. Henze, M. C. M. van Loosdrecht, G. A. Ekama and D. Brdjamovic (eds.), IWA Publishing, London, UK.

van Lier J. B., Vashi A., van der Lubbe J. and Heffernan B. (2010). Anaerobic sewage treatment using UASB reactors: engineering and operational aspects. In: Environmental Anaerobic Technology: Applications and New Developments, H. H. P. Fang (ed.), Imperial College Press, London, United Kingdom, pp. 59–89.

van Lier J. B., van der Zee F. P., Frijters C. T. M. J. and Ersahin M. E. (2015). Celebrating 40 years anaerobic sludge bed reactors for industrial wastewater treatment. *Reviews in Environmental Science and Bio/technology*, **14**, 681–702.

van Linden N., Spanjers H. and van Lier J. B. (2018). Efficient ammonium recovery by electrodialysis for power production. Nutrient Removal and Recovery Conference 2018, Brisbane, Australia.

Viruela A., Murgui M., Gómez-Gil T., Durán F., Robles Á, Ruano M. V., Ferrer J. and Seco A. (2016). Water resource recovery by means of microalgae cultivation in outdoor photobioreactors using the effluent from an anaerobic membrane bioreactor fed with pre-treated sewage. *Bioresource Technology*, **218**, 447–454.

Viruela A., Robles Á, Durán F., Ruano M. V., Barat R., Ferrer J. and Seco A. (2018). Performance of an outdoor membrane photobioreactor for resource recovery from anaerobically treated sewage. *Journal of Cleaner Production*, **178**, 665–674.

Vyrides I. and Stuckey D. C. (2009). Saline sewage treatment using a submerged anaerobic membrane reactor (SAMBR): effects of activated carbon addition and biogas-sparging time. *Water Research*, **43**, 933–942.

Waeger F., Delhaye T. and Fuchs W. (2010). The use of ceramic microfiltration and ultrafiltration membranes for particle removal from anaerobic digester effluents. *Separation and Purification Technology*, **73**, 271–278.

Wang K. M., Cingolani D., Eusebi A. L., Soares A., Jefferson B. and McAdam E. J. (2018a). Identification of gas sparging regimes for granular anaerobic membrane bioreactor to enable energy neutral municipal wastewater treatment. *Journal of Membrane Science*, **555**, 125–133.

Wang K. M., Jefferson B., Soares A. and McAdam E. J. (2018b). Sustaining membrane permeability during unsteady state operation of anaerobic membrane bioreactor for municipal sewage treatment following peak flow. *Journal of Membrane Science*, **654**, 289–297.

Wang X., Wang C., Tang C. Y., Hu T., Li X. and Ren Y. (2017). Development of a novel anaerobic membrane bioreactor simultaneously integrating microfiltration and forward osmosis membranes for low strength wastewater treatment. *Journal of Membrane Science*, **527**, 1–7.

Watanabe R., Nie Y., Wakahara S., Komori D. and Li Y. Y. (2017). Investigation on the response of anaerobic membrane bioreactor to temperature decrease from 25°C to 10°C in sewage treatment. *Bioresource Technology*, **243**, 747–754.

Wei C., Harb M., Amy G., Hong P. and Leiknes T. (2014). Sustainable organic loading rate and energy recovery potential of mesophilic anaerobic membrane bioreactor for municipal wastewater treatment. *Bioresource Technology*, **166**, 326–334.

WssTP The European Platform (2016). Water Vision 2030 – The Value of Water: Towards a future proof model for a European water-smart society. http://wsstp.eu/wp-content/uploads/sites/102/2017/11/WssTP-Water-Vision_ english_2edition_online.pdf (accessed: 18 November 2018)

Wu B., An Y., Li Y. and Wong F. S. (2009). Effect of adsorption/coagulation on membrane fouling in microfiltration process post-treating anaerobic digestion effluent. *Desalination*, **242**, 183–192.

Xiao Y., Yaohari H., De Araujo C., Sze C. C. and Stuckey D. C. (2017). Removal of selected pharmaceuticals in an anaerobic membrane bioreactor (AnMBR) with/without powdered activated carbon (PAC). *Chemical Engineering Journal*, **321**, 335–345.

Xie K., Lin H. J., Mahendran B., Bagley D. M., Leung K. T., Liss S. N. and Liao B. Q. (2010). Performance and fouling characteristics of a submerged anaerobic membrane bioreactor for evaporator condensate treatment. *Environmental Technology*, **31**, 511–521.

Zhang X., Wang Z., Wu Z., Lu F., Tong J. and Zang L. (2010). Formation of dynamic membrane in an anaerobic membrane bioreactor for municipal wastewater treatment. *Chemical Engineering Journal*, **165**, 175–183.

Zhang X., Wang Z., Wu Z., Wei T., Lu F., Tong J. and Mai S. (2011). Membrane fouling in an anaerobic dynamic membrane bioreactor (andmbr) for municipal wastewater treatment: characteristics of membrane foulants and bulk sludge. *Process Biochemistry*, **46**, 1538–1544.

Zhi-Guo M., Feng-lin Y. and Xing-wen Z. (2005). MBR Focus: do non-wovens offer a cheaper option? *Filtration and Separation*, **42**, 28–30.

Chapter 12

Closing cycles in anaerobic-based sewage treatment systems

Ana Soares[1], César R. Mota Filho[2], Fernando Fdz-Polanco[3], Xiangjun Huang[1], Fabiana Passos[2], Thiago Bressani-Ribeiro[2] and Carlos A. L. Chernicharo[2]

[1]Cranfield Water Science Institute, Vincent Building, Cranfield University, Cranfield, Beds, MK43 0AL, UK
[2]Departament of Sanitary and Environmental Engineering (DESA), School of Engineering, Universidade Federal de Minas Gerais (UFMG), Av. Antônio Carlos 6.627, Belo Horizonte/MG, 31270-901 – Brazil
[3]Chemical Engineering and Environmental Technology Department, University of Valladolid, EII (sede Mergelina), 47011 Valladolid, Spain
Corresponding author: a.soares@cranfield.ac.uk

12.1 INTRODUCTION

During the last two decades, anaerobic-based sewage treatment has evolved considerably leading to a significant increase in the number of full-scale treatment facilities, as well as the development of new reactor configurations and post-treatment solutions. Anaerobic processes, as secondary treatment, are at the centre of a paradigm shift in sewage treatment, moving from waste treatment facilities to resource recovery factories. The ability to treat sewage coupled with recovery of energy and other resources is currently being explored worldwide. Nonetheless, in most cases, solutions have been focused on high-tech processes for resource recovery in centralised sewage treatment plants (STPs). Generally, little effort has been directed to small and medium STPs in developing countries (Bressani-Ribeiro *et al.*, 2018). The context in which such plants are inserted is usually limited by capacity and skills for operation and maintenance, hence mechanised and complex technologies should be avoided.

Editor(s): Carlos Augusto de Lemos Chernicharo and Thiago Bressani-Ribeiro
doi: 10.2166/9781780409238_0367

Additionally, investment is also a major barrier and resource recovery can significantly increase capital (CAPEX) and operational expenditure (OPEX) of STPs.

Particularly in developing countries, most with average temperatures above 25°C and low-cost workforce for maintenance and operation, anaerobic sewage treatment is a sustainable solution for closing cycles and producing enhanced-value sewage treatment. In these systems, biosolids and gas could be recovered to generate valuable by-products, such as soil conditioners and bioenergy. However, in most cases such products are disposed in landfills (e.g., sludge and scum) or flared to the atmosphere (e.g., biogas). This represents both a negative environmental impact and loss of possible revenue for the water industry and nearby communities. For instance, a recent study comparing the carbon footprint of different scenarios on biogas management showed how a positive balance was achieved when biogas was used for sludge sanitization and at household level for cooking or heating (16.5 kg CO_2 PE^{-1} $year^{-1}$) (Bressani-Ribeiro *et al.*, 2018). In contrast, typical scenarios in which the biogas produced in the anaerobic reactor was flared and excess sludge was not recovered, the balance of the carbon footprint was negative (−54.4 kg CO_2 PE^{-1} $year^{-1}$) (Bressani-Ribeiro *et al.*, 2018). The use of biogas and sludge as added-value by-products can avoid CO_2 emissions that take place when biogas is flared or when sludge is transported to landfills (Chernicharo *et al.*, 2017).

Overall, the technology used for resource recovery will depend on each particular case. There are many important aspects to be considered, such as STP scale, existing infrastructure, legislation, land availability and agricultural fields, market demands, public perception and acceptance, among others. In this manner, as stated in Chapter 10 regarding post-treatment options for UASB reactor effluents, there is also no single ideal resource recovery solution that could be applied in all cases, nor models to define a best arrangement. In reality, recovering resources and closing cycles in STPs will demand a reasonable balance between economic benefits, technical aspects and social acceptance. In developing countries, in which most rural communities still lack sanitation, aspects such as implementation and maintenance costs, operational simplicity and process robustness are usually given more importance than aspects such as process efficiency or reliability. Understanding local priorities is critical for evaluating and identifying appropriate technologies for sustainable STPs in particular contexts.

In this manner, it is crucial to consider new STPs as part of a socio-economic context, by minimising energy and external inputs, reducing or eliminating wastes and recovering resources when technically and economically feasible. In this context, STPs may be proposed as resource recovery units, converting sewage into energy and high-valuable products by using sophisticated physical separation and biotechnological processes in the developed world. While, on the other hand, small decentralised anaerobic-based STPs in Latin America, Africa or Southeast Asia, should integrate the sewage treatment process into the vicinity and local economic activities. Resource recovery in STPs could therefore serve as powerful leverage to increase the historically lagging coverage of sanitation infrastructure in the developing world.

As expressed in Chapters 4, 7 and 10, UASB reactors without post-treatment allow COD removal efficiencies in the range from 60 to 75% in warm climate regions. As also stated in Chapter 10, nitrogen and phosphorus removal in UASB reactors are negligible. Due to organic solids hydrolysis, there is an increase in ammonia concentrations by 20–40% during anaerobic sewage treatment, whereas typical phosphorus removals range from 0 to 5%, due to biomass assimilation. Therefore, effluents from anaerobic STPs offer opportunities for water reuse and nutrients recovery. In addition, produced biogas can be recovery for energy purposes and biosolids can be applied to land. This Chapter focuses on exploring resource recovery options for closing cycles in anaerobic-based sewage treatment systems.

12.2 TREATED AND UNTREATED SEWAGE FOR WATER REUSE

It is estimated that, globally, more than 80% of all wastewater is discharged without treatment. High-income countries treat approximately 70% of the wastewater they generate (municipal and industrial). Upper middle-income, lower middle-income and low-income countries treat only 38%, 28% and 8% of wastewater they generate, respectively (Sato *et al.*, 2013). The widespread discharge of untreated wastewater has contributed significantly to exacerbating problems related to water scarcity. That is due to the fact that water availability is intrinsically linked to water quality, as the pollution of water sources may prohibit different types of water uses. Two thirds of the world's population currently live in areas that experience water scarcity for at least one month a year, and approximately 500 million people live in areas where water consumption exceeds the locally renewable water resources by a factor of two. Problems related to water scarcity are predicted to worsen over the coming decades, as also addressed in Chapter 11. In addition to the agricultural sector, which is responsible for 70% of water abstractions worldwide, large increases in water demand are predicted for industry and energy production. Accelerated urbanization and the expansion of municipal water supply and sanitation systems also contribute to rising water demand (UN Water Report, 2017).

If adequately treated and safely applied, sewage can enhance food security and have significant health benefits, including improved nutrition. The use of sewage for agricultural irrigation is a common pattern in countries of the Middle East and North Africa, Australia and the Mediterranean, as well as in China, Mexico and the USA. The practice has been most successful in urban and peri-urban areas, where wastewater is easily available, generally free of charge, and where there is a market for agricultural products (UN Report, 2017). There are other uses for sewage besides agriculture, including landscape irrigation, groundwater recharge, industrial water reuse etc. Figure 12.1 shows common types of water reuse, following tertiary sewage treatment. It describes only a portion of the total reuse practiced worldwide.

Currently, there is limited reliable data on the amount of water reuse and raw sewage use in the agricultural sector. The limited evidence available suggests that the area of land irrigated with untreated

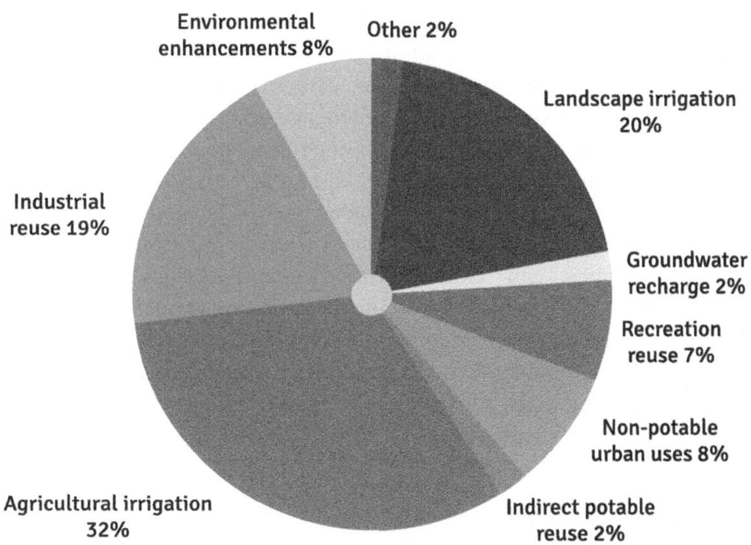

Figure 12.1 Global water reuse after tertiary sewage treatment: market share by application. (Source: Adapted from Lautze *et al.* 2014).

sewage is more than 10 times as great as the area irrigated with reclaimed water (Scott *et al.*, 2010). Rough estimations suggest that, globally, approximately 20 million ha of agricultural land is irrigated with mostly untreated sewage (Lautze *et al.*, 2014). As such, the proportion of sewage used in agriculture may be far greater than that, as only documented cases were reported. Growth in the global water reuse sector is expected to migrate from being dominated by agriculture towards higher-value applications, mostly in municipal applications, such as potable, industrial, and landscape irrigation reuse (Lautze *et al.*, 2014). China, the United States, Spain, Mexico, India, Australia, Israel, Kuwait, Japan, and Singapore lead the world in total installed advanced water reuse capacity to date (GWI, 2009).

Although the use of sewage is an ancient habit, it has not always been properly practiced. The risks of untreated sewage use in agriculture are extensive, ranging from changes to physical-chemical properties of soils to impacts on human health. In unfavourable economic conditions, the search for alternative irrigation sources, such as raw or inadequately treated sewage may result in avoidable risks (Jaramillo & Restrepo, 2017). The concentration levels and types of pathogens and chemical substances present in sewage vary by region, according to the sanitary and socio-economic conditions of a particular community (Gerba & Rose, 2003). The concentration of viruses, protozoan parasites and helminths in wastewater can be 10–1000 times higher in developing countries than in developed ones (Jiménez *et al.*, 2010).

The World Health Organization (WHO), in order to tackle the health and environmental risks caused by microbiological and chemical contaminants potentially present in water, recommends the implementation of a risk management framework. A risk management approach provides the conceptual framework for the *WHO Guidelines for Drinking Water Quality* (WHO, 2011), and the *Guidelines for the Safe Use of Wastewater, Excreta and Greywater* (WHO, 2006).

Sewage treatment technologies play a key role in making wastewater safe for reuse. There is a wide range of sewage treatment options for safe water, nutrient recovery, and irrigation with particular relevance for low-income countries. Several experts have summarised appropriate treatment options, including Mara (2004), Laugesen *et al.* (2010), von Sperling and Chernicharo (2005), and Libhaber and Orozco-Jaramillo 2013. This was addressed in Chapter 2 and in Section 12.5 of this Chapter, which highlights some of the available anaerobic-based sewage treatment technologies and flowsheets that can be used in the context of a circular economy. Such options are increasingly being used in developing countries and can be considered sustainable solutions for sewage treatment, as they are capable of producing effluents, sludge and other resources that can be suitable for reuse in agriculture and/or generate energy, at low cost, while requiring low-maintenance and relatively simplified operation.

The type of use intended for the treated sewage will determine the technologies that should be applied. For instance, if the objective is to use treated sewage for aquifer recharge, treatment technologies capable of removing organic matter, nutrients, solids and pathogens should be applied to avoid contamination of groundwater. Whereas if the objective is to use treated sewage for crop irrigation, technologies that remove excess solids (as solids can clog irrigation apparatus) and pathogens (e.g. anaerobic membrane-based alternatives – see Chapter 11), but retain nutrients in the treated effluent should be prefered. As such, anaerobic-based sewage treatment systems are uniquely capable of producing safe (if coupled with an operational unit for disinfection) water for reuse in agriculture as their treated effluent retain part of the organic matter and the most part of valuable macro- (nitrogen and phosphorus) and micro-nutrients (Ca, Mg, B, Mg, Fe, Mn or Zn).

12.3 NUTRIENT RECOVERY

As previously discussed in this Chapter, nitrogen and phosphorus removal in UASB reactors is negligible. Therefore, these nutrients may be recovered and reused, adding value to the sewage treatment process and minimising the environmental burden related to mineral extraction. There are many technological mechanisms for nutrient recovery in anaerobic-based STPs which can be applied to solid (Ponsa *et al.*, 2017) or liquid phases. In addition to the discussion previously presented in section 11.3.7 of Chapter 11, this section aims at further presenting and discussing examples of some technological approaches for nutrient recovery. Firstly, recovery from digested sludge will be described for understanding the reuse of nutrients from the solid phase after anaerobic digestion. Afterwards, microalgae-based treatment ponds are shown as alternatives for recovering nutrients from effluents at the post-treatment stage. Therefore, microalgae assimilate nutrients and, following, biomass can be processed for recovering them. Finally, solutions for nutrient capture from the liquid phase are presented. In this instance, fertirrigation and ion-exchange (IEX) processes are commented as possibilities in rural and urban contexts, respectively.

12.3.1 Recovery of nutrients from sludge

Sludge composition varies according to several factors, including sewage origin (e.g. only domestic effluents, industrial contributions etc.), health of the population (e.g. infestation by helminth eggs), and treatment process. Therefore, it is important to perform a full characterisation of the sludge generated in each process in order to evaluate possible management strategies. Table 12.1 shows the composition of sludge from UASB reactors treating sewage reported in the literature.

Table 12.1 Composition of sludge from UASB reactors treating sewage reported in the literature.

Parameter	Value	References
Organic carbon (dag kg^{-1})	20.1	Andreoli *et al.* (2001)
Total nitrogen (dag kg^{-1})	2.2	Andreoli *et al.* (2001)
Total phosphorus (dag kg^{-1})	0.67	Andreoli *et al.* (2001)
Total potassium (dag kg^{-1})	0.95	Andreoli *et al.* (2001)
Helminth eggs (eggs gST^{-1})	0.9–48.0	Fernandes and Silva (1999); Passamani *et al.* (2002)
Viable helminth eggs (eggs gST^{-1})	0.4–2.67	Fernandes and Silva (1999)
Thermotolerant coliforms (MPN gST^{-1})	10^5–10^7	Andreoli *et al.* (2001); Passamani *et al.* (2002)

Source: Adapted from Bortolini (2018).

Amongst the several possible routes for final disposal of sludge (e.g. landfills, ocean discharge, incineration, recovery of degraded areas and the like), its use in agriculture fits well into the circular economy concept, as sludge returns to the productive system as resource, instead of being handled as waste. This reduces pressure for the exploitation of natural resources and minimises waste generation. For example, when phosphorus in the sewage sludge is returned to the soil, it minimises pressure for the exploitation of phosphate rocks in the production of phosphate fertilizers. Nutrients present in sewage originate mainly from undigested or partially digested food. Crops that are used for food production incorporate such nutrients from soil.

Usually, in low-income countries small amounts of sewage sludge are generated, as there is limited sanitation infrastructure. Middle-income countries usually dispose sewage sludge in landfills, practice incineration or land disposal. High-income countries have increasingly used sewage sludge to generate biosolids that can be used in agriculture or otherwise (UN-HABITAT, 2008). Table 12.2 summarises the

Table 12.2 Sewage sludge destination and regulatory aspect of sludge disposal in selected countries.

Destination	Region/Country							
	Japan	**USA**	**Australia**	**New Zealand**	**EU27**	**United Kingdom**	**Germany**	**Greece**
Ocean disposal	Banned (2004)	Banned (1982)	Being banned (1% in 2010)	Discouraged (10% in 2010)	Banned (1999)	Banned (1998)	Banned	Banned (2000)
Landfill	Permitted only for incineration ash	Discouraged	Abandoned in most states	Main method of disposal	Restricted and about to be banned	Discouraged (<1% em 2010)	Applicable only if organic matter content <3%	Discouraged
Agriculture use	Sparsely used	Main alternative	Main alternative	Encouraged but still scarce use	Main alternative	Main alternative	Significant but declining	Encouraged and on the rise
Other forms of land application	None	Biosolids (compost) marketed for landscaping	Forestry, reclamation of degraded areas, mining	Forestry, reclamation of degraded areas, mining	Biosolids (compost) marketed for landscaping	Forestry, reclamation of degraded areas	Biosolids (compost) marketed for landscaping	Biosolids fertilizer
Thermal treatment	Main use (>70%)	Moderate use	Not preferred, limited application	Not applied	Increasing	Not preferred	Main alternative (almost >50%)	Recently the main alternative (>50%)
Possible upcoming changes in legislation	Inclusion of phosphorus recovery and energy recovery	Restricted landfill disposal rules. Promotion of thermal methods and energy generation	Minimization of landfill and ocean disposal. Increase of agricultural use and energy use	Minimization of landfill and ocean disposal. Increase of agricultural use and energy use	Restriction of landfill disposal. Promotion of energy generation	Phosphorus recovery from the liquid effluent	Restricted use in agriculture. Increased phosphorus recovery and energy generation	Minimization of landfill disposal. Increase of agricultural use and energy generation

Source: Bortolini (2018) – adapted from Christodoulou and Stamatelatou (2016).

main routes used for sludge disposal by different countries, as well as possible upcoming changes in regulation on sewage sludge disposal. Agricultural use of sewage sludge is the main route adopted by the USA, Australia, EU and United Kingdom. In New Zealand and Greece, agricultural use of sewage sludge has gained ground. Regulation in several countries has increasingly restricted the disposal of sewage sludge in landfills and promoted sewage sludge for use in agriculture, phosphorus recovery or as biomass for energy generation.

12.3.2 Microalgal biomass recovery

Another option for nutrient recovery in anaerobic-based STPs is by producing a high-value biomass through an aerobic biological post-treatment. An example of this approach is high rate algal ponds (HRAPs), in which not only effluent is treated, but microalgal biomass is recovered. HRAPs consist of an open pond with closed-loop flow channels with a shallow depth (~0.3–0.4 m) and slowly rotating paddlewheels. Mechanical stirring keeps the biomass in suspension and contributes to a decrease in light photosynthesis limitation, subsequently maximising microalgae activity (Díez-Monteiro *et al.*, 2018). HRAPs have been studied since the 1950s as a cost-effective alternative to conventional activated sludge systems, due to their low energy demand and simplicity of operation (Oswald & Gotaas, 1957).

In HRAPs two main mechanisms are involved: (i) direct or indirect transformation of pollutants by microalgae (e.g. nutrients assimilation and precipitation); and (ii) enhancement of bacterial biodegradation by oxygen generated through microalgae photosynthesis (Rawat *et al.*, 2013). Both mechanisms take place simultaneously through the so-called "algae-bacteria symbiosis" (Oswald & Gotaas, 1957). Since oxygen needed for organic matter removal is provided by microalgae photosynthesis, there is no need for mechanical aeration, as in conventional activated sludge processes. This is a major advantage, since aeration is responsible for 60 to 80% of total energy demand in activated sludge (Chachuat *et al.*, 2005; Longo *et al.*, 2016). However, maybe the most important benefit of microalgae-based systems is that generated biomass can be recovered and valorised for different purposes such as animal feeding, biomolecules extraction and biofuel production. In recent years, bioenergy generation through microalgae has been intensively studied and is nowadays a promising topic. However, several processes for biomass production and harvesting must still be optimised for cost-effective full-scale applications (Pittman *et al.*, 2011).

Among downstream options for microalgal biomass, anaerobic digestion for biogas production seems to currently be the most viable process in developing countries. Most ongoing research on microalgae production for biofuel purposes in the developed world has been focused on biodiesel. However, full-scale biodiesel production still has many limitations; for instance, the necessity of energy-consuming technologies for biomass harvesting and processing. As an alternative, biogas generation has been identified as a more energy-favorable process (Wiley *et al.*, 2011). In this case, wet biomass can be used and, therefore, no extensive drying methods are required.

The energy input for operating anaerobic reactors is very low, meaning that net energy production may be positive. For instance, a recent study estimated the energy balance of a hypothetical 10,000 population equivalent STP combining HRAPs and anaerobic co-digestion with primary sludge. This work showed how the system was able to keep the reactor at mesophilic conditions during the whole year when biogas produced was converted into heat using a boiler. The net heat production was further increased by 30–40% after microalgae thermal pretreatment to improve biomass biodegradability and increase biogas production. In this case, it reached from 2.4 to 5.3 net energy ratio during the experimental year (Passos *et al.*, 2017).

Overall, biomass pre-treatment methods aim to increase solubilisation of organic matter and, therefore, make organic compounds more readily available to anaerobic digestion, which would ultimately increase the methanogenesis rates and methane yield (Passos *et al.*, 2014). Particularly, the main reason why microalgae have slow and/or low biodegradability is due to the nature of their cell wall structure and composition. Most species have a complex cell wall composed of recalcitrant components, especially those grown in open ponds treating wastewater (Park *et al.*, 2011). Nonetheless, the characteristics of these cell walls may vary depending on the strain and environmental/operational conditions. A recent study found that, although proteins, lipids and a considerable amount of carbohydrates were present in the cell walls of microalgae species, microalgae resistance was not correlated to the presence of a unique monomer (Montingelli *et al.*, 2015). It is hypothesised that the cross-link of several compounds into a complex network of building layers around the cell could eventually work as a barrier to anaerobic microbial community (Klassen *et al.*, 2016). Such pre-treatment techniques may be classified into four main categories: mechanical, thermal, chemical and biological methods. These methods are based on different mechanisms and, therefore, support different disruption efficiencies. Most recent reviews in microalgae pre-treatment concluded that thermal pretreatment is the best method, by combining the highest methane improvement and the lowest energy input (Jankowska *et al.*, 2017; Passos *et al.*, 2014; Rodriguez *et al.*, 2015).

In order to employ HRAP in anaerobic-based STP, a possible system configuration could be as that shown in Figure 12.2. Following effluent treatment in the HRAPs, recovered biomass would be co-digested with sewage in the anaerobic reactor, the digested sludge reused, and the produced biogas converted into energy. Other possibilities may include anaerobic co-digestion of microalgal biomass in a continuous stirred tank reactor (CSTR) with agricultural and/or animal waste, since it has a higher solid content compared to sludge. Currently, there are very few full-scale STPs using HRAP. An example is the Aqualia STP in Cadiz municipality (Spain), where 100 tons per hectare per year of microalgae are recovered after treating 100 cubic metres of sewage. Subsequently, harvested biomass is anaerobically digested, biogas is upgraded and used as biofuel, which is enough for approximately 40 cars running 18,000 kilometres each (All-gas, 2017; De Godos *et al.*, 2013).

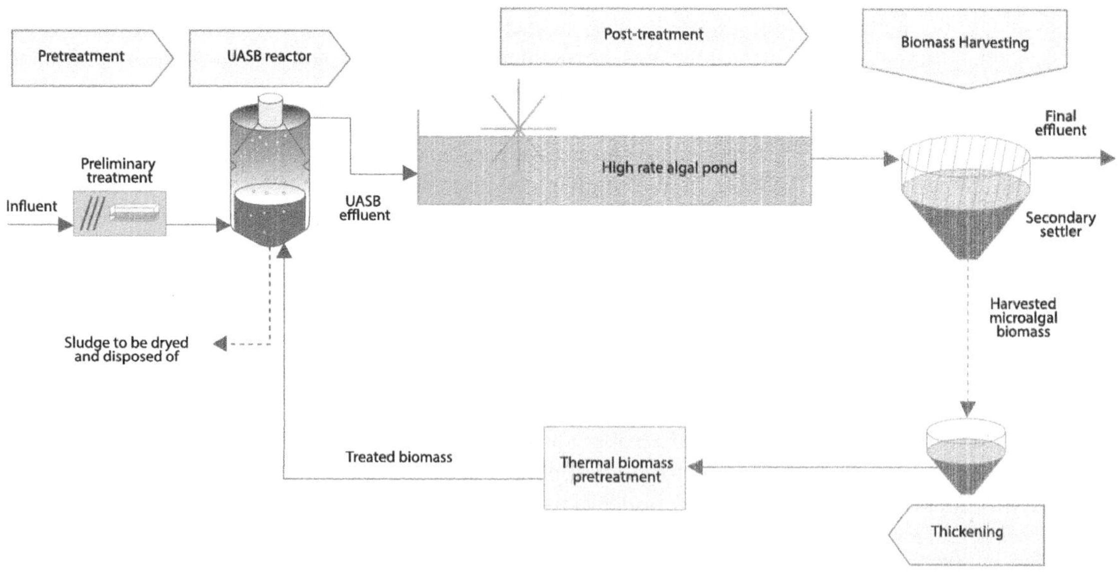

Figure 12.2 Sustainable STP based on UASB reactor and HRAP.

12.3.3 Fertirrigation

The nutrient content in UASB effluents can be attractive for agricultural irrigation, in a process called fertirrigation. In this case, the effluent provides water and nutrients to the soil and plants, which is important especially for countries with a scarce amount of water or frequent droughts. The potential of soil to degrade sewage is even greater if compared to water bodies, since the soil behaves as a filter besides promoting biodegradation (Bertoncini, 2008).

A recent study used life cycle assessment (LCA) for comparing different strategies in STPs and showed how the application of the effluent for irrigation and fertilization of willow-cropped land led to non-negligible energy savings and contributed to renewable energy generation. In this case, fertirrigation decreased the impact in freshwater eutrophication (Freshwater Eutrophication Potential-FEP) by 60% when the effluent was reused instead of being discharged in water bodies. Nonetheless, the amount of sewage recycled into the land would be approximately 50% of the total annual volume produced in the STP. The authors emphasised that this strategy would only be feasible if agricultural land is available at a short distance from the STP and that the plant design considers this option since the very beginning (Buonocore *et al.*, 2016). Another detailed study on LCA comparing 18 different technologies for recovering phosphorus indicated that recycling sewage for land application showed positive impacts in the environment, such as reduction in eutrophication potential by reducing phosphate rock mining, mitigation of toxic effects of heavy metals input from commercial fertilizers, among others (Amann *et al.*, 2018). The authors suggested that recovering phosphorus from the liquid phase had mostly positive or comparable little impacts on emissions and energy demand in respect to sewage sludge. Nonetheless, there is still a low recovery potential that contradicts the demand for efficient recycling rates (Amann *et al.*, 2018).

In anaerobic-based STPs in developing countries, fertirrigation may be an important approach for closing cycles. Generally, there are a great number of decentralised STPs in the rural area, where agricultural and/or pasture land are present nearby. In Brazil, few studies have been conducted with sewage recycling in agricultural fields with most research focusing on reuse of vinasse from sugarcane plantations. A research on UASB reactor followed by post-treatment with anaerobic filter, planted wetlands, non-planted wetlands and polishing ponds showed that for all cases effluent had suitable nutrient concentrations for most crops in the Brazilian northeast region. However, concentration of faecal indicators was lower than recommended by the World Health Organization (WHO) for unrestricted irrigation only for polishing ponds (Sousa *et al.*, 2006).

There are still political, technical and social obstacles for using treated sewage for agricultural purposes, especially in developing countries. There are often very restrictive laws and lack of investment on studies that subsidise better reuse regulations (Bertoncini, 2008).

12.3.4 Ion exchange (IEX) processes

Ion exchange (IEX) processes can be an alternative option for nutrient recovery from the soluble phase, where there is full control of effluent quality and the nutrients are selectively removed and concentrated in the regenerant. Nevertheless, IEX has been sparsely applied in STPs for nutrient removal. The main reason is that frequent regenerations with new chemical solution entail high costs for disposal of spent regenerant (Martin *et al.*, 2009; Miladinovic & Weatherley, 2008). Recent studies have indicated that a promising solution is to clean up and reuse the spent regenerant and recover the captured nutrients at the same time. It was demonstrated that ammonium-rich spent regenerant can be cleaned up through hollow fibre membrane contactor (HFMC). By adding sulphuric acid, ammonium can be recovered in the form of ammonium sulphate (Canellas, 2018). Phosphate-rich spent regenerant can be cleaned up through a filter after adding hydrated lime, and phosphorous can be recovered in the form of hydroxyapatite (Muhammad, 2018). The media used in the studies has high selectivity and capacity, for instance, Mesolite for ammonia adsorption up to 4.6 mg g^{-1} in a hybrid ion exchange resin (HAIX) with ferric oxide nanoparticles embedded in the base structure for phosphorus adsorption up to 4.9–6.2 mg g^{-1} with an empty bed contact time (EBCT) of 5 minutes (Canellas, 2018; Gonzales, 2012; Muhammad, 2018; Williams & Mayer, 2013).

12.4 BIOGAS USE

Although the fundamental components of biogas are CH_4 and CO_2, in practice, its composition is very variable and depends on the sewage composition. In conventional sewage sludge anaerobic digestion, the standard biogas composition is 65% CH_4, 34% CO_2, 1% N_2, <0.5% O_2, up to 1,500 ppm H_2S and different compounds such as siloxanes and volatile organic compounds (VOC's) are also present in the gaseous stream. On the other side, in UASB-based sewage treatment plants (mainstream treatment) the biogas composition is somewhat different from sludge digesters, more specifically in terms of CH_4, CO_2, N_2 and H_2S concentrations, which are usually in the range of 70–85%, 5–15%, 2–25% and 233–5,000 ppm, respectively, as discussed in Chapter 8. The major impurities present in the biogas, such as moisture, siloxanes and particularly H_2S, must be removed according to its final use, as also addressed in Chapter 8.

This section aims to present complementary perspectives to those discussed in Chapter 8, considering an expanded view of the energy recovery possibilities from biogas. Overall technological arrangements for biogas use are addressed considering the dissolved methane recovery from the liquid phase.

12.4.1 Energy recovery, odour control, and carbon neutrality in anaerobic-based STPs

The potential application and end use of the biogas depends on the biogas flow generated, which is primarily a direct function of the scale of the STP. However, a series of other factors can significantly affect the amount of biogas that is effectively collected and made available for the proposed end use. In order to analyse in a systematic way the possible uses of biogas and the different technologies for its exploitation, it is therefore necessary to take into account the following aspects:

- Methane is lost solubilised in the effluent (30 to 40% according to Souza *et al.*, 2011 – see further in Chapters 8 and 9). If methane is to be recovered, different degassing technologies provide different methane yields (see Chapter 9).
- Sewage dilution due to the presence of rain water can significantly reduce biogas production during and immediately after storm events (see Chapter 8).
- Major biogas losses can occur in badly designed, constructed and operated reactors (see Chapter 8).
- Hydrogen sulphide must be removed in any situation (see Chapter 8).

Furthermore, it is important to understand that sulphide and methane can simultaneously occur in different points of the anaerobic reactor (in the biogas, in the waste gas and also dissolved in the effluent). Therefore, there is a need for considering gaseous management alternatives that can simultaneously tackle *energy recovery, odour control and carbon neutrality* in the STP. As presented in Chapter 9, different management alternatives can be considered for the abatement of sulphide and methane dissolved in the anaerobic effluent, as well as for the abatement of the diffuse emissions from the settler compartment of UASB reactors. Such alternatives are depicted in Figure 12.3, considering the following conditions (Chernicharo *et al.*, 2017):

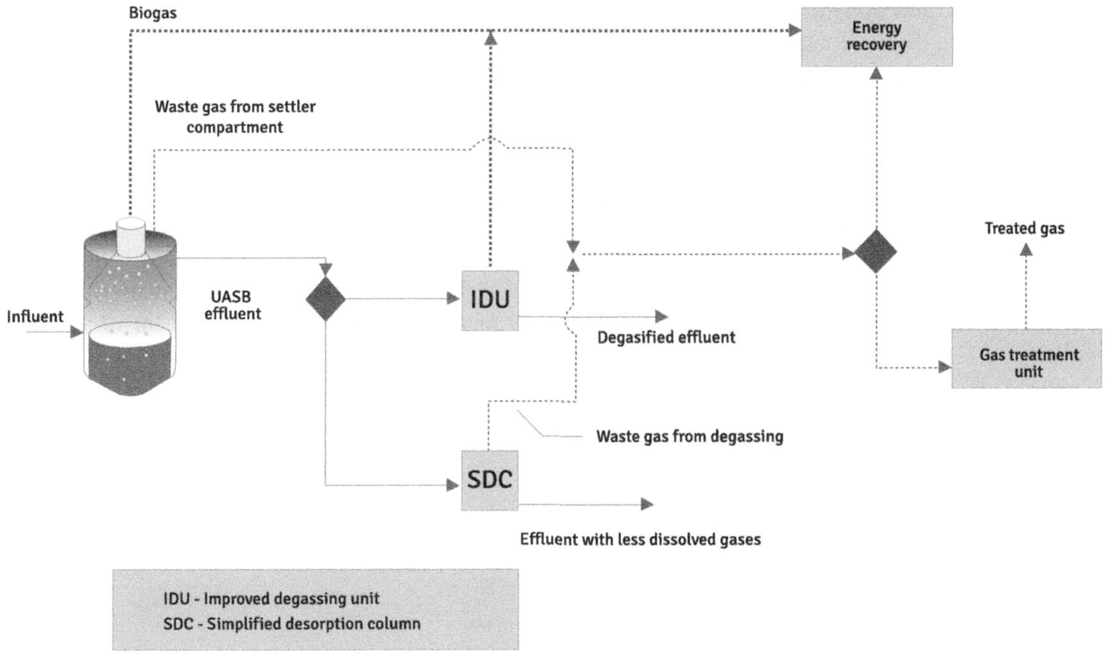

Figure 12.3 Schematic representation of possible management alternatives for gaseous emissions in UASB-based sewage treatment plants.

- The use of UASB reactors with enclosed settler compartments, targeting downstream management of the emissions.
- The use of simplified desorption columns to remove dissolved sulphide and methane, followed by their oxidation in biofilters.
- The use of improved degassing units (such as membrane contactors) to recover dissolved CH_4 for energy purposes, followed by the oxidation of sulphide.

For the purpose of this Chapter, emphasis is given to the energetic management approach, although some aspects regarding the reduction of equivalent emissions of CO_2 ($CO_{2equiv.}$) are also pointed out. Most recently, an environmental tool was developed specifically for improving the management of gaseous emissions in UASB-based STPs, so-called "Sulphide and Carbon Emission Avoidance and Energy Recovery in STPs" (Chernicharo *et al.*, 2017). The methodology considers different scenarios for the management of sludge, biogas and gaseous emissions (in particular sulphide and methane). The tool comprises simple alternatives for the use of biogas (e.g. for excess sludge sanitization and household usages) for small STPs (<10,000 PE) and, other biogas uses (e.g. sludge dryers, boilers or CHP engines for electricity generation and heat recovery) for medium- (>10,000 PE; <100,000 PE) and large-scale (>100,000 PE) STPs. To this aim, authors performed an extensive literature review on the main advantages and disadvantages of treatment technologies for gaseous emissions in anaerobic reactors treating domestic sewage. Afterwards, the most suitable technological arrangements for sulphide and methane diffuse emissions management were consolidated. The tool calculates the corresponding amounts of sulphide and methane in three different emission compartments (internal and external surface of the GLS separator, and the degassing unit). Afterwards, the total amount of methane supposed to be emitted into the atmosphere is estimated, which is then converted to CO_2 equivalent, allowing the assessment of the carbon footprint of the STP. The results indicated how the carbon footprint should be the main decision factor when evaluating alternatives for the management of gaseous emissions in UASB-based STPs.

12.4.2 Biogas management options aiming at energy recovery

Figure 12.4 shows a general overview of the potential uses of biogas for various STP scales, expressed in population equivalent (PE). For the three arbitrarily selected STP sizes, i.e. 5,000; 50,000 and 500,000 PE, the biogas use includes heat and/or electricity generation and the main difference derives from the application of different technologies to recover the dissolved methane in the effluent, as follows:

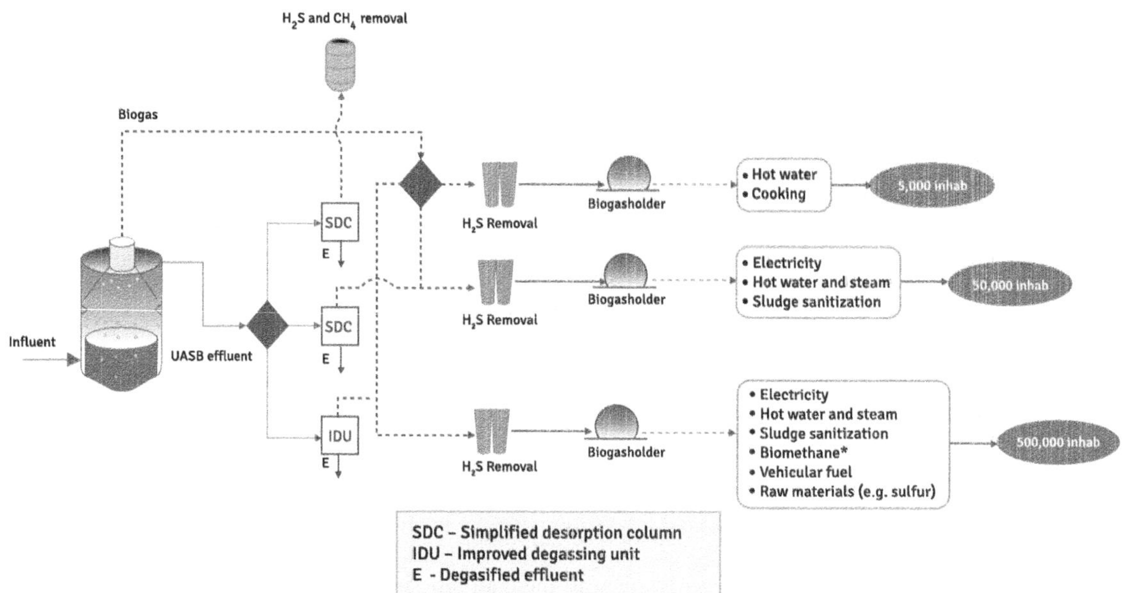

Figure 12.4 Potential uses of the biogas generated in different sizes of anaerobic-based sewage treatment plants.
**See Section 12.4.2.5.*

- For a small-scale STP (5,000 PE), a degassing unit was considered only for the purpose of H_2S and CH_4 abatement, therefore no energy gains were achieved.
- For a medium-scale STP (50,000 PE), a simplified degassing unit (such as spray aeration column) was considered, allowing energy gains.
- For a large-scale STP (500,000 PE), an improved degassing unit (such as membrane contactor) was considered, allowing energy gains.

Based on the biogas proposed uses depicted in Figure 12.4, mass and energy balances were made for the three aforementioned different STP scales, as shown in Examples 12.1 to 12.4. The standard values of the parameters used in the calculations are shown in Table 12.3. It should be emphasised that the results presented in the examples below are an estimate. For more accurate energy balances it is necessary to use real and specific data from STPs.

Table 12.3 Standard parameters for mass and energy balances.

Parameter	Unit	Value	References
Per capita sewage generation	$L\ PE^{-1}\ d^{-1}$	160	Adopted arbitrarily
Biogas consumption for cooking ($BC_{cooking}$)	Nm^3 biogas $family^{-1}\ d^{-1}$	0.25	Adopted arbitrarily
Unitary methane yield	$NL\ m^{-3}$sewage	64	Lobato *et al.* (2012)
Methane content in biogas	%	75	Cabral *et al.* (2016)
Dissolved CH_4 in UASB effluent (liquid temperature: 22–25°C)	$gCH_4\ m^{-3}$	20	Souza *et al.* (2011)
CH_4 recovery in simplified degassing unit (spray aeration)	$NL\ m^{-3}$sewage	14	Glória *et al.* (2016)
CH_4 recovery in improved degassing unit (membrane contactor)	$NL\ m^{-3}$sewage	19	Brown (2006)
Net calorific value for methane	$kWh\ Nm^{-3}CH_4{}^{-1}$	9.97	Moran *et al.* (2014)
Lower calorific value of methane	$kJ\ Nm^{-3}CH_4{}^{-1}$	35,892	Moran *et al.* (2014)
Lower calorific value of biogas	$kJ\ Nm^{-3}CH_4{}^{-1}$	26,919	Moran *et al.* (2014)
Per capita dry sludge generation in UASB reactors	$gDS\ PE^{-1}\ d^{-1}$	15	Andreoli *et al.* (2001)
Water specific heat	$kJ\ kg^{-1}\ °C^{-1}$	4.18	Moran *et al.* (2014)
Drinking water temperature	°C	20	Adopted arbitrarily
Temperature of a bath	°C	40	Adopted arbitrarily
Thermal efficiency of boilers	%	90	Standard engine
Overall efficiency of CHP engines	%	84	Standard engine
Electric efficiency of CHP engines	%	35	Standard engine
Overall thermal efficiency of CHP engines	%	45	Standard engine
Hot water efficiency	%	27	Standard engine
Exhaust gases efficiency	%	18	Standard engine
Exhaust gases temperature	°C	450	Standard engine
Specific electricity consumption of an aerobic post-treatment unit (activated sludge or aerated biofilter)	$kWh\ PE^{-1}\ y^{-1}$	17	von Sperling and Chernicharo (2005)

12.4.2.1 Biogas uses in small-scale STPs to produce thermal energy for domestic use

When small volumes of biogas (1–5 $Nm^3\ d^{-1}$) are produced, the most traditional use is directly burning biogas after H_2S removal for domestic applications, such as heat production and especially for cooking, as illustrated in Figure 12.5. Example 12.1 shows the mass and energy balances for this scenario of biogas use.

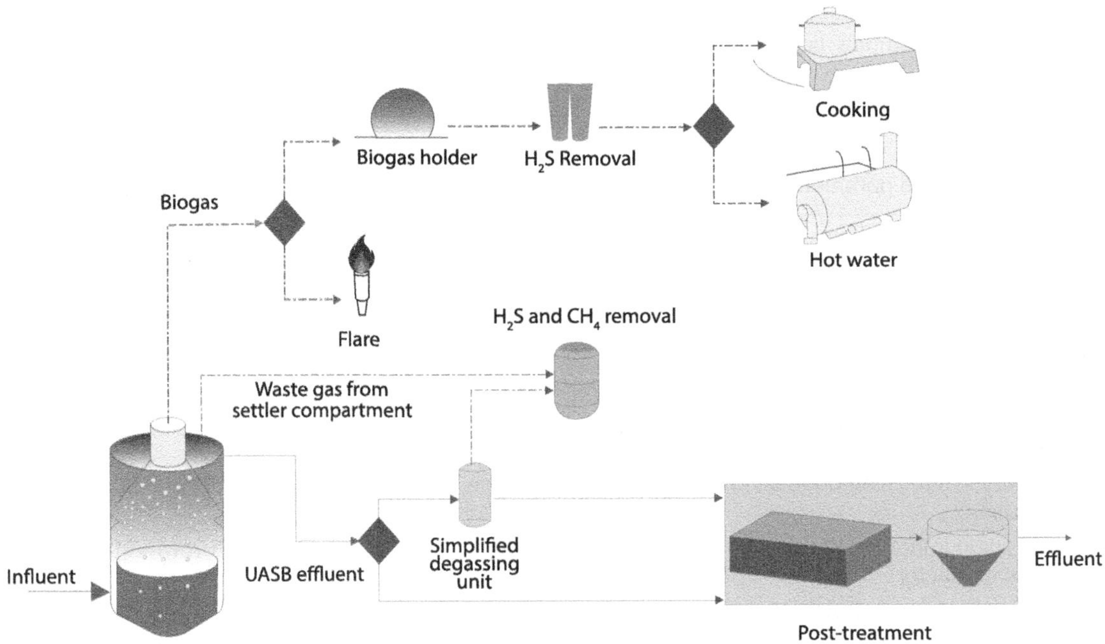

Figure 12.5 Flowsheet of biogas uses in small-scale STPs.

EXAMPLE 12.1

Mass and energy balance for a small-scale STP (5,000 PE): Conversion of biogas into heat. *Although the dissolved methane recovery was excluded from this scenario, its recovery can have a significant impact reducing STP greenhouse gas emissions.*

Option 1: Use of biogas to heat water (as illustrated in Figure 12.6)

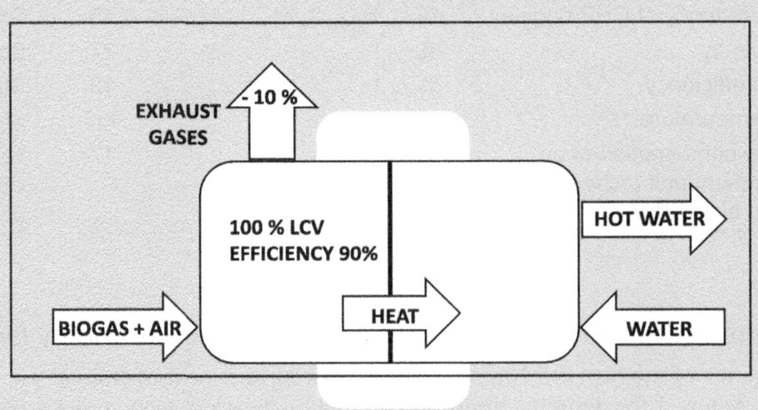

Figure 12.6 Schematic of a domestic heater producing hot water.

Input data

- Population equivalent: 5,000 PE
- Average influent sewage flowrate: $Q_{s\text{-}avg} = 8.4\ L\ s^{-1}$ ($729.6\ m^3\ d^{-1}$)
- Methane yield: $Y_{CH4} = 64\ NL\ m^{-3}sewage$
- Lower calorific value of methane: $LCV_{CH4} = 35{,}892\ kJ\ Nm^{-3}CH_4{}^{-1}$
- Thermal efficiency of boiler: $\eta_{boiler} = 90\%$
- Water specific heat: $SH_{water} = 4.18\ kJ\ kg^{-1}\ °C^{-1}$
- Drinking water temperature: $T_{water} = 20°C$
- Hot water temperature: $T_{hot\text{-}water} = 60°C$

(1) Estimate the daily methane production (Q_{CH4})

$$Q_{CH4} = Q_{s\text{-}avg} \times Y_{CH4} = 729.6\ m^3\ d^{-1} \times 64\ NL\ CH_4\ m^{-3}\ sewage \times (10^{-3}\ NL\ Nm^{-3}) = 46.7\ Nm^3CH_4\ d^{-1}$$

(2) Estimate the amount of chemical energy in the biogas (Ebiogas)

$$E_{biogas} = Q_{CH4} \times LCV_{CH4} = 46.7\ Nm^3CH_4\ d^{-1} \times 35{,}892\ kJ\ Nm^{-3}CH_4 = 1.68 \times 10^6\ kJ\ d^{-1}$$

(3) Estimate the amount of heat generated in the boiler (Eth-boiler)

$$E_{th\text{-}boiler} = E_{biogas} \times \eta_{boiler} = 1.68 \times 10^6\ kJ\ d^{-1} \times 0.90 = 1.51 \times 10^6\ kJ\ d^{-1}$$

(4) Estimate the amount of hot water ($Q_{hot\text{-}water}$)

$$Q_{hot\text{-}water} = (E_{th\text{-}boiler})/[SH_{water} \times (T_{hot\text{-}water} - T_{water})] = 1.51 \times 10^6\ kJ\ d^{-1}/[4.18\ kJ\ kg^{-1}\ °C^{-1} \times (60°C\text{–}20\ °C)] = 9{,}031\ kg\ hot\ water\ d^{-1}\ or\ 9{,}031\ L\ hot\ water\ d^{-1}$$

Comments

- As noticed, the biogas is sufficient to produce almost 10 m^3 per day of hot water (60°C). Considering a family with four persons and a consumption of 50 litres of hot water per bath/shower, the amount of produced hot water would be sufficient to supply approximately 200 persons per day (around 50 families).

Option 2: Direct use of biogas for cooking

Input data

- Methane production: $Q_{CH4} = 46.7\ Nm^3\ CH_4\ d^{-1}$
- Methane content in the biogas: $\%CH_4 = 75\%$
- Standard biogas consumption: $BC_{cooking} = 0.25\ Nm^3$ biogas $family^{-1}\ d^{-1}$
- Estimate the number of families attended ($N_{families}$):

$$N_{families} = (Q_{CH4}/\%CH_4)/BC_{cooking} = (46.7\ Nm^3CH_4\ d^{-1}/0.75)/0.25\ Nm^3\ biogas\ family^{-1}\ d^{-1} = 249\ families$$

Comments

- The direct use of biogas allows 250 families to cook without the need for another external source of heat. This means that approximately 1,000 PE (or 20% of the STP population equivalent) could be supplied with the generated biogas.

12.4.2.2 Use of biogas from medium-scale STPs as fuel for CHP engines

Recovering energy from biogas using CHP engines is by far the most developed and used technology in anaerobic sludge digesters in developed countries, mainly in Europe and USA. Gas engines, gas turbines and microturbines are commonly considered combined heat and power (CHP) technologies to convert biogas to electricity and heat, as illustrated in Figure 12.7.

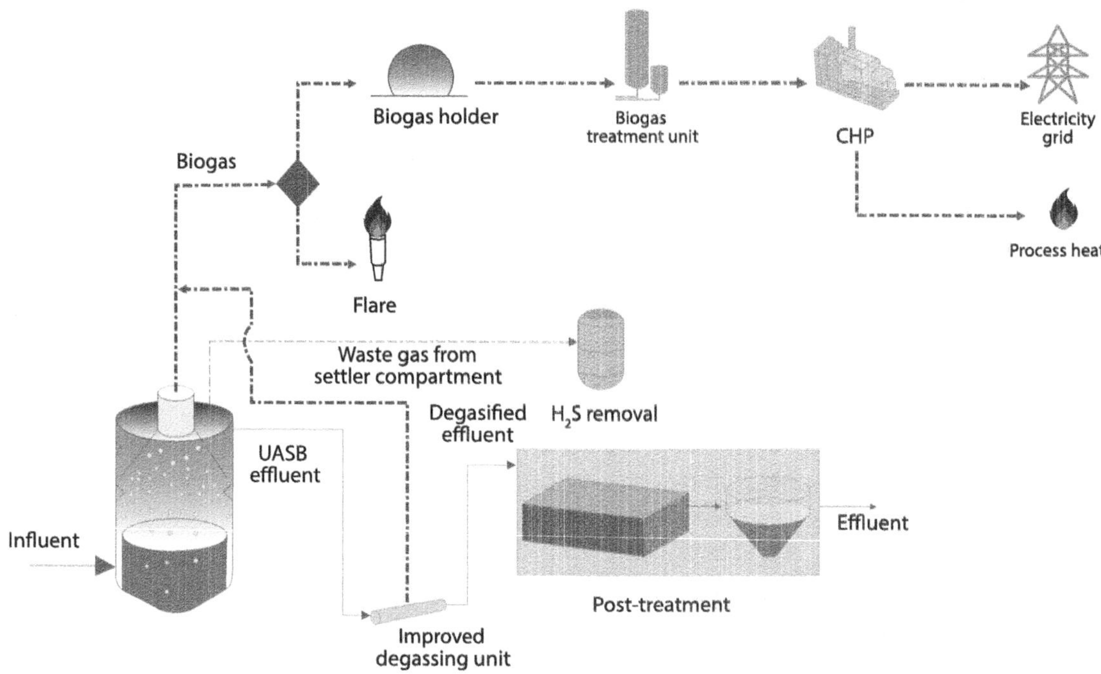

Figure 12.7 Flowsheet of biogas uses in medium- and large-scale STPs.

According to Figure 12.8, most of the heat, the primary by-product of the mechanical power generation, is recoverable in two different forms: continuous engine cooling with jacket water and hot exhaust gases. Standard values for the efficiencies of the different flows of energy are depicted in the aforementioned Table 12.3. Example 12.2 shows mass and energy balances for this scenario of use of biogas.

12.4.2.3 Use of biogas from large-scale STPs as fuel for CHP engines aiming at the electricity self-sufficiency of the plant

In large-scale STPs with high biogas production, virtually all the technological possibilities depicted in Figure 12.4 are economically feasible to be implemented. Example 12.3 shows mass and energy balances for a scenario in which a mechanised aerobic post-treatment is used. Possible applications of hot currents are also presented.

12.4.2.4 Use of thermal energy produced by the CHP engine for sludge sanitization

Currently, the sanitization of sludge to meet the requirements for safe agricultural use is acquiring great interest. According to the United States Environmental Protection Agency (USEPA) rules, sludge must be

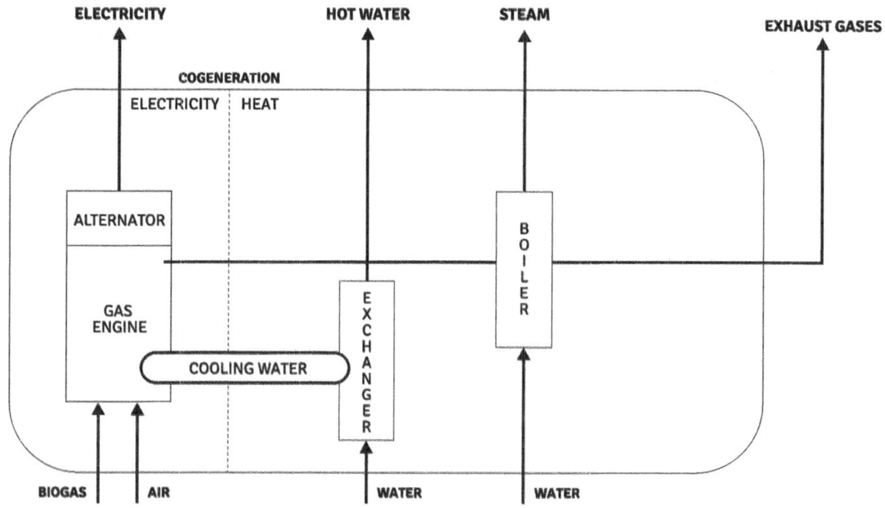

Figure 12.8 Heat recovery from combined heat and power (CHP) cogeneration.

EXAMPLE 12.2

Mass and energy balance for a medium-scale STP (50,000 PE): conversion of biogas into heat and power in CHP engines including a simplified degassing unit to recover dissolved methane.

Input data

- Population equivalent: 50,000 PE
- Average influent sewage flowrate: $Q_{s\text{-}avg} = 96.8\ L\ s^{-1}$ ($8{,}364\ m^3\ d^{-1}$)
- Unitary methane yield: $Y_{CH4} = 64\ NL\ m^{-3}$ sewage
- Methane recovery in simplified degassing unit: $Y_{CH4\text{-}recov\text{-}sdu} = 14\ NL\ m^{-3}$ sewage
- Lower calorific value of methane: $LCV_{CH4} = 35{,}892\ kJ\ Nm^{-3}CH_4^{-1}$ ($9{,}97\ kWh\ Nm^{-3}CH_4^{-1}$)
- Electric efficiency of CHP: $\eta_{el} = 35\%$
- Thermal efficiency CHP (cooling water): $\eta_{th\text{-}cool\text{-}water} = 27\%$
- Thermal efficiency CHP (exhaust gases): $\eta_{th\text{-}exh\text{-}gas} = 18\%$
- Water specific heat: $SH_{water} = 4.18\ kJ\ kg^{-1}\ °C^{-1}$
- Steam (6 bar) condensation enthalpy: $\Delta H_c = 2{,}085\ kJ\ kg^{-1}$
- Drinking water temperature: $T_{water} = 20°C$
- Hot water temperature: $T_{hot\text{-}water} = 90°C$

(1) Estimate the daily methane production (Q_{CH4})

$$Q_{CH4} = Q_{s\text{-}avg} \times (Y_{CH4} + Y_{CH4\text{-}recov\text{-}sdu}) = 8{,}364\ m^3\ d^{-1} \times (64 + 14)\ NL\ CH_4\ m^{-3}\ \text{sewage} \times (10^{-3}\ NL\ Nm^{-3}) = 652\ Nm^3\ CH_4\ d^{-1}$$

(2) Estimate the amount of chemical energy in the biogas (E_{biogas})

$$E_{biogas\text{-}kJ} = Q_{CH4} \times LCV_{CH4} = 652\ Nm^3CH_4\ d^{-1} \times 35{,}892\ kJ\ Nm^{-3}CH_4^{-1} = 23.4 \times 10^6\ kJ\ d^{-1}$$

$E_{biogas\text{-}kWh} = Q_{CH4} \times LCV_{CH4} = 652\ Nm^3CH_4\ d^{-1} \times 9.97\ kWh\ Nm^{-3}CH_4^{-1} = 6{,}504.3\ kWh\ d^{-1}$

(3) Estimate the amount of electricity generation (E_{el})

$E_{el} = E_{biogas\text{-}kWh} \times \eta_{el} = 6{,}504.3\ kWh\ d^{-1} \times 0.35 = 2{,}276.5\ kWh\ d^{-1}$ (95 kW power)

(4) Estimate the amount of thermal energy in cooling water ($E_{th\text{-}cool\text{-}water}$)

$E_{th\text{-}cool\text{-}water} = E_{biogas\text{-}kJ} \times \eta_{th\text{-}cool\text{-}water} = 23.4 \times 10^6\ kJ\ d^{-1} \times 0.27 = 6.32 \times 10^6\ kJ\ d^{-1}$

(5) Estimate the amount of hot water produced ($Q_{hot\text{-}water}$)

$$
\begin{aligned}
Q_{hot\text{-}water} &= (E_{th\text{-}cool\text{-}water})/[SH_{water} \times (T_{hot\text{-}water} - T_{water})] \\
&= 6.32 \times 10^6\ kJ\ d^{-1}/[4.18\ kJ\ kg^{-1}\ °C^{-1} \times (90°C - 20°C)] \\
&= 21{,}607\ kg\ water\ d^{-1} \text{ or } 21{,}607\ L\ d^{-1}
\end{aligned}
$$

(6) Estimate the amount of thermal energy in exhaust gases: ($E_{th\text{-}exh\text{-}gas}$)

$(E_{th\text{-}exh\text{-}gas}) = E_{biogas} \times \eta_{th\text{-}exh\text{-}gas} = 23.4 \times 10^6\ kJ\ d^{-1} \times 0.18 = 4.21 \times 10^6\ kJ\ d^{-1}$

(7) Estimate the amount of steam generated: ($Q_{steam\text{-}exh\text{-}gas}$)

Accepting that steam is produced with exhaust gases that are cooled from 450 to 150°C, the global heat transfer efficiency (η_{ht}) is 60%.

$Q_{steam\text{-}exh\text{-}gas} = (E_{th\text{-}exh\text{-}gas} \times \eta_{ht})/\Delta H_c = (4.21 \times 10^6\ kJ\ d^{-1} \times 0.60)/2{,}085\ kJ\ kg^{-1} = 1{,}213\ kg\ steam\ d^{-1}$

Comments

- The heat released by the cooling water of the CHP engine is sufficient to produce approximately 22 m^3 per day of hot water at 90°C.
- The heat contained in the exhaust gases of the CHP engine is sufficient to produce more than one ton per day of steam at 6 bar.
- The use of CHP engines allows a reasonable conversion of biogas in energy, however, for medium-scale STPs, a rigorous feasibility analysis should be carried out to guarantee that the investment is justified, as discussed in Chapter 8.

EXAMPLE 12.3

Mass and energy balance for a large-scale STP (500,000 PE): Conversion of biogas into heat and power including an improved degassing unit to recover dissolved methane.

Input data

- Population equivalent: 500,000 PE
- Average influent sewage flowrate: $Q_{s\text{-}avg} = 943.1\ L\ s^{-1}$ ($81{,}480\ m^3\ d^{-1}$)
- Unitary methane yield: $Y_{CH4} = 64\ NL\ m^{-3}$ sewage
- Methane recovery in improved degassing unit: $Y_{CH4\text{-}recov\text{-}idu} = 19\ NL\ m^{-3}$ sewage
- Lower calorific value of methane: $LCV_{CH4} = 35{,}892\ kJ\ Nm^{-3}CH_4^{-1}$ ($9.97\ kWh\ Nm^{-3}CH_4^{-1}$)
- Electric efficiency of CHP: $\eta_{el} = 35\%$
- Thermal efficiency CHP (cooling water): $\eta_{th\text{-}cool\text{-}water} = 27\%$
- Thermal efficiency CHP (exhaust gases): $\eta_{th\text{-}exh\text{-}gas} = 18\%$

- Water specific heat: $SH_{water} = 4.18$ kJ kg^{-1} °C^{-1}
- Steam (6 bar) condensation enthalpy: $\Delta H_c = 2{,}085$ kJ kg^{-1}
- Drinking water temperature: $T_{water} = 20$°C
- Hot water temperature: $T_{hot\text{-}water} = 90$°C
- Specific electricity consumption of a post-treatment unit: $SE_{consump} = 17$ kWh PE^{-1} y^{-1}

(1) Estimate the methane production (Q_{CH4})

$$Q_{CH4} = Q_s \times (Y_{CH4} + Y_{CH4\text{-recov-idu}}) = 81{,}480\ m^3\ d^{-1} \times (64 + 19)\ NL\ CH_4\ m^{-3}\ sewage \times (10^{-3}\ NL\ Nm^{-3}) = 6{,}762.8\ Nm^3\ CH_4\ d^{-1}$$

(2) Estimate the amount of chemical energy in the biogas (E_{biogas})

$$E_{biogas\text{-}kJ} = Q_{CH4} \times LCV_{CH4} = 6{,}762.8\ Nm^3\ CH_4\ d^{-1} \times 35.892\ kJ\ Nm^{-3} CH_4^{-1} = 242.7 \times 10^6\ kJ\ d^{-1}$$

$$E_{biogas\text{-}kWh} = Q_{CH4} \times LCV_{CH4} = 6{,}762.8\ Nm^3 CH_4\ d^{-1} \times 9.97\ kWh\ Nm^{-3} CH_4^{-1} = 67{,}425.5\ kWh\ d^{-1}$$

(3) Estimate the amount of electricity generation (E_{el})

$$E_{el} = E_{biogas} \times \eta_{el} = 67{,}425.5\ kWh\ d^{-1} \times 0.35 = 23{,}598.9\ kWh\ d^{-1}\ (983\ kW\ power)$$

(4) Estimate the amount of thermal energy in cooling water ($E_{th\text{-cool-water}}$)

$$E_{th\text{-cool-water}} = E_{biogas\text{-}kJ} \times \eta_{th\text{-cool-water}} = 242.7 \times 10^6\ kJ\ d^{-1} \times 0.27 = 65.5 \times 10^6\ kJ\ d^{-1}$$

(5) Estimate the amount of hot water generated ($Q_{hot\text{-water}}$)

$$Q_{hot\text{-water}} = (E_{th\text{-cool-water}})/[SH_{water} \times (T_{hot\text{-water}} - T_{water})] = 65.5 \times 10^6\ kJ\ d^{-1}/[4.18\ kJ\ kg^{-1}\ °C^{-1} \times (90°C - 20°C)] = 223{,}954\ kg\ water\ d^{-1}\ or\ 223{,}954\ L\ water\ d^{-1}$$

(6) Estimate the amount of thermal energy in exhaust gases ($E_{th\text{-exh-gas}}$)

$$E_{th\text{-exh-gas}} = E_{biogas\text{-}kJ} \times \eta_{th\text{-exh-gas}} = 242.7 \times 10^6\ kJ\ d^{-1} \times 0.18 = 43.7 \times 10^6\ kJ\ d^{-1}$$

(7) Estimate the amount of steam generated ($Q_{steam\text{-exh-gas}}$)
Accepting that steam is produced with exhaust gases that are cooled from 450 to 150°C, the global heat transfer efficiency (η_{ht}) is 60%.

$$Q_{steam\text{-exh-gas}} = (E_{th\text{-exh-gas}} \times \eta_{ht})/\Delta H_c = (43.7 \times 10^6\ kJ\ d^{-1} \times 0.60)/2{,}085\ kJ\ kg^{-1} = 12{,}573\ kg\ steam\ d^{-1}$$

(8) Electricity consumed by the aerobic post-treatment unit (e.g. activated sludge) ($E_{el\text{-consumed}}$)

$$E_{el\text{-consumed}} = PE \times SE_{consump} = (500{,}000\ PE \times 17\ kWh\ PE^{-1}\ y^{-1})/(365\ d\ y^{-1} \times 24\ h\ d^{-1}) = 970\ kW$$

Comments

- The heat contained in the cooling water of the CHP engine is sufficient to produce approximately 220 m^3 per day of hot water at 90°C.
- The heat contained in the exhaust gases of the CHP engine is sufficient to produce more than 12 ton per day of steam at 6 bar.
- Under normal operating conditions, the STP can be considered energetically self-sufficient, since the electricity produced by the CHP engine (983 kW) is slightly higher than the consumed by an aerobic post-treatment unit such as an activated sludge process (970 kW).

maintained for at least 30 minutes at the pasteurisation temperature of 70°C. Hot water and steam produced by the CHP engine are perfect energy vectors for this purpose. Example 12.4 shows mass and energy balances for such a scenario. The feasibility of sludge sanitization through pasteurisation of sludge using hot water or steam produced as residual energy in the CHP engine is analysed according to two scenarios:

EXAMPLE 12.4

Mass and energy balance for sludge sanitization using the residual energy of a CHP engine.

Input data

- Population equivalent: 500,000 PE
- Sludge yield: $Y_{sludge} = 15$ g DS PE^{-1} d^{-1}
- DS concentration of the sludge: $C_{sludge} = 6\% = 0.06$ kg DS kg sludge^{-1}
- Temperature of sludge entering the heat exchanger: $ST_{in} = 25°C$
- Sludge specific heat: $SH_{sludge} = 4.18$ kJ.kg^{-1} °C^{-1} (considered the same as water)
- Flow of hot water from CHP (Example 12.3): $Q_{hot\text{-}water} = 223{,}984$ kg d^{-1}
- Temperature of water entering the heat exchanger: $WT_{in} = 90°C$
- Temperature of water leaving the heat exchanger: $WT_{out} = 50°C$
- Water heat exchanger efficiency: $\eta_{he} = 90\%$
- Steam heat exchanger efficiency: $\eta_{st} = 60\%$
- Flow of steam (6 bar) from CHP (Example 12.3): $Q_{steam\text{-}exh\text{-}gas} = 12{,}573$ kg d^{-1}
- Steam condensation enthalpy: $\Delta H_c = 2{,}085$ kJ kg^{-1}

(1) Estimate the daily mass of sludge production (P_{sludge})

$$P_{sludge} = PE \times Y_{sludge} = 500{,}000\ PE \times 15 \times 10^{-3}\ kg\ DS\ PE^{-1}\ d^{-1} = 7{,}500\ kg\ DS\ d^{-1}$$

(2) Estimate the daily volumetric sludge production (6% DS) (Q_{sludge})

$$Q_{sludge} = P_{sludge}/C_{sludge} = (7{,}500\ kg\ DS\ d^{-1})/(0.06\ kg\ DS\ kg\ sludge^{-1}) = 125{,}000\ kg\ sludge\ d^{-1}$$

(3) Estimate the sludge temperature out of the heat exchanger (heating sludge with hot water) (ST_{out})

$$Q_{hot\text{-}water} \times (WT_{in} - WT_{out})\ (\eta_{he}) = Q_{sludge} \times (ST_{out} - ST_{in})$$

$$(223{,}984\ kg\ d^{-1}) \times (90\text{–}50°C) \times 0.90 = (125{,}000\ kg\ d^{-1}) \times (ST_{out} - 25°C)$$

$ST_{out} = 89.5°C > 70°C \rightarrow$ The process is feasible.

(4) Estimate the sludge temperature out of the heat exchanger (heating sludge with steam) (ST_{out})

$$Q_{steam\text{-}exh\text{-}gas} \times \Delta H_c \times \eta_{st} = Q_{sludge} \times (ST_{out} - ST_{in})$$

$$(12{,}573\ kg\ d^{-1}) \times (2{,}085\ kJ\ kg^{-1}) \times (0.60) = (125{,}000\ kg\ d^{-1}) \times (ST_{out} - 25°C)$$

$ST_{out} = 151°C > 70°C \rightarrow$ The process is feasible.

Comments

- Sludge pasteurisation can be achieved using as heating fluid both hot water or steam produced as residual heat in the CHP engine.

- By using hot water and maintaining in the heat exchanger reasonably temperature differences between hot and cold fluids (exchange surface is not very high), the sanitization temperature is easily reached.
- The direct injection of less than 20% of the steam generated in the CHP engine makes possible to reach and maintain the sanitization temperature.

Note: for sludge thermal drying, please consult the design example in Chapter 8.

(i) Through a heat exchanger operating in counter-current using hot water to pre-heat the sludge.
(ii) By direct injection of steam to maintain 70°C in a tank dimensioned for 30 minutes of hydraulic residence time of the sludge.

Sludge concentration is a controlling parameter, as sludge behaves like a non-Newtonian fluid. To ensure an adequate sludge flow, in the Example 12.4, a concentration of 6% dry solids (DS) is adopted. For higher DS values it is necessary to access the rheological behavior of the sludge.

12.4.2.5 Biogas upgrading

Strict environmental policies and incentives for renewable energy production in Europe have pushed forward the transformation of biogas from sludge digesters into biomethane. Conceptually, there are two groups of technologies: the conventional ones that remove CO_2 from biogas (physical or chemical processes) and those that convert CO_2 into CH_4 (chemical or biological processes). Both of them allow biomethane to be produced with CH_4 concentrations higher than 90%, which can be used as an automotive fuel or injected into the natural gas grid, after being pressurised.

Nonetheless, as addressed in Chapter 8, for UASB reactors applied to the mainstream sewage treatment, transformation of biogas into biomethane is typically not feasible, because of the low amount of produced biogas compared to sidestream processes, such as sludge digestion. Additionally, biogas purification to biomethane requires the removal of the high N_2 concentrations in the biogas from UASB reactors treating sewage, which appears to be economically unfeasible.

12.5 SUSTAINABLE ANAEROBIC-BASED SEWAGE TREATMENT PLANTS: FLOWSHEET PROPOSITIONS FOR CLOSING CYCLES

Section 2.5 of Chapter 2 presents different sanitation concepts for anaerobic-based systems for warm and low/moderate climates. Following a similar approach, this section 12.5 proposes sanitation solutions for warm climate regions.

Based on a research project conducted in Brazil, schematic diagrams of sustainable STPs were designed, considering an integrated approach for management of liquid, solid and gaseous phases, as well as by-product recovery alternatives applicable to different STP scales (small-, medium- and large-scale STPs). Such conceptual diagrams considered the socioeconomic and environmental context of the study area (two watersheds in the state of Minas Gerais). Therefore, different local realities would benefit from different combinations of UASB reactor and post-treatment systems, as well as different technological arrangements for by-products recovery. Further details can be found on Bressani-Ribeiro *et al.* (in press).

Selected technological processes for small- and medium-scale STPs (2,000 < PE < 100,000) intended to integrate treatment plants into local communities / economic activities. On the other hand, large-scale STPs

(PE > 100,000) were conceived as industries, where the input (raw sewage) can be converted to a wide range of by-products. The main intent to use sewage treatment by-products were:

- *Liquid phase (treated sewage)*: fertirrigation, use of treated sewage within the STP itself and use in industrial processes;
- *Solid phase (sludge)*: agricultural use and recovery of degraded areas (e.g. areas degraded by mining activities);
- *Gas phase (biogas from anaerobic processes)*: sludge drying, water heating, cooking and combined heat and power generation.

12.5.1 Proposed flowsheets for small-scale STPs

For small-scale STPs (2,000 < PE < 10,000), a UASB + controlled land application (e.g. overland flow) flowsheet is proposed, as depicted in Figure 12.9.

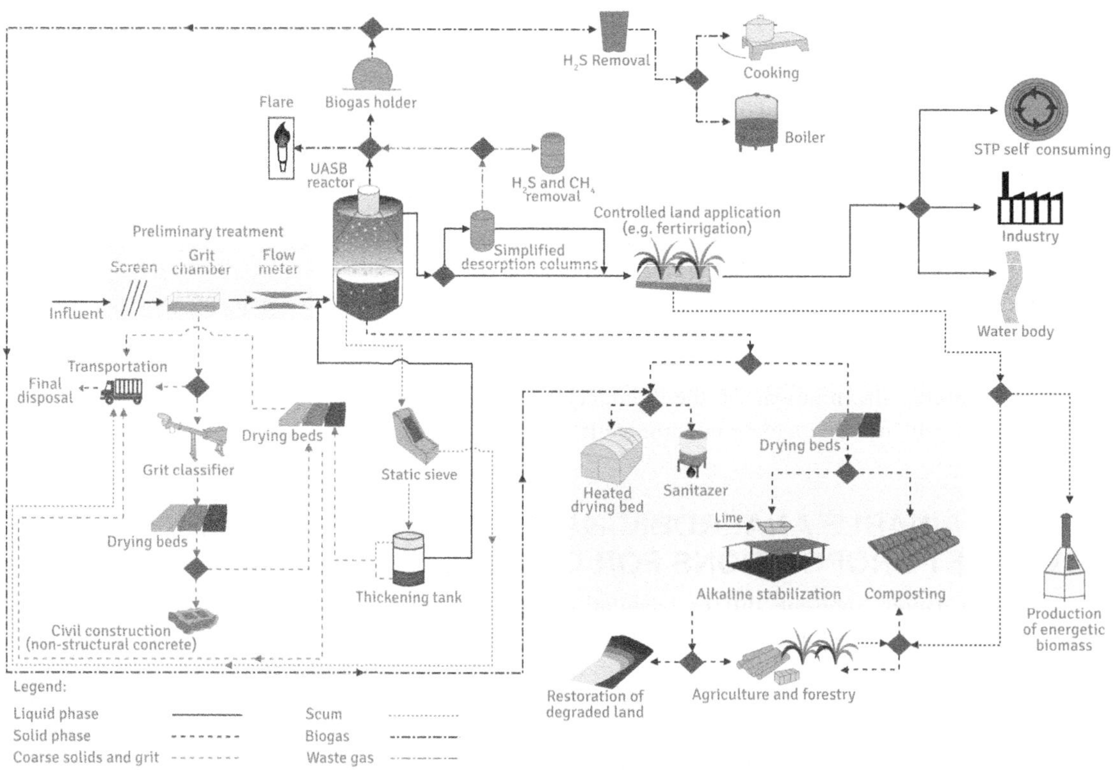

Figure 12.9 Schematic diagram of the UASB + controlled land application system for a small-scale STP.

For the preliminary treatment step, the possibility of energy recovery through processing the retained solid material was not considered, therefore, the traditional alternative of dispose the retained material in landfill was indicated. It is suggested that the grit removed is reused in the construction and building industries due to its suitable granulometry and properties. For this, the grit must be further processed in sequential classification, washing and drying steps. Another possible utilizations include: (i) replacement material in the sludge drying beds of the STP itself or in other treatment plants nearby or (ii) as a small aggregate in the incorporation in coating mortars and preparation of non-structural concrete for civil construction (Borges, 2014).

For the sludge management, the following technological alternatives were considered: (i) drying beds; (ii) thermal drying beds; or (iii) sanitizing tanks. After drying beds, the removed sludge has a solid content of approximately 35% and could be directly routed to liming/maturation or composting. For thermal drying beds, a decrease of the sludge drying time is expected, allowing the reduction of the required area and the inactivation of pathogenic organisms. Regarding sanitizers, the sludge volume remains constant, however a complete sanitization is possible to achieve with an increase in temperature, as previously discussed in this Chapter. Low-cost thermal drying systems using solar energy and thermal energy from biogas could also be an interesting alternative for small and medium-scale STPs, as addressed in Chapter 8.

For the biogas management, two technological alternatives were considered for biogas uses: (i) cooking or as fuel for heating water systems in the vicinity of the STP, benefiting the surrounding community and; (ii) used to produce heat for sludge drying (thermal drying bed) or sanitizing processes (e.g. sanitizing tank). As previously presented in this Chapter, such biogas uses allow the thermal energy recovery in simplified facilities and with low operational requirements.

12.5.2 Proposed flowsheets for medium-scale STPs

For medium-scale STPs (10,000 < PE < 100,000 - see Chapter 1), a UASB + polishing ponds flowsheet is proposed, as depicted in Figure 12.10.

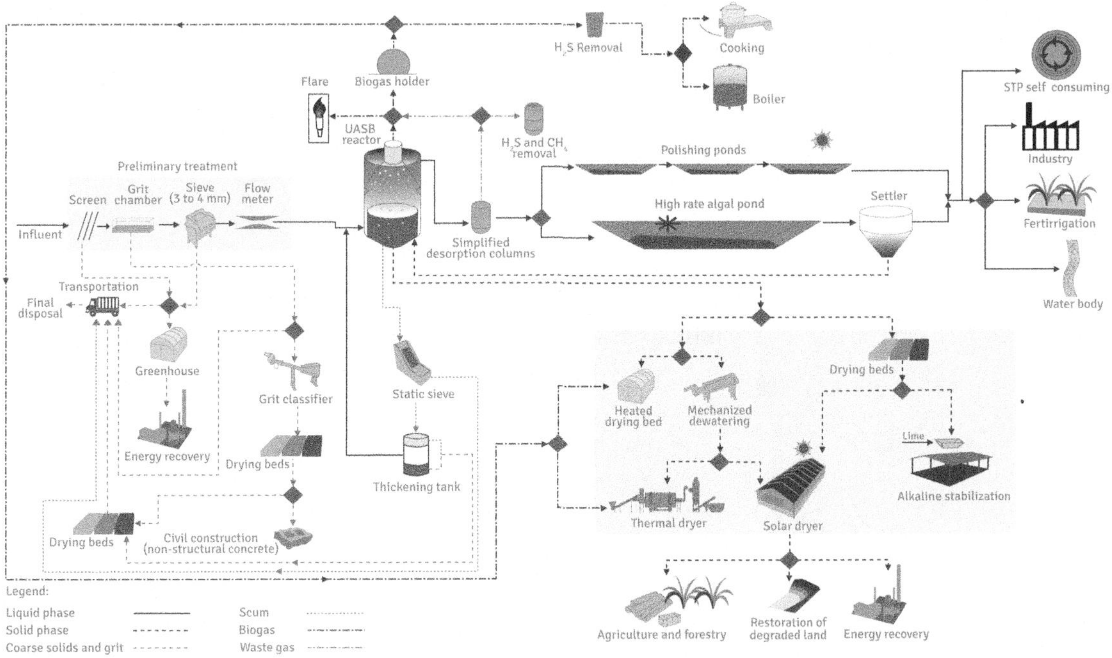

Figure 12.10 Schematic diagram of the UASB + polishing ponds system for a medium-scale STP.

For such STPs, especially those with >50,000 PE, it was considered the possibility of energy recovery through processing the solid material retained in the preliminary treatment units. Taking into account the typical lower calorific value (LCV) of the retained material in the screening step of medium and large-scale STPs (23 MJ kg^{-1}) (Borges, 2014), it can be considered for co-processing in cement kilns as secondary fuel, for instance. Occasionally, after a previous drying step, such solid material could also feed sludge thermal dryers in the STP, if these units are adapted for receiving solid fuel.

Regarding grit processing, the proposed management and reuse options are similar to that previously described for small-scale STPs.

For the sludge management, considering the possibility of biogas recovery for heat generation, sludge thermal dryers seem to be an interesting alternative. These dryers would be fed with sludge from a previous mechanized dewatering step (e.g. centrifuges or filter press), allowing the production of a biosolid with a solid content up to 80%, as addressed in Chapter 8.

Alternatively, the use of solar dryers also leads to a biosolid with high solid content (Tchobanoglous *et al.*, 2016). Moreover, if the reduction of sludge volumes or even the production of energetic biosolid (sludge with total solids concentration higher than 90%) is not a target, liming or composting would be sufficient for the production of a class A biosolid, which could be used for agricultural purposes.

12.5.3 Proposed flowsheets for large-scale STPs

For large-scale STPS (PE > 100,000), a UASB + trickling filter is proposed, as depicted in Figure 12.11.

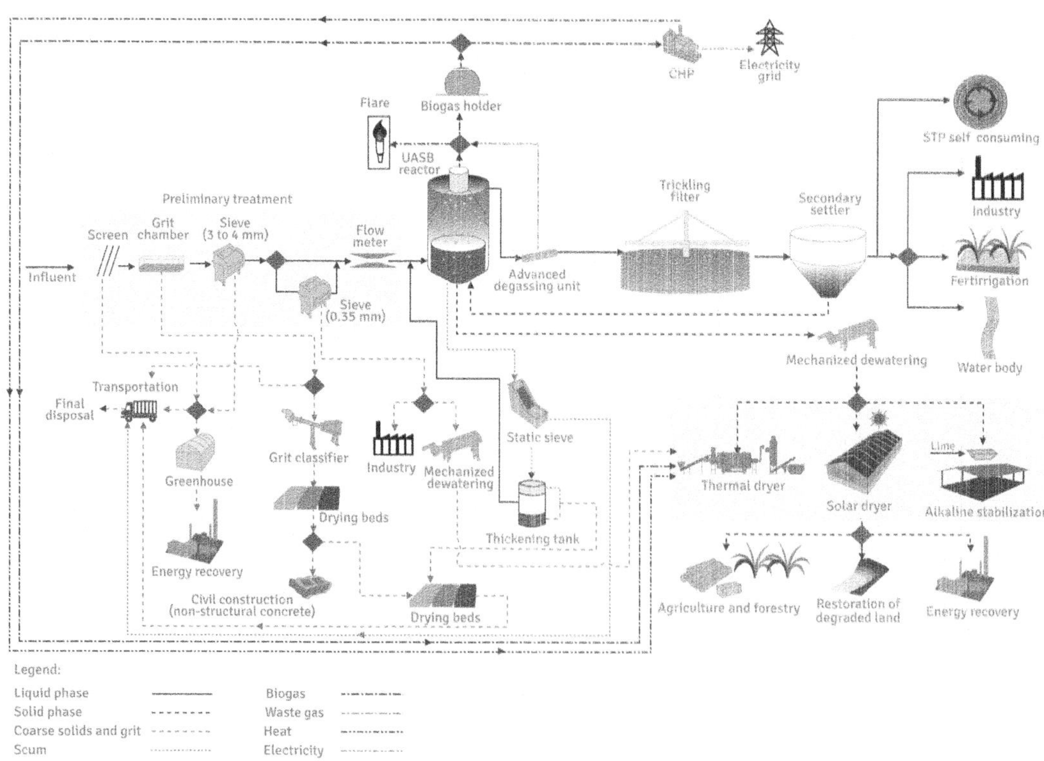

Figure 12.11 Schematic diagram of the UASB + trickling filter system for a large-scale STP.

As for medium-scale STPs, the possibility of energy recovery by processing the solid material retained in the preliminary treatment units was also considered. Additionally, the proposed management and reuse options are similar to that previously described for small-scale STPs. Moreover, cellulose recovery can be considered. An assessment of the technical-economic feasibility suggested that micro screening (0.35 mm) after a previous fine screening step could retain significant amounts of cellulose fibers with payback time of approximately 7 years (Ruiken *et al.*, 2013). The alternatives for reusing this by-product can be: (i) biomass for energy (i.e. co-processing) or as solid fuel for sludge dryers; (ii) production of second-generation ethanol; and (iii) production of toilet paper. The latter alternative is technically possible (Ruiken *et al.*, 2013), although sociocultural aspects should be taken into account.

For large-scale STPs, biogas use unfolds in several alternatives, which comprise its use as thermal or electric energy, mainly using CHP engines, as previously discussed in this Chapter. An important interface is verified with the solid phase management, in which thermal drying seems to be an adequate alternative for sludge handling, as addressed in Chapter 8.

12.6 FINAL REMARKS

Closing cycles by recovering and reusing by-products is a sustainable solution for encouraging the implementation of sanitation facilities and engagement with local populations. Anaerobic-based STPs have the benefit of generating several valuable products that can bring economic income. Nonetheless, commonly these products are still not fully utilised or are perceived as wastes. In UASB reactors treating sewage, it is very usual that dissolved methane is not recovered from the effluent, the produced biogas is flared to the atmosphere and digested sludge is disposed in landfills.

An integrated approach for liquid, solid and gaseous phases management should be pursued. Especially in developing countries, important challenges remain for the implementation of by-products recovery from anaerobic-based STPs. These challenges include: (i) the need to evaluate different business models (e.g. supply, pre-operation, assisted operation); (ii) the need for policies regarding energy efficiency in the sanitation sector and use of alternative energy sources (e.g. through tariff subsidies); (iii) the legal framework regarding environmental permits and use of sewage treatment by-products; and (iv) social perception.

Apart from those challenges, the depletion of non-renewable natural resources (e.g. phosphorus) will certainly motivate the finding for new alternatives that rely on a more integrated view of the biogeochemical cycles. The recovery of by-products from anaerobic-based STPs is aligned to this perspective and it can trigger improvements in sewage collection and treatment infrastructure, especially in developing countries.

REFERENCES

All-gas (2017). All-gas Newsletter 11/2017. http://www.all-gas.eu/documents/1509955/1514023/Newsletter_2017.11_eng.pdf/629a9f4c-17e4–4cca-0b8f-b1b8bbe855ca (acessed 23 October 2018).

Amann A, Zoboli O., Krampe J., Rechberger H., Zessner M. and Egle L. (2018). Environmental impacts of , phosphorus recovery from municipal wastewater. *Resources, Conservation and Recycling*, **130**, 129–139.

Andreoli C. V., von Sperling M. and Fernandes F. (2001). Lodo de esgoto: tratamento e disposição final (Sewage sludge: treatment and final disposal), Department of Sanitary and Environmental Engineering of UMG (ed.), 1st edn, Curitiba, Paraná, Brazil, 6, pp. 1–484.

Bertoncini E. I. (2008). Tratamento de efluentes e reuso da água no meio agrícola (Treatment of effluents and water reuse in agriculture). *Revista Tecnologia e Inovação Agropecuária*, **1**(1), 152–169.

Bortolini R. L. R. (2018). Composto orgânico de lodo de reator UASB e casca de café: qualidade e oportunidades para produção e uso agrícola em Minas Gerais (Composting of UASB reactor excess sludge with coffe fruit peel: quality and opportunities for agricultural production and use in Minas Gerais state). MSc thesis, Graduate Programme on Sanitation, Environmental and Water Resources, Federal University of Minas Gerais, Belo Horizonte, Minas Gerais.

Borges N. B. (2014) Aproveitamento dos resíduos gerados no tratamento preliminar de estações de tratamento de esgoto (Waste recovery from preliminary treatment of sewage treatment plants). PhD thesis, Universidade de São Paulo (USP), São Paulo, 2014.

Bressani-Ribeiro T., Brandt E. M. F., Melo V. R., Bianchetti F. J., McAdam E., Mota Filho C. R. and Chernicharo C. A. L. (2018). Carbon footprint assessment and management of sludge and gaseous emissions in decentralized anaerobic-based sewage treatment plants. In: Proceedings of the 14th International IWA Conference on Sustainable Solutions for Small Water and Wastewater Treatment Systems, Nantes, France.

Bressani-Ribeiro T., Mota C.R., Melo V.R., Bianchetti F.J. and Chernicharo C.A.L. (in press). Planning for Achieving Low Carbon and Integrated Resources Recovery from Sewage Treatment Plants for the State of Minas Gerais, Brazil. *Journal of Environmental Management*.

Brown N. (2006) Methane Dissolved in Wastewater Exiting UASB Reactors: Concentration Measurement and Methods for Neutralisation. MSc thesis. Department of Energy Technology, Royal Institute of Technology (KTH), Stockholm.

Buonocore E., Mellino S., De Angelis G., Liu G. and Ulgiati S. (2016). Life cycle assessment indicators of urban wastewater and sewage sludge treatment. *Ecological Indicators*, **94**(3), 13–23.

Cabral C. B. G., Chernicharo C. A. L., Hoffmann H., Neves P. N. P., Platzer C., Bressani-Ribeiro T. and Rosenfeldt S. (2016). Resultados do projeto de medições de biogás em reatores anaeróbios/Probiogás; organizadores, Ministério das Cidades (Results of the biogas measurement program in UASB reactors/Probiogás; organizers, Ministry of Cities). Deutsche Gesellschaf für Internationale Zusammenarbeit GmbH (ed.), Ministério das Cidades, Brasília, Brazil.

Canellas J. (2018). Tertiary ammonium removal with zeolites. PhD thesis. Cranfield University, Cranfield, United Kingdom.

Chachuat B., Roche N. and Latifi M. A. (2005). Long-term optimal aeration strategies for small-size alternating activated sludge treatment plants. *Chemical Engineering and Processing: Process Intensification*, **44**, 591–604.

Chernicharo C. A. L., Brandt E. M. F., Bressani-Ribeiro T., Melo V. R., Bianchetti F. J., Mota Filho C. R. and McAdam E. (2017). Development of a tool for improving the management of gaseous emissions in UASB-based sewage treatment plants. *Water Practice and Technology*, **12**, 917–926.

Christodoulou A. and Stamatelatou K. (2016). Overview of legislation on sewage sludge management in developed countries worldwide. *Water Science and Technology*, **73**, 453–462.

De Godos I., Arbib Z., Feria M. J., Lara E., Santiago J. R., Rogalla F., Fernández M. and De la Rubia M. A. (2013). Proyecto All-Gas. Cultivo de microalgas com producción de biocombustibles y eliminación de nutrientes (Cultivation of microalgae with production of biofuels and nutrients removal). *RETEMA Especial Bioenergía*, **169**, 56–62.

Díez-Monteiro R., Solimeno A., Uggetti E., García-Galán M. J. and García J. (2018). Feasibility assessment of energy-neutral microalgae-based wastewater treatment plants under Spanish climatic conditions. *Process Safety and Environmental Protection*, **119**, 242–252.

Elmitwalli T. A., Zandvoort M. H., Zeeman G., Bruning H. and Lettinga G. (1999). Low temperature treatment of domestic sewage in upflow anaerobic sludge blanket and anaerobic hybrid reactors. *Water Science Technology*, **39**(5), 177185.

Fernandes F. and Silva S. M. C. P. (1999). Manual Prático para Compostagem de Biossólidos (Practical Manual for Composting of Biosolids). Brazilian Research Program on Basic Sanitation – PROSAB, State University of Londrina, Londrina, Paraná, Brazil.

Gerba C. and Rose J. (2003). International guidelines for water recycling: Microbiological considerations. *Water Science and Technology*, **3**(4), 311–316.

Glória R. M., Motta T. M., Silva P. V. O., Costa P., Brandt E. M. F., Souza C. L. and Chernicharo C. A. L. (2016). Stripping and dissipation techniques for the removal of dissolved gases from anaerobic effluents. *Brazilian Journal of Chemical Engineering*, **33**(4), 713–721.

Gonzales A. M. (2012). Nutrient recovery from ion exchange regenarants. PhD thesis. Cranfield University, Cranfield, United Kingdom.

GWI - Global Water Intelligence (2009). Municipal Water Reuse Markets. Media Analytics, Oxford, United Kingdom.

Jankowska E., Sahu A. K. and Oleskowicz-Popiel P. (2017). Biogas from microalgae: Review on microalgae's cultivation, harvesting and pretreatment for anaerobic digestion. *Renewable & Sustainable Energy Reviews*, **75**, 692–709.

Jaramillo M. F. and Restrepo I. (2017). Wastewater reuse in agriculture: a review about its limitations and benefits. *Sustainability*, **9**(10), 1734.

Jiménez B., Mara D., Carr R. and Brissaud F. (2010). Wastewater treament for pathogen removal and nutrient conservation: Suitable systems for use in developing countries. IWMI Books, Reports H042608, International Water Management Institute.

Klassen V., Blifernez-Klassen O., Wobbe L., Schlüter A., Kruse O. and Mussgnug J. H. (2016). Efficiency and biotechnological aspects of biogas production from microalgal substrates. *Journal of Biotechnology*, **234**, 7–26.

Laugesen C., Fryd O., Brix H. and Koottatep T. (eds.) (2010). Sustainable Wastewater Management in Developing Countries: New Paradigms and Case Studies from the Field. ASCE Press, Reston, Virginia, United States.

Lautze J., Stander E., Drechsel P., da Silva A. K. and Keraita B. (2014). Global Experiences in Water Reuse. Resource Recovey and Reuse Series 4, International Water Management Institute – IWMI, CGIAR Research Program on Water, Land and Ecosystems – WLE, Pelawatta Battaramulla, Sri Lanka, pp. 1–23.

Libhaber M. and Orozco-Jaramillo A. (2013). Sustainable treatment of municipal wastewater. *Water21*, **15**(5), 25–28.

Lobato L. C. S., Chernicharo C. A. L. and Souza C. L. (2012). Estimates of methane loss and energy recovery potential in anaerobic reactors treating domestic wastewater. *Water Science and Technology*, **66**(12), 2745–2753.

Longo S., Benedetto Mirko A. B., Bongards M., Chaparro A., Cronrath A., Fatone F., Lema J. L., Mauricio-Iglesias M., Soares A. and Hospido A. (2016). Monitoring and diagnosis of energy consumption in wastewater treatment plants. A state of the art and proposals for improvement. *Applied Energy*, **179**, 1251–1268.

Mara D. D. (2004). Domestic Wastewater Treatment in Developing Countries. Earthscan, London, UK, pp. 1–293.

Martin B. D., Parsons S. A. and Jefferson B. (2009). Removal and recovery of phosphate from municipal wastewaters using a polymeric anion exchanger bound with hydrated ferric oxide nanoparticles. *Water Science and Technology*, **60**, 2637–2645.

Miladinovic N. and Weatherley L. R. (2008). Intensification of ammonia removal in a combined ion-exchange and nitrification column. *Chemical Engineering Journal*, **135**(1–2), 15–24.

Montingelli M. E., Tedesco S. and Olabi A. G. (2015). Biogas production from algal biomass: A review. *Renewable and Sustainable Energy Reviews*, **43**(C), 961–972.

Moran M. J., Shapiro H. N., Boettner D. D. and Bailey M. B. (2014). Fundamentals of Engineering Thermodynamics, 8th edn. Wiley, New Jersey.

Muhammad A. (2018). Hybrid ion exchange resins for phosphorus removal from wastewater. PhD thesis. Cranfield University, Cranfield, United Kingdom.

Oswald W. J. and Gotaas H. B. (1957). Photosynthesis in sewage treatment. *Transactions of the American Society of Civil Engineers*, **122**, 73–105.

Park J. B. K., Craggs R. J. and Shilton A. N. (2011). Wastewater treatment high rate algal ponds for biofuel production. *Bioresource Technology*, **102**, 35–42.

Passamani F. R. F., Keller R. and Gonçalves R. F. (2002). Higienização de lodo utilizando caleagem e pasteurização em uma pequena estação de tratamento de esgoto combinando reator UASB e biofiltro aerado submerso (Sludge sanitization using alcaline stabilization and pasteurization in a small sewage treatment plant combining UASB reactor and submerged aerated biofilter). In: Congreso Interamericano de Ingeriería Sanitaria y Ambiental, 28th edn. Cancun, México.

Passos F., Uggetti E., Carrère H. and Ferrer I. (2014). Pretreatment of microalgae to improve biogas production: a review. *Bioresource Technology*, **172**, 403–412.

Passos F., Gutiérrez R., Uggetti E., Garfí M., García J. and Ferrer I. (2017). Towards energy neutral microalgae-based wastewater treatment plants. *Algal Research*, **28**, 235–243.

Pittman J. K., Dean A. P. and Osundeko O. (2011). The potential of sustainable algal biofuel production using wastewater resources. *Bioresource Technology*, **102**(1), 17–25.

Ponsa S., Bolzonella D., Deshusses J. C. M. A., Fonts I., Gil-Lalaguna N., Komilis D., Lyberatos G., Pérez-Elvira S. I. and Sánchez J. L. (2017). Recovering energy from sludge. In: Innovative Wastewater Treatment & Resource Recovery Technologies: Impacts on Energy, Economy and Environment, J. M. Lema and S. Sayarez (eds), IWA Publishing, London, UK.

Rawat I., Kumar R. R., Mutanda T. and Bux F. (2013). Biodiesel from microalgae: A critical evaluation from laboratory to large scale production. *Applied Energy*, **103**, 444–467.

Rodriguez C., Alaswad A., Mooney J., Prescott T. and Olabi A. G. (2015). Pre-treatment techniques used for anaerobic digestion of algae. *Fuel Processing Technology*, **138**, 765–779.

Ruiken C. J., Breuer G., Klaversma E., Santiago T. and van Loosdrecht M. C. M. (2013). Sieving wastewater - Cellulose recovery, economic and energy evaluation. *Water Research*, **47**(1), 43–48.

Sato T., Qadir M., Yamamoto S., Endo T. and Zahoor A. (2013). Global, regional, and country level need for data on wastewater generation, treatment, and use. *Agricultural Water Management*, **130**, 1–13.

Scott C., Drechsel P., Raschid-Sally L., Bahri A., Mara D., Redwood M. and Jimenez B. (2010). Wastewater irrigation and health: Challenges and outlook for mitigating risks in low-income countries. In: Wastewater irrigation and health: Assessing and mitigating risk in low-income countries. International Development Research Centre (IDRC). London, UK: Earthscan; Ottawa, Canada; International Water Management Institute (IWMI), Colombo, Sri Lanka.

Sousa J. T., Henrique I. N., Leite V. D. and Lopes W. S. (2006). Tratamento de águas residuárias: uma proposta para a sustentabilidade ambiental (Wastewater treatment: a proposal for environmental sustainability). *Revista de Biologia e Ciências da Terra*, **1**, 90–97.

Souza C. L., Chernicharo C. A. L. and Aquino S. F. (2011). Quantification of dissolved methane in UASB reactors treating domestic wastewater under different operating conditions. *Water Science and Technology*, **64**, 2259–2264.

Tchobanoglous G., Stensel H. D., Tssuchihashi R., Burton F. L., Abu-Orf M., Bowden G. and Pfrang W. (2013). Wastewater engineering, treatment and reuse. 5th edn, McGraw Hill, New York.

The United Nations World Water Development Report (2017). Wastewater: The Untapped Resource. United Nations Education, Scientific, and Cultural Organization, Paris, France.

Uemura S. and Harada H. (2000). Treatment of sewage by a UASB reactor under moderate to low temperature conditions. *Bioresource Technology*, **72**(3), 275–282.

UN-HABITAT (2008). Global Atlas of Excreta, Wastewater Sludge, and Biosolids Management: Moving Forward the Sustainable and Welcome Uses of a Global Resource, United Nations Human Settlements Programme, Nairobi, Kenya.

Von Sperling M. and Chernicharo C. A. L. (2005). Biological Wastewater Treatment in Warm Climate Regions. Two volumes. IWA Publishing, London, UK. Available at http://www.iwapublishing.com/open-access-ebooks/3567.

Wiley P. E., Campbell J. E. and McKuin B. (2011). Production of biodiesel and biogas from algae: a review of process train options. *Water Environment Research*, **83**, 326–338.

Williams A. T. and Mayer B. K. (2013). Advancement in ion exchange processes for municipal wastewater nutrient recovery. *Proceedings of the Water Environment Federation*, **12**, 6474–6485.

World Health Organization (2006). Guidelines for the Safe Use of Wastewater. Excreta and Greywater in Agriculture - Wastewater Use in Agriculture, United Nations Environment Programme, World Health Organization, Geneva, Switzerland.

World Health Organization (2011). Guidelines for Drinking Water Quality, 4th edn, World Health Organization, Geneva, Switzerland.

Index

C

D

E

F

IWA Publishing's authorised EU representative for General Product Safety Regulations is Diane D'Arras, 15 rue Duret, 75116 Paris, France, e-mail: safety@iwap.co.uk.

Printed and bound by CPI Group (UK) Ltd, Croydon, CR0 4YY
12/05/2026
02108870-0014